# HSPA Evolution

# HSPA Evolution

## The Fundamentals for Mobile Broadband

**Thomas Chapman**

**Erik Larsson**

**Peter von Wrycza**

**Erik Dahlman**

**Stefan Parkvall**

**Johan Sköld**

AMSTERDAM • BOSTON • HEIDELBERG • LONDON
NEW YORK • OXFORD • PARIS • SAN DIEGO
SAN FRANCISCO • SINGAPORE • SYDNEY • TOKYO

Academic Press is an Imprint of Elsevier

Academic Press is an imprint of Elsevier
32 Jamestown Road, London NW1 7BY, UK
525 B Street, Suite 1800, San Diego, CA 92101-4495, USA
225 Wyman Street, Waltham, MA 02451, USA
The Boulevard, Langford Lane, Kidlington, Oxford OX5 1GB, UK

First published 2015



**British Library Cataloguing in Publication Data**
A catalogue record for this book is available from the British Library

**Library of Congress Cataloguing in Publication Data**
A catalog record for this book is available from the Library of Congress

ISBN: 978-0-08-099969-2

For information on all Academic Press publications
visit our website at **http://store.elsevier.com/**

Printed and bound in the UK

# Contents

# List of Figures

# List of Tables

# Preface

From the early deployments at the beginning of the twenty-first century to the powerful multi-carrier networks today serving millions of smartphones and laptops worldwide, HSPA technology has risen to become the engine of the mobile broadband revolution. During these years, a rapid evolution has occurred, transforming the initial HSDPA/HSUPA specifications into even more powerful tools for serving an expanding and diverse range of traffic to ever more complex devices and networks.

This book describes HSPA technology in detail from release 99 to release 13 enabling the reader to gain a complete picture of the toolbox of features and build a sound understanding of the 3GPP specifications, their background, and practical issues in building and deploying networks and devices. This description of the latest HSPA features is built upon an outline of the basic technologies and concepts that lie behind the standardization.

The book is divided into four parts. The first part provides an overview of the origins of the 3G standards and the processes behind the evolution of 3G and 4G and licensing of spectrum. The motivations behind continuing the evolution of HSPA are examined in the context of current market trends, such that an insight can be gained into the reasons for introducing the various features.

In Part II, a deeper insight is provided into some of the basic technologies that have been embedded into HSPA. The description is generic in nature, such that it enables not only an understanding of the technology drivers for HSPA but also similar use of the technologies in other specifications such as LTE and WiMAX.

In Part III, a core description of the HSPA feature set is provided. The underlying WCDMA, HSDPA, and HSUPA specifications are reviewed in detail. This is followed by detailed examination of the full set of HSPA features, including MIMO in downlink and uplink, higher-order modulations, multicarrier, multiflow, and traffic handling schemes such as CPC and enhanced CELL_FACH. These detailed chapters both trace the development of features in the standardization and provide a deep insight into their operation.

To build 3GPP-compliant equipment, certain radio-related performance criteria must be met. Part IV reviews how the RF performance specifications for HSPA are built. Furthermore, the definition and structure of HSPA frequency bands are discussed. Band definitions and appropriate RF criteria are essential to allow global mobility and operate in a radio environment in which other types of systems are ready.

Although the standards provide a specification of interfaces and signals, it is important to understand what factors drive performance in different types of network environments. Part IV also provides an insight into techniques for capturing HSPA performance using simulations and the key factors that drive performance, in addition to the throughput and capacity that can actually be provided.

Despite the growth of LTE, HSPA technology will remain deployed for many years to come, and the final chapter of the book provides an outlook on the challenges faced in the evolving mobile broadband environment and the likely feature developments that will be needed to meet these challenges.

# Acknowledgments

We thank all our colleagues at Ericsson for assisting in this project by helping with contributions to the book, giving suggestions, and comments on the contents, and taking part in the huge team effort of developing HSPA.

The standardization process involves people from all parts of the world, and we acknowledge the efforts of our colleagues in the wireless industry in general and in 3GPP RAN in particular. Without their work and contributions to the standardization, this book would not have been possible.

Finally, we are immensely grateful to our families for bearing with us and supporting us during the long process of writing this book.

PART

# Introduction

# I

CHAPTER

# From 3G to 4G: background and motivation of 3G evolution

1

## CHAPTER OUTLINE



From the first experiments with radio communication by Guglielmo Marconi in the 1890s, the road to truly mobile radio communication has been quite long. To understand the complex 3G mobile-communication systems of today, it is also important to understand where they came from and how cellular systems have evolved from an expensive technology for a few selected individuals to today's global mobile-communication systems used by almost half of the world's population. Developing mobile technologies has also changed, from being a national or regional concern to becoming a very complex task undertaken by global standards-developing organizations, such as the *Third-Generation Partnership Project* (3GPP), and involving thousands of people.

## 1.1 HISTORY AND BACKGROUND OF 3G

The cellular technologies specified by 3GPP are the most widely deployed in the world, with more than 6.6 billion users in 2013 [1]. A very significant step forward was taken with the development of *High-Speed Downlink Packet Access* (HSPA) in


HSPA Evolution: The Fundamentals for Mobile Broadband. DOI: 10.1016/B978-0-08-099969-2.00001-6

the first few years of the 2000s, which was followed both by the continuing evolution of HSPA and the standardization of an advanced radio access referred to as the *Long-Term Evolution* (LTE) and an evolved packet access core network in the *System Architecture Evolution* (SAE). Both HSPA and LTE networks are now widely deployed and together have continued to evolve.

Looking back to when it all started, it began several decades ago with early deployments of analog cellular services.

### 1.1.1 BEFORE 3G

The US *Federal Communications Commission* (FCC) approved the first commercial car-borne telephony service in 1946, operated by AT&T. In 1947, AT&T also introduced the cellular concept of reusing radio frequencies, which became fundamental to all subsequent mobile-communication systems. Commercial mobile telephony continued to be car-borne for many years because of bulky and power-hungry equipment. In spite of the limitations of the service, there were systems deployed in many countries during the 1950s and 1960s, but the users counted only in thousands at the most.

These first steps toward mobile communication were taken by the monopoly telephone administrations and wire-line operators. The big uptake of subscribers and usage came when mobile communication became an international concern and the industry was invited into the process. The first international mobile communication system was the analog *Nordic Mobile Telephony* (NMT) system, which was introduced in the Nordic countries in 1981, at the same time as analog *Advanced Mobile Phone Service* (AMPS) was introduced in North America. Other analog cellular technologies deployed worldwide were *Total Access Communications System* (TACS) and J-TACS. They all had in common that equipment was still bulky, mainly car-borne, and voice quality was often inconsistent, with "cross-talk" between users being a common problem.

With an international system such as NMT came the concept of "roaming," giving a service for users traveling outside the area of their "home" operator. This also gave a larger market for the mobile phones, which attracted additional companies into the mobile communication business.

The analog cellular systems supported "plain old telephony services," that is, voice with some related supplementary services. With the advent of digital communication during the 1980s, the opportunity to develop a second generation of mobile-communication standards and systems based on digital technology surfaced. With digital technology came an opportunity to increase the capacity of the systems, to give a more consistent quality of service, and to develop much more attractive truly mobile devices.

In Europe, the telecommunication administrations in CEPT[1] initiated the *Global System for Mobile communications* (GSM) project to develop a pan-European mobile-telephony system. The GSM activities were in 1989 continued within the newly formed *European Telecommunication Standards Institute* (ETSI). After

[1]The European Conference of Postal and Telecommunications Administrations (CEPT) consists of the telecom administrations from 48 countries.

evaluations of TDMA-, CDMA-, and FDMA-based proposals in the mid-1980s, the final GSM standard was built on TDMA. Development of a digital cellular standard was simultaneously done by *Telecommunications Industry Association* (TIA) in the USA, resulting in the TDMA-based IS-54 standard, later simply referred to as US-TDMA. A somewhat later development of a CDMA standard called IS-95 was completed by TIA in 1993. In Japan, a second-generation TDMA standard was also developed, usually referred to as *Personal Digital Cellular* (PDC).

All of these standards were "narrowband" in the sense that they targeted "low-bandwidth" services, such as voice. With the second-generation digital mobile communications came also the opportunity to provide data services over the mobile-communication networks. The primary data services introduced in 2G were text messaging (SMS) and circuit-switched data services enabling e-mail and other data applications. The peak data rate in 2G was initially 9.6 kbps. Higher data rates were introduced later in evolved 2G systems by assigning multiple time slots to a user and by modified coding schemes.

Packet data over cellular systems became a reality during the second half of the 1990s, with *General Packet Radio Services* (GPRS) introduced in GSM and packet data also added to other cellular technologies, such as the Japanese PDC standard. These technologies are often referred to as 2.5G. The success of the wireless data service iMode in Japan gave a very clear indication of the potential for applications over packet data in mobile systems, in spite of the fairly low data rates supported at the time.

With the advent of 3G and the higher-bandwidth radio interface of *Universal Terrestrial Radio Access* (UTRA) came possibilities for a range of new services that were only hinted at with 2G and 2.5G. The 3G radio access development is today handled in 3GPP. However, the initial steps for 3G were taken in the early 1990s, long before 3GPP was formed.

What also set the stage for 3G was the internationalization of cellular standardization. GSM was a pan-European project, but quickly attracted worldwide interest when the GSM standard was deployed in a number of countries outside Europe. A global standard gains in economy of scale, since the market for products becomes larger. This has driven a much tighter international cooperation around 3G cellular technologies than for the earlier generations.

### 1.1.2 RESEARCH ON 3G

In parallel with the widespread deployment and evolution of 2G mobile-communication systems during the 1990s, substantial efforts were put into 3G research activities. In Europe, the partially EU-funded project *Research into Advanced Communications in Europe* (RACE) carried out initial 3G research in its first phase. 3G in Europe was named *Universal Mobile Telecommunications Services* (UMTS). In the second phase of RACE, the *Code Division Test bed* (CODIT) project and the *Advanced TDMA Mobile Access* (ATDMA) project further developed 3G concepts based on *Wideband CDMA* (WCDMA) and Wideband TDMA technologies. The next phase of related European research was *Advanced Communication Technologies and Services* (ACTS), which included the UMTS-related project *Future Radio Wideband Multiple*

*Access System* (FRAMES). The FRAMES project resulted in a multiple access concept that included both WCDMA and Wideband TDMA components.

At the same time, parallel 3G activities were going on in other parts of the world. In Japan, the *Association of Radio Industries and Businesses* (ARIB) was in the process of defining a 3G wireless communication technology based on WCDMA. Also in the United States, a WCDMA concept called WIMS was developed within the T1.P1[2] committee. Korea started work on WCDMA at this time.

The FRAMES concept was submitted to the standardization activities for 3G in ETSI,[3] where other multiple-access proposals were also introduced by the industry, including the WCDMA concept from the ARIB standardization in Japan. The ETSI proposals were merged into five concept groups, which also meant that the WCDMA proposals from Europe and Japan were merged.

### 1.1.3 3G STANDARDIZATION ROOTS

The outcome of the ETSI process in early 1998 was the selection of WCDMA as the technology for UMTS in the paired spectrum (FDD) and *Time Division CDMA* (TD-CDMA) for the unpaired spectrum (TDD). There was also a decision to harmonize the parameters between the FDD and the TDD components.

The standardization of WCDMA went on in parallel in ETSI and ARIB until the end of 1998 when the 3GPP was formed by standards-developing organizations from all regions of the world. This solved the problem of trying to maintain parallel development of aligned specifications in multiple regions. The present organizational partners of 3GPP are ARIB (Japan), CCSA (China), ETSI (Europe), ATIS (USA), TTA (Korea), and TTC (Japan).

The WCDMA and HSPA standards continue to be developed within 3GPP, which is currently (that is in 2014) working on release 12 of the specifications.

## 1.2 STANDARDIZATION

### 1.2.1 THE STANDARDIZATION PROCESS

Setting a standard for mobile communication is not a one-time job, it is an ongoing process. The standardization forums are constantly evolving their standards trying to meet new demands for services and features. The standardization process is different in the different forums but typically includes the four phases illustrated in Figure 1.1:

1. *Requirements*, where it is decided what is to be achieved by the standard.
2. *Architecture*, where the main building blocks and interfaces are decided.
3. *Detailed specifications*, where every interface is specified in detail.
4. *Testing and verification*, where the interface specifications are proven to work with real-life equipment.

[2]The T1.P1 committee was part of T1 which presently has joined the ATIS standardization organization.
[3]The TDMA part of the FRAMES project was also fed into 2G standardization as the evolution of GSM into EDGE (Enhanced Data rates for GSM Evolution).

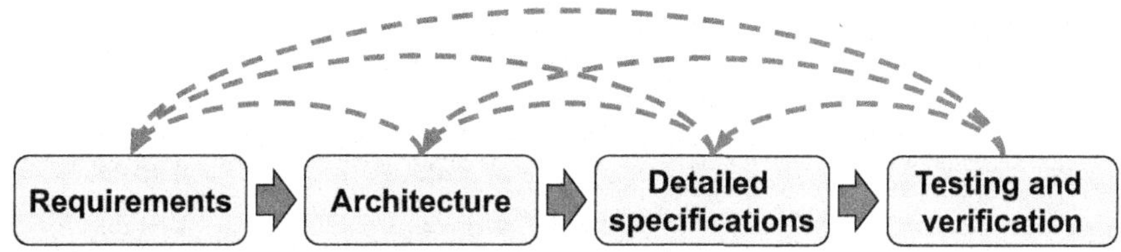

**FIGURE 1.1 The standardization phases and iterative process.**

These phases are overlapping and iterative. As an example, requirements can be added, changed, or dropped during the later phases if the technical solutions call for it. Likewise, the technical solution in the detailed specifications can change due to problems found in the testing and verification phase.

Standardization starts with the *requirements* phase, where the standards body decides what should be achieved with the standard. This phase is usually relatively short.

In the *architecture* phase, the standards body decides about the architecture, that is, the principles of how to meet the requirements. The architecture phase includes decisions about reference points and interfaces to be standardized. This phase is usually quite long and may change the requirements.

After the architecture phase, the *detailed specification* phase starts. In this phase the details for each of the identified interfaces are specified. During the detailed specification of the interfaces, the standards body may find that it has to change decisions made either in the architecture or even in the requirements phases.

Finally, the *testing and verification* phase starts. It is usually not a part of the actual standardization in the standards bodies but takes place in parallel through testing by vendors and interoperability testing between vendors. This phase is the final proof of the standard. During the testing and verification phase, errors in the standard may still be found and those errors may change decisions in the detailed standard. Albeit not common, changes may need to be done also to the architecture or the requirements. To verify the standard, products are needed. Hence, the implementation of the products starts after (or during) the detailed specification phase. The testing and verification phase ends when there are stable test specifications that can be used to verify that the equipment is fulfilling the standard.

### 1.2.2 3GPP

The 3GPP is the standards-developing body that specifies the LTE, 3G UTRA and GSM systems. 3GPP is a partnership project formed by the standards bodies ETSI, ARIB, TTC, TTA, CCSA, and ATIS. 3GPP consists of several *Technical Specifications Groups* (TSGs); see Figure 1.2.

A parallel partnership project called 3GPP2 was formed in 1999. It also developed 3G specifications, but for cdma2000, which is the 3G technology developed from the 2G CDMA-based standard IS-95. It is also a global project, and the organizational partners are ARIB, CCSA, TIA, TTA, and TTC.

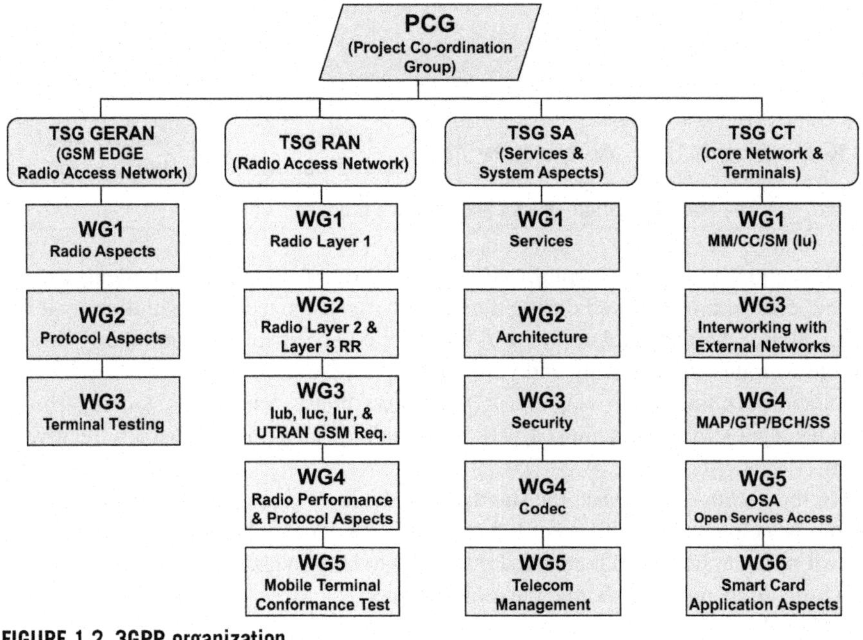


**FIGURE 1.2 3GPP organization.**

3GPP TSG RAN is the technical specification group that has developed WCDMA and its evolution HSPA and also LTE and is in the forefront of the technology. TSG RAN consists of five *working groups* (WGs):

1. RAN WG1 deals with the physical layer specifications.
2. RAN WG2 deals with the layer 2 and layer 3 radio interface specifications.
3. RAN WG3 deals with the fixed RAN interfaces, for example interfaces between nodes in the RAN, but also the interface between the RAN and the core network.
4. RAN WG4 deals with the *radio frequency* (RF) and *radio resource management* (RRM) performance requirements.
5. RAN WG5 deals with the *User Equipment* (UE) conformance testing.

The scope of 3GPP when it was formed in 1998 was to produce global specifications for a 3G mobile system based on an evolved GSM core network, including the WCDMA-based radio access of the UTRA FDD and the TD-CDMA-based radio access of the UTRA TDD mode. The task to maintain and develop the GSM/EDGE specifications was added to 3GPP at a later stage. The UTRA (and GSM/EDGE) specifications are developed, maintained, and approved in 3GPP. After approval, the organizational partners transpose them into appropriate deliverables as standards in each region.

In parallel with the initial 3GPP work, a 3G system based on TD-SCDMA was developed in China. TD-SCDMA was eventually merged into release 4 of the 3GPP specifications as an additional TDD mode.

The work in 3GPP is carried out with relevant ITU recommendations in mind and the result of the work is also submitted to ITU. The organizational partners are obliged to identify regional requirements that may lead to options in the standard. Examples are regional frequency bands and special protection requirements local to a region. The specifications are developed with global roaming and circulation of UEs in mind. This implies that many regional requirements in essence will be global requirements for all UEs, since a roaming UE has to meet the strictest of all regional requirements. Regional options in the specifications are thus more common for base stations than for UEs.

The specifications of all releases can be updated after each set of TSG meetings, which occur four times a year. The 3GPP documents are divided into releases, where each release has a set of features added compared to the previous release. The features are defined in *Work Items* (WIs) agreed and undertaken by the TSGs. The contents of releases 99 to 12 are described in Chapters 6–15 and 19. The date shown for each release is the day the content of the release was frozen. For historical reasons, the first release is numbered by the year it was frozen (1999), whereas the following releases are numbered 4, 5, etc.

For the WCDMA radio access developed in TSG RAN, release 99 contains all of the features needed to meet the IMT-2000 requirements as defined by ITU. There are circuit-switched voice and video services and data services over both packet-switched and circuit-switched bearers. The first major addition of radio access features to WCDMA is release 5 with *High Speed Downlink Packet Access* (HSDPA) and release 6 with *High Speed Uplink Packet Access* (HSUPA), also known as *Enhanced Uplink*. These two are together referred to as HSPA and are described in more detail in Part III of this book. With HSPA, UTRA became the first mass market mobile broadband technology.

HSPA has evolved to incorporate technology advances such that its current evolution state, release 11, is technologically capable of meeting the most significant IMT-Advanced requirements.

With the inclusion of an Evolved UTRAN (LTE) and the related SAE in release 8 and its subsequent evolution, further steps are taken in terms of broadband capabilities. LTE evolution is described in more detail in the companion book to this one [2].

Table 1.1 and Figure 1.3 illustrate some of the most important of the specifications maintained by each 3GPP working group.

### 1.2.3 IMT-2000 AND IMT-ADVANCED ACTIVITIES IN ITU

The ITU work on 3G took place in ITU-R Working Party 5D (WP5D),[4] where 3G systems were referred to as IMT-2000. WP5D does not write technical specifications for IMT-2000, but kept the role of defining IMT-2000, cooperating with the regional standardization bodies to maintain a set of recommendations for IMT-2000 and

[4]The work on IMT-2000 was moved from Working Party 8F to Working Party 5D in 2008.

Table 1.1 Illustration of Working Group Specification (FDD) Responsibility

| Working Group | Area | Specification | Content |
|---|---|---|---|
| RAN 1 | Layer 1 (physical layer) | 25.211 | **Physical channels and mapping of transport channels onto physical channels** |
| | | 25.212 | **Multiplexing and channel coding** |
| | | 25.213 | **Spreading and modulation** |
| | | 25.214 | **Physical layer procedures**<br>*Power control, power settings, random access, synchronization, DRX and DTX, etc.* |
| | | 25.215 | **Physical layer measurements** |
| RAN 2 | Layer 2 (data link) | 25.321 | **Medium Access Control (MAC) protocol**<br>*TFC selection procedures, priority handling, multiplexing/demultiplexing of PDUs, hybrid-ARQ functionality, etc.* |
| | | 25.322 | **Radio Link Controller (RLC) protocol**<br>*Transparent/acknowledged/ unacknowledged data transfer, QoS and error management, flow control, segmentation, padding and concatenation, ciphering, etc.* |
| | | 25.306 | **UE Radio Access capabilities** |
| | Layer 3 (network layer) | 25.331 | **Radio Resource Control (RRC)**<br>*Control plane signaling, broadcast, establishment, reconfiguration and release of RRC connection and radio bearers, outer-loop control, mobility functions, radio resource allocation, cell search / paging, etc.* |
| RAN 3 | Fixed interfaces | 25.420-25.426 | **Iur interface** |
| | | 25.427-25.435 | **Iub Interface** |
| | | 25.410-25.419 | **Iu interface** |
| RAN 4 | RF and RRM performance requirements | 25.101 | **UE radio transmission and reception** |
| | | 25.104 | **BS radio transmission and reception** |
| | | 25.133 | **Requirements for support of radio resource management** |

IMT-Advanced. Recently, ITU-R has initiated an activity, sometimes referred to as IMT-2020, to look at the development of IMT technologies beyond what is defined for IMT-2000 and IMT-Advanced.

The main IMT-2000 recommendation is ITU-R M.1457 [3], which identifies the IMT-2000 *radio interface specifications* (RSPC). The recommendation contains a "family" of radio interfaces, all included on an equal basis. The original family of six

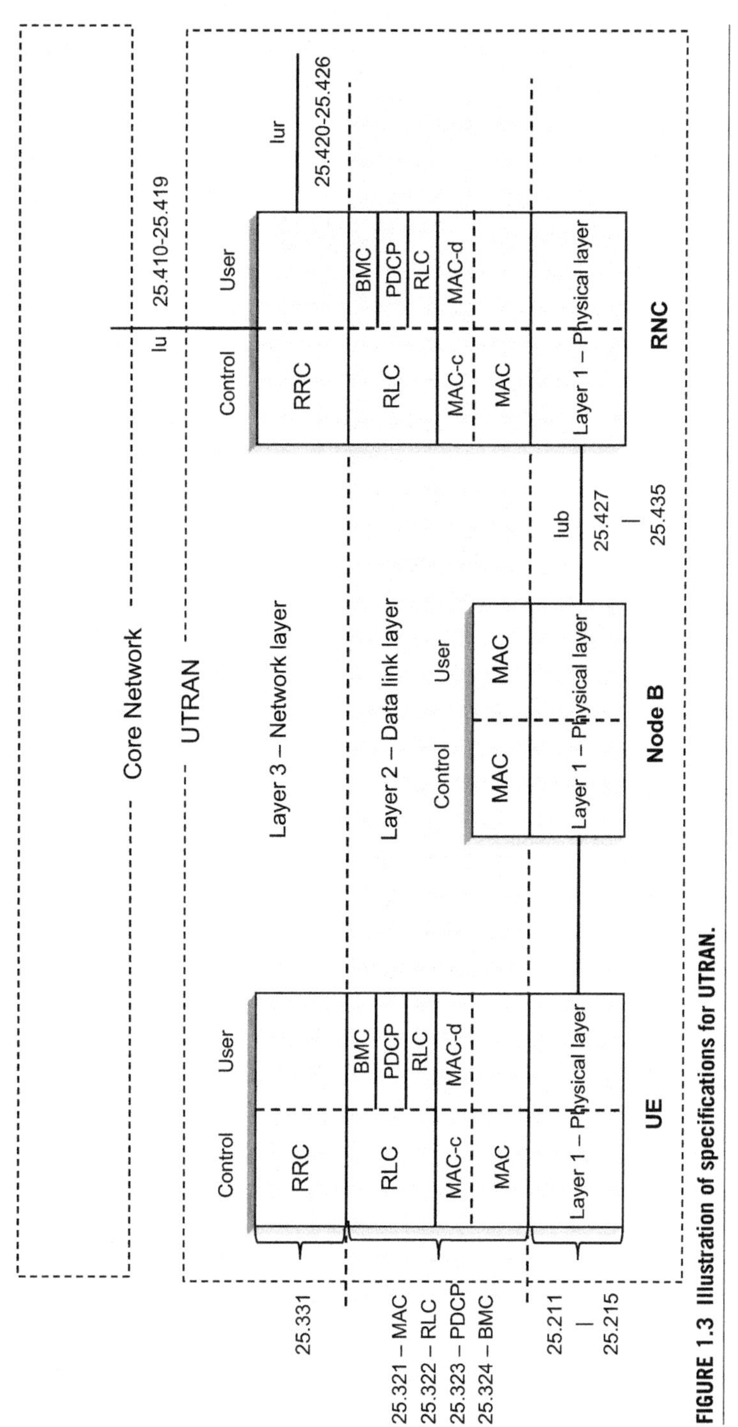


**FIGURE 1.3 Illustration of specifications for UTRAN.**

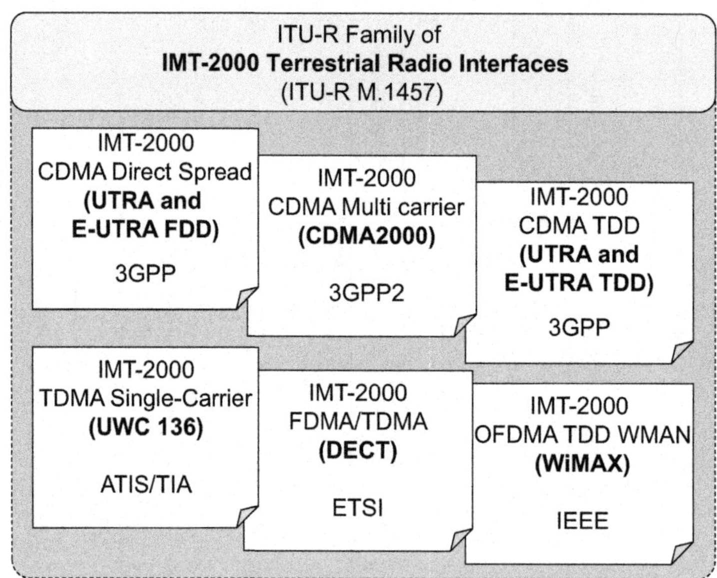


**FIGURE 1.4 The definition of IMT-2000 in ITU-R.**

terrestrial radio interfaces of IMT-2000 is illustrated in Figure 1.4, which also shows what *Standards Developing Organizations* (SDOs) or Partnership Projects produce the specifications. In addition, there are several IMT-2000 satellite radio interfaces defined, not illustrated in Figure 1.4.

For each radio interface, M.1457 contains an overview of the radio interface, followed by a list of references to the detailed specifications. The actual specifications are maintained by the individual SDOs, and M.1457 provides URLs locating the specifications at each SDO's web archive.

With the continuing development of the IMT-2000 radio interfaces, including the evolution of UTRA to Evolved UTRA, the ITU recommendations also need to be updated. ITU-R WP5D continuously revises recommendation M.1457 and at the time of writing, it is in its 11th version. Input to the updates is provided by the SDOs and Partnership Projects writing the standards. In ITU-R M.1457, LTE (or E-UTRA) is included in the family through the 3GPP family members for UTRA FDD and TDD, whereas *Ultra Mobile Broadband* (UMB) is included through CDMA2000, as shown in the figure (development of UMB stopped in 2008 and the technology is not commercially available). WiMAX is also included as the sixth family member for IMT-2000.

IMT-Advanced is the term used for systems that include new radio interfaces supporting the new capabilities of systems beyond IMT-2000, as demonstrated with the "van diagram" in Figure 1.5. The step into IMT-Advanced capabilities is seen by ITU-R as the step into 4G, the next generation of mobile technologies after 3G.

The process for defining IMT-Advanced was set by ITU-R WP5D [4] and was quite similar to the process used in developing the IMT-2000 recommendations.

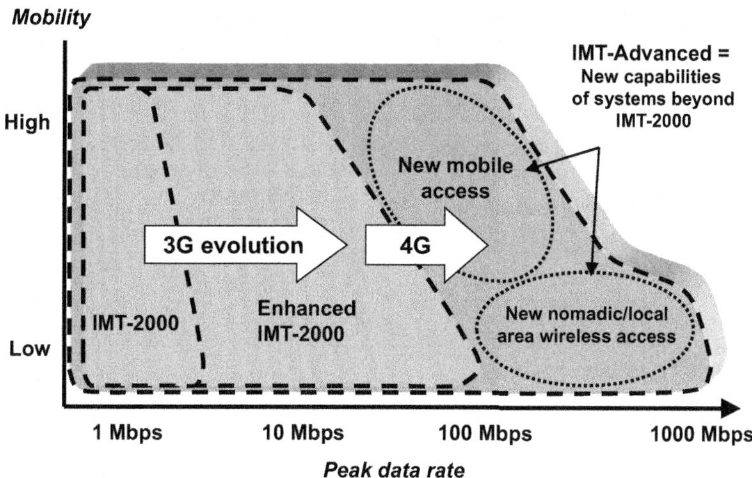


**FIGURE 1.5 Technology evolution toward IMT-A.**

ITU-R first concluded studies for IMT-Advanced of services and technologies, market forecasts, principles for standardization, estimation of spectrum needs, and identification of candidate frequency bands [5]. Evaluation criteria were agreed, where proposed technologies were to be evaluated according to a set of minimum technical requirements. All ITU members and other organizations were then invited to the process through a circular letter [6] in March 2008. After submission of candidate technologies in 2009, an evaluation was performed in cooperation with external bodies such as standards-developing organizations, industry fora, and national groups.

An evolution of LTE as developed by 3GPP was submitted as one candidate to the ITU-R evaluation. Although it actually was a new release (release 10) of the LTE system and thus an integral part of the continuing LTE development, the candidate was named LTE-Advanced for the purpose of ITU submission. 3GPP also set up its own set of technical requirements for LTE-Advanced, with the ITU-R requirements as a basis. HSPA was also developed into release 11 and, though not formally submitted to ITU, meets all of the IMT-Advanced requirements. The target of the process was always harmonization of the candidates through consensus building. ITU-R determined in October 2010 that two technologies will be included in the first release of IMT-Advanced, those two being LTE from release 10 ("LTE-Advanced") and WirelessMAN-Advanced [7] based on the IEEE 802.16m specification. The two can be viewed as the "family" of IMT-Advanced technologies as shown in Figure 1.6. The main IMT-Advanced recommendation, identifying the IMT-Advanced *radio interface specifications*, is ITU-R M.2012 [8]. As for the corresponding IMT-2000 specification, it contains an overview of each radio interface, followed by a list of references to the detailed specifications.

Looking further into the future, a new recommendation, "Framework and Overall Objectives of Future Development of IMT for 2020 and Beyond," is initiated within ITU-R WP5D. The recommendation is a first step for defining developments of IMT

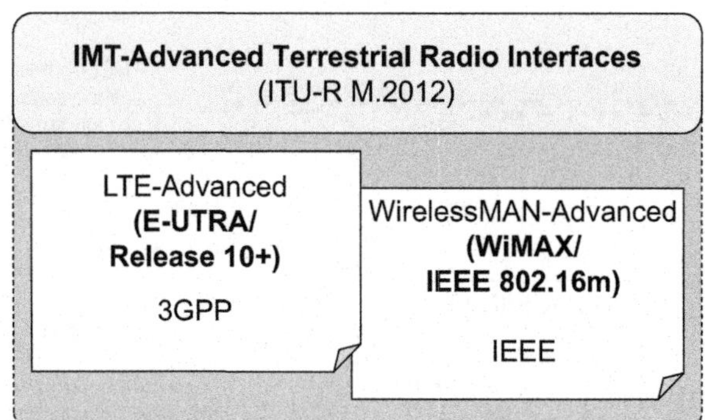


**FIGURE 1.6 IMT-A technology family.**

in the future, looking at the future roles of IMT and how it can serve the society, considering market, user, and technology trends and spectrum implications. As a parallel activity, ITU-R WP5D is also developing a report on "Future technology trends of terrestrial IMT systems," with a focus on the time period 2015–2020. The report will cover trends of future IMT technology aspects by looking at the technical and operational characteristics of IMT systems and how they are improved with the evolution of IMT technologies.

## 1.3 SPECTRUM FOR 3G AND SYSTEMS BEYOND 3G

Another major activity within ITU-R has been to identify globally available spectrum, suitable for IMT systems. The spectrum work has involved studies of spectrum sharing between IMT and other technologies. Adequate spectrum availability and globally harmonized spectrum are identified as essential for IMT-Advanced.

Spectrum for 3G was first identified at the *World Administrative Radio Congress* WARC-92, where 230 MHz was identified as intended for use by national administrations that want to implement IMT-2000. The so-called IMT-2000 "core band" at 2 GHz is in this frequency range and was the first band where 3G systems were deployed.

Additional spectrum was identified for IMT-2000 at later World Radio communication conferences. WRC-2000 identified the existing 2G bands at 800/900 MHz and 1800/1900 MHz plus an additional 190 MHz of spectrum at 2.6 GHz, all for IMT-2000. As additional spectrum for IMT-2000, WRC'07 identified a band at 450 MHz, the so-called "digital dividend" at 698–806 MHz, plus an additional 300 MHz of spectrum at higher frequencies. The applicability of these new bands varies on a regional and national basis. WRC'12 did not identify any additional spectrum allocations for IMT, but the issue was put on the agenda for WRC'15. It was also determined to study the use of the band 694–790 MHz for mobile services in Region 1 (Europe, Middle East, and Africa).

The worldwide frequency arrangements for IMT are outlined in ITU-R recommendation M.1036 [9], which is presently being updated with the arrangements for the most recent frequency bands added. The recommendation outlines the regional variations in how the bands are implemented and also identifies which parts of the spectrum are paired and which are unpaired. For the paired spectrum, the bands for uplink (mobile transmit) and downlink (base station transmit) are identified for *Frequency-Division Duplex* (FDD) operation. The unpaired bands can, for example, be used for *Time-Division Duplex* (TDD) operation. Note that the band that is most globally deployed for HSPA is still 2 GHz.

The same bands that were originally defined for IMT-2000 and used for 3G deployment are also being used for 4G, including LTE-Advanced deployment. The latest version of the ITU-R spectrum recommendation M.1036 [9] is renamed with a more generic title having the identifier "IMT" instead of "IMT-2000." Some regions and countries have issued licenses for new spectrum identified at WRC-2000 and WRC'07 with the intention of allowing 4G deployments, but in many cases, new spectrum is licensed on a technology-neutral basis. Many new deployments are either 4G or 3G, but some are also a mix of 4G and 3G, and some more also include 2G, since significant parts of the spectrum still used for 2G.

The somewhat diverging arrangement between regions of the frequency bands assigned to 3G means that there is not a single band that can be used for 3G roaming worldwide. Large efforts have, however, been put into defining a minimum set of bands that can be used to provide roaming. In this way, multi-band devices can provide efficient worldwide roaming for 3G.

3GPP first defined UTRA in release 99 for the 2-GHz bands, with $2 \times 60$ MHz for UTRA FDD and $20 + 15$ MHz of unpaired spectrum for UTRA TDD. A separate definition was also made for the use of UTRA in the US PCS bands at 1900 MHz. The concept of frequency bands with separate and release-independent requirements was defined in release 5 of the 3GPP specifications. The release-independence implies that a new frequency band added at a later release also can be implemented for earlier releases. All bands are also defined with consideration of what other bands may be deployed in the same region through special coexistence requirements for both base stations and UEs. These tailored requirements enable coexistence between 4G and 3G (and 2G) deployments in different bands in the same geographical area and even for colocation of base stations at the same sites using different bands.

## 1.4 THE MOTIVATIONS BEHIND CONTINUING HSPA EVOLUTION

Before entering the detailed discussion on technologies being used or considered for the evolution of HSPA mobile communication, it is important to understand the motivation for this evolution, that is, understanding the underlying driving forces. This section will try to highlight some of the driving forces, giving the reader an understanding of where the technical requirements and solutions are coming from.

### 1.4.1 GENERAL DRIVING FORCES

A key factor for success in any business is to understand the forces that will drive the business in the future. This is in particular true for the mobile-communication industry, where the rapid growth of the number of subscribers and the global presence of the technologies have attracted several new players that want to be successful. Both new operators and new vendors try to compete with the existing operators and vendors by adopting new technologies and standards to provide new and existing services better and at a lower cost than earlier systems. The existing operators and vendors will, of course, also follow or drive new technologies to stay ahead of competition. Thus, there is a key driving force in staying competitive or becoming competitive.

#### *1.4.1.1 Technology advancements*

Technology advancements in many areas make it possible to build devices that were not possible 20, 10, or even 5 years ago. Even though Moore's law[5] is not a law of physics, it gives an indication of the rapid technology evolution for integrated circuits. This evolution enables faster processing/computing and more memory in smaller devices at lower cost. Similarly, the rapid development of color touch-screens, small digital cameras, etc. makes it possible to envisage services to a device that were seen as utopian 10 years ago.

The size and weight of UEs have reduced dramatically during the past 20 years. The standby and talk times have also been extended significantly and the end users do not need to re-charge their devices every day. Simple black-and-white (or brown-and-gray) numerical screens have evolved into color screens capable of showing digital photos at good quality and further into the touch-screens common on today's smartphones and tablets. Mega-pixel-capable digital cameras have been added, making the device more attractive to use. Thus, the mobile device has become a multi-purpose device, not only a mobile phone for voice communications.

On the network side, a number of technologies have emerged that have altered the ways in which networks can be built. Base station miniaturization enables low cost, small cell deployments, while faster and more cost-effective backhaul mechanisms enable more varied deployments. In larger sites, advanced designs that integrate transceiver electronics and antenna systems reduce losses, energy costs, form factors, and site costs. In the longer term, network coordination and relocation of network functionality have the potential to change the topology of networks.

[5]Moore's law is an empirical observation, and states that with the present rate of technological development, the complexity of an integrated circuit, with respect to minimum component cost, will double in about 18 months.

#### *1.4.1.2 Services*

Delivering services to the end users is the fundamental goal of any mobile-communication system. Because it is impossible to know what services will be popular and because service possibilities and offers will differ with time and possibly also among countries, the future mobile-communication systems will need to be adaptive to the changing service environment. Luckily, there are a few known key services that span the technology space:

- *Real-time gaming applications*: These have the characteristic to require small amounts of data (game update information) relatively frequently with short delays.[6] Only a limited delay jitter is tolerable.
- *Voice*: This has the characteristic to require small amounts of data (voice packets) frequently with low delay jitter. The end-to-end delay has to be small enough not to be noticeable.[7]
- *Interactive file download and upload applications*: These have the characteristics of requiring low delay and high data rates.
- *Background file download and upload applications*: These have the characteristics of accepting lower bit rates and longer delays. E-mail (mostly) is an example of background file download and upload.
- *App operation and synchronization*: The development of the App ecosystem leads to a demand for downloading, synchronizing, and maintaining Apps.
- *Broadcast*: This has the characteristics of streaming downlink to many users at the same time requiring low delay jitter. The service can tolerate delays as long as it is approximately the same delay for all users in the neighborhood.
- *Machine type communications (MTC):* Machine-to-machine services are likely to expand enormously in coming years and in reality represent a plethora of different requirement types. Some general needs, however, include low cost, ultra-long battery life, improved link budget, and potentially high reliability. MTC will be further discussed in Chapter 19.
- *Device to device (D2D).* This very new type of traffic is one in which data are transmitted directly between devices, mainly under the control of the network. The principal application is support of specific groups, such as emergency services. High reliability and the ability to operate when a network is lacking capacity or not operating correctly are key needs of this type of service.

A mobile-communication system is designed to handle these services and the services in between (see Figure 1.7). The demands of these services and the mixture of services vary and evolve, such that even if general categories of services are known, the specific demands on the networks are difficult to predict.

[6]The faster the data is delivered the better. Expert Counter Strike players look for game servers with a ping time of less than 50 ms.

[7]In 3G systems the end-to-end delay requirement for circuit-switched voice is approximately 400 ms. This delay is not disturbing humans in voice communications.

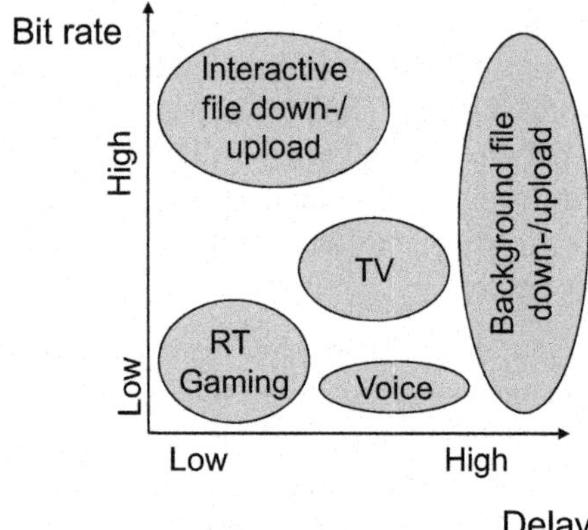


FIGURE 1.7 The bit rate–delay service space that is important to cover when designing a new cellular system.

### 1.4.1.3 User expectations

In addition to the services themselves, user expectations evolve, too. In the first days of mobile communications, voice quality and dropped calls were key issues. As the quality of voice calls improved and mobile data began to play a larger role, peak rates became of greater importance. The development of the "App" ecosystem has significantly changed user expectations. With today's devices, universal App availability, latency, and battery life play as important a role as data rate in shaping users' expectations of services.

### 1.4.1.4 Operator business models

In the first days of mobile communications, operators could make money by providing voice services and charging per time unit. This business model began to shift with the introduction of flat rate charging for data usage. Flat rate charging led to the explosive success of mobile broadband but also broke the link between the cost of a service charged to a user and the cost of providing services. Furthermore, making Internet connectivity available to end users devalued the service offering from operators, reducing it toward being a bit-pipe provider.

Operators need to respond by providing differentiation to users by means of high-quality networks, service differentiation, and innovative charging structures. Mobile technologies, such as QoS-related traffic steering, play a key role in enabling operators to maintain a viable business in a rapidly evolving and challenging environment.

### 1.4.1.5 Cost

Another important driving factor for future mobile-communication systems is the cost of the service provisioning. Cost takes two forms: the so-called CAPEX, which is related to the cost of buying and owning the networks, and OPEX, which is the running cost for the networks. Factors contributing to CAPEX include the cost of the base stations, the efficiency of the base stations (that is, how much capacity each base station can serve), and the cost of devices (which are often subsidized by operators). Factors that contribute to OPEX include energy costs, site rental costs, and site maintenance costs.

### 1.4.2 HSPA-SPECIFIC EVOLUTION DRIVERS AND PHILOSOPHY

WCDMA, HSDPA, and HSUPA are in commercial operation throughout the world with established network node sites, especially the base station sites with their antenna arrangements and hardware[8]. This equipment is serving millions of UEs with different characteristics and supported 3GPP releases. These UEs need to be supported by the WCDMA operators for many more years.

The philosophy of the HSPA Evolution work is to continue to add new and more advanced technical features for the reasons outlined in the previous sections and at the same time be able to serve the already existing UEs. This is the successful strategy of GSM that has added new features constantly since its introduction in the early 1990s. The success stems from the fact that there are millions of existing UEs at the launch time of the new features that can take the cost of the upgrade of the network for the initially few new UEs before the UE fleet is upgraded. The time it takes to upgrade the UE fleet is different from country to country, but a rule of thumb is that a UE is used for two years before it is replaced. For HSPA Evolution, that means that millions of HSPA-capable UEs need to be supported at launch. In other words, HSPA Evolution needs to be backward compatible with the previous releases in the sense that it is possible to serve UEs of earlier releases of WCDMA on the same carrier as HSPA-Evolution-capable UEs.

The backward compatibility requirement on the HSPA Evolution puts certain constraints on the technology; for example, the physical layer fundamentals need to be the same as for WCDMA release 99.

## REFERENCES


[1] Ericsson Mobility report, November 2013 <www.ericsson.com/ericsson-mobility-report>.

[2] Dahlman, Parkvall, Sköld, '4G: LTE/LTE-Advanced for mobile broadband. 2nd ed.'. Academic Press, London; 2013.

[3] 'Detailed Specifications of the Radio Interfaces of International Mobile Telecommunications-2000 (IMT-2000)', ITU-R, Recommendation ITU-R M. 1457-11, February 2013.

[4] ITU-R, 'Principles for the process of development of IMT Advanced', Resolution ITU-R 57 October 2007.

[5] ITU-R, 'Framework and overall objectives of the future development of IMT-2000 and systems beyond IMT-2000', Recommendation ITU-R M. 1645 June 2003.

[6] ITU-R, 'Invitation for submission of proposals for candidate radio interface technologies for the terrestrial components of the radio interface(s) for IMT advanced and invitation to participate in their subsequent evaluation', ITU-R SG5 Circular letter 5/LCCE/2, March 2008.


[8]When operating with HSDPA and Enhanced UL, the system is known as HSPA.

[7] ITU-R, 'ITU paves the way for next generation 4G mobile technologies; ITU-R IMT-advanced 4G standards to usher new era of mobile broadband communications', ITU press release, October 2010.

[8] ITU-R WP5D, 'Recommendation ITU-R M. 2012. Detailed specifications of the terrestrial radio interfaces of International Mobile Telecommunications Advanced (IMT-Advanced), January 2012.

[9] ITU-R, 'Frequency arrangements for implementation of the terrestrial component of international mobile telecommunications 2000 (IMT-2000) in the bands 806–960 MHz, 1710–2025 MHz, 2110–2200 MHz, 2500–2690 MHz, Recommendation ITU-R M1036-4, March 2012.

PART

# Technologies for HSPA

CHAPTER

# High data rates in mobile communication

2

## CHAPTER OUTLINE



As discussed in Chapter 1, one main target for the evolution of HSPA mobile communication is to provide the possibility for significantly higher end user data rates compared to what is achievable with, for example, the first releases of the 3G standards. This includes the possibility for higher peak data rates but, as pointed out in the previous chapter, even more so the possibility for significantly higher data rates over the entire cell area, also including, for example, users at the cell-edge. The initial part of this chapter will briefly discuss some of the more fundamental constraints that exist in terms of what data rates can actually be achieved in different scenarios. This will provide a background to subsequent discussions in the latter part of the chapter, as well as in the following chapters, concerning different means to increase the achievable data rates in different mobile communication scenarios.

## 2.1 HIGH DATA RATES: FUNDAMENTAL CONSTRAINTS

In [1], Shannon provided the basic theoretical tools needed to determine the maximum rate, also known as the *channel capacity*, by which information can be transferred over a given communication channel. Although relatively complicated in the general case, for the special case of communication over a channel, for example a radio link, only impaired by additive white Gaussian noise, the channel capacity $C$ is given by the relatively simple expression [2]

$$C = BW \cdot \log_2\left(1+\frac{S}{N}\right), \tag{2.1}$$


**HSPA Evolution: The Fundamentals for Mobile Broadband. DOI: 10.1016/B978-0-08-099969-2.00002-8**

where $BW$ is the bandwidth available for the communication, $S$ denotes the received signal power, and $N$ denotes the power of the white noise impairing the received signal.

Already from (2.1) it should be clear that the two fundamental factors limiting the achievable data rate are the available received signal power, or more generally the available signal-power-to-noise-power ratio $S/N$, and the available bandwidth. To further clarify how and when these factors limit the achievable data rate, assume communication with a certain information rate $R$. The received signal power can then be expressed as

$$S = E_b \cdot R, \tag{2.2}$$

where $E_b$ is the received energy per information bit. Furthermore, the noise power can be expressed as

$$N = N_0 \cdot BW, \tag{2.3}$$

where $N_0$ is the constant noise power spectral density measured in W/Hz.

Clearly, the information rate can never exceed the channel capacity. Together with the above expressions for the received signal power and noise power, this leads to the inequality

$$R \le C = BW \cdot \log_2\left(1+\frac{S}{N}\right) = BW \cdot \log_2\left(1+\frac{E_b \cdot R}{N_0 \cdot BW}\right), \tag{2.4}$$

or, by defining the radio link *bandwidth utilization* $\gamma = R/BW$,

$$\gamma \le \log_2\left(1+\gamma \cdot \frac{E_b}{N_0}\right). \tag{2.5}$$

This inequality can be reformulated to provide a lower bound on the required received energy per information bit, normalized to the noise power density, for a given bandwidth utilization $\gamma$

$$\frac{E_b}{N_0} \ge \min\left\{\frac{E_b}{N_0}\right\} = \frac{2^\gamma - 1}{\gamma}. \tag{2.6}$$

The rightmost expression, that is, the minimum required $E_b/N_0$ at the receiver as a function of the bandwidth utilization, is illustrated in Figure 2.1. As can be seen, for bandwidth utilizations significantly less than one, that is, for information rates substantially smaller than the utilized bandwidth, the minimum required $E_b/N_0$ is relatively constant, regardless of $\gamma$. For a given noise power density, any increase of the information data rate then implies a similar relative increase in the minimum required signal power $S = E_b \cdot R$ at the receiver. On the other hand, for bandwidth utilizations larger than one, the minimum required $E_b/N_0$ increases rapidly with $\gamma$. Thus, in case of data rates in the same order as or larger than the communication

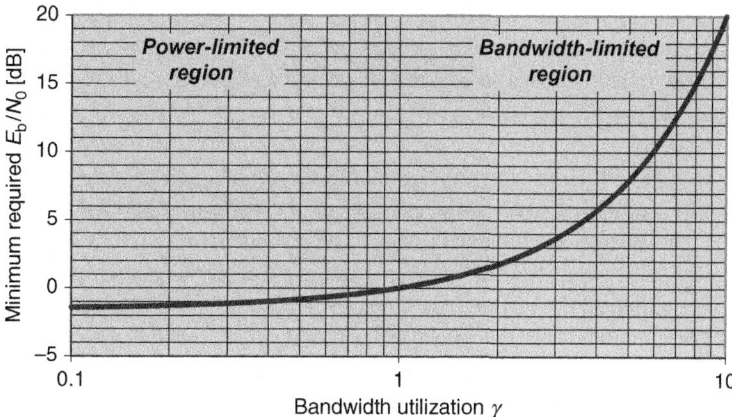


**FIGURE 2.1 Minimum required $E_b/N_0$ at the receiver as a function of bandwidth utilization.**

bandwidth, any further increase of the information data rate, without a corresponding increase in the available bandwidth, implies a larger, eventually much larger, relative increase in the minimum required received signal power.

## 2.1.1 HIGH DATA RATES IN NOISE-LIMITED SCENARIOS

From the discussion above, some basic conclusions can be drawn regarding the provisioning of higher data rates in a mobile communication system when the thermal noise level is greater than the level of other sources of interference, and hence noise is the main source of radio link impairment (a *noise-limited* scenario):

- The data rates that can be provided in such scenarios are always limited by the available received signal power or, in the general case, the received signal-power-to-noise-power ratio. Furthermore, any increase of the achievable data rate within a given bandwidth will require at least the same relative increase of the received signal power. At the same time, if sufficient received signal power can be made available, basically any data rate can, at least in theory, be provided within a given limited bandwidth.
- In case of low bandwidth utilization, that is, as long as the radio link data rate is substantially lower than the available bandwidth, any further increase of the data rate requires *approximately the same* relative increase in the received signal power. This can be referred to as *power-limited* operation (in contrast to *bandwidth-limited* operation, see below) as, in this case, an increase in the available bandwidth does not substantially impact what received signal power is required for a certain data rate.
- On the other hand, in case of high bandwidth utilization, that is, in case of data rates in the same order as or exceeding the available bandwidth, any further increase in the data rate requires *a much larger* relative increase in the received

signal power unless the bandwidth is increased in proportion to the increase in data rate. This can be referred to as *bandwidth-limited operation* as, in this case, an increase in the bandwidth will reduce the received signal power required for a certain data rate.

Thus, to make efficient use of the available received signal power or, in the general case, the available signal-to-noise ratio, the transmission bandwidth should at least be of the same order as the data rates to be provided.

Assuming a constant transmit power, the received signal power can always be increased by reducing the distance between the transmitter and the receiver, thereby reducing the attenuation of the signal as it propagates from the transmitter to the receiver. Thus, in a noise-limited scenario, it is at least in theory always possible to increase the achievable data rates, assuming that one is prepared to accept a reduction in the transmitter/receiver distance, that is, a reduced range. In a mobile communication system, this would correspond to a reduced cell size and thus the need for more cell sites to cover the same overall area. Providing data rates in the same order as or larger than the available bandwidth, that is, with a high bandwidth utilization, would require a significant cell size reduction. Alternatively, one has to accept that the high data rates are only available for *User Equipments* (UEs) in the center of the cell, that is, not over the entire cell area.

Another means to increase the overall received signal power for a given transmit power is the use of additional antennas at the receiver side, also known as *receive antenna diversity*. Multiple receive antennas can be applied at the base station (that is, for the uplink) or at the UE (that is, for the downlink). By proper combining of the signals received at the different antennas, the signal-to-noise ratio after the antenna combining can be increased in proportion to the number of receive antennas, thereby allowing for higher data rates for a given transmitter/receiver distance.

Multiple antennas can also be applied at the transmitter side, typically at the base station, and be used to focus a given total transmit power in the direction of the receiver, that is, toward the target UE. This will increase the received signal power and thus, once again, allow for higher data rates for a given transmitter/receiver distance.

However, providing higher data rates by the use of multiple transmit or receive antennas is only efficient up to a certain level, that is, as long as the data rates are power limited rather than bandwidth limited. Beyond this point, the achievable data rates start to saturate and any further increase in the number of transmit or receive antennas, although leading to a correspondingly improved signal-to-noise ratio at the receiver, will only provide a marginal increase in the achievable data rates. This saturation in achievable data rates can be avoided though, by the use of multiple antennas at both the transmitter *and* the receiver, enabling what can be referred to as *spatial multiplexing*, often also referred to as *Multiple Input Multiple Output* (MIMO). Different types of multi-antenna techniques, including spatial multiplexing, will be discussed in more detail in Chapters 4 and 11.

An alternative to increasing the received signal power is to reduce the noise power, or more exactly the noise power density, at the receiver. This can, at least to some extent, be achieved by more advanced receiver RF design, allowing for a reduced receiver noise figure.

### 2.1.2 HIGHER DATA RATES IN INTERFERENCE-LIMITED SCENARIOS

The discussion above assumed communication over a radio link only impaired by noise. However, in actual mobile communication scenarios, interference from transmissions in neighbor cells, also referred to as *inter-cell interference*, is often the dominating source of radio link impairment, more so than noise. This is especially the case in small cell deployments with a high traffic load. Furthermore, in addition to inter-cell interference, there may in some cases also be interference from other transmissions *within the current cell*, also referred to as *intra-cell interference*.

In many respects, the impact of interference on a radio link is similar to that of noise. The basic principles discussed above apply also to a scenario where interference is the main radio link impairment:

- The maximum data rate that can be achieved in a given bandwidth is limited by the available signal-power-to-interference-power ratio.
- Providing data rates larger than the available bandwidth (high bandwidth utilization) is costly in the sense that it requires a disproportionally high signal-to-interference ratio.

Also similar to a scenario where noise is the dominating radio link impairment, reducing the cell size as well as the use of multiantenna techniques are key means to increase the achievable data rates also in an interference-limited scenario:

- Reducing the cell size will obviously reduce the number of users, and thus also the overall traffic, per cell. This will reduce the relative interference level and thus allow for higher data rates.
- Similar to the increase in signal-to-noise ratio, proper combining of the signals received at multiple antennas will also increase the signal-to-interference ratio after the antenna combining.
- The use of beamforming by means of multiple transmit antennas will focus the transmit power in the direction of the target receiver, leading to reduced interference to other radio links and thus improving the overall signal-to-interference ratio in the system.

One important difference between interference and noise is that interference, in contrast to noise, typically has a certain structure that makes it, at least to some extent, predictable and thus possible to further suppress or even remove completely. As an example, a dominant interfering signal may arrive from a certain direction in which case the corresponding interference can be further suppressed, or even completely removed, by means of *spatial processing* using multiple antennas at the receiver. This will be further discussed in Chapter 4. Also any differences in the spectrum properties between the target signal and an interfering signal can be used to suppress the interferer and thus reduce the overall interference level.

## 2.2 HIGHER DATA RATES WITHIN A LIMITED BANDWIDTH: HIGHER-ORDER MODULATION

As discussed in the previous section, providing data rates larger than the available bandwidth is fundamentally inefficient in the sense that it requires disproportionally high signal-to-noise and signal-to-interference ratios at the receiver. Still, bandwidth is often a scarce and expensive resource and, at least in some mobile communication scenarios, high signal-to-noise and signal-to-interference ratios can be made available, for example, in small cell environments with a low traffic load or for UEs close to the cell site. Future mobile communication systems, including the evolution of 3G mobile communication, should be designed to be able to take advantage of such scenarios; that is, they should be able to offer very high data rates within a limited bandwidth when the radio conditions so allow.

A straightforward means to provide higher data rates within a given transmission bandwidth is the use of *higher-order modulation*, implying that the modulation alphabet is extended to include additional signaling alternatives, thus allowing for more bits of information to be communicated per modulation symbol.

In case of QPSK modulation, which was the modulation scheme used for the downlink in the first releases of the 3G mobile communication standards (WCDMA and CDMA2000), the modulation alphabet consists of four different signaling alternatives. These four signaling alternatives can be illustrated as four different points in a two-dimensional plane (see Figure 2.2a). With four different signaling alternatives, QPSK allows for up to two bits of information to be communicated during each modulation symbol interval. By extending to 16-QAM modulation (Figure 2.2b), 16 different signaling alternatives are available. The use of 16-QAM thus allows for up to four bits of information to be communicated per symbol interval. Further extension to 64-QAM (Figure 2.2c), with 64 different signaling alternatives, allows for up to six bits of information to be communicated per symbol interval. At the same time, the bandwidth of the transmitted signal is, at least in principle, independent of the size of the modulation alphabet and mainly depends on the modulation rate, that is, the number of modulation symbols per second. The maximum bandwidth utilization, expressed in bits/s/Hz, of 16-QAM and 64-QAM are thus, at least in principle, two and three times that of QPSK, respectively.

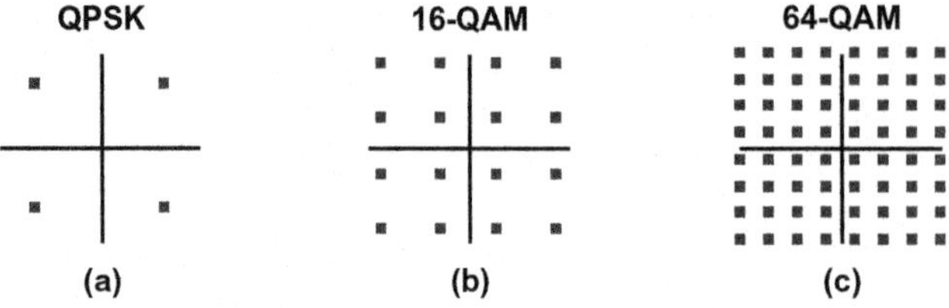


**FIGURE 2.2** Signal constellations for (a) QPSK, (b) 16-QAM, and (c) 64-QAM.

It should be pointed out that there are many other possible modulation schemes, in addition to those illustrated in Figure 2.2. One example is 8-PSK, consisting of eight signaling alternatives and thus providing up to three bits of information per modulation symbol. Readers are referred to [1] for a more thorough discussion on different modulation schemes.

The use of higher-order modulation provides the possibility for higher bandwidth utilization, that is, the possibility to provide higher data rates within a given bandwidth. However, the higher bandwidth utilization comes at the cost of reduced robustness to noise and interference. Alternatively expressed, higher-order modulation schemes, such as 16-QAM or 64-QAM, require a higher $E_b/N_0$ at the receiver for a given bit error probability, compared to QPSK. This is in line with the discussion in the previous section where it was concluded that high bandwidth utilization, that is, a high information rate within a limited bandwidth, in general requires a higher receiver $E_b/N_0$.

### 2.2.1 HIGHER-ORDER MODULATION IN COMBINATION WITH CHANNEL CODING

Higher-order modulation schemes such as 16-QAM and 64-QAM require, in themselves, a higher receiver $E_b/N_0$ for a given error rate, compared to QPSK. However, in combination with channel coding, the use of higher-order modulation will sometimes be more efficient, that is, require a lower receiver $E_b/N_0$ for a given error rate, compared to the use of lower-order modulation such as QPSK. This may, for example, occur when the target bandwidth utilization implies that, with lower-order modulation, no or very little channel coding can be applied. In such a case, the additional channel coding that can be applied by using a higher-order modulation scheme such as 16-QAM may lead to an overall gain in power efficiency compared to the use of QPSK.

As an example, if a bandwidth utilization of close to two information bits per modulation symbol is required, QPSK modulation would allow for very limited channel coding (channel coding rate close to one). On the other hand, the use of 16-QAM modulation would allow for a channel coding rate in the order of one half. Similarly, if a bandwidth efficiency close to four information bits per modulation symbol is required, the use of 64-QAM may be more efficient than 16-QAM modulation, taking into account the possibility for lower rate channel coding and corresponding additional coding gain in case of 64-QAM. It should be noted that this does not speak against the general discussion in Section 2.1 where it was concluded that transmission with high bandwidth utilization is inherently power inefficient. The use of rate 1/2 channel coding for 16-QAM obviously reduces the information data rate, and thus also the bandwidth utilization, to the same level as uncoded QPSK.

From the discussion above it can be concluded that, for a given signal-to-noise/interference ratio, a certain combination of modulation scheme and channel coding rate is optimal in the sense that it can deliver the highest bandwidth utilization (the highest data rate within a given bandwidth) for that signal-to-noise/interference ratio.

### 2.2.2 VARIATIONS IN INSTANTANEOUS TRANSMIT POWER

A general drawback of higher-order modulation schemes such as 16-QAM and 64-QAM, where information is encoded also in the instantaneous amplitude of the modulated signal, is that the modulated signal will have larger variations, and thus also larger peaks, in its instantaneous power. This can be seen from Figure 2.3, which illustrates the distribution of the instantaneous power, more specifically the probability that the instantaneous power is above a certain value, for QPSK, 16-QAM, and 64-QAM, respectively. Clearly, the probability for large peaks in the instantaneous power is higher in case of higher-order modulation.

Larger peaks in the instantaneous signal power imply that the transmitter power amplifier must be over-dimensioned to avoid the power amplifier nonlinearities, occurring at high instantaneous power levels, that cause corruption to the signal to be transmitted. As a consequence, the power amplifier efficiency will be reduced, leading to increased power consumption. In addition, there will be a negative impact on the power amplifier cost. Alternatively, the average transmit power must be reduced, implying a reduced range for a given data rate. High power amplifier efficiency is especially important for the UE, that is, in the uplink direction, because of the importance of low UE power consumption and cost. For the base station, high power amplifier efficiency, though far from irrelevant, is still somewhat less important. Thus, large peaks in the instantaneous signal power are less of an issue for the downlink compared to the uplink and, consequently, higher-order modulation is more suitable for the downlink compared to the uplink.

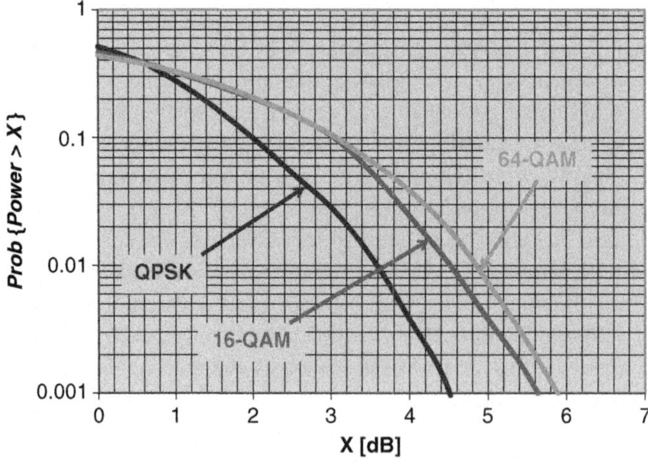


**FIGURE 2.3** Distribution of instantaneous power for different modulation schemes. Average power is the same in all cases.

## 2.3 WIDER BANDWIDTH INCLUDING MULTI-CARRIER TRANSMISSION

As was shown in Section 2.1, transmission with a high bandwidth utilization is fundamentally power inefficient in the sense that it will require disproportionally high signal-to-noise and signal-to-interference ratios for a given data rate. Providing very high data rates within a limited bandwidth, for example by means of higher-order modulation, is thus only possible in situations where relatively high signal-to-noise and signal-to-interference ratios can be made available, for example, in small cell environments with low traffic load or for UEs close to the cell site.

Instead, to provide high data rates as efficiently as possible in terms of required signal-to-noise and signal-to-interference ratios, implying as good coverage as possible for high data rates, the transmission bandwidth should be at least of the same order as the data rates to be provided.

Having in mind that the provisioning of higher data rates with good coverage is one of the main targets for the evolution of HSPA mobile communication, it can thus be concluded that support for even wider transmission bandwidth is an important part of this evolution.

However, there are several critical issues related to the use of wider transmission bandwidths in a mobile communication system:

- Spectrum is, as already mentioned, often a scarce and expensive resource, and it may be difficult to find spectrum allocations of sufficient size to allow for very wideband transmission, especially at lower frequency bands.
- The use of wider transmission and reception bandwidths has an impact on the complexity of the radio equipment, both at the base station and at the UE. As an example, a wider transmission bandwidth has a direct impact on the transmitter and the receiver sampling rates, and thus on the complexity and power consumption of digital-to-analog and analog-to-digital converters as well as front-end digital signal processing. *Radio frequency* (RF) components are also, in general, more complicated to design and more expensive to produce, the wider the bandwidth they are to handle.

The two issues above are mainly outside the scope of this book. However, a more specific technical issue related to wider band transmission is the increased corruption of the transmitted signal due to time dispersion on the radio channel. Time dispersion occurs when the transmitted signal propagates to the receiver via multiple paths with different delays (see Figure 2.4a). In the frequency domain, a time dispersive channel corresponds to a non-constant channel frequency response, as illustrated in Figure 2.4b. This radio channel *frequency selectivity* will corrupt the frequency domain structure of the transmitted signal and lead to higher error rates for given signal-to-noise/interference ratios. Every radio channel is subject to frequency selectivity, at least to some extent. However, the extent to which the frequency selectivity impacts the radio communication depends on the bandwidth of the transmitted signal with, in general, larger impact for wider band transmission. The amount of radio channel

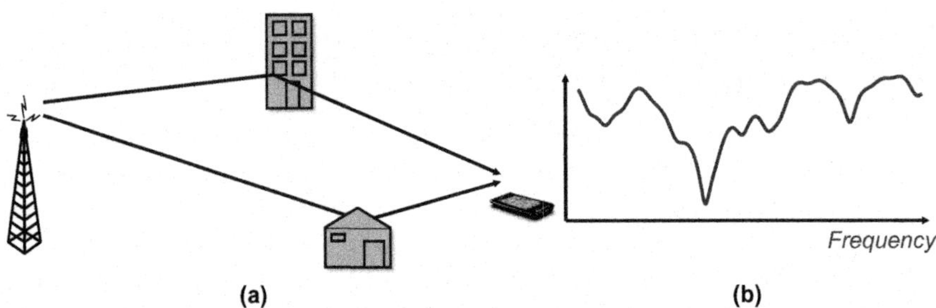


**FIGURE 2.4 Multipath propagation causing time dispersion (a) and radio channel frequency selectivity (b).**

frequency selectivity also depends on the environment, with typically less frequency selectivity (less time dispersion) in case of small cells and in environments with few obstructions and potential reflectors, such as rural environments.

It should be noted that Figure 2.4b illustrates a "snapshot" of the channel frequency response. As a UE is moving through the environment, the detailed structure of the multipath propagation, and thus also the detailed structure of the channel frequency response, may vary rapidly in time. The rate of the variations in the channel frequency response is related to the channel *Doppler spread*, $f_D$, defined as $f_D = v/c \cdot f_c$, where $v$ is the speed of the UE, $f_c$ is the carrier frequency (for example, 2 GHz), and $c$ is the speed of light.

Receiver side *equalization* [1] has for many years been used to counteract signal corruption due to radio channel frequency selectivity. Equalization has been shown to provide satisfactory performance with reasonable complexity at least up to bandwidths corresponding to the WCDMA bandwidth of 5 MHz (see, for example, [3]). However, if the transmission bandwidth is further increased, the complexity of straightforward high performance equalization starts to become a serious issue. One option is then to apply less optimal equalization, with a corresponding negative impact on the equalizer capability to counteract the signal corruption due to radio channel frequency selectivity and thus a corresponding negative impact on the radio link performance.

An alternative approach is to consider specific transmission schemes and signal designs that allow for good radio link performance also in case of substantial radio channel frequency selectivity without a prohibitively large receiver complexity. One good example of this is the use of different types of *multi-carrier transmission*, that is, transmitting an overall wider bandwidth signal as several frequency multiplexed signals.

### 2.3.1 MULTI-CARRIER TRANSMISSION

One way to increase the overall transmission bandwidth, without suffering from increased signal corruption due to radio channel frequency selectivity, is the use of so-called *multi-carrier transmission*.

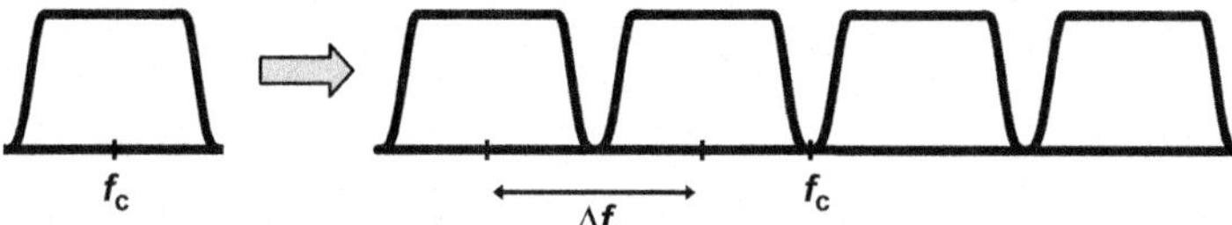


**FIGURE 2.5 Extension to wider transmission bandwidth by means of multi-carrier transmission.**

A drawback of the kind of multi-carrier evolution outlined in Figure 2.5, where an existing more narrowband radio access technology is extended to a wider overall transmission bandwidth by the parallel transmission of $M$ more narrowband carriers, is that the spectrum of each subcarrier typically does not allow for very tight subcarrier "packing." This is illustrated by the "valleys" in the overall multi-carrier spectrum outlined in Figure 2.5. This has a somewhat negative impact on the overall bandwidth efficiency of this kind of multi-carrier transmission.

As an example, consider a WCDMA multi-carrier evolution toward wider bandwidth. WCDMA has a modulation rate, also referred to as the *WCDMA chip rate*, of $f_{cr}$ = 3.84 Mchips/s. However, because of spectrum shaping, even the theoretical WCDMA spectrum, not including spectrum widening due to transmitter imperfections, has a bandwidth that significantly exceeds 3.84 MHz. More specifically, as can be seen in Figure 2.6, the theoretical WCDMA spectrum has a *raised cosine* shape with roll-off $\alpha = 0.22$. As a consequence, the bandwidth outside of which the WCDMA theoretical spectrum equals zero is approximately 4.7 MHz (see right part of Figure 2.6).

For a straightforward multi-carrier extension of WCDMA, the subcarriers must thus be spaced approximately 4.7 MHz from each other to completely avoid inter-subcarrier interference. It should be noted though that a smaller subcarrier spacing can be used with only limited inter-subcarrier interference.

A second drawback of multi-carrier transmission is that, similar to the use of higher-order modulation, the parallel transmission of multiple carriers will lead to

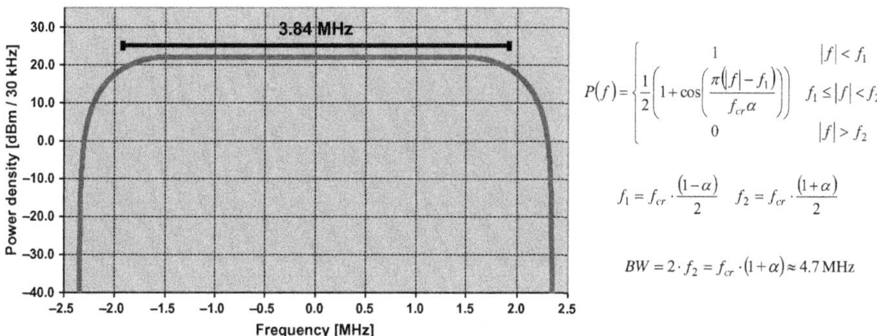


**FIGURE 2.6 Theoretical WCDMA spectrum. Raised cosine shape with roll-off $\alpha = 0.22$.**

larger variations in the instantaneous transmit power. Thus, similar to the use of higher-order modulation, multi-carrier transmission will have a negative impact on the transmitter power amplifier efficiency, implying increased transmitter power consumption and increased power amplifier cost. Alternatively, the average transmit power must be reduced, implying a reduced range for a given data rate. For this reason, similar to the use of higher-order modulation, multi-carrier transmission is more suitable for the downlink (base-station transmission), compared to the uplink (UE transmission), because of the higher importance of high power amplifier efficiency at the UE.

The main advantage with the kind of multi-carrier extension outlined in Figure 2.5 is that it provides a very smooth evolution, in terms of both radio equipment and spectrum, of an already existing radio-access technology to wider transmission bandwidth and a corresponding possibility for higher data rates, especially for the downlink. In essence, this kind of multi-carrier evolution to wider bandwidth can be designed so that, for legacy UEs not capable of multi-carrier reception, each downlink "subcarrier" will appear as an original, more narrowband carrier, whereas, for a multi-carrier-capable UE, the network can make use of the full multi-carrier bandwidth to provide higher data rates. Furthermore, the UE can take advantage of carriers that are not contiguous in spectrum, or even not in the same frequency bands.

## REFERENCES

[1] C.E. Shannon, A mathematical theory of communication, Bell Syst. Tech. J. 27 (July and October) (1948) 379–423, 623–656.

[2] J.G. Proakis, Digital Communications, McGraw-Hill, New York, (2001).

[3] G. Bottomley, T. Ottosson, Y.-P. Eric Wang, A generalized RAKE receiver for interference suppression, IEEE J. Select. Areas Commun. 18 (8) (2000) 1536–1545.

CHAPTER

# CDMA transmission principles

3

## CHAPTER OUTLINE



HSPA is built upon release 99 Wideband CDMA, which is a technology based on the concept of spread spectrum transmission. The spread spectrum technique emerged out of military research as a basis for communications that aim to flexibly multiplex multiple users and services and to provide robustness against external interference. This chapter reviews some of the basics of spread spectrum techniques and building blocks that are required for a successful communications link.

## 3.1 SPREAD SPECTRUM BASICS

A spread spectrum communications system is one that is built upon the principle of transmitting information signals over a much wider bandwidth than is strictly necessary for transferring the information. By transmitting over a larger bandwidth, robustness against external narrowband interference is increased, since the wider the bandwidth of any transmitted signal the lower will be the relative influence of interference over a small part of the bandwidth. Although from a single-link point of view, spread spectrum transmission may seem like very inefficient use of spectrum, this is not the case on a system level as spread spectrum techniques allow for simultaneous multiplexing of multiple transmissions in the same bandwidth. Thus, if transmissions to/from $K$ users are multiplexed using $K$ times the bandwidth that would be required without spreading, then the bandwidth utilization is not compromised (Figure 3.1). Furthermore, if the number $K$ of users varies, the spreading applied to each individual user can be varied in such a manner as to ensure that all users are served while bandwidth utilization is maintained.[1]

[1]Wideband CDMA, in particular in the uplink, initially multiplexed many users together with high so-called spreading factor; however, in recent years there has been a trend to time multiplex users with lower values of the spreading factor because TDM generally provides a higher degree of orthogonality than CDM.

HSPA Evolution: The Fundamentals for Mobile Broadband. DOI: 10.1016/B978-0-08-099969-2.00003-X

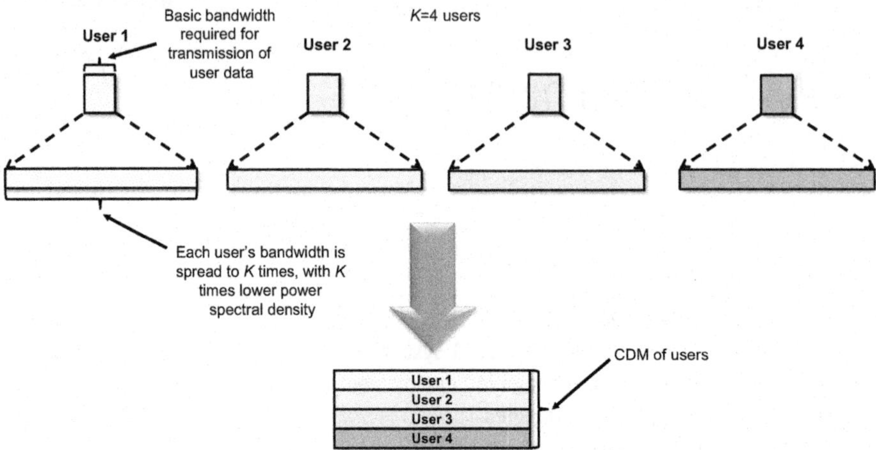


**FIGURE 3.1 The principle of spread spectrum based user multiplexing.**

A variety of spread spectrum techniques exist, such as *Frequency Hopping Spread Spectrum* (FHSS) and *Direct Sequence Spread Spectrum* (DSSS). The focus of this chapter is on DSSS.

Consider an information sequence coded/modulated into a sequence of $N$ modulation symbols. In order to spread the sequence using DSSS, an $M$-chip-long spreading sequence is selected. The $M$ chips of the spreading sequence are replicated $N$ times, with each of the replications being multiplied by one of the $N$ symbols (see Figure 3.2). The symbols of this spread sequence are referred to as "chips."

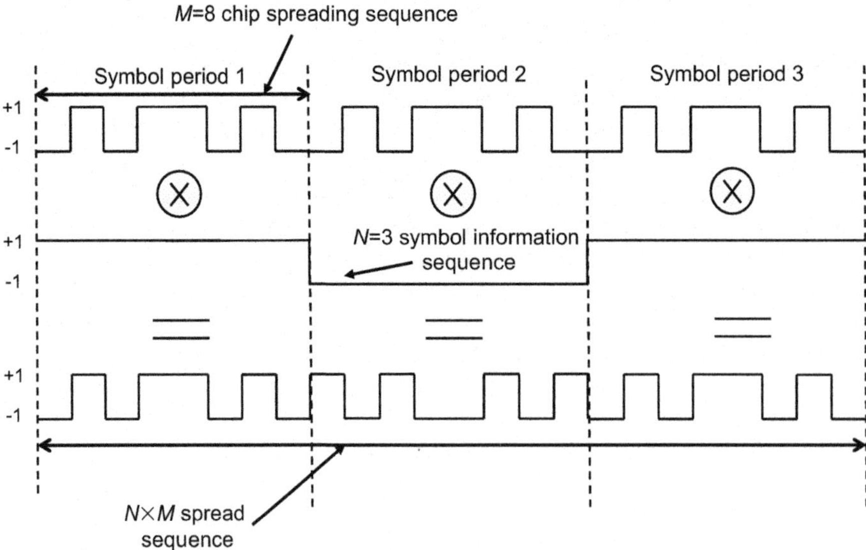


**FIGURE 3.2 Spreading of an information sequence.**

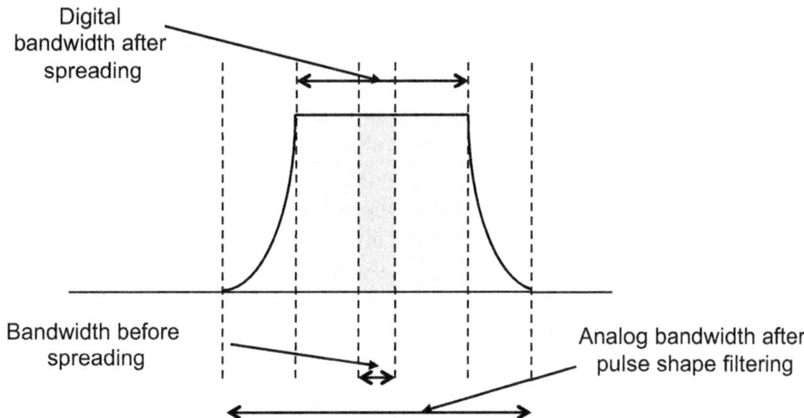


**FIGURE 3.3 The impact of pulse shape filtering to the occupied bandwidth.**

The resulting sequence of $N \times M$ chips must be transmitted within the same time as the original $N$ symbol message in order to maintain the data rate. In the frequency domain, this implies an increase in the occupied spectrum for the signal of $M$ times. Hence, $M$ is referred to as the "spreading factor." In practice, a pulse-shaping filter is then applied, such as a *root raised cosine* (RRC) filter. The overall occupied bandwidth of the transmitted signal is somewhat larger than $M$ times the bandwidth of the information sequence, as illustrated in Figure 3.3.

In the absence of other types of channel effects, the receiver can detect the information sequence by means of a synchronized correlation of the spreading sequence with the received sequence (Figure 3.4).

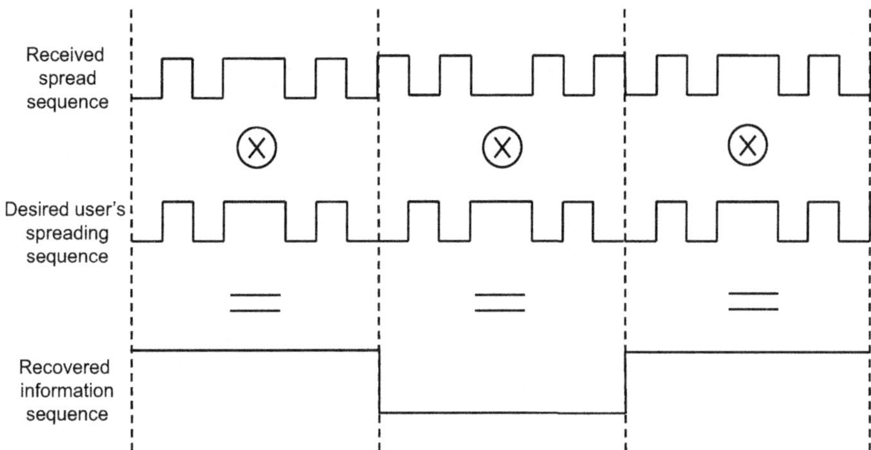


**FIGURE 3.4 Simple despreading of a spread sequence.**

The spreading sequence may in principle be a sequence of random chips. If this is the case, then transmissions to/from different users may be assigned different random sequences. In order to detect a sequence transmitted to or from a particular user, the total received signal, which will consist of aggregation of all users' sequences, can be correlated with the users' spreading code at the receiver side. In this case, the target user's signal will experience a combining gain across the $M$ chips, whereas the other users' signals will not and hence appear as uncorrelated noise. This combining gain is referred to as spreading gain. This principle is illustrated in Figure 3.5. In Figure 3.5a, two users' signals are spread with different random sequences and combined. In Figure 3.5b, the combined signal is correlated (that is, multiplied and accumulated) with the same random sequence used for the first user. In this case, the first user's signal is recovered with a (power) gain factor of $M$, whereas the multiplication of the second user's sequence with the first user's random spreading sequence results in a further random sequence with no gain factor.

In the following paragraphs, a situation in which two users have information to transmit is considered, in order to simplify the discussion and equations. The discussion may, however, be generalized to consider any number of users.

Spreading of a single modulation symbol is equivalent to

$$S(k) = s_1 c_1(k) + s_2 c_2(k), \quad k = 1, \ldots, M, \tag{3.1}$$

where $s_1$ is the first user's modulation symbol, $s_2$ the second user's modulation symbol, $c_i(k)$ the random spreading sequence used for the $i$th user, and $S(k)$ the combined spread sequence. It is assumed that the elements of the spreading sequences

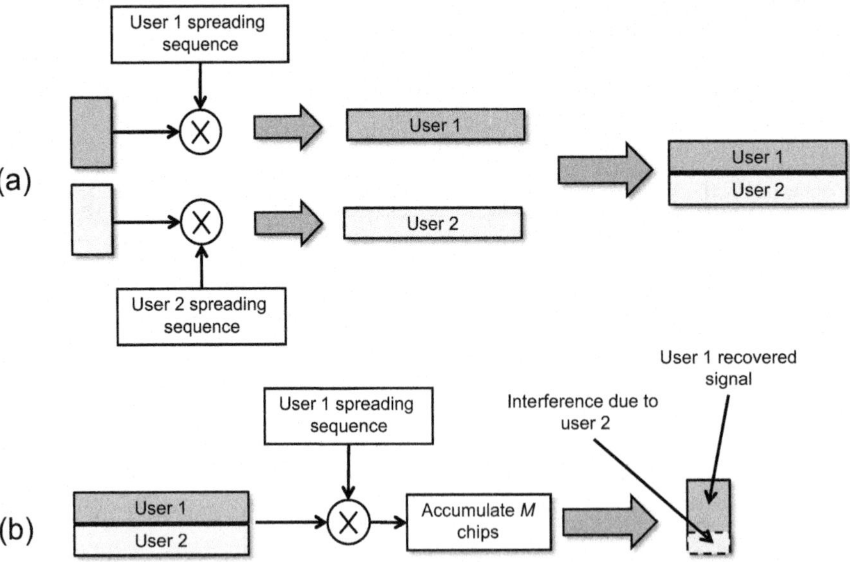


**FIGURE 3.5 Pseudo-noise (PN) code based spreading (a) and despreading (b).**

are normalized according to $|c_i(k)| = 1$. The process of recovering the wanted sequence from the spread sequence is known as despreading. Despreading user 1's signal then becomes

$$\tilde{s}_1 = \sum_{k=1}^{M} S(k)c_1^*(k) = \sum_{k=1}^{M} (s_1c_1(k) + s_2c_2(k))c_1^*(k) = Ms_1 + \sum_{k=1}^{M} s_2c_2(k)c_1^*(k). \quad (3.2)$$

In equation (3.2), the second term in the right-hand side has a power level $M$ times lower than the first term (note that $c_x(k) \cdot c_x^*(k)$ is always equal to unity).

A problem can arise when the signals from different users are received at very different power levels, for example, because in the uplink one user is close to the base station whereas a second user is not, as shown in Figure 3.6. In this case, the spreading gain will not be able to suppress the interference from the (strong) second user's transmission sufficiently and the reception of the first user will experience a poor signal-to-interference ratio because of interference from the second user. This problem is known as the "near-far" problem and is mitigated by the use of power control that keeps the receive power from different transmissions at appropriate levels (depending on the users' relative data rates).

There exist better choices of spreading sequence than a random sequence. One example is the Walsh-Hadamard sequences [1], also known as *Orthogonal Variable Spreading Factor* (OVSF) sequences. These sequences are available as different lengths and form a hierarchical tree structure, as shown in Figure 3.7. The sequences have the property that if a sequence from one part of the tree is correlated with another sequence that comes from a different branch of the tree, the resulting cross-correlation is zero

$$\sum_{k=1}^{M} s_2c_1(k)c_2^*(k) = 0. \quad (3.3)$$

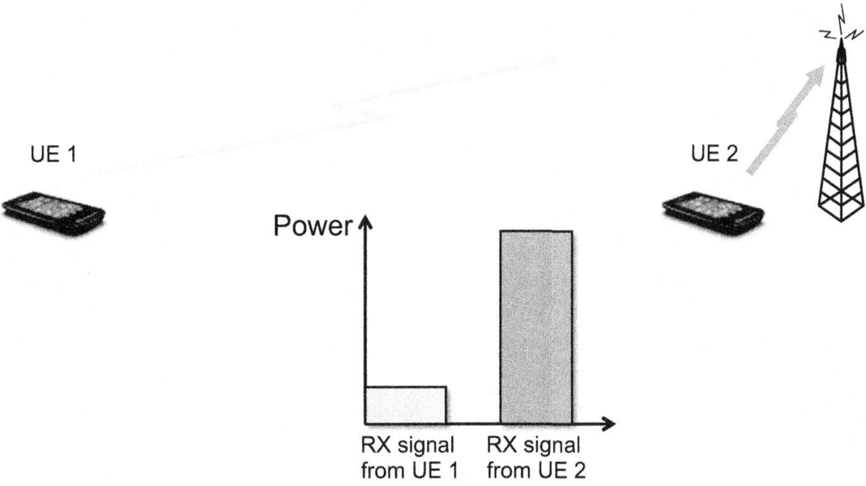


**FIGURE 3.6 Near-far problem with DSSS.**

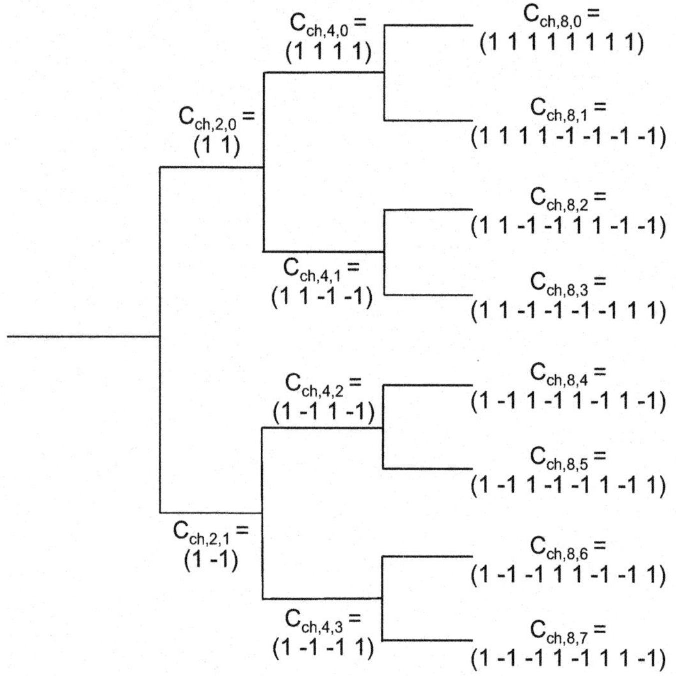


**FIGURE 3.7 A three-level example of a Walsh code.**

The cross code orthogonality also holds true in the general case that OVSF codes of differing lengths (i.e., different *M*) from different branches of the OVSF tree (see Figure 3.7 in which $C_{ch,x,y}$ refers to channelization code *y* with spreading factor *x*) are applied; thus, it is possible to apply different spreading factors and transmit data with different symbol rates on different codes. If codes of differing length from the same branch are selected, they will, however, not be orthogonal.

This property of OVSF sequences makes them useful in DSSS systems, because the use of different OVSF sequences for different users allows for different information sequences to be transmitted, at least in principle, without experiencing any cross-user interference when correlation is applied at the receiver. This is different from the case of random codes in which the cross-user interference is only suppressed by the spreading gain. A key requirement, though, for maintaining orthogonality between OVSF codes is that the codes are exactly time synchronous. In some situations, in particular uplink, it may not be feasible to time synchronize transmissions and thus the orthogonality between the codes is not preserved.

The use of orthogonal spreading sequences, such as Walsh-Hadamard, is a powerful technique, because (in the absence of channel effects) users do not interfere with one another. However, the amount of available codes is limited and may not be sufficient to support all users and all cells in a cellular communications system. Typically, spreading will involve a mixture of orthogonal and non-orthogonal codes. Groups of users may share orthogonal codes in a Walsh-Hadamard tree. After spreading, a process called *scrambling* may be applied in which each group is further multiplied with another random sequence, but not spread. Assuming that all codes at the transmitter are time aligned, then codes that are spread with the same sequence will remain orthogonal, but codes that are spread using different sequences are no longer orthogonal. The scrambling sequences may be random, but more usefully they can be selected to obtain good cross-correlation properties or to facilitate more efficient receiver implementations.

Consider a scenario with three users. In this example, the first two users' modulation symbols are spread with different orthogonal codes but the same scrambling code $v_1(k)$. The third user's sequence is spread with the same spreading sequence as user 1 but a different scrambling code $v_2(k)$. It is assumed that the elements of the scrambling sequences are normalized to unity, $|v_i(k)| = 1$. The combined spread sequence becomes

$$S(k) = s_1c_1(k)v_1(k) + s_2c_2(k)v_1(k) + s_3c_1(k)v_2(k). \tag{3.4}$$

The symbol $s_1$ may be recovered by correlation with both the spreading and scrambling code

$$\begin{aligned}\tilde{s}_1 &= \sum\nolimits_{k=1}^{M} S(k)c_1^*(k)v_1^*(k) = \sum\nolimits_{k=1}^{M}\left(s_1c_1(k)v_1(k)+s_2c_2(k)v_1(k)+s_3c_1(k)v_2(k)\right)c_1^*(k)v_1^*(k)\\ &= Ms_1 + \sum\nolimits_{k=1}^{M} s_2c_2(k)c_1^*(k) + \sum\nolimits_{k=1}^{M} s_3c_1(k)c_1^*(k)v_2(k)v_1^*(k)\\ &= Ms_1 + \sum\nolimits_{k=1}^{M} s_3v_2(k)v_1^*(k).\end{aligned} \tag{3.5}$$

The first user experiences interference from the third user but not from the second user.

## 3.2 BASEBAND TRANSMITTER MODEL FOR A CDMA SYSTEM

The simple example of spreading of a single symbol outlined in Section 3.1 is now developed into a multiple-user baseband model. Figure 3.8 shows the transmitter chain for a CDMA system, such as WCDMA. The $i$th modulated symbol corresponding to a user-assigned spreading code $k$, denoted $s_k(i)$, is spread by the mutually orthogonal spreading sequences $\{c_k(l)\}$. The resulting $k = 1,\ldots,K$ sequences $\{d_k(l)\}$ are summed together, the output is then multiplied by the node- and symbol period-dependent scrambling sequence $\{v_i(l)\}$, and finally filtered by the pulse-shaping filter $p(t)$. For simplicity, it is assumed that all spreading codes have the same spreading

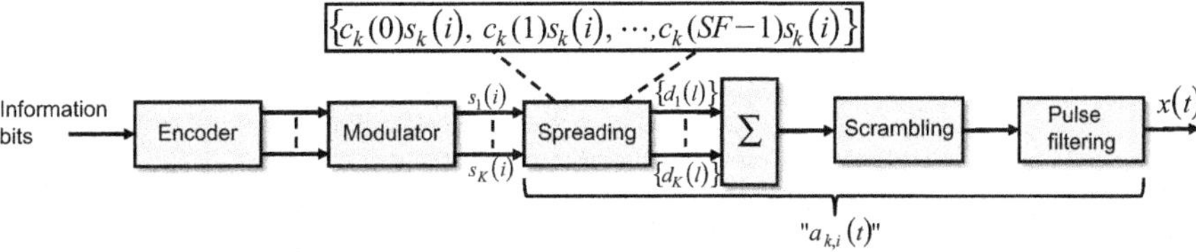


**FIGURE 3.8 Basic spread spectrum transmit signal processing chain.**

factor (*SF*). Typically, in the downlink, different codes are used for carrying information for different users. In the uplink, if there is a lack of time synchronization, each user transmits using a different scrambling code. In this model, different codes under the same scrambling code are likely to be different physical channels transmitted by the same user. The baseband model for the transmitted signal $x(t)$ can be described as

$$x(t)=\sum_{k=1}^{K}\sqrt{E_k}\sum_{i=-\infty}^{\infty}s_k(i)a_{k,i}(t-iT), \tag{3.6}$$

where $E_k$ is the symbol energy for the $k$th symbol, $T$ the symbol duration ($T = SF \cdot T_c$), $T_c$ denotes the chip rate, and $a_{k,i}(t)$ is the spreading waveform given by

$$a_{k,i}(t)=\sum_{l=0}^{SF\text{-}1}u_{k,i}(l)p(t-lT_c), \tag{3.7}$$

where the sequence $\{u_{k,i}(l)\}$ represents the combined spreading and scrambling sequence for spreading code $k$ and symbol period $i$. The model (3.6) is described as a function of the symbols but can just as well be represented as a function of the chip values, depending on what form is most convenient.

## 3.3 SPREAD SPECTRUM IN A REAL PROPAGATION ENVIRONMENT

In a real propagation environment, a signal propagates along multiple paths, such that the receiver experiences multiple time-delayed versions of the transmitted signal. Furthermore, if the user is moving, a Doppler spectrum arises, causing uncorrelated fading between the different received paths. If a simple correlation receiver is applied to the received signal, delayed versions of the transmitted signal will not correlate properly and thus cause self-interference. For example, suppose that a spread signal $S(k) = s_1c_1(k)$ is transmitted over a channel with two multi-paths, the amplitude of each channel being $a_i$ and the delay profile $\{0, \tau\}$. At the receiver

$$\sum\nolimits_{k=1}^{M}(a_1s_1c_1(k)+a_2s_1c_1(k-\tau))c_1^*(k)=Ma_1s_1+\sum\nolimits_{k=1}^{M}a_2s_1c_1(k-\tau)c_1^*(k). \tag{3.8}$$

The second term in equation (3.8) represents self-interference whose magnitude depends on $a_2c_1(k-\tau)c_1^*(k)$.

Furthermore, if information is transmitted to or from multiple users using Walsh-Hadamard codes as described in the previous section, then delayed versions of other users' signals will not sum to zero in the correlator, and thus signals intended for different users will mutually interfere

$$Ma_1s_1+\sum\nolimits_{k=1}^{M}a_2s_1c_1(k-\tau)c_1^*+\sum\nolimits_{k=1}^{M}a_2s_2c_2(k-\tau)c_1^*(k), \tag{3.9}$$

where the third term in equation (3.9) represents inter-user interference whose magnitude depends on the channel amplitude $a_2$ and the cross correlation $c_2(k-\tau)c_1^*(k)$.

In general, the multipath channel will consist of more than two paths, and it is instructive to develop a more generic baseband model for the received signal. Consider the signal $x(t)$ as derived in Section 3.2. $x(t)$ is transmitted over the air, where the propagation channel is described by a baseband equivalent impulse response according to

$$g(t)=\sum_{p=1}^{P} g_p\delta(t-\tau_p), \tag{3.10}$$

where $P$ denotes the number of channel taps, and $g_p$ and $\tau_p$ are the gain and delay of the $p$th tap (assumed to be time-invariant for simplicity). The received signal $y(t)$ is then given by

$$y(t)=\sum_{p=1}^{P} g_p x(t-\tau_p)+n(t), \tag{3.11}$$

where the noise term $n(t)$ accounts for impairments (interference and thermal noise).

A receiver that despreads a received signal that has experienced multipath fading using a set of correlators is known as a RAKE receiver. To despread the symbol of interest, $s_k(i)$, a bank of $Q$ correlators (Rake fingers) tuned to the spreading waveform of the desired symbol (see Figure 3.9) can be used. The output of correlator $q$ (delay $d_q$) can be written as

$$y_{ki}(d_q)=\sum_{n=1}^{K}\sqrt{E_n}\sum_{j=-\infty}^{\infty} s_n(i-j)\sum_{p=1}^{P} g_p r_{knij}(jT+d_q-\tau_p)+\tilde{n}(t), \tag{3.12}$$

where $r_{knij}(t)$ is the waveform correlation function describing the correlation between the $k$th spreading waveform during the $i$th symbol period and the $n$th spreading waveform during the $(i-j)$th symbol period, which can be expressed as

$$r_{knij}\tau=\frac{1}{SF}\sum_{u=1-SF}^{SF-1} C_{knij}(u)f(\tau-uT_c), \tag{3.13}$$

where $f(t)$ is a raised cosine pulse filter and

$$C_{knij}(u)=\begin{cases}\sum_{v=0}^{SF-1-u} u_{k,i}^{*}(v)u_{n,i-j}(u+v), & 0\le u\le SF-1\\ \sum_{v=0}^{SF-1+u} u_{k,i}^{*}(v-u)u_{n,i-j}(v), & 1-SF\le u<0\end{cases}, \tag{3.14}$$

where the sequence $\{u_{k,i}(l)\}$ is the combined spreading and scrambling sequence for scrambling code $k$ and time period $i$. Because $f(t)$ satisfies the Nyquist criterion, so does $r(t)$, that is, it is zero at multiples of the chip rate $t=kT_c$, $k\neq 0$. Hence, if there is no multipath ($P=1$ and $\tau_q=\tau_p=0$), the only contribution from $r_{knij}(jT)$ comes when $j=0$ and equals 1 if $n=k$ and zero otherwise because of the spreading sequence properties. If there is multipath present (as usually is the case), then there will be ISI because $r(t)$ will not be sampled exactly at multiples of the chip rate. In the frequency domain, this can be viewed as the effective combined receiver waveform equals

$$A(w)G(w)A(w), \tag{3.15}$$

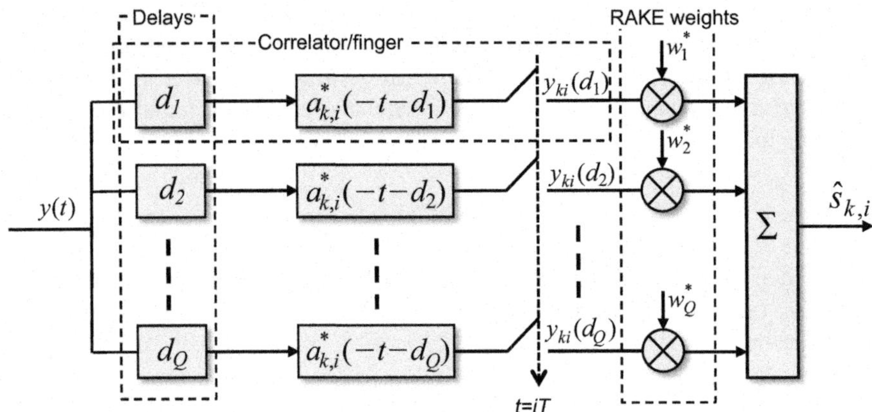


**FIGURE 3.9 General time–domain linear equalization.**

where $A(w)$ and $G(w)$ are the frequency domain representation of the spreading waveform $a_{k,i}(t)$ and the channel impulse response $g(t)$, respectively. This quantity should ideally satisfy the Nyquist criterion, but in general this is only true if $G(w) = 1$, in which case the resulting filter satisfies the Nyquist criterion.

Suppression or removal of negative impact from the propagation channel, $G(w)$, is highly desirable and is commonly referred to as *equalization*.

## 3.4 RECEIVER AND EQUALIZATION STRATEGIES

Multipath propagation in the time domain as discussed in the previous section is equivalent to frequency selectivity in the frequency domain. Traditionally, the main method to handle signal corruption due to radio-channel frequency selectivity has been to apply different forms of *equalization* [2–5] at the receiver side. The aim of equalization is to, by different means, compensate for the channel frequency selectivity and thus, at least to some extent, restore the original signal shape. In the time domain, this is equivalent to resolving the multipath components into a single signal, such that the self- and multiuser interference is reduced.

### 3.4.1 TIME–DOMAIN LINEAR EQUALIZATION

The most simple approach to equalization is the time–domain linear equalizer, consisting of a linear filter with an impulse response $w(\tau)$ applied to the received signal $r(t)$. The time-discrete representation becomes (see Figure 3.10)

$$\hat{s}_n = \sum_{l=0}^{L-1} w_l r_{(n-l)}. \tag{3.16}$$

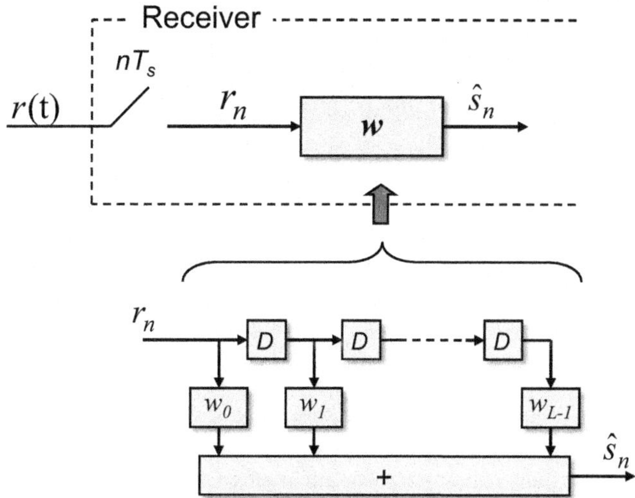


**FIGURE 3.10 Linear equalization implemented as a time-discrete FIR filter. $r_n$ represents $r(t)$ sampled at $nT_s$, where $T_s$ is the sampling rate. $D$ represents the delay operator.**

By selecting different filter impulse responses, different receiver/equalizer strategies can be implemented. As an example, in Direct Sequence CDMA-based systems, a so-called RAKE receiver structure has historically often been used. The RAKE receiver is simply the receiver structure of Figure 3.10 where the filter impulse response has been selected to provide *channel-matched filtering*

$$w(\tau) = h^*(-\tau), \tag{3.17}$$

that is, the filter response has been selected as the complex conjugate of the time-reversed channel impulse response. This is also often referred to as a *Maximum-Ratio Combining* (MRC) filter setting [2].

Selecting the receiver filter according to the MRC criterion, that is, as a channel-matched filter, maximizes the post-filter signal-to-noise ratio (assuming that the interference is white, that is, spectrally flat), hence the term maximum-ratio combining. However, MRC-based filtering does not provide any compensation for any radio-channel frequency selectivity, that is, no equalization. Thus, MRC-based receiver filtering is appropriate when the received signal is mainly impaired by noise or interference from other transmissions but not when a main part of the overall signal corruption is due to the radio-channel frequency selectivity, that is multipath, causing self-interference.

Another alternative is to select the receiver filter to fully compensate for the radio-channel frequency selectivity. This can be achieved by selecting the receiver-filter impulse response to fulfill the relation

$$h(\tau) \otimes w(\tau) = 1, \tag{3.18}$$

where "⊗" denotes linear convolution. This selecting of the filter setting, also known as *Zero-Forcing* (ZF) equalization [2], provides full compensation for any radio-channel frequency selectivity (complete equalization) and thus full suppression of any related signal corruption. However, zero-forcing equalization may lead to a large, potentially very large, increase in the noise level after equalization and thus to an overall degradation in the link performance. This will especially be the case when the channel has large variations in its frequency response.

A third and, in most cases, better alternative is to select a filter setting that provides a trade-off between signal corruption due to radio-channel frequency selectivity and noise/interference. This can, for example, be done by selecting the filter to minimize the mean-square error between the equalizer output $\hat{s}(k)$ and the transmitted signal $s(k)$, that is, to minimize

$$\varepsilon = E\left\{\left|\hat{s}(t) - s(t)\right|^2\right\} \tag{3.19}$$

where $E$ represents the expectation operator.[2] This is also referred to as a *Minimum Mean-Square Error* (MMSE) equalizer setting [2].

In practice, the linear equalizer has most often been implemented as a *time-discrete FIR filter* [6] with $L$ filter taps applied to the sampled received signal, as illustrated in Figure 3.10. In general, the complexity of such a time-discrete equalizer grows relatively rapidly with the bandwidth of the signal to be equalized:

- A more wideband signal is subject to relatively more radio-channel frequency selectivity or, equivalently, relatively more time dispersion. This implies that the equalizer needs to have a larger span (larger length $L$, that is, more filter taps) to be able to properly compensate for the channel frequency selectivity.
- A more wideband signal leads to a correspondingly higher sampling rate for the received signal. Thus also the receiver-filter processing needs to be carried out with a correspondingly higher rate.

It can be shown [7] that the time-discrete MMSE equalizer setting

$$\boldsymbol{w} = \begin{bmatrix} w_0 & w_1 & \cdots & w_{L-1} \end{bmatrix}^T, \tag{3.20}$$

corresponding to (3.19) is given by the expression

$$\boldsymbol{w} = \boldsymbol{R}^{-1}\boldsymbol{p}. \tag{3.21}$$

In this expression, $\boldsymbol{R}$ is the *channel-output auto-correlation matrix* of size $L \times L$, which depends on the channel impulse response as well as on the noise level, and $\boldsymbol{p}$ is the *channel-output/channel-input cross-correlation vector* of size $L \times 1$ that depends on the channel impulse response.

In order to calculate the autocorrelation matrix at the receiver, it is necessary to know the transmitted signal. Furthermore, if there are multiple transmitted signals, calculation of the full autocorrelation matrix becomes non-trivial. In the downlink of a cellular system, multiple transmitted signals are received from multiple base

[2]The expectation operator can operate over different ensembles, for example, the noise ensemble or the ensemble of scrambling sequences, or a combination of these.

stations, whereas in the uplink, multiple transmitted signals are received from multiple UEs.

Typically, the receiver will contain algorithms that effectively make an estimate for the autocorrelation and cross-correlation matrices/vectors based on a known sequence sent by the transmitter from which the UE tends to receive. The known transmitter sequence is often known as a pilot sequence. Thermal noise, in addition to interference from all other transmitters, is estimated and treated as being spectrally flat, that is, AWGN. Making these assumptions, equation (3.21) becomes effectively

$$\boldsymbol{w} = (\boldsymbol{H}\boldsymbol{H}^{H} + N_0 I)^{-1}\boldsymbol{H}\delta_D, \tag{3.22}$$

where $N_0$ is the noise power, $\boldsymbol{H}$ the channel matrix and $\delta_D$ sets the delay of the FIR filter (selects column $D$ of $\boldsymbol{H}$).

Equation (3.22) is only an approximation of the autocorrelation and cross correlation of the full received signal, because it has approximated interference from other transmitters to be white. However, it may be the case that the other transmitters also transmit known pilot sequences such that the receiver can estimate their interference characteristics. In this case, the equalization can be improved by taking the other transmitters into account

$$\boldsymbol{w} = (\boldsymbol{H}\boldsymbol{H}^{H} + \sum\nolimits_{i=1}^{N} \boldsymbol{H}_i\boldsymbol{H}_i^{H} + N_0 I)^{-1}\boldsymbol{H}\delta_D, \tag{3.23}$$

where $\boldsymbol{H}_i$ is the channel matrix of the $i$th transmitted signal and $N$ is the number of interfering receive signals that are considered.

Many receivers implement receive diversity, and in this case, the most optimal means to calculate the receiver weights is to jointly calculate the weights that should be applied in the first and second receiver branch

$$\begin{bmatrix} \boldsymbol{w}^{(1)} \\ \boldsymbol{w}^{(2)} \end{bmatrix} = \left( \begin{bmatrix} \boldsymbol{H}^{(1)} \\ \boldsymbol{H}^{(2)} \end{bmatrix} \begin{bmatrix} \boldsymbol{H}^{(1)} \\ \boldsymbol{H}^{(2)} \end{bmatrix} + \sum\nolimits_{i=1}^{N} \left( \begin{bmatrix} \boldsymbol{H}_i^{(1)} \\ \boldsymbol{H}_i^{(2)} \end{bmatrix} \begin{bmatrix} \boldsymbol{H}_i^{(1)} \\ \boldsymbol{H}_i^{(2)} \end{bmatrix}^{H} \right) + N_0 I \right)^{-1} \begin{bmatrix} \boldsymbol{H}^{(1)} \\ \boldsymbol{H}^{(2)} \end{bmatrix} \delta_D, \tag{3.24}$$

where $x^{(k)}$ represents $x$ for the $k$th receiver branch.

An equalizer that is applied across receive diversity branches and aims to mitigate the most significant interferers is often known as an interference mitigation receiver.

Especially in case of a large equalizer span (large $L$), the time–domain MMSE equalizer may be of relatively high complexity:

- The equalization itself (the actual filtering) may be of relatively high complexity according to the above.
- Calculation of the MMSE equalizer setting, especially the calculation of the inverse of the size $L \times L$ correlation matrix $\boldsymbol{R}$, may be of relatively high complexity.

Receive diversity and interference mitigation increase complexity further, necessitating the need to carefully select a trade-off between receiver complexity and performance.

Apart from the above approaches to linear equalization as a means to counteract signal corruption of a wideband signal due to radio-channel frequency selectivity, other more complex approaches to equalization, such as *Decision-Feedback Equalization* (DFE) [2], *Maximum-Likelihood* (ML) detection, also known as *Maximum Likelihood Sequence Estimation* (MLSE) [4], and the *Viterbi algorithm*[3] [6], have been the subject of research; however, they are not practical for 3G systems because of a prohibitive complexity level at 3G chip and data rates.

## REFERENCES

[1] H.F. Harmuth, Transmission of Information by Orthogonal Functions, Springer-Verlag, Berlin, 1970.
[2] J.G. Proakis, Digital Communications, McGraw-Hill, New York, 2001.
[3] D. Falconer, et al. Frequency Domain Equalization for Single-Carrier Broadband Wireless Systems, IEEE Commun Magazine 40 (4) (April 2002) 58–66.
[4] G. Forney, Maximum likelihood sequence estimation of digital sequences in the presence of intersymbol interference, IEEE Trans Information Theory IT-18 (May 1972) 363–378.
[5] G. Forney The Viterbi algorithm. Proceedings of the IEEE, IEEE, Piscataway, NJ, USA, Vol. 61, No. 3, March 1973, p. 268–278.
[6] A. Oppenheim Schafer RW. Digital Signal Processing. Prentice-Hall International (2000), Englewood Cliffs, NJ. ISBN 0-13-214107-8 01.
[7] S. Haykin, Adaptive Filter Theory, Prentice-Hall International, NJ, USA (1986), Upper Saddle River, NJ ISBN 0-13-004052-5 025.

[3]Thus the term "*Viterbi equalizer*" is sometimes used for the ML detector.

CHAPTER

# Multi-antenna techniques

# 4

## CHAPTER OUTLINE



Multi-antenna techniques can be seen as a joint name for a set of techniques with the common theme that they rely on the use of multiple antennas at the receiver and/or the transmitter, in combination with more or less advanced signal processing. Multi-antenna techniques can be used to achieve improved system performance, including improved system capacity (more users per cell) and improved coverage (possibility for larger cells), as well as improved service provisioning, for example, higher per-user data rates. This chapter will provide a general overview of different multi-antenna techniques applicable to the 3G evolution. How multi-antenna techniques are specifically applied to HSPA is discussed in somewhat more detail in Chapter 11.

## 4.1 MULTI-ANTENNA CONFIGURATIONS

An important characteristic of any multi-antenna configuration is the distance between the different antenna elements, to a large extent because of the relation between the antenna distance and the mutual correlation between the radio-channel fading experienced by the signals at the different antennas.

The antennas in a multi-antenna configuration can be located relatively far from each other, typically implying a relatively low mutual correlation. Alternatively, the antennas can be located relatively close to each other, typically implying a high mutual fading correlation, that is, the different antennas experience the same, or at


HSPA Evolution: The Fundamentals for Mobile Broadband. DOI: 10.1016/B978-0-08-099969-2.00004-1

least very similar, instantaneous fading. Whether high or low correlation is desirable depends on what is to be achieved with the multi-antenna configuration (diversity, beamforming, or spatial multiplexing) as discussed further below.

What actual antenna distance is needed for low, alternatively high, fading correlation depends on the wavelength or, equivalently, the carrier frequency used for the radio communication. However, it also depends on the deployment scenario.

In case of base station antennas in typical macro-cell environments (relatively large cells, relatively high base station antenna positions, etc.), an antenna distance in the order of 10 wavelengths is typically needed to ensure a low mutual fading correlation. At the same time, for a UE in the same kind of environment, an antenna distance in the order of only half a wavelength ($0.5\lambda$) is often sufficient to achieve relatively low mutual correlation [1]. The reason for the difference between the base station and the UE in this respect is that, in the macro-cell scenario, the multipath reflections that cause the fading occur mainly in the near-zone around the UE. Thus, as seen from the UE, the different paths will typically arrive from a wide angle, implying a low fading correlation already with a relatively small antenna distance. At the same time, as seen from the (macro-cell) base station, the different paths will typically arrive within a much smaller angle, implying the need for significantly larger antenna distance to achieve low fading correlation.[1]

On the other hand, in other deployment scenarios, such as micro-cell deployments with base station antennas below roof-top level and indoor deployments, the environment as seen from the base station is more similar to the environment as seen from the UE. In such scenarios, a smaller base-station antenna distance is typically sufficient to ensure relatively low mutual correlation between the fading experienced by the different antennas.

The above discussion assumed antennas with the same polarization direction. Another means to achieve low mutual fading correlation is to apply different polarization directions for the different antennas [1]. The antennas can then be located relatively close to each other, implying a compact antenna arrangement, while still experiencing low mutual fading correlation.

## 4.2 BENEFITS OF MULTI-ANTENNA TECHNIQUES

The availability of multiple antennas at the transmitter and/or the receiver can be utilized in different ways to achieve different aims:

- Multiple antennas at the transmitter and/or the receiver can be used to provide additional diversity against fading on the radio channel. In this case,

[1]Although the term *arrive* is used above, the situation is exactly the same in case of multiple *transmit* antennas. Thus, in case of multiple base station transmit antennas in a macro-cell scenario, the antenna distance typically needs to be a number of wavelengths to ensure low fading correlation while, in case of multiple transmit antennas at the UE, an antenna distance of a fraction of a wavelength is sufficient.

the channels experienced by the different antennas should have low mutual correlation, implying the need for a sufficiently large inter-antenna distance (*spatial diversity*), alternatively the use of different antenna polarization directions (*polarization diversity*).

- Multiple antennas at the transmitter and/or the receiver can be used to "shape" the overall antenna beam (transmit beam and receive beam, respectively) in a certain way, for example, to maximize the overall antenna gain in the direction of the target receiver/transmitter or to suppress specific dominant interfering signals. Such *beamforming* can be based either on high or low fading correlation between the antennas, as is further discussed in Section 4.4.2.
- The simultaneous availability of multiple antennas at the transmitter and the receiver can be used to create what can be seen as multiple parallel communication "channels" over the radio interface. This provides the possibility for very high bandwidth utilization without a corresponding reduction in power efficiency or, in other words, the possibility for very high data rates within a limited bandwidth without an un-proportionally large degradation in terms of coverage. Herein, we will refer to this as *spatial multiplexing*. It is often also referred to as *Multiple-Input Multiple-Output* (MIMO) antenna processing.

## 4.3 MULTIPLE RECEIVE ANTENNAS

Perhaps the most straightforward and historically the most commonly used multi-antenna configuration is the use of multiple antennas at the receiver side. This is often referred to as *receive diversity* or *Rx diversity* even if the aim of the multiple receive antennas is not always to achieve additional diversity against radio-channel fading.

Figure 4.1 illustrates the basic principle of linear combining of signals $r_1, \ldots, r_{N_R}$ received at $N_R$ different antennas, with the received signals being multiplied by complex weight factors $w_1^* \ldots w_{N_R}^*$ before being added together. In vector notation, this *linear receive-antenna combining* can be expressed as[2]

$$\hat{s} = \begin{bmatrix} w_1^* & \cdots & w_{N_R}^* \end{bmatrix} \begin{bmatrix} r_1 \\ \vdots \\ r_{N_R} \end{bmatrix} = \boldsymbol{w}^H \boldsymbol{r}. \tag{4.1}$$

What is outlined in (4.1) and Figure 4.1 is linear receive-antenna combining in general. Different specific antenna-combining approaches then differ in the exact choice of the weight vector $\boldsymbol{w}$.

[2]Note that the weight factors are expressed as complex conjugates of $w_1, \ldots, w_{N_R}$.

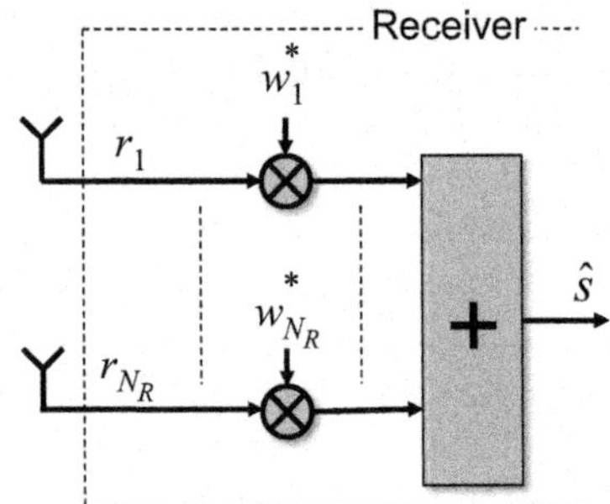


FIGURE 4.1 Linear receive-antenna combining.

Assuming that the transmitted signal is only subject to non-frequency-selective fading and (white) noise, that is, there is no radio-channel time dispersion, the signals received at the different antennas in Figure 4.1 can be expressed as

$$\boldsymbol{r} = \begin{bmatrix} r_1 \\ \vdots \\ r_{N_R} \end{bmatrix} = \begin{bmatrix} h_1 \\ \vdots \\ h_{N_R} \end{bmatrix} s + \begin{bmatrix} n_1 \\ \vdots \\ n_{N_R} \end{bmatrix} = \boldsymbol{h}s + \boldsymbol{n}, \tag{4.2}$$

where $s$ is the transmitted signal, the vector $\boldsymbol{h}$ consists of the $N_R$ complex channel gains, and the vector $\boldsymbol{n}$ consists of the noise impairing the signals received at the different antennas (see also Figure 4.2).

One can easily show that to maximize the signal-to-noise ratio after linear combining, the weight vector $\boldsymbol{w}$ should be selected as [2]

$$\boldsymbol{w}_{MRC} = \boldsymbol{h}. \tag{4.3}$$

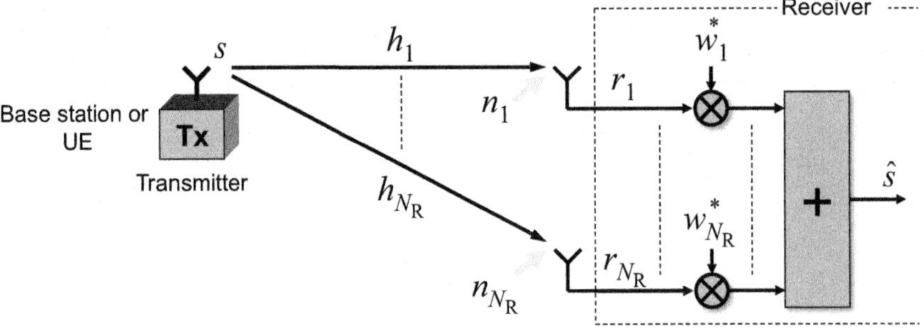


FIGURE 4.2 Linear receive-antenna combining.

This is also known as *Maximum-Ratio Combining* (MRC). The MRC weights fulfill two purposes:

- phase rotate the signals received at the different antennas to compensate for the corresponding channel phases and ensure that the signals are phase aligned when added together (coherent combining); and
- weight the signals in proportion to their corresponding channel gains, that is, apply higher weights for stronger received signals.

In case of mutually uncorrelated antennas, that is, sufficiently large antenna distances or different polarization directions, the channel gains $h_1,\ldots, h_{N_R}$ are uncorrelated and the linear antenna combining provides diversity of order $N_R$. In terms of receiver-side beamforming, selecting the antenna weights according to (4.3) corresponds to a receiver beam with maximum gain $N_R$ in the direction of the target signal. Thus, the use of multiple receive antennas may increase the post-combined signal-to-noise ratio in proportion to the number of receive antennas.

MRC is an appropriate antenna-combining strategy when the received signal is mainly impaired by noise. However, in many cases of mobile communication the received signal is mainly impaired by interference from other transmitters within the system, rather than by noise. In a situation with a relatively large number of interfering signals of approximately equal strength, MRC is typically still a good choice as, in this case, the overall interference will appear relatively "noise-like" with no specific direction of arrival. However, in situations where there is a single dominating interferer (or, in the general case, a limited number of dominating interferers), as illustrated in Figure 4.3, improved performance can be achieved if, instead of selecting the antenna weights to maximize the received signal-to-noise ratio after antenna combining (MRC), the antenna weights are selected so that the interferer is suppressed. In terms of receiver-side beamforming, this corresponds to a receiver beam with high attenuation in the direction of the interferer, rather than focusing the receiver beam in the direction of the target signal. Applying receive-antenna combined with a target to suppress specific interferers is often referred to as *Interference Rejection Combining* (IRC) [3].

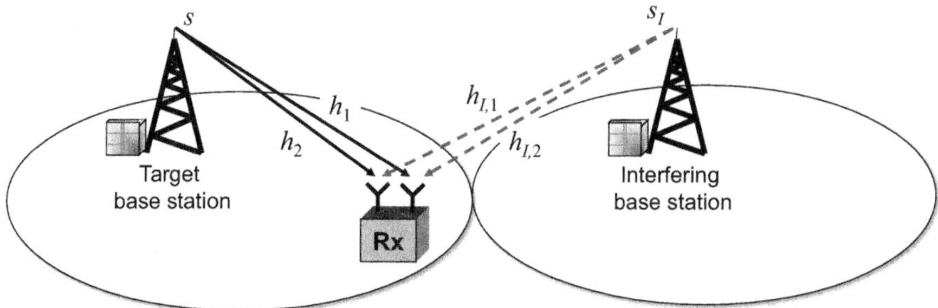


**FIGURE 4.3 Downlink scenario with a single dominating interferer (special case of only two receive antennas).**

In the case of a single dominating interferer as outlined in Figure 4.3, expression (4.2) can be extended according to

$$\boldsymbol{r} = \begin{bmatrix} r_1 \\ \vdots \\ r_{N_R} \end{bmatrix} = \begin{bmatrix} h_1 \\ \vdots \\ h_{N_R} \end{bmatrix} s + \begin{bmatrix} h_{I,1} \\ \vdots \\ h_{I,N_R} \end{bmatrix} s_I + \begin{bmatrix} n_1 \\ \vdots \\ n_{N_R} \end{bmatrix} = \boldsymbol{h}s + \boldsymbol{h}_I s_I + \boldsymbol{n}, \tag{4.4}$$

where $s_I$ is the transmitted interferer signal and the vector $\boldsymbol{h}_I$ consists of the complex channel gains from the interferer to the $N_R$ receive antennas. By applying expression (4.1) to (4.4), it becomes clear that the interfering signal will be completely suppressed if the weight vector $\boldsymbol{w}$ is selected to fulfill the expression

$$\boldsymbol{w}^H \boldsymbol{h}_I = 0. \tag{4.5}$$

In the general case, (4.5) has $N_R - 1$ non-trivial solutions, indicating flexibility in the weight-vector selection. This flexibility can be used to suppress additional dominating interferers. More specifically, in the general case of $N_R$ receive antennas there is a possibility to, at least in theory, completely suppress up to $N_R - 1$ separate interferers. However, such a choice of antenna weights, providing complete suppression of a number of dominating interferers, may lead to a large, potentially very large, increase in the noise level after the antenna combining. This is similar to the potentially large increase in the noise level in case of a *Zero-Forcing* equalizer, as discussed in Chapter 3.

Thus, similar to the case of linear equalization, a better approach is to select the antenna weight vector $\boldsymbol{w}$ to minimize the mean square error

$$\varepsilon = E\{|\hat{s} - s|^2\}, \tag{4.6}$$

also known as the *Minimum Mean Square Error* (MMSE) combining [4].

Although Figure 4.3 illustrates a downlink scenario with a dominating interfering base station, IRC can also be applied to the uplink to suppress interference from specific UEs. In this case, the interfering UE may either be in the same cell as the target UE (*intra-cell interference*) or in a neighbor cell (*inter-cell interference*) (see Figure 4.4a and b, respectively). Suppression of intra-cell interference is relevant in case of a non-orthogonal uplink, that is, when multiple UEs are transmitting simultaneously using the same time-frequency resource. Uplink intra-cell-interference suppression by means of IRC is sometimes also referred to as *Spatial Division Multiple Access* (SDMA) [5,6].

As discussed in Chapter 2, in practice a radio channel is always subject to at least some degree of time dispersion or, equivalently, frequency selectivity, causing corruption to a wide-band signal. As discussed in Chapter 3, one method to counteract such signal corruption is to apply equalization, either in the time or frequency domain.

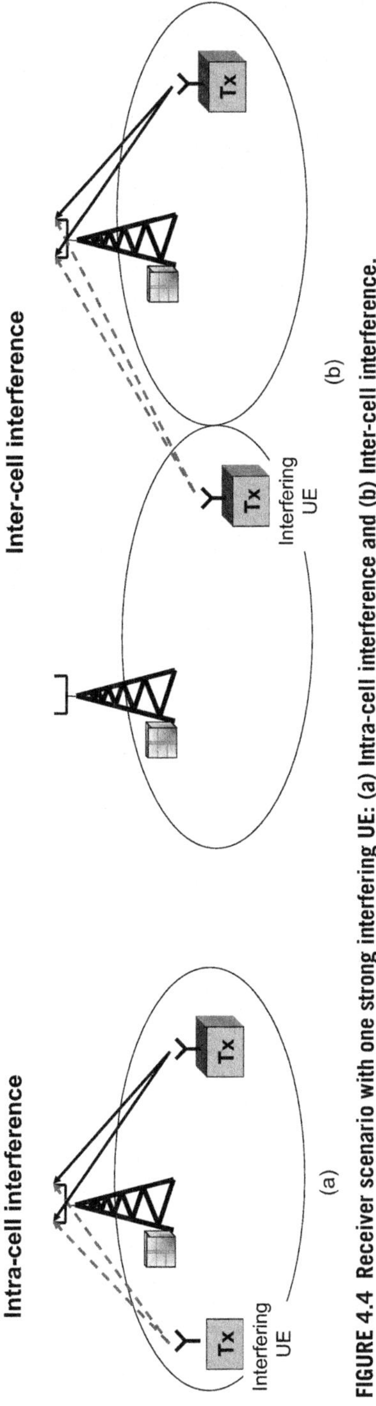


**FIGURE 4.4** Receiver scenario with one strong interfering UE: (a) Intra-cell interference and (b) Inter-cell interference.

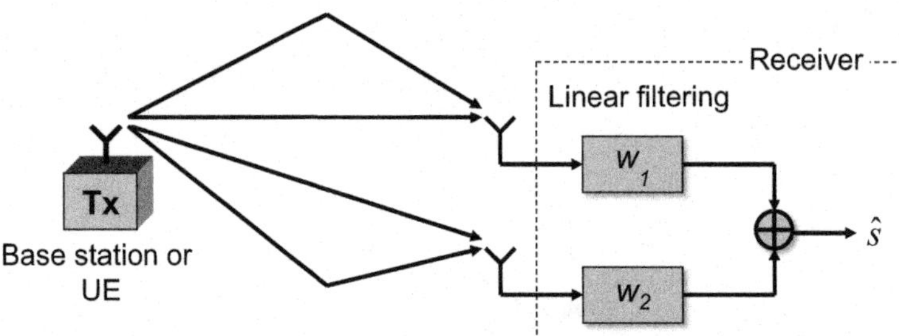


**FIGURE 4.5 Two-dimensional space–time linear processing (two receive antennas).**

It should be clear from the discussion above that linear receive-antenna combining has many similarities to linear equalization:

- Linear time-domain (frequency-domain) filtering/equalization as described in Chapter 3 implies that linear processing is applied to signals received at different time instances (different frequencies) with a target to maximize the post-equalizer SNR (MRC-based equalization), alternatively to suppress signal corruption due to radio-channel frequency selectivity (zero-forcing equalization, MMSE equalization, etc.).
- Linear receive-antenna combining implies that linear processing is applied to signals received at different antennas, that is, processing in the spatial domain, with a target to maximize the post-combined SNR (MRC-based combining), alternatively to suppress specific interferers (IRC based on, for example, MMSE).

Thus, in the general case of frequency-selective channel and multiple receive antennas, two-dimensional time/space linear processing/filtering can be applied as illustrated in Figure 4.5 where the linear filtering can be seen as a generalization of the antenna weights of Figure 4.1. The filters should be jointly selected to minimize the overall impact of noise, interference, and signal corruption due to radio-channel frequency selectivity. Chapter 3 describes how interference rejection is performed in a direct sequence spread spectrum system by means of jointly calculating the receiver filter weights on both antenna branches.

Note that although Figure 4.5 assumes two receive antennas, the corresponding receiver structures can straightforwardly be extended to more than two antennas.

## 4.4 MULTIPLE TRANSMIT ANTENNAS

As an alternative or complement to multiple receive antennas, diversity and beamforming can also be achieved by applying multiple antennas at the transmitter side. The use of multiple transmit antennas is primarily of interest for the downlink, that is, at the base station. In this case, the use of multiple transmit antennas provides an opportunity for diversity and beamforming without the need for additional receive antennas and corresponding additional receiver chains at the UE. On the other hand, because of complexity reasons, the use of multiple transmit antennas for the uplink, that is, at the UE, is less attractive. In this case, it is typically preferable to apply additional receive antennas and corresponding receiver chains at the base station.

### 4.4.1 TRANSMIT-ANTENNA DIVERSITY

If no knowledge of the downlink channels of the different transmit antennas is available at the transmitter, multiple transmit antennas cannot provide beamforming but only diversity. For this to be possible, there should be low mutual correlation between the channels of the different antennas. As discussed in Section 4.1, this can be achieved by means of sufficiently large distance between the antennas, alternatively by the use of different antenna polarization directions. Assuming such antenna configurations, different approaches can be taken to realize the diversity offered by the multiple transmit antennas.

#### 4.4.1.1 Delay diversity

A radio channel subject to time dispersion, with the transmitted signal propagating to the receiver via multiple, independently fading paths with different delays, provides the possibility for multi-path diversity or, equivalently, frequency diversity. Thus, multi-path propagation is actually beneficial in terms of radio-link performance, assuming that the amount of multi-path propagation is not too extensive and that the transmission scheme includes tools to counteract signal corruption due to the radio-channel frequency selectivity, for example, by means of advanced receiver-side equalization.

If the channel in itself is not time dispersive, the availability of multiple transmit antennas can be used to create *artificial time dispersion* or, equivalently, *artificial frequency selectivity* by transmitting identical signals with different relative delays from the different antennas. In this way, the antenna diversity, that is, the fact that the fading experienced by the different antennas has low mutual correlation, can be transformed into frequency diversity. This kind of *delay diversity* is illustrated in Figure 4.6 for the special case of two transmit antennas. The relative delay $T$ should be selected to ensure a suitable amount of frequency selectivity over the bandwidth of the signal to be transmitted. It should be noted that although Figure 4.6 assumes

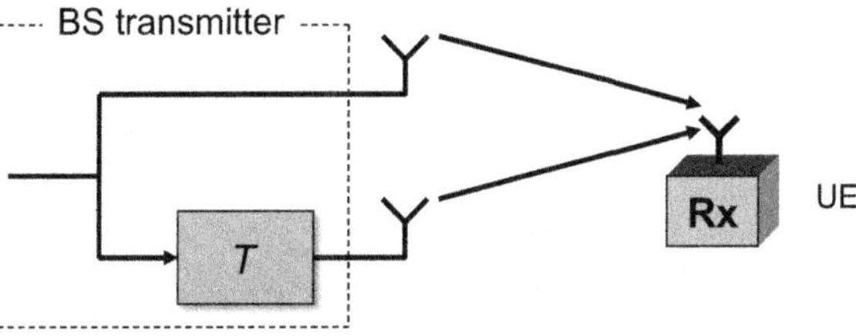


**FIGURE 4.6 Two-antenna delay diversity.**

two transmit antennas, delay diversity can straightforwardly be extended to more than two transmit antennas with different relative delays for each antenna.

Delay diversity is in essence invisible to the UE, which will simply see a single radio-channel subject to additional time dispersion. Delay diversity can thus straightforwardly be introduced in an existing mobile-communication system without requiring any specific support in a corresponding radio-interface standard. Delay diversity is also applicable to basically any kind of transmission scheme that is designed to handle and benefit from frequency-selective fading, including, for example, WCDMA and CDMA2000.

#### 4.4.1.2 Diversity by means of space–time coding

Space-time coding is a general term used to indicate multi-antenna transmission schemes where modulation symbols are mapped in the time and spatial (transmit-antenna) domain to capture the diversity offered by the multiple transmit antennas. Two-antenna *Space–Time Block Coding* (STBC), more specifically a scheme referred to as *Space–Time Transmit Diversity* (STTD), has been part of the 3G WCDMA standard from its first release [7].

As shown in Figure 4.7, STTD operates on pairs of modulation symbols. The modulation symbols are directly transmitted on the first antenna. However, on the second antenna the order of the modulation symbols within a pair is reversed. Furthermore, the modulation symbols are sign-reversed and complex-conjugated, as illustrated in Figure 4.7.

In vector notation, STTD transmission can be expressed as

$$\boldsymbol{r} = \begin{bmatrix} r_{2n} \\ r_{2n+1}^{*} \end{bmatrix} = \begin{bmatrix} h_1 & -h_2 \\ h_2^{*} & h_1^{*} \end{bmatrix} \begin{bmatrix} s_{2n} \\ s_{2n+1}^{*} \end{bmatrix} = \boldsymbol{Hs}, \tag{4.7}$$

where $r_{2n}$ and $r_{2n+1}$ are the received symbols during the symbol intervals $2n$ and $2n + 1$, respectively.[3] It should be noted that this expression assumes that the channel coefficients $h_1$ and $h_2$ are constant over the time corresponding to two consecutive

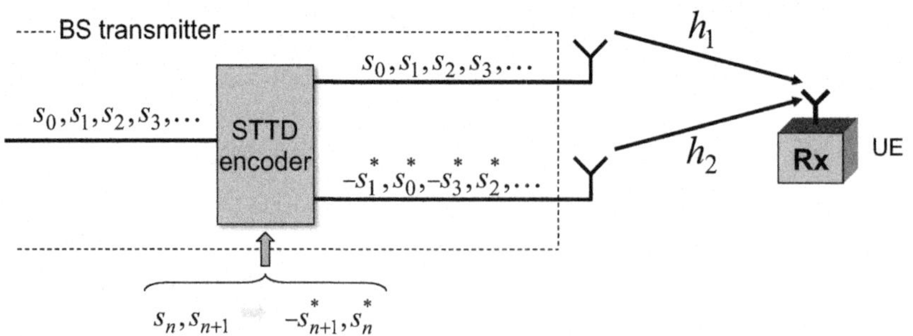


**FIGURE 4.7 WCDMA Space–Time Transmit Diversity (STTD).**

[3]Note that, for convenience, complex-conjugates have been applied for the second elements of $\boldsymbol{r}$ and $\boldsymbol{s}$.

symbol intervals, an assumption that is typically valid. As the matrix $\boldsymbol{H}$ is a scaled unitary matrix, the sent symbols $s_{2n}$ and $s_{2n+1}$ can be recovered from the received symbols $r_{2n}$ and $r_{2n+1}$, without any interference between the symbols, by applying the matrix $\boldsymbol{W} = \boldsymbol{H}^{-1}$ to the vector $\boldsymbol{r}$.

The two-antenna space–time coding of Figure 4.7 can be said to be of rate one, implying that the input symbol rate is the same as the symbol rate at each antenna, corresponding to a bandwidth utilization of one. Space–time coding can also be extended to more than two antennas. However, in case of complex-valued modulation, such as QPSK or 16/64-QAM, space–time codes of rate one without any intersymbol interference (*orthogonal space–time codes*) only exist for two antennas [8]. If intersymbol interference is to be avoided in case of more than two antennas, space–time codes with rate less than one must be used, corresponding to reduced bandwidth utilization.

### 4.4.2 TRANSMITTER-SIDE BEAMFORMING

If some knowledge of the downlink channels of the different transmit antennas, more specifically some knowledge of the relative channel phases, is available at the transmitter side, multiple transmit antennas can, in addition to diversity, also provide beamforming, that is, the shaping of the overall antenna beam in the direction of a target receiver. In general, such beamforming can increase the signal strength at the receiver with up to a factor $N_T$, that is, in proportion to the number of transmit antennas. When discussing transmission schemes relying on multiple transmit antennas to provide beamforming, one can distinguish between the cases of *high* and *low mutual antenna correlation*, respectively.

High mutual antenna correlation typically implies an antenna configuration with a small inter-antenna distance, as illustrated in Figure 4.8a. In this case, the channels between the different antennas and a specific receiver are essentially the same, including the same radio-channel fading, except for a direction-dependent phase

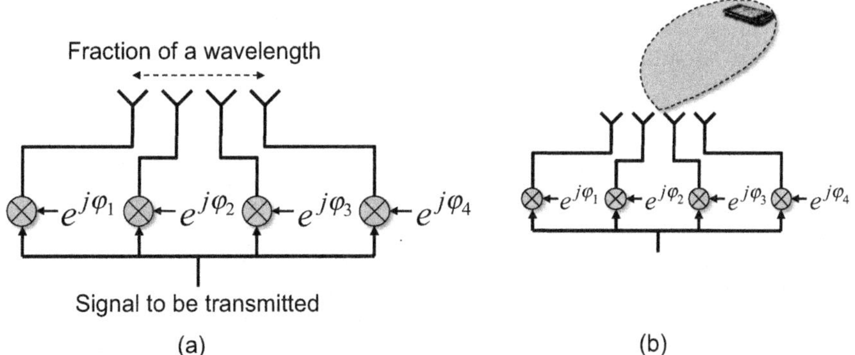


**FIGURE 4.8 Classical beamforming with high mutual antennas correlation: (a) antenna configuration and (b) beam-structure.**

difference. The overall transmission beam can then be steered in different directions by applying *different phase shifts* to the signals to be transmitted on the different antennas, as illustrated in Figure 4.8b.

This approach to transmitter side beamforming, with different phase shifts applied to highly correlated antennas, is sometimes referred to as "classical" beamforming. Because of the small antenna distance, the overall transmission beam will be relatively wide and any adjustments of the beam direction, in practice adjustments of the antenna phase shifts, will typically be carried out on a relatively slow basis. The adjustments could, for example, be based on estimates of the direction to the target UE derived from uplink measurements. Furthermore, because of the assumption of high correlation between the different transmit antennas, classical beamforming cannot provide any diversity against radio-channel fading but only an increase of the received signal strength.

Low mutual antenna correlation typically implies either a sufficiently large antenna distance, as illustrated in Figure 4.9, or different antenna polarization directions. With low mutual antenna correlation, the basic beamforming principle is similar to that of Figure 4.8, that is, the signals to be transmitted on the different antennas are multiplied by different complex weights. However, in contrast to classical beamforming, the antenna weights should now take general complex values; that is, both the phase and the amplitude of the signals to be transmitted on the different antennas can be adjusted. This reflects the fact that because of the low mutual antenna correlation, both the phase and the instantaneous gain of the channels of each antenna may differ.

Applying different complex weights to the signals to be transmitted on the different antennas can be expressed, in vector notation, as applying a *precoding vector* $\mathbf{v}$ to the signal to be transmitted according to

$$\mathbf{s} = \begin{bmatrix} s_1 \\ \vdots \\ s_{N_T} \end{bmatrix} = \begin{bmatrix} v_1 \\ \vdots \\ v_{N_T} \end{bmatrix} s = \mathbf{v}s. \tag{4.8}$$

It should be noted that classical beamforming according to Figure 4.8 can also be described according to (4.8), that is as *transmit-antenna precoding*, with the

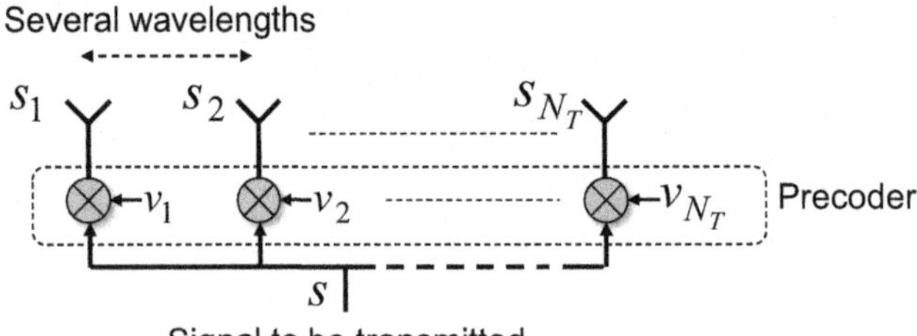


**FIGURE 4.9 Precoder-based beamforming in case of low mutual antenna correlation.**

constraint that the antenna weights are limited to unit gain and only provide phase shifts to the different transmit antennas.

Assuming that the signals transmitted from the different antennas are only subject to non-frequency-selective fading and white noise, that is, there is no radio-channel time dispersion, it can be shown [9] that, in order to maximize the received signal power, the precoding weights should be selected according to

$$v_i = \frac{h_i^*}{\sqrt{\sum_{k=1}^{N_T} |h_k|^2}}, \tag{4.9}$$

that is, as the complex conjugate of the corresponding channel coefficient $h_i$ and with a normalization to ensure a fixed overall transmit power. The precoding vector thus

- phase rotates the transmitted signals to compensate for the instantaneous channel phase and thereby ensures that the received signals are received phase aligned;
- allocates power to the different antennas, with, in general, more power being allocated to antennas with good instantaneous channel conditions (high channel gain $|h_i|$); and
- ensures an overall unit (or any other constant) transmit power.

A key difference between classical beamforming according to Figure 4.8, assuming high mutual antenna correlation, and beamforming according to Figure 4.9, assuming low mutual antenna correlation, is that, in the latter case, there is a need for more detailed channel knowledge, including estimates of the instantaneous channel fading. Updates to the precoding vector are thus typically done on a relatively short time scale to capture the fading variations. As the adjustment of the precoder weights takes into account also the instantaneous fading, including the instantaneous channel gain, fast beamforming according to Figure 4.9 also provides diversity against radio-channel fading.

Furthermore, at least in case of communication based on *Frequency-Division Duplex* (FDD), with uplink and downlink communication taking place in different frequency bands, the fading is typically uncorrelated between the downlink and uplink. Thus, in case of FDD, only the UE can determine the downlink fading. The UE would then report an estimate of the downlink channel to the base station by means of uplink signaling. Alternatively, the UE may, in itself, select a suitable precoding vector from a limited set of possible precoding vectors, the so-called precoder codebook, and report this to the base station.

The above discussion assumed that the channel gain was constant in the frequency domain, that is, there was no radio-channel frequency selectivity. In case of a frequency-selective channel, there is obviously not a single channel coefficient per antenna based on which the antennas' weights can be selected according to (4.9).

It should be pointed out that in case of single-carrier transmission, such as WCDMA, the one-weight-per-antenna approach outlined in Figure 4.9 can be extended to take into account also a time-dispersive/frequency-selective channel [10].

## 4.5 SPATIAL MULTIPLEXING

The use of multiple antennas at both the transmitter and the receiver can simply be seen as a tool to further improve the signal-to-noise/interference ratio and/or achieve additional diversity against fading, compared to the use of only multiple receive antennas or multiple transmit antennas. However, in case of multiple antennas at both the transmitter and the receiver, there is also the possibility for so-called *spatial multiplexing*, allowing for more efficient utilization of high signal-to-noise/interference ratios and significantly higher data rates over the radio interface.

### 4.5.1 BASIC PRINCIPLES

It should be clear from the previous sections that multiple antennas at the receiver and the transmitter can be used to improve the receiver signal-to-noise ratio in proportion to the number of antennas by applying beamforming at the receiver and the transmitter. In the general case of $N_T$ transmit antennas and $N_R$ receive antennas, the receiver signal-to-noise ratio can be made to increase in proportion to the product $N_T \times N_R$. As discussed in Chapter 2, such an increase in the receiver signal-to-noise ratio allows for a corresponding increase in the achievable data rates, assuming that the data rates are power limited rather than bandwidth limited. However, once the bandwidth-limited range-of-operation is reached, the achievable data rates start to saturate unless the bandwidth is also allowed to increase.

One way to understand this saturation in achievable data rates is to consider the basic expression for the normalized channel capacity

$$\frac{C}{BW} = \log_2\left(1 + \frac{S}{N}\right), \tag{4.10}$$

where, by means of beamforming, the signal-to-noise ratio $S/N$ can be made to grow proportionally to $N_T \times N_R$. In general, $\log_2(1 + x)$ is proportional to $x$ for small $x$, implying that, for low signal-to-noise ratios, the capacity grows approximately proportionally to the signal-to-noise ratio. However, for larger $x$, $\log_2(1 + x) \approx \log_2(x)$, implying that, for larger signal-to-noise ratios, capacity grows only logarithmically with the signal-to-noise ratio.

However, in the case of multiple antennas at the transmitter *and* the receiver, it is, *under certain conditions*, possible to create up to $N_L = \min\{N_T, N_R\}$ parallel "channels" each with $N_L$ times lower signal-to-noise ratio (the signal power is "split" between the channels), that is, with a channel capacity

$$\frac{C}{BW} = \log_2\left(1 + \frac{N_R}{N_L} \cdot \frac{S}{N}\right). \tag{4.11}$$

As there are now $N_L$ parallel channels, each with a channel capacity given by (4.11), the overall channel capacity for such a multi-antenna configuration is thus given by

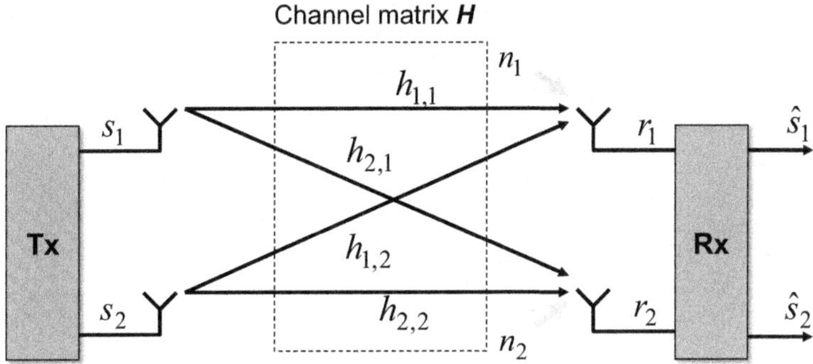


**FIGURE 4.10 2 × 2-antenna configuration.**

$$\begin{aligned}\frac{C}{BW} &= N_L \cdot \log_2\left(1+\frac{N_R}{N_L}\cdot\frac{S}{N}\right)\\ &= \min\{N_T,N_R\}\cdot\log_2\left(1+\frac{N_R}{\min\{N_T,N_R\}}\cdot\frac{S}{N}\right).\end{aligned} \tag{4.12}$$

Thus, *under certain conditions*, the channel capacity can be made to grow essentially linearly with the number of antennas, avoiding the saturation in the data rates. We will refer to this as *Spatial Multiplexing*. The term *Multiple-Input Multiple-Output* antenna processing is also very often used, although the term strictly speaking refers to all cases of multiple transmit antennas and multiple receive antennas, including also the case of combined transmit and receive diversity.[4]

To understand the basic principles of how multiple parallel channels can be created in case of multiple antennas at the transmitter and the receiver, consider a 2 × 2 antenna configuration, that is, two transmit antennas and two receive antennas, as outlined in Figure 4.10. Furthermore, assume that the transmitted signals are only subject to non-frequency-selective fading and white noise; that is, there is no radio-channel time dispersion.

Based on Figure 4.10, the received signals can be expressed as

$$\boldsymbol{r} = \begin{bmatrix} r_1 \\ r_2 \end{bmatrix} = \begin{bmatrix} h_{1,1} & h_{1,2} \\ h_{2,1} & h_{2,2} \end{bmatrix}\begin{bmatrix} s_1 \\ s_2 \end{bmatrix} + \begin{bmatrix} n_1 \\ n_2 \end{bmatrix} = \boldsymbol{Hs} + \boldsymbol{n}, \tag{4.13}$$

where $\boldsymbol{H}$ is the 2 × 2 *channel matrix*. This expression can be seen as a generalization of (4.2) in Section 4.3 to cover multiple transmit antennas with different signals being transmitted from the different antennas.

[4]The case of a single transmit antenna and multiple receive antennas is, consequently, often referred to as *SIMO* (Single-Input/Multiple-Output). Similarly, the case of multiple transmit antennas and a single receiver antenna can be referred to as *MISO* (Multiple-Input/Single-Output).

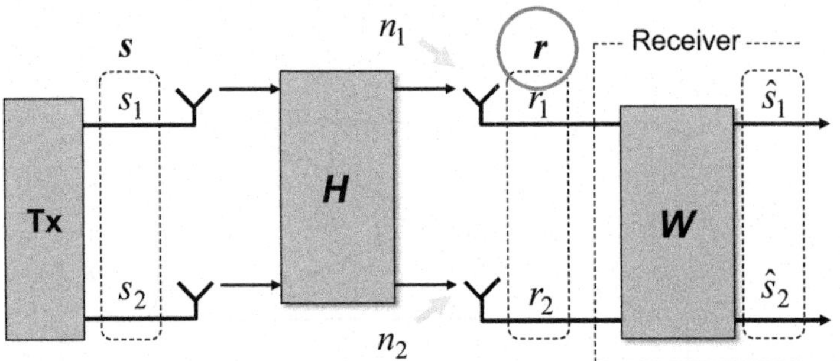


**FIGURE 4.11 Linear reception/demodulation of spatially multiplexed signals.**

Assuming no noise and that the channel matrix $\boldsymbol{H}$ is invertible, the vector $\boldsymbol{s}$, and thus both signals $s_1$ and $s_2$, can be perfectly recovered at the receiver, with no residual interference between the signals, by multiplying the received vector $\boldsymbol{r}$ with a matrix $\boldsymbol{W} = \boldsymbol{H}^{-1}$ (see also Figure 4.11).

$$\begin{bmatrix} \hat{s}_1 \\ \hat{s}_2 \end{bmatrix} = \boldsymbol{W}\boldsymbol{r} = \begin{bmatrix} s_1 \\ s_2 \end{bmatrix} + \boldsymbol{H}^{-1}\boldsymbol{n}. \tag{4.14}$$

Although the vector $\boldsymbol{s}$ can be perfectly recovered in case of no noise, as long as the channel matrix $\boldsymbol{H}$ is invertible, (4.14) also indicates that the properties of $\boldsymbol{H}$ will determine to what extent the joint demodulation of the two signals will increase the noise level. More specifically, the closer the channel matrix is to being a singular matrix, the larger the increase in the noise level.

One way to interpret the matrix $\boldsymbol{W}$ is to realize that the signals transmitted from the two transmit antennas are two signals causing interference to each other. The two receive antennas can then be used to carry out IRC, in essence completely suppressing the interference from the signal transmitted on the second antenna when detecting the signal transmitted at the first antenna and vice versa. The rows of the receiver matrix $\boldsymbol{W}$ simply implement such IRC.

In the general case, a multiple-antenna configuration will consist of $N_T$ transmit antennas and $N_R$ receive antennas. As discussed above, in such a case, the number of parallel signals that can be spatially multiplexed is, at least in practice, upper limited by $N_L = \min\{N_T, N_R\}$. This can intuitively be understood from the fact that

- obviously no more than $N_T$ different signals can be transmitted from $N_T$ transmit antennas, implying a maximum of $N_T$ spatially multiplexed signals; and
- with $N_R$ receive antennas, a maximum of $N_R - 1$ interfering signals can be suppressed, implying a maximum of $N_R$ spatially multiplexed signals.

However, in many cases, the number of spatially multiplexed signals, or the *order of the spatial multiplexing*, will be less than $N_L$ given above:

- In case of very bad channel conditions (low signal-to-noise ratio), there is no gain of spatial multiplexing as the channel capacity is anyway a linear function of the signal-to-noise ratio. In such a case, the multiple transmit and receive antennas should be used for beamforming to improve the signal-to-noise ratio, rather than for spatial multiplexing.
- In a more general case, the spatial-multiplexing order should be determined based on the properties of the size $N_R \times N_T$ channel matrix. Any excess antennas should then be used to provide beamforming. Such combined beamforming and spatial multiplexing can be achieved by means of *precoder-based* spatial multiplexing, as discussed below.

## 4.5.2 PRECODER-BASED SPATIAL MULTIPLEXING

Linear precoding in case of spatial multiplexing implies that linear processing by means of a size $N_T \times N_L$ precoding matrix is applied at the transmitter side, as illustrated in Figure 4.12. In line with the discussion above, in the general case $N_L$ is equal to or smaller than $N_T$, implying that $N_L$ signals are spatially multiplexed and transmitted using $N_T$ transmit antennas.

It should be noted that precoder-based spatial multiplexing can be seen as a generalization of precoder-based beamforming, as described in Section 4.4.2, with the precoding vector of size $N_T \times 1$ replaced by a precoding matrix of size $N_T \times N_L$.

The precoding of Figure 4.12 can serve two purposes:

- In the case when the number of signals to be spatially multiplexed equals the number of transmit antennas ($N_L = N_T$), the precoding can be used to "orthogonalize" the parallel transmissions, allowing for improved signal isolation at the receiver side.
- In the case when the number of signals to be spatially multiplexed is less than the number of transmit antennas ($N_L < N_T$), the precoding in addition provides

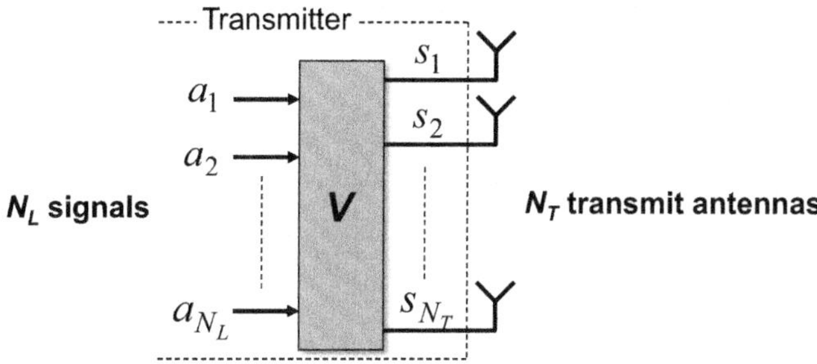


**FIGURE 4.12 Precoder-based spatial multiplexing.**

the mapping of the $N_L$ spatially multiplexed signals to the $N_T$ transmit antennas, including the combination of spatially multiplexing and beamforming.

To confirm that precoding can improve the isolation between the spatially multiplexed signals, express the channel matrix $\boldsymbol{H}$ as its singular-value decomposition [11]

$$\boldsymbol{H} = \boldsymbol{W} \cdot \Sigma \cdot \boldsymbol{V}^H, \tag{4.15}$$

where $\boldsymbol{W}$ and $\boldsymbol{V}$ are unitary matrices of dimension $N_R \times N_R$ and $N_T \times N_T$, respectively, and $\boldsymbol{\Sigma}$ is an $N_R \times N_T$ rectangular diagonal matrix, where the diagonal elements of $\boldsymbol{\Sigma}$ equal the square root of the eigenvalues of $\boldsymbol{H}^H\boldsymbol{H}$ (arranged in decreasing order). Assuming that $\boldsymbol{H}$ has rank $r$ ($r<=N_R$, $N_T$), (4.15) can be composed as

$$\boldsymbol{H} = \boldsymbol{W}_1\Sigma_1\boldsymbol{V}_1^H + \boldsymbol{W}_2\Sigma_2\boldsymbol{V}_2^H = \boldsymbol{W}_1\Sigma_1\boldsymbol{V}_1^H, \tag{4.16}$$

where the matrices $\boldsymbol{W}_1$, $\boldsymbol{V}_1$, $\boldsymbol{W}_2$, and $\boldsymbol{V}_2$ are of dimension $N_R \times r$, $N_T \times r$, $N_R \times (N_R - r)$, and $N_T \times (N_T - r)$, and where the columns of $[\boldsymbol{W}_1\ \boldsymbol{W}_2]$ and $[\boldsymbol{V}_1\ \boldsymbol{V}_2]$ each form an orthonormal set. Thus, by assuming that $N_L = r$ and applying the matrix $\boldsymbol{V}_1$ as precoding matrix at the transmitter side and the matrix $\boldsymbol{W}_1^H$ (or $\boldsymbol{W}^H$) at the receiver side, one arrives at an equivalent channel matrix $\boldsymbol{H}' = \Sigma_1$ (see Figure 4.13). As $\boldsymbol{H}'$ is a diagonal matrix, there is thus no interference between the spatially multiplexed signals at the receiver. At the same time, as both $\boldsymbol{V}_1$ and $\boldsymbol{W}_1$ have orthonormal columns, the transmit power as well as the demodulator noise level (assuming spatially white noise) are unchanged.

Clearly, in case of precoding, each received signal will have a certain "quality," depending on the eigenvalues of the channel matrix (see right part of Figure 4.13). This indicates potential benefits of applying dynamic link adaptation in the *spatial domain*, that is, adaptively select the coding rates and/or modulation schemes for each signal to be transmitted.

As the precoding matrix will never perfectly match the channel matrix in practice, there will always be some residual interference between the spatially multiplexed signals. This interference can be taken care of by means of additional receiver-size linear processing, according to Figure 4.11, or nonlinear processing, as discussed in Section 4.5.3.

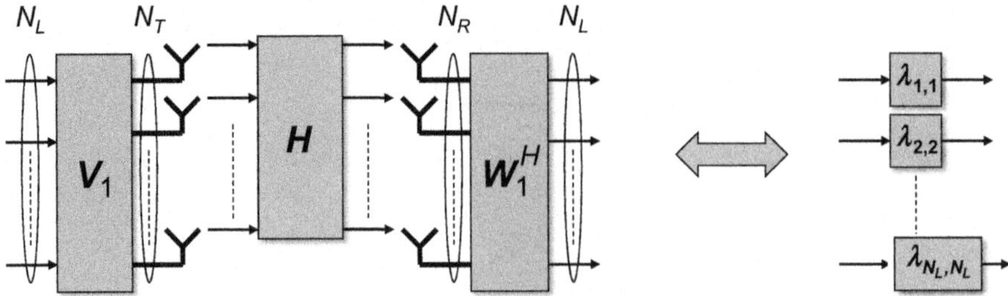


**FIGURE 4.13 Orthogonalization of spatially multiplexed signals by means of precoding. $\lambda_{i,i}$ is the $i$th eigenvalue of the matrix $H^HH$.**

To determine the precoding matrix $\boldsymbol{V}$ (or $\boldsymbol{V}_l$), knowledge about the channel matrix $\boldsymbol{H}$ is needed. Similar to precoder-based beamforming, a common approach is to have the receiver estimate the channel and decide on a suitable precoding matrix from a set of available precoding matrices (the precoder *codebook*). The receiver then feeds back information about the selected precoding matrix to the transmitter.

### 4.5.3 NONLINEAR RECEIVER PROCESSING

The previous sections discussed the use of linear receiver processing to jointly recover spatially multiplexed signals. However, improved demodulation performance can be achieved if nonlinear receiver processing can be applied in case of spatial multiplexing.

The "optimal" receiver approach for spatially multiplexed signals is to apply *Maximum-Likelihood* (ML) detection [12]. However, in many cases, ML detection is too complex to use. Thus, several different proposals have been made for reduced complexity of most ML schemes (see, e.g., [13]).

Another non-linear approach to the demodulation of spatially multiplexed signals is to apply *Successive Interference Cancellation* (SIC) [14]. SIC is based on an assumption that the spatially multiplexed signals are separately coded before the spatial multiplexing. This is often referred to as *Multi-codeword* transmission, in contrast to *Single-codeword* transmission where the spatially multiplexed signals are assumed to be jointly coded (Figure 4.14). It should be understood that, also in the case of multi-codeword transmission, the data may originate from the same source but then be de-multiplexed into different signals to be spatially multiplexed before channel coding.

As shown in Figure 4.15, in case of SIC, the receiver first demodulates and decodes one of the spatially multiplexed signals. The corresponding decoded data is then, if correctly decoded, re-encoded and subtracted from the received signals. A second spatially multiplexed signal can then be demodulated and decoded without, at least in the ideal case, any interference from the first signal, that is, with an improved

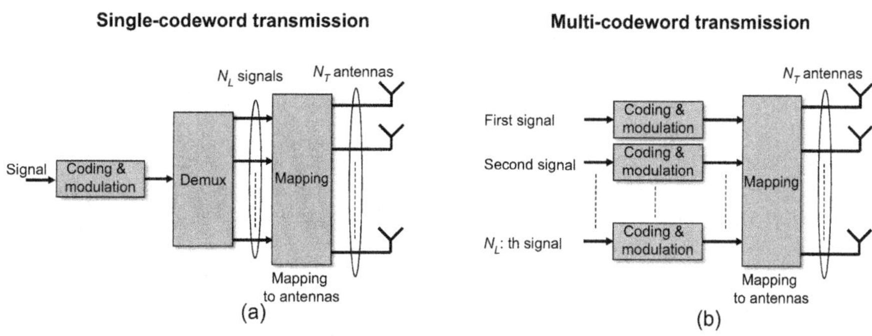


**FIGURE 4.14 Single-codeword transmission (a) versus multi-codeword transmission (b).**

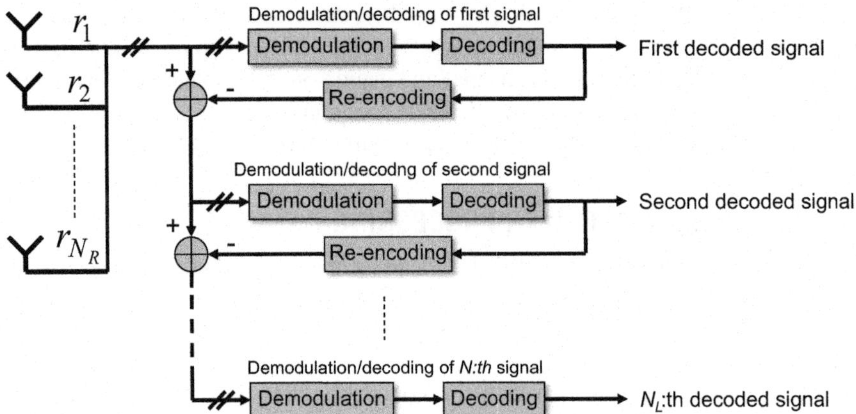


**FIGURE 4.15 Demodulation/decoding of spatially multiplexed signals based on *Successive Interference Cancellation* (SIC).**

signal-to-interference ratio. The decoded data of the second signal is then, if correctly decoded, re-encoded and subtracted from the received signal before decoding of a third signal. These iterations continue until all spatially multiplexed signals have been demodulated and decoded.

Clearly, in case of SIC, the first signals to be decoded are subject to a higher interference level, compared to later decoded signals. To work properly, there should thus be a differentiation in the robustness of the different signals, with, at least in principle, the first signal to be decoded being more robust than the second signal, the second signal being more robust than the third signal, etc. Assuming multi-codeword transmission according to Figure 4.14b, this can be achieved by applying different modulation schemes and coding rates to the different signals with, typically, lower-order modulation and lower coding rate, implying a lower data rate, for the first signals to be decoded. This is often referred to as *Per-Antenna Rate Control* (PARC) [15].

## REFERENCES

[1] Lee W. Mobile communications engineering. New York, NY, USA; 1997 McGraw-Hill; ISBN 0-07-037039-7.

[2] J.G. Proakis, Digital communications, McGraw-Hill, New York, (2001).

[3] Karlsson J, Heinegard J. Interference rejection combining for GSM. In: Proceedings of the 5th IEEE International Conference on Universal Personal Communications, Cambridge, MA, USA; 1996. p. 433–437.

[4] Oppenheim A, Schafer RW. Digital signal processing. Prentice-Hall International, Englewood Cliffs, NJ; 1975 ISBN 0-13-214107-8 01.

[5] Godara LC. Applications of antenna arrays to mobile communications, part I: beamforming and direction-of-arrival considerations. In: Proceedings of the IEEE, Vol. 85, No. 7, July 1997. p. 1029–1030, New York, NY.

[6] Godara LC. Applications of antenna arrays to mobile communications, part II: beam-forming and direction-of-arrival considerations. In: Proceedings of the IEEE, Vol. 85, No. 7, July 1997. p. 1031–1060, New York, NY.
[7] 3rd Generation Partnership Project; Technical Specification Group Radio Access Network; Physical Channels and Mapping of Transport Channels onto Physical Channels (FDD). 3GPP, TS 25.211.
[8] V. Tarokh, N. Seshadri, A. Calderbank, Space–time block codes from orthogonal design, IEEE Trans Information Theory 45 (5) (July 1999) 1456–1467.
[9] Hottinen A, Tirkkonen O, Wichman R. Multi-antenna transceiver techniques for 3G and beyond. Chichester, UK: 2003 John Wiley & Sons; ISBN 0470 84542 2.
[10] C. Kambiz, L. Krasny, Capacity-achieving transmitter and receiver pairs for dispersive MISO channels, IEEE Trans Commun 42 (April 1994) 1431–1440.
[11] Horn R, Johnson C. Matrix analysis. In: Proceedings of the 36th Asilomar Conference on Signals, Systems and Computers, Pacific Grove, CA, USA, November 2002.
[12] G. Forney, Maximum likelihood sequence estimation of digital sequences in the presence of intersymbol interference, IEEE Trans Information Theory IT-18 (May 1972) 363–378.
[13] Kim KJ. Channel estimation and data detection algorithms for MIMO–OFDM systems. In: Proceedings of the 36th Asilomar Conference on Signals, Systems and Computers, Pacific Grove, CA, USA, November 2002.
[14] Varanasi MK, Guess T. Optimum decision feedback multi-user equalization with successive decoding achieves the total capacity of the Gaussian multiple-access channel. In: Proceedings of the Asilomar Conference on Signals, Systems, and Computers, Monterey, CA, November 1997.
[15] Grant S, et al. Per-antenna-rate-control (PARC) in frequency selective fading with SIC-GRAKE receiver. In: 60th IEEE Vehicular Technology Conference, Vol. 2. Los Angeles, CA, USA; September 2004. p. 1458–1462.

CHAPTER

# Scheduling, link adaptation, and hybrid-ARQ

# 5

## CHAPTER OUTLINE



One key characteristic of mobile radio communication is the typically rapid and significant variations in the instantaneous channel conditions. There are several reasons for these variations. Frequency-selective fading will result in rapid and random variations in the channel attenuation. Shadow fading and distance-dependent path loss will also affect the average received signal strength significantly. Finally, the interference at the receiver due to transmissions in other cells and by other UEs will also impact the interference level. Hence, to summarize, there will be rapid, and to some extent random, variations in the experienced quality of each radio link in a cell, variations that must be taken into account and preferably exploited.

In this chapter, some of the techniques for handling variations in the instantaneous radio-link quality will be discussed. *Channel-dependent scheduling* in a mobile-communication system deals with the question of how to share, between different users (different UEs), the radio resource(s) available in the system to achieve as efficient resource utilization as possible. Typically, this implies minimizing the amount of resources needed per user and thus allowing for as many users as possible in the system, while still satisfying whatever quality-of-service requirements that may exist. Closely related to scheduling is *link adaptation*, which deals with how to set the transmission parameters of a radio link to handle variations of the radio-link quality.

Both channel-dependent scheduling and link adaptation try to exploit the channel variations through appropriate processing *prior to* transmission of the data. However, because of the random nature of the variations in the radio-link quality, perfect adaptation to the instantaneous radio-link quality is never possible. *Hybrid-ARQ*, which requests retransmission of erroneously received data packets, is therefore

**HSPA Evolution: The Fundamentals for Mobile Broadband. DOI: 10.1016/B978-0-08-099969-2.00005-3**

useful. This can be seen as a mechanism for handling variations in the instantaneous radio-link quality *after* transmission and nicely complements channel-dependent scheduling and link adaptation. Hybrid-ARQ also serves the purpose of handling random errors due to, for example, noise in the receiver.

## 5.1 LINK ADAPTATION: POWER AND RATE CONTROL

Historically, *dynamic transmit-power control* has been used in CDMA-based mobile-communication systems, such as WCDMA and cdma2000, to compensate for variations in the instantaneous channel conditions. As the name suggests, dynamic power control dynamically adjusts the radio-link transmit power to compensate for variations and differences in the instantaneous channel conditions. The aim of these adjustments is to maintain a (near) constant $E_b/N_0$ at the receiver to successfully transmit data without a too high error probability. In principle, transmit-power control increases the power at the transmitter when the radio link experiences poor radio conditions (and vice versa). Thus, the transmit power is in essence inversely proportional to the channel quality, as illustrated in Figure 5.1a. This results in a basically constant data rate, regardless of the channel variations. For services such as circuit-switched voice, this is a desirable property. Transmit-power control can be seen as one type of link adaptation, that is, the adjustment of transmission parameters, in this case the transmit power, to adapt to differences and variations in the instantaneous channel conditions to maintain the received $E_b/N_0$ at a desired level.

However, in many cases of mobile communication, especially in case of packet-data traffic, there is not a strong need to provide a certain constant data rate over a radio link. Rather, from a user perspective, the data rate provided over the radio interface should simply be as "high as possible." Actually, even in case of typical "constant-rate" services, such as voice and video, (short-term) variations in the data rate are often not an issue, as long as the average data rate remains constant, assuming averaging over some relatively short time interval. In such cases, that is, when a constant data rate is not required, an alternative to transmit-power control is link adaptation by means of *dynamic rate control*. Rate control does not aim at keeping the instantaneous radio-link data rate constant, regardless of the instantaneous channel conditions. Instead, with rate control, the data rate is dynamically adjusted to compensate for the varying channel conditions. In situations with advantageous channel conditions, the data rate is increased and vice versa. Thus, rate control maintains the $E_b/N_0 \sim P/R$ at the desired level, not by adjusting the transmission power $P$, but rather by adjusting the data rate $R$. This is illustrated in Figure 5.1b.

It can be shown that rate control is more efficient than power control [1,2]. Rate control in principle implies that the power amplifier is always transmitting at full power and therefore efficiently utilized. Power control, on the other hand, results in the power amplifier not being efficiently utilized in most situations as the transmission power is less than its maximum.

In practice, the radio-link data rate is controlled by adjusting the modulation scheme and/or the channel coding rate. In case of advantageous radio-link conditions,

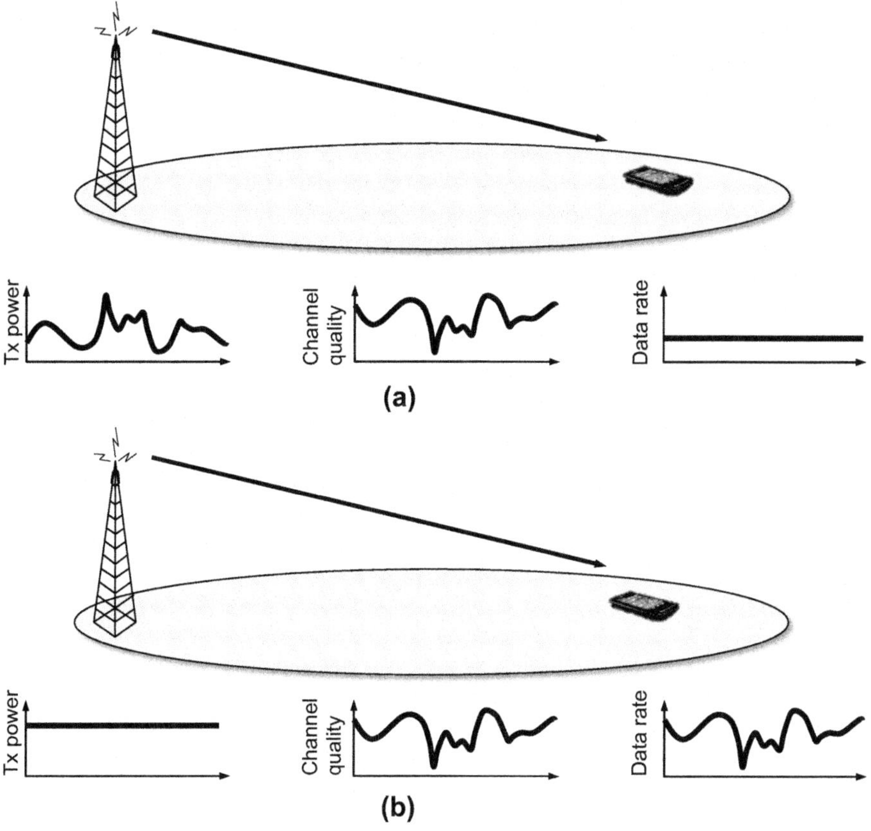


**FIGURE 5.1 (a) Power control and (b) rate control.**

the $E_b/N_0$ at the receiver is high and the main limitation of the data rate is the bandwidth of the radio link. Hence, in such situations higher-order modulation, for example, 16-QAM or 64-QAM, together with a high code-rate is appropriate, as discussed in Chapter 2. Similarly, in case of poor radio-link conditions, QPSK and low-rate coding is used. For this reason, link adaptation by means of rate control is sometimes also referred to as *Adaptive Modulation and Coding* (AMC).

## 5.2 CHANNEL-DEPENDENT SCHEDULING

Scheduling controls the allocation of the shared resources among the users at each time instant. It is closely related to link adaptation, and often scheduling and link adaptation are seen as one joint function. The scheduling principles, as well as which resources are shared between users, differ depending on the radio-interface characteristics, for example, whether uplink or downlink is considered and whether different users' transmissions are mutually orthogonal or not.

### 5.2.1 DOWNLINK SCHEDULING

In the downlink, transmissions to different UEs within a cell are typically mutually orthogonal, implying that, at least in theory, there is no interference between the transmissions (no intra-cell interference). Downlink intra-cell orthogonality can be achieved in time domain, *Time Division Multiplexing* (TDM); in the frequency domain, *Frequency-Domain Multiplexing* (FDM); or in the code domain, *Code Domain Multiplexing* (CDM). In addition, the spatial domain can also be used to separate users, at least in a quasi-orthogonal way, through different antenna arrangements. This is sometimes referred to as *Spatial Division Multiplexing* (SDM), although in most cases it is used in combination with one or several of the above multiplexing strategies. It is not discussed further in this chapter.

For packet data, where the traffic often is very bursty, it can be shown that TDM is preferable from a theoretical point of view [3,4] and is therefore typically the main component in the downlink [5,6]. However, the TDM component is often combined with sharing of the radio resource also in the frequency domain (FDM) or in the code domain (CDM). For example, in case of HSDPA (see Chapter 8), downlink multiplexing is a combination of TDM and CDM. HSDPA offers little scope for FDM, apart from multicarrier HSDPA, which offers a scope for course FDM. The reasons for sharing the resources not only in the time domain will be further elaborated on later in this section.

When transmissions to multiple users occur in parallel, using CDM, there is also an instantaneous sharing of the total available cell transmit power. In other words, not only are the time/code resources shared resources but also the power resource in the base station. In contrast, in case of sharing only in the time domain, there is, per definition, only a single transmission at a time and thus no instantaneous sharing of the total available cell transmit power.

For the purpose of discussion, assume initially a TDM-based downlink with a single user being scheduled at a time. In this case, the utilization of the radio resources is maximized if, at each time-instant, all resources are assigned to the user with the best instantaneous channel condition:

- In case of link adaptation based on power control, this implies that the lowest possible transmit power can be used for a given data rate and thus minimizes the interference to transmissions in other cells for a given link utilization.
- In case of link adaptation based on rate control, this implies that the highest data rate is achieved for a given transmit power, or, in other words, for a given interference to other cells, the highest link utilization is achieved.

However, if applied to the downlink, transmit-power control in combination with TDM scheduling implies that the total available cell transmit power will, in most cases, not be fully utilized. Thus, rate control is generally preferred [1,3,6,7].

The strategy outlined above is an example of channel-dependent scheduling, where the scheduler takes the instantaneous radio-link conditions into account. Scheduling the user with the instantaneously best radio link conditions is often referred to as *max-C/I*

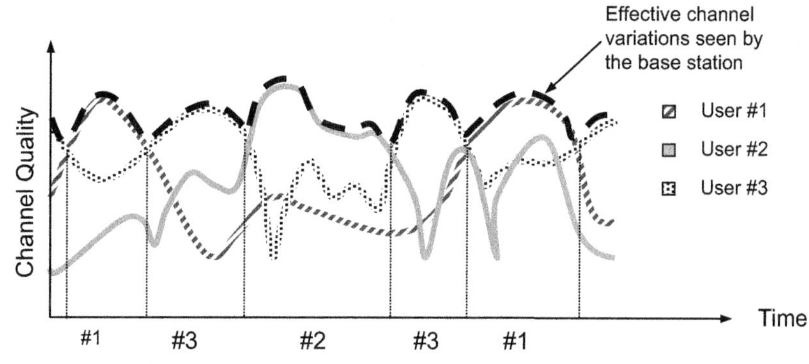


**FIGURE 5.2 Channel-dependent scheduling.**

(or *maximum rate*) scheduling. Because the radio conditions for the different radio links within a cell typically vary independently, at each point in time there is almost always a radio link whose channel quality is near its peak (see Figure 5.2). Thus, the channel eventually used for transmission will typically have a high quality and, with rate control, a correspondingly high data rate can be used. This translates into a high system capacity. The gain obtained by transmitting to users with favorable radio-link conditions is commonly known as multi-user diversity; the gains are larger, the larger the channel variations and the larger the number of users in a cell. Hence, in contrast to the traditional view that fast fading, that is rapid variations in the radio-link quality, is an undesirable effect that has to be combated, with the possibility for channel-dependent scheduling, *fading is in fact potentially beneficial and should be exploited.*

Mathematically, the max-C/I (maximum rate) scheduler can be expressed as scheduling user $k$ given by

$$k = \arg\max_i R_i, \tag{5.1}$$

where $R_i$ is the instantaneous data rate for user $i$. Although from a system capacity perspective max-C/I scheduling is beneficial, this scheduling principle will not be fair in all situations. If all UEs are, on average, experiencing similar channel conditions and large variations in the instantaneous channel conditions are only due to, for example, fast multipath fading, all users will experience the same average data rate. Any variations in the instantaneous data rate are rapid and often not even noticeable by the user. However, in practice different UEs will experience also differences in the (short-term) average channel conditions, for example, due to differences in the distance and shadow fading between the base station and the UE. In this case, the channel conditions experienced by one UE may, for a relatively long time, be worse than the channel conditions experienced by other UEs. A pure max-C/I-scheduling strategy may then, in essence, "starve" the UE with the bad channel conditions, and the UE with bad channel conditions will never be scheduled. This is illustrated in Figure 5.3a, where a max-C/I scheduler is used to schedule between two different users with different average channel quality. Virtually all the time, the same user is

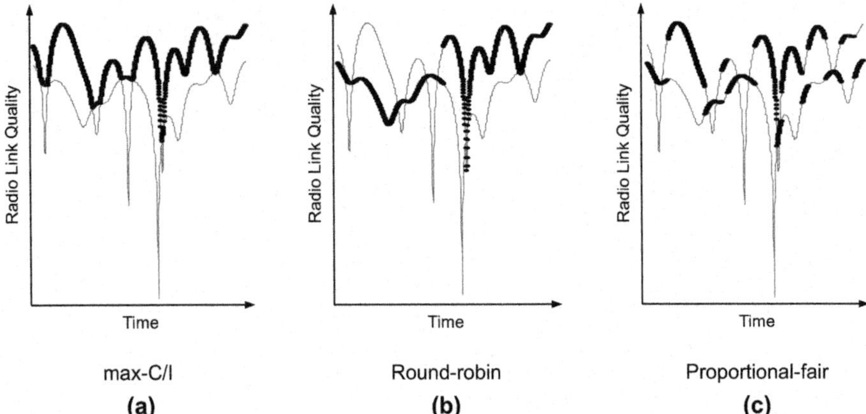


**FIGURE 5.3 Examples of three different scheduling behaviors for two users with different average channel quality: (a) max-C/I, (b) round-robin, and (c) proportional-fair. The selected user is shown with bold lines.**

scheduled. Although resulting in the highest system capacity, this situation is often not acceptable from a quality-of-service point of view.

An alternative to the max-C/I scheduling strategy is the so-called *round-robin* scheduling, illustrated in Figure 5.3b. This scheduling strategy lets the users take turns in using the shared resources without taking the instantaneous channel conditions into account. Round-robin scheduling can be seen as fair scheduling in the sense that the same amount of radio resources (the same amount of time) is given to each communication link. However, round-robin scheduling is not fair in the sense of providing the same service quality to all communication links. In that case, more radio resources (more time) must be given to communication links with bad channel conditions. Furthermore, as round-robin scheduling does not take the instantaneous channel conditions into account in the scheduling process, it will lead to lower overall system performance but more equal service quality between different communication links, compared to max-C/I scheduling.

Thus, what is needed is a scheduling strategy that is able to utilize the fast channel variations to improve the overall cell throughput while still ensuring the same average user throughput for all users or at least a certain minimum user throughput for all users. When discussing and comparing different scheduling algorithms, it is important to distinguish between different types of variations in the service quality:

- Fast variations in the service quality corresponding to, for example, fast multipath fading and fast variations in the interference level. For many packet-data applications, relatively large short-term variations in service quality are often acceptable or not even noticeable to the user.
- More long-term differences in the service quality between different communication links corresponding to, for example, differences in the distance

to the cell-site and shadow fading. In many cases there is a need to limit such long-term differences in service quality.

A practical scheduler should thus operate somewhere in between the max-C/I scheduler and the round-robin scheduler, that is, try to utilize fast variations in channel conditions as much as possible while still satisfying some degree of fairness between users.

One example of such a scheduler is the proportional-fair scheduler [8,9,10], illustrated in Figure 5.3c. In this strategy, the shared resources are assigned to the user with the *relatively* best radio-link conditions, that is, at each time instant user $k$ is selected for transmission according to

$$k = \arg\max_{i} \frac{R_i}{\overline{R}_i}, \tag{5.2}$$

where $R_i$ is the instantaneous data rate for user $i$ and $\overline{R}_i$ the average data rate for user $i$. The average is calculated over a certain averaging period $T_{PF}$. To ensure efficient usage of the short-term channel variations and, at the same time, limit the long-term differences in service quality to an acceptable level, the time constant $T_{PF}$ should be set longer than the time constant for the short-term variations. At the same time, $T_{PF}$ should be sufficiently short so that quality variations within the interval $T_{PF}$ are not strongly noticed by a user. Typically, $T_{PF}$ can be set to be in the order of one sec.

In the above discussion, it was assumed that all the radio resources in the downlink were assigned to a single user at a time; that is, scheduling was done purely in the time domain using TDM between users. However, in several situations, TDM is complemented by CDM or (to a limited extent with MC-HSDPA) FDM. In principle, there are two reasons for not relying solely on TDM in the downlink:

- in case of insufficient payload, that is, when the amount of data to transfer to a user is not sufficiently large to utilize the full channel capacity, and a fraction of resources could be assigned to another user, either through FDM or CDM; and
- in case channel variations in the frequency domain are exploited through FDM, as discussed further below.

The scheduling strategies in these cases can be seen as generalizations of the schemes discussed for the TDM-only cases above. For example, to handle small payloads, a greedy filling approach can be used, where the scheduled user is selected according to max-C/I (or any other scheduling scheme). Once this user has been assigned resources matching the amount of data awaiting transmission, the second best user according to the scheduling strategy is selected and assigned (a fraction of) the residual resources, and so on.

Finally, it should also be noted that the scheduling algorithm typically is a base station implementation issue and nothing that is normally specified in any standard. What needs to be specified in a standard to support channel-dependent scheduling are channel-quality measurements/reports and the signaling needed for dynamic resource allocation.

### 5.2.2 UPLINK SCHEDULING

The previous section discussed scheduling from a downlink perspective. However, scheduling is equally applicable to uplink transmissions, and to a large extent the same principles can be reused although there are some differences between the two.

Fundamentally, the uplink power resource is *distributed* among the users, while in the downlink the power resource is *centralized* within the base station. Furthermore, the maximum uplink transmission power of a single UE is typically significantly lower than the output power of a base station. This has a significant impact on the scheduling strategy. Unlike the downlink, where pure TDMA often can be used, uplink scheduling typically may have to rely on sharing in the code domain in addition to the time domain as a single UE may not have sufficient power for efficiently utilizing the link capacity.

Channel-dependent scheduling is, similar to the downlink case, also beneficial in the uplink case. However, the characteristics of the underlying radio interface, most notably whether the uplink relies on orthogonal or nonorthogonal multiple access and the type of link adaptation scheme used, also have a significant impact on the uplink scheduling strategy.

In case of a nonorthogonal multiple-access scheme, such as CDMA, power control is typically essential for proper operation. As discussed earlier in this chapter, the purpose of power control is to control the received $E_b/N_0$ such that the received information can be recovered. However, in a non-orthogonal multiple-access setting, power control also serves the purpose of controlling the amount of interference affecting *other* users that are sharing the same code resources. This can be expressed as the *maximum tolerable interference level* at the base station, which is a shared resource. Even if, from a single user's perspective, it would be beneficial to transmit at full power to maximize the data rate, this may not be acceptable from an interference perspective as other UEs in this case may not be able to successfully transfer any data. Thus, with non-orthogonal multiple access, scheduling a UE when the channel conditions are favorable may not directly translate into a higher data rate as the interference generated to other simultaneously transmitting UEs in the cell must be taken into account. Stated differently, the *received* power (and thus the data rate) is, thanks to power control, in principle constant, regardless of the channel conditions at the time of transmission, while the *transmitted* power depends on the channel conditions at the time of transmission. Hence, even though channel-dependent scheduling in this example does not give a direct gain in terms of a higher data rate from the UE, channel-dependent scheduling will still provide a gain for the system in terms of lower inter-cell interference.

The above discussion on non-orthogonal multiple access was simplified in the sense that no bounds on the UEs' transmission power were assumed. In practice, the transmission power of a UE is upper-bounded, due to both implementation and regulatory reasons, and scheduling a UE for transmission in favorable channel conditions decreases the probability that the UE has insufficient power to utilize the channel capacity.

The same basic scheduling principles as for the downlink can be used. If UEs are not power limited, then it is preferable to schedule users only in the time domain in the WCDMA uplink, because TDM scheduling maintains orthogonality between users. However, in large cells, the transmission power in a UE may be limited and therefore additional sharing of the uplink resources in code domain may be required.

A max-C/I scheduler would assign all the uplink resources to the UE with the best uplink channel conditions. Neglecting any power limitations in the UE, this would result in the highest capacity (in an isolated cell) [3].

In case of a non-orthogonal multiple-access scheme, *greedy filling* is one possible scheduling strategy [7]. With greedy filling, the UE with the best radio conditions is assigned as high a data rate as possible. If the interference level at the receiver is smaller than the maximum tolerable level, the UE with the second best channel conditions is allowed to transmit as well, continuing with more and more UEs until the maximum tolerable interference level at the receiver is reached. This strategy maximizes the air interface utilization but is achieved at the cost of potentially large differences in data rates between users. In the extreme case, a user at the cell border with poor channel conditions may not be allowed to transmit at all.

Strategies between greedy filling and max-C/I can also be envisioned as, for example, different proportional-fair strategies. This can be achieved by including a weighting factor for each user, proportional to the ratio between the instantaneous and average data rates, into the greedy filling algorithm.

The schedulers above all assume knowledge of the instantaneous radio-link conditions–knowledge that can be hard to obtain in the uplink scenario, as discussed in Section 5.2.4 below. In situations when no information about the uplink radio-link quality is available at the scheduler, *round-robin* scheduling can be used. Similar to the downlink, round-robin implies UEs taking turns in transmitting, thus creating a TDMA-like operation with interuser orthogonality in the time domain. Although the round-robin scheduler is simple, it is far from the optimal scheduling strategy.

### 5.2.3 LINK ADAPTATION AND CHANNEL-DEPENDENT SCHEDULING IN THE FREQUENCY DOMAIN

In the previous section, for the downlink, TDM-based scheduling was assumed and it was explained how, in this case, channel variations in the time domain could be used to improve system performance by applying channel-dependent scheduling, especially in combination with dynamic rate control. However, if the scheduler has access to the frequency domain (e.g., can schedule across multiple HSPA carriers), scheduling and link adaptation can also take place in the frequency domain.

Link adaptation in the frequency domain implies that the data modulation and coding is adjusted based on knowledge about the instantaneous channel conditions also in the frequency domain; that is, knowledge about the attenuation as well as the noise/interference level of every carrier.

Similarly, channel-dependent scheduling in the frequency domain implies that, based on knowledge about the instantaneous channel conditions also in the frequency domain, different carriers are used for transmission to or from different UEs. The scheduling gains from exploiting variations in the frequency domain are similar to those obtained from time domain variations. Obviously, in situations where the channel quality varies significantly with the frequency while the channel quality only varies slowly with time, channel-dependent scheduling in the frequency domain can enhance system capacity. An example of such a situation is an indoor system with low mobility, where the quality only varies slowly with time.

### 5.2.4 ACQUIRING CHANNEL-STATE INFORMATION

To select a suitable data rate, in practice a suitable modulation scheme and channel-coding rate, the transmitter needs information about the radio-link channel conditions. Such information is also required for the purpose of channel-dependent scheduling. In case of a system based on *Frequency-Division Duplex* (FDD), only the receiver can accurately estimate the radio-link channel conditions.

For the downlink, most systems provide a downlink signal of a predetermined structure, known as the downlink pilot or the downlink reference signal. This reference signal is transmitted from the base station with a constant power and can be used by the UE to estimate the instantaneous downlink channel conditions. Information about the instantaneous downlink conditions can then be reported to the base station.

Basically, what is relevant for the transmitter is an estimate reflecting the channel conditions *at the time of transmission*. Hence, in principle, the UE could apply a predictor, trying to predict the future channel conditions and report this predicted value to the base station. However, as this would require specification of prediction algorithms and how they would operate when the UE is moving at different speeds, most practical systems simply report the measured channel conditions to the base station. This can be seen as a very simple predictor, basically assuming the conditions in the near future will be similar to the current conditions. Thus, the more rapid the time domain channel variations are, the less efficient link adaptation is.

As there inevitably will be a delay between the point in time when the UE measured the channel conditions and the application of the reported value in the transmitter, channel-dependent scheduling and link adaptation typically operates at its best at low UE mobility. If the UE starts to move at a high speed, the measurement reports will be outdated once reported to the base station. In such cases, it is often preferable to perform link adaptation on the long-term average channel quality and rely on hybrid-ARQ with soft combining for the rapid adaptation.

For the uplink, estimation of the uplink channel conditions is not as straightforward as there is typically not any reference signal transmitted with constant power from each UE.

### 5.2.5 TRAFFIC BEHAVIOR AND SCHEDULING

It should be noted that there is little difference between different scheduling algorithms at low system load, that is, when only one or, in some cases, a few users have data waiting for transmission at the base station at each scheduling instant.

The differences between different scheduling algorithms are primarily visible at high load. However, not only the load but also the traffic behavior affects the overall scheduling performance.

As discussed above, channel-dependent scheduling tries to exploit short-term variations in radio quality. Generally speaking, a certain degree of long-term fairness in service quality is desirable, which should be accounted for in the scheduler design. However, because system throughput decreases the more fairness is enforced, a trade-off between fairness and system throughput is necessary. In this trade-off, it is important to take traffic characteristics into account as they have a significant influence on the trade-off between system throughput and service quality.

To illustrate this, consider three different downlink schedulers:

1. *Round-robin (RR) scheduler*, where channel conditions are not taken into account.
2. *Proportional-fair (PF) scheduler*, where short-term channel variations are exploited while maintaining the long-term average user data rate.
3. *Max-C/I scheduler*, where the user with the best instantaneous channel quality in absolute terms is scheduled.

For a full buffer scenario when there are always data available at the base station for all UEs in the cell, a max-C/I scheduler will result in no, or a very low, user throughput for users at the cell-edge with a low average channel quality. The reason is the fundamental strategy of the max-C/I scheduler—all resources are allocated for transmission to the UE whose channel conditions support the highest data rate. Only in the rare, not to say unlikely, case of a cell-edge user having better conditions than a cell-center user, for example, due to a deep fading dip for the cell-center user, will the cell-edge user be scheduled. A proportional-fair scheduler, on the other hand, will ensure some degree of fairness by selecting the user supporting the highest data rate relative to its average data rate. Hence, users tend to be scheduled on their fading peaks, regardless of the absolute quality. Thus, users on the cell edge will also be scheduled, thereby resulting in some degree of fairness between users.

For a scenario with bursty packet data, the situation is different. In this case, the users' buffers will be finite and in many cases also empty. For example, a web page has a certain size and, after transmitting the page, there are no more data to be transmitted to the UE in question until the users request a new page by clicking on a link. In this case, a max-C/I scheduler can still provide a certain degree of fairness. Once the buffer for the user with the highest C/I has been emptied, another user with non-empty buffers will have the highest C/I and be scheduled, and so on. This is the reason for the difference between full buffer and web-browsing traffic illustrated in Figure 5.4. The proportional-fair scheduler has similar performance in both scenarios.

Clearly, the degree of fairness introduced by the traffic properties depends heavily on the actual traffic; a design made with certain assumptions may be less desirable in an actual network where the traffic pattern may be different from the assumptions made during the design. Therefore, relying solely on the traffic properties for fairness is not a good strategy, but the discussion above emphasizes the need to not merely design the scheduler for the full buffer case.

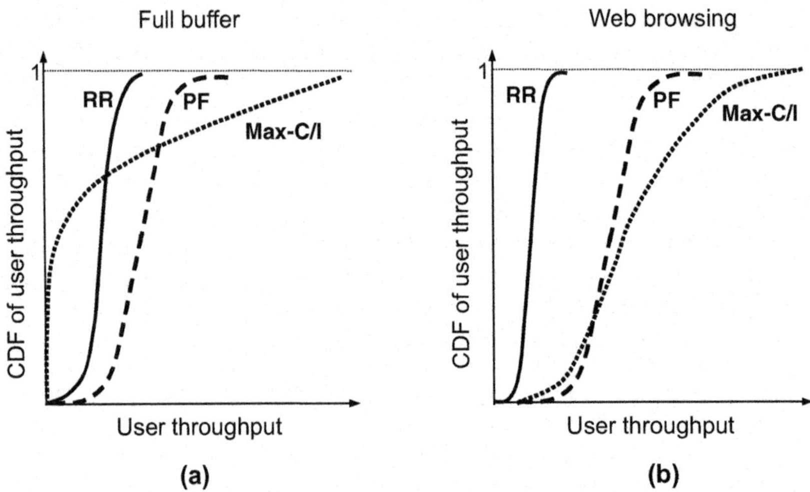


**FIGURE 5.4 Illustration of the principal behavior of different scheduling strategies: (a) for full buffers and (b) for web browsing traffic model.**

## 5.3 RETRANSMISSION SCHEMES

Transmissions over wireless channels are subject to errors, for example, due to variations in the received signal quality. To some degree, such variations can be counteracted through link adaptation, as discussed above. However, receiver noise and unpredictable interference variations cannot be counteracted. Therefore, virtually all wireless communication systems employ some form of *Forward Error Correction* (FEC), tracing its roots to the pioneering work by Claude Shannon in 1948 [11]. There is a rich literature in the area of error-correcting coding (see, e.g., [12,13] and the references therein), and a detailed description is beyond the scope of this book. In short, the basic principle beyond forward error-correcting coding is to introduce redundancy in the transmitted signal. This is achieved by adding *parity bits* to the information bits prior to transmission (alternatively, the transmission could consist of parity bits alone, depending on the coding scheme used). The parity bits are computed from the information bits using a method given by the coding structure used. Thus, the number of bits transmitted over the channel is larger than the number of original information bits, and a certain amount of *redundancy* has been introduced in the transmitted signal.

Another approach to handle transmissions errors is to use *Automatic Repeat Request* (ARQ). In an ARQ scheme, the receiver uses an error-detecting code, typically a *Cyclic Redundancy Check* (CRC), to detect if the received packet is in error or not. If no error is detected in the received data packet, the received data is declared error-free and the transmitter is notified by sending a positive *acknowledgment* (ACK). On the other hand, if an error is detected, the receiver discards the received data and notifies the transmitter via a return channel by sending a *negative acknowledgment* (NACK). In response to a NACK, the transmitter retransmits the same information.

Virtually all modern communication systems, including WCDMA and cdma2000, employ a combination of forward error-correcting coding and ARQ, a combination known as *hybrid-ARQ*. Hybrid-ARQ uses forward error-correcting codes to correct a subset of all errors and relies on error detection to detect uncorrectable errors. Erroneously received packets are discarded and the receiver requests retransmissions of corrupted packets. Thus, it is a combination of FEC and ARQ, as described above. Hybrid-ARQ was first proposed in [14] and numerous publications on hybrid-ARQ have appeared since (see [12] and references therein). Most practical hybrid-ARQ schemes are built around a CRC code for error detection and convolutional or Turbo codes for error correction, but in principle any error-detecting and error-correcting code can be used.

## 5.4 HYBRID-ARQ WITH SOFT COMBINING

The hybrid-ARQ operation described above discards erroneously received packets and requests retransmission. However, despite that the packet was not possible to decode, the received signal still contains information, which is lost by discarding erroneously received packets. This short-coming is addressed by *hybrid-ARQ with soft combining*. In hybrid-ARQ with soft combining, the erroneously received packet is stored in a buffer memory and later combined with the retransmission to obtain a single, combined packet that is more reliable than its constituents. Decoding of the error-correcting code operates on the combined signal. If the decoding fails (typically a CRC code is used to detect this event), a retransmission is requested.

Retransmission in any hybrid-ARQ scheme must, by definition, represent the same set of information bits as the original transmission. However, the set of coded bits transmitted in each retransmission may be selected differently as long as they represent the same set of information bits. Hybrid-ARQ with soft combining is therefore usually categorized into *Chase combining* and *Incremental Redundancy*, depending on whether the retransmitted bits are required to be identical to the original transmission or not.

*Chase combining*, where the retransmissions consist of the same set of coded bits as the original transmission, was first proposed in [9]. After each retransmission, the receiver uses maximum-ratio combining to combine each received channel bit with any previous transmissions of the same bit and the combined signal is fed to the decoder. As each retransmission is an identical copy of the original transmission, retransmissions with Chase combining can be seen as additional repetition coding. Therefore, as no new redundancy is transmitted, Chase combining does not give any additional coding gain but only increases the accumulated received $E_b/N_0$ for each retransmission (Figure 5.5).

Several variants of Chase combining exist. For example, only a subset of the bits transmitted in the original transmission might be retransmitted, so-called partial

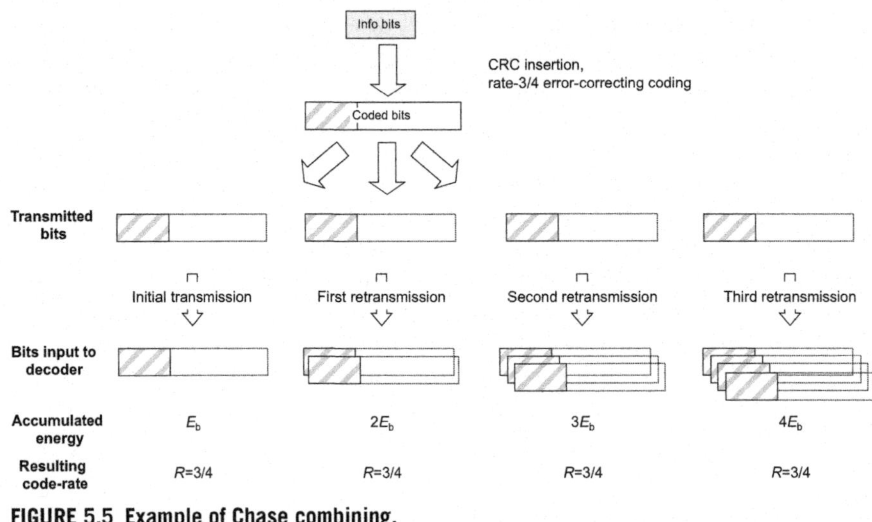


**FIGURE 5.5 Example of Chase combining.**

Chase combining. Furthermore, although combining is often done after demodulation but before channel decoding, combining can also be done at the modulation symbol level before demodulation, as long as the modulation scheme is unchanged between transmission and retransmission.

With *Incremental Redundancy* (IR), each retransmission does not have to be identical to the original transmission. Instead, *multiple sets* of coded bits are generated, each of them representing the same set of information bits [15,16]. Whenever a retransmission is required, the retransmission typically uses a different set of coded bits than the previous transmission. The receiver combines the retransmission with previous transmission attempts of the same packet. As the retransmission may contain additional parity bits not included in the previous transmission attempts, the resulting code-rate is generally lowered by a retransmission. Furthermore, each retransmission does not necessarily have to consist of the same number of coded bits as the original and, in general, the modulation scheme as well can be different for different retransmissions. Hence, incremental redundancy can be seen as a generalization of Chase combining or, stated differently, Chase combining is a special case of incremental redundancy.

Typically, incremental redundancy is based on a low-rate code and the different redundancy versions are generated by puncturing the output of the encoder. In the first transmission, only a limited number of the coded bits are transmitted, effectively leading to a high-rate code. In the retransmissions, additional coded bits are transmitted. As an example, assume a basic rate-1/4 code. In the first transmission, only every third coded bit is transmitted, effectively giving a rate-3/4 code, as illustrated in Figure 5.6. In case of a decoding error and a subsequent request for a retransmission, additional bits are transmitted, effectively leading to a rate-3/8 code. After a second retransmission, the code-rate is 1/4. In case of more than two

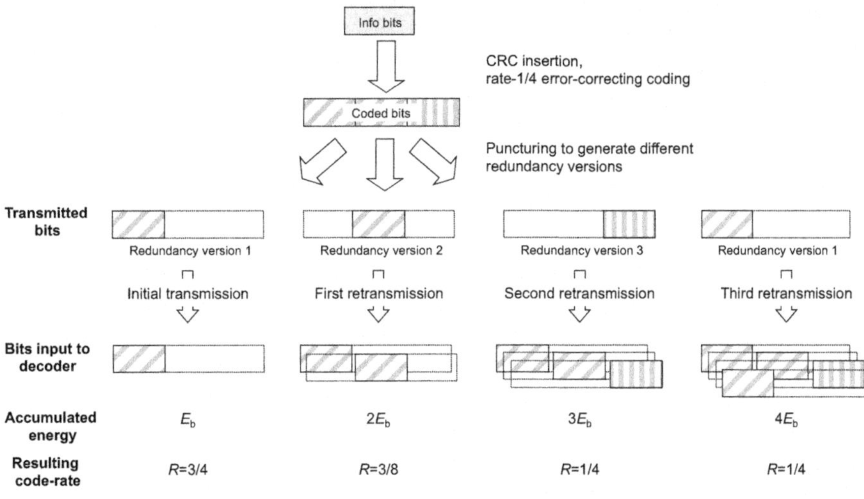


**FIGURE 5.6 Example of incremental redundancy.**

retransmissions, already transmitted coded bits would be repeated. In addition to a gain in accumulated received $E_b/N_0$, incremental redundancy also results in a coding gain for each retransmission. The gain with IR compared to Chase is larger for high initial code rates while at lower initial coding rates, Chase combining is almost as good as IR [10]. Furthermore, as shown in [17], the performance gain of incremental redundancy compared to Chase combining can also depend on the relative power difference between the transmission attempts.

With incremental redundancy, the code used for the first transmission should provide good performance not only when used alone but also when used in combination with the code for the second transmission. The same holds for subsequent retransmissions. Thus, as the different redundancy versions typically are generated through puncturing of a low-rate mother code, the puncturing patterns should be defined such that all the code bits used by a high-rate code should also be part of any lower-rate codes. In other words, the resulting code-rate $R_i$ after transmission attempt $i$, consisting of the coded bits from redundancy versions $RV_k$, $k = 1,\ldots, i$, should have similar performance as a good code designed directly for rate $R_i$. Examples of this for convolutional codes are the so-called rate-compatible convolutional codes [18].

In the discussion so far, it has been assumed that the receiver has received all of the previously transmitted redundancy versions. If all redundancy versions provide the same amount of information about the data packet, the order of the redundancy versions is not critical. However, for some code structures, not all redundancy versions are of equal importance. One example hereof is Turbo codes, where the systematic bits are of higher importance than the parity bits. Hence, the initial transmission should at least include all the systematic bits and some parity bits. In the retransmission(s), parity bits not part of the initial transmission can be included. However, if the initial transmission was received with poor quality or not at all, a retransmission with only parity bits is not appropriate, as a retransmission of (some of) the systematic bits provides better performance. Incremental redundancy with Turbo codes can therefore benefit from multiple levels of feedback, for example, by using two different negative acknowledgments (NACK) to request additional parity bits and LOST to request a retransmission of the systematic bits. In general, the problem of determining the amount of systematic and parity bits in a retransmission based on the signal quality of previous transmission attempts is non-trivial.

Hybrid-ARQ with soft combining, regardless of whether Chase or incremental redundancy is used, leads to an implicit reduction of the data rate by means of retransmissions and can thus be seen as *implicit* link adaptation. However, in contrast to link adaptation based on explicit estimates of the instantaneous channel conditions, hybrid-ARQ with soft combining implicitly adjusts the coding rate based on the result of the decoding. In terms of overall throughput, this kind of implicit link adaptation can be superior to explicit link adaptation as additional redundancy is only added *when needed*, that is, when previous higher rate transmissions were not possible to decode correctly. Furthermore, as it does not try to predict any channel variations, it works equally well, regardless of the speed at which the UE is moving. Because implicit link adaptation can provide a gain in system throughput, a valid question is why explicit link adaptation at all is necessary. One major reason for having explicit link adaptation is the reduced delay. Although relying on implicit link adaptation alone is sufficient from a system throughput perspective, the end-user service quality may not be acceptable from a delay perspective.

## REFERENCES


[1] S.T. Chung, A.J. Goldsmith, Degrees of freedom in adaptive modulation: a unified view, IEEE Trans Commun 49 (9) (September 2001) 1561–1571.

[2] A.J. Goldsmith, P. Varaiya, Capacity of fading channels with channel side information, IEEE Trans Information Theory 43 (November 1997) 1986–1992.

[3] Knopp R, Humblet PA. Information capacity and power control in single-cell multi-user communications. In: Proceedings of the IEEE International Conference on Communications, Vol. 1, Seattle, WA, USA, 1995, pp. 331–335.

[4] Tse D. Optimal power allocation over parallel Gaussian broadcast channels. In: Proceedings of the International Symposium on Information Theory, Ulm, Germany; June 1997, p. 7.

[5] Honig ML, Madhow U. Hybrid intra-cell TDMA/inter-cell CDMA with inter-cell interference suppression for wireless networks. In: Proceedings of the IEEE Vehicular Technology Conference, Secaucus, NJ, USA; 1993. p. 309–312.

[6] S. Ramakrishna, J.M. Holtzman, A scheme for throughput maximization in a dual-class CDMA system, IEEE J Selected Areas Commun 16 (6) (1998) 830–844.

[7] Oh SJ, Wasserman KM. Optimality of greedy power control and variable spreading gain in multi-class CDMA mobile networks. In: Proceedings of the AMC/IEEE MobiComp, Seattle, Washington, USA, 1999, pp. 102–112.

[8] Holtzman JM. CDMA forward link waterfilling power control. In: Proceedings of the IEEE Vehicular Technology Conference, Vol. 3, Tokyo, Japan; May 2000, pp. 1663–1667.

[9] Holtzman JM. Asymptotic analysis of proportional fair algorithm. In: Proceedings of the IEEE Conference on Personal Indoor and Mobile Radio Communications, Vol. 2, San Diego, CA, USA, 2001, pp. 33–37.

[10] P. Viswanath, D. Tse, R. Laroia, Opportunistic beamforming using dumb antennas, IEEE Trans Information Theory 48 (6) (2002) 1277–1294.

[11] C.E. Shannon, A mathematical theory of communication, Bell System Tech J 27 (July and October 1948) 379–423 623–656.

[12] Lin S, Costello D. Error control coding (1983). Upper Saddle River, NJ, USA: Prentice-Hall.

[13] C. Schlegel, Trellis and turbo coding, Wiley – IEEE Press, Chichester, UK, (March 2004).

[14] Wozencraft JM, Horstein M. Digitalised communication over two-way channels. In: Fourth London Symposium on Information Theory, London, UK; September 1960.

[15] M.B. Pursley, S.D. Sandberg, Incremental-redundancy transmission for meteor-burst communications, IEEE Trans Commun 39 (May 1991) 689–702.

[16] S.B. Wicker, M. Bartz, Type-I hybrid ARQ protocols using punctured MDS codes, IEEE Trans Commun 42 (April 1994) 1431–1440.

[17] Frenger P, Parkvall S, Dahlman E. Performance comparison of HARQ with chase combining and incremental redundancy for HSDPA. In: Proceedings of the IEEE Vehicular Technology Conference, Atlantic City, NJ, USA; October 2001, pp. 1829–1833.

[18] J. Hagenauer, Rate-compatible punctured convolutional codes (RCPC Codes) and their applications, IEEE Trans Commun 36 (April 1988) 389–400.

PART

# HSPA and its evolution

CHAPTER

# Overview of release 99 WCDMA

# 6

## CHAPTER OUTLINE



This section provides a brief overview of WCDMA release 99 to serve as a background to subsequent chapters. As a thorough walkthrough of WCDMA is a topic of its own and beyond the scope of this book, the reader is referred to other books and articles, for example [1–3], for a detailed description. WCDMA is a versatile and highly flexible radio interface that can be configured to meet the requirements from a large number of services, but the focus for the description below is the functionality commonly used to support packet-data transmissions. The main purpose is to provide a brief background to WCDMA to put the enhancements described later on into a perspective.

## 6.1 SYSTEM ARCHITECTURE

WCDMA is based on a hierarchical architecture [4] with different nodes and interfaces, as illustrated in Figure 6.1. A terminal, also referred to as *User Equipment* (UE) in 3GPP terminology, communicates with one (or several) Node Bs. In the WCDMA architecture, the term *Node B* refers to a logical node, responsible for physical-layer processing such as error-correcting coding, modulation and spreading, as well as conversion from baseband to the radio-frequency signal transmitted from the antenna(s). A Node B is handling transmission and reception in one or several cells. Three-sector sites are common, where each Node B is handling transmissions in three cells, although other arrangements of the cells belonging to one Node B can be thought of, for example, a large number of indoor cells or several cells along a highway belonging to the same Node B. Thus, a base station is a possible implementation of a Node B.

The *Radio Network Controller* (RNC) controls multiple Node Bs. The number of Node Bs connected to one RNC varies depending on the implementation and deployment, but up to a few hundred Node Bs per RNC is not uncommon. The


**HSPA Evolution: The Fundamentals for Mobile Broadband. DOI: 10.1016/B978-0-08-099969-2.00006-5**

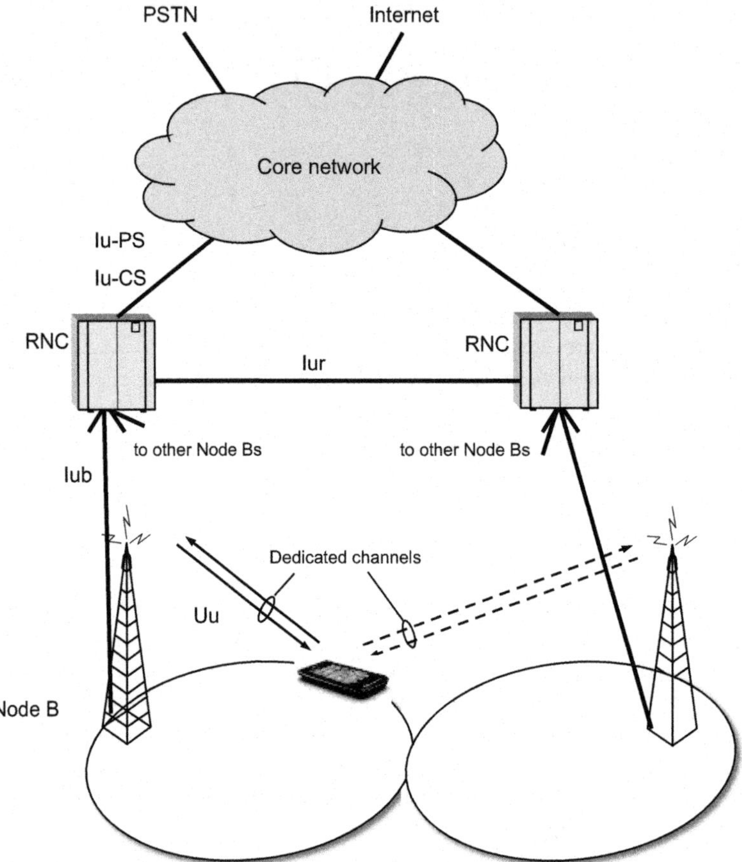


**FIGURE 6.1 WCDMA radio-access network architecture.**

RNC is, among other things, in charge of call setup, quality-of-service handling, and management of the radio resources in the cells for which it is responsible. The ARQ protocol, handling retransmissions of erroneous data, is also located in the RNC. Thus, in release 99, most of the "intelligence" in the radio-access network resides in the RNC, whereas the Node Bs mainly act as modems.

The RNC functionality can be partitioned into functionality that is responsible for management of radio resources and functionality that is responsible for the management of connections to individual UEs. In some circumstances, it can be the case that a UE is located in a cell served by a Node B that is controlled from one RNC, whereas the data management for the UE link is managed by a different RNC. Under such circumstances, the RNC that is responsible for radio resource management in the cell in which the UE resides is known as the "controlling" RNC, whereas the RNC that is responsible for data link management is known as the "serving" RNC.

Obviously, user plane data needs to be routed from the core network to the serving RNC and then from the serving RNC to the controlling RNC.

Finally, the RNCs are connected to the Internet and the traditional wired telephony network through the core network.

In the 3GPP specifications, a number of interfaces between network elements are defined:

- *Uu* connects the UE with the Node B;
- *Iub* connects the Node B with the RNC;
- *Iur* inter-connects RNCs; and
- *Iu-PS* and *Iu-CS* are interfaces connecting RNCs with the core network in the circuit-switched and packet-switched domains, respectively.

## 6.2 PROTOCOL ARCHITECTURE

Most modern communication systems structure the processing into different *layers*, and WCDMA is no exception. The layered approach is beneficial as it provides a certain structure to the overall processing where each layer is responsible for a specific part of the radio-access functionality. The different protocol layers used in WCDMA are illustrated in Figure 6.2 and briefly described below (see also [5]).

User data from the core network, for example in the form of IP packets, are first processed by the *Packet Data Convergence Protocol* (PDCP) which performs (optional) header compression. IP packets have a relatively large header, 40 bytes for IPv4 and 60 bytes for IPv6, and to save radio-interface resources, header compression is beneficial.

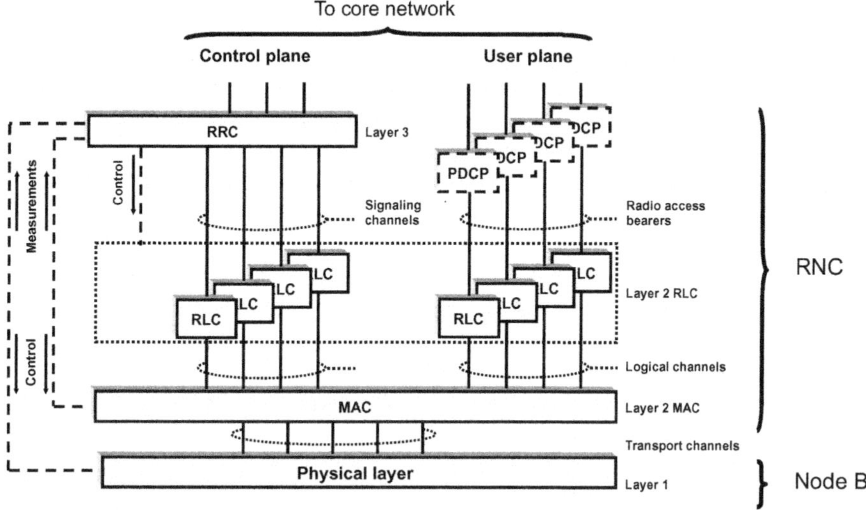


**FIGURE 6.2 WCDMA protocol architecture.**

The *Radio Link Control* (RLC) protocol is, among other things, responsible for segmentation of the IP packets into smaller units known as RLC *Protocol Data Units* (RLC PDUs). At the receiving end, the RLC performs the corresponding reassembly of the received segments. The RLC also handles the ARQ protocol. For packet-data services, error-free delivery of data is often essential, and the RLC can therefore be configured to request retransmissions of erroneous RLC PDUs in this case. For each incorrectly received PDU, the RLC requests a retransmission. The need for a retransmission is indicated by the RLC entity at the receiving end to its peer RLC entity at the transmitting end by means of status reports.

The *Medium Access Control* (MAC) layer offers services to the RLC in the form of so-called *Logical Channels*. The MAC layer can multiplex data from multiple logical channels. It is also responsible for determining the *Transport Format* of the data sent to the next layer, the *physical* layer. In essence, the transport format is the instantaneous data rate used over the radio link. The interface between MAC and PHY carries so-called *Transport Channels* over which data in the form of *transport blocks* are transferred. There are several transport channels defined in WCDMA; for an overview and the allowed mappings of logical channels to transport channels, the reader is referred to [5].

In each *Transmission Time Interval* (TTI), one or several transport blocks are fed from the MAC layer to the physical layer, which performs coding, interleaving, multiplexing, spreading, etc., prior to data transmission. Thus, for WCDMA, the TTI is the time that the interleaver spans and the time it takes to transmit the transport block over the radio interface. A larger TTI implies better time diversity but also a longer delay. In the first release, WCDMA relies on TTI lengths of 10, 20, 40, and 80 ms. As will be seen later, HSPA introduces additional 2 ms TTI to reduce the latency.

To support different data rates, the MAC can vary the transport format between consecutive TTIs. The transport format consists of several parameters describing how the data shall be transmitted in a TTI. By varying the transport-block size and/or the number of transport blocks, different data rates can be realized.

At the bottom of the protocol stack is the *physical layer*. The physical layer is responsible for operations such as coding, spreading, and data modulation, as well as modulation of the radio-frequency carrier.

The PDCP, RLC, MAC, and physical layer are configured by the *Radio Resource Control* (RRC) protocol. RRC performs admission control, handover decisions, and active set management for soft handover. By setting the parameters of the RLC, MAC, and physical layers properly, the RRC can provide the necessary quality-of-service (QoS) requested by the core network for a certain service.

On the network side, the MAC, RLC, and RRC entities in release 99 are all located in the RNC whereas the physical layer is mainly located in the Node B. The same entities also exist in the UE. For example, the MAC in the UE is responsible for selecting the transport format for uplink transmissions from a set of formats configured by the network. However, the handling of the radio resources in the cell is controlled by the RRC entity in the network and the UE obeys the RRC decisions taken in the network.

## 6.3 PHYSICAL LAYER; DOWNLINK

The basis for the physical layer of WCDMA is spreading of the data to be transmitted to the chip rate corresponding to 3.84 Mchip/s using channelization codes. It is also responsible for coding, transport-channel multiplexing, and modulation of the radio-frequency carrier. A simplified illustration of the physical layer processing is provided in Figure 6.3.

The channelization codes are so-called *Orthogonal Variable Spreading Factor* (OVSF) codes, defined by the tree structure, as shown in Figure 6.4. As discussed in Chapter 3, a key property of the OVSF codes is the mutual orthogonality between data streams spread with different OVSF codes, even if the data rate, and therefore the spreading factors, are different. This holds as long as the different OVSF codes are selected from different branches of the channelization code tree and transmitted

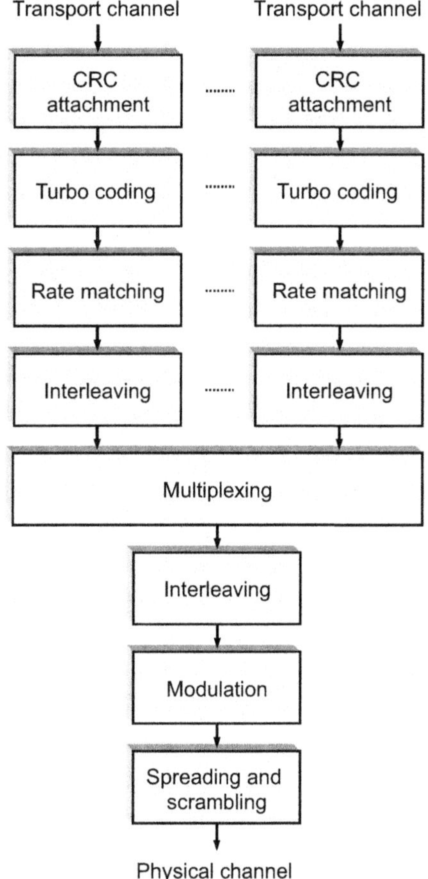


**FIGURE 6.3 Simplified view of physical layer processing in WCDMA.**

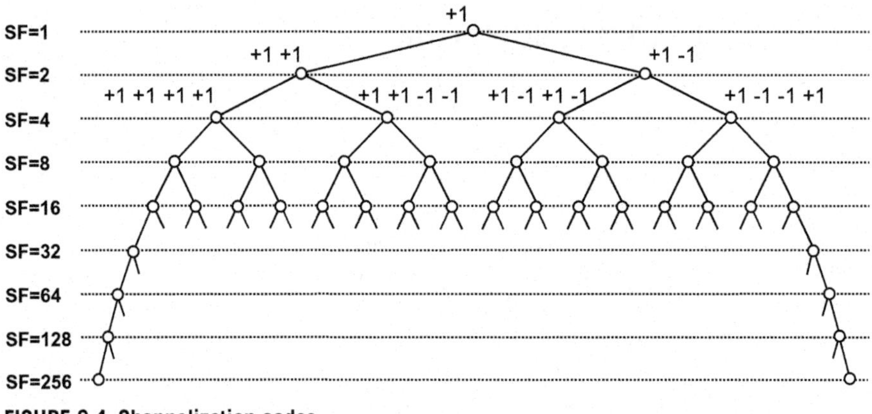


**FIGURE 6.4 Channelization codes.**

with the same timing. By selecting different spreading factors, physical channels with different data rates can be provided.

Because physical channels are separated by OVSF codes, transmissions on different physical channels are orthogonal and will not interfere with each other. The WCDMA downlink is therefore often said to be *orthogonal*. However, at the receiver side, orthogonality will partly be lost in case of a frequency-selective channel. As discussed in Chapter 3, this will cause signal corruption to the received signal and result in interference between different codes used for downlink transmission. Such signal corruption can be counteracted by an equalizer.

The spreading operation actually consists of two steps: spreading to the chip rate by using orthogonal *channelization codes* with a length equal to the symbol time followed by cell-specific scrambling using non-orthogonal scrambling sequences with a length equal to the 10 ms radio frame. The purpose of the scrambling step is to randomize interference between different cells. Transmissions in neighboring cells will cause interference, interference that will be suppressed by the UE receiver with a factor proportional to the processing gain (the processing gain is given by the spreading factor divided by the code rate after rate matching). The scrambling sequences are designed such that the cross-correlation, and thereby the interference after despreading in the UE, is minimized irrespective of the time difference between different cells. Hence, WCDMA does not require multiple cells to be time synchronized.

Downlink channelization codes are used for carrying physical channels and physical signals. Some channels (such as Primary CPICH, Primary CCPCH, and SCH) are assigned to fixed channelization codes, whereas for other channels the channelization codes are assigned by RRC.

The downlink physical channels are as follows:

- The *Primary Common Pilot Channel* (P-CPICH) contains a known sequence on a known OVSF code and is used as a reference for downlink channel estimation by all UEs in the cell.
- The *Primary Common Control Physical Channel* (P-CCPCH) carries the *Broadcast* transport channel, BCH. The P-CCPCH is transmitted on a known

and fixed code such that UEs involved in handover or registering to the network can find it. Information is carried on the broadcast channel in a series of so-called *System Information Blocks* (SIBs).

- *Secondary Common Control Physical Channels* (S-CCPCHs) can be configured and can carry common transport channels such as the *Forward Access Channel* (FACH) and *Paging Channel* (PCH).
- *Secondary Common Pilot Channels* (S-CPICHs) may be configured such that additional common pilots can be provided. These pilots can, for example, be provided over a subsection of the cell area in order to enable certain types of beamforming.
- *Synchronization Channel* (SCH) is a fixed sequence that is time multiplexed with P-CCPCH on a known channelization code in order to allow the UE to perform synchronization with the network.
- The *Paging Indicator Channel* (PICH) is used for carrying indications to UEs in low connectivity states that do not have DCH allocated that the network wishes to communicate with them.
- The *Acquisition Indicator Channel* (AICH) is used for the network to indicate that it has detected *Physical Random Access Channel* (PRACH) preambles in uplink.
- *Dedicated Physical Channels* (DPCHs) are channels that are assigned to specific users.

Data to a specific user, including the necessary control information, is carried on a DPCH corresponding to one channelization code. By varying the spreading factor, different DPCH symbol rates can be provided. The spreading factor is determined by the transport format selected by the MAC layer. There is also a possibility to use multiple physical channels to a single user for the highest data rates. The DPCH is subdivided into two channels; the DPCCH, which is a control channel carrying *Transmit Power Control* (TPC) bits relating to uplink, pilots, and optionally *Transport Format Combination Indicator* (TFCI) bits and the DPDCH, carrying encoded transport blocks. The DPDCH and DPCCH channels are time multiplexed.

The coding and modulation is broadly similar for DPCH and other common channels that carry data, such as P-CCPCH and S-CCPCH. To each transport block to be transmitted, a *Cyclic Redundancy Check* (CRC) is added for the purpose of error detection. If the receiver detects an error, the RLC protocol in the receiver is informed and requests a retransmission. After CRC attachment, the data are encoded using a rate-1/3 Turbo coder (convolutional coding is also supported by WCDMA but typically not used for packet-data services on DPDCH). Rate matching by means of puncturing or repetition of the coded bits is used to fine-tune the code-rate, and in the case of dedicated channels, multiple coded and interleaved transport channels can be multiplexed together, forming a single stream of bits to be spread to the chip rate and subsequently modulated. For the downlink, QPSK modulation is used, whereas the uplink uses BPSK. The resulting stream of modulation symbols is mapped to a *physical channel* and subsequently digital-to-analog converted and modulated onto a radio-frequency carrier. In principle, each physical channel corresponds to a unique spreading code, used to separate transmissions to/from different users.

## 6.4 PHYSICAL LAYER; UPLINK

In the uplink, there are two physical channel types. Most user and control plane data is carried on a BPSK-modulated *Dedicated Physical Data Channel* (DPDCH), whereas the *Physical Random Access Channel* (PRACH) is used, as its name suggests, for random access.

For the DPDCH, similar to the downlink, different data rates are realized by using different spreading factors. The uplink data rate is determined by the transport block size. Similarly to downlink, uplink transport blocks are transmitted within TTIs. The transport block size and coding parameters that are used for a particular TTI are known as the *Transport Format Combination* (TFC). Coherent demodulation is used also in the uplink, which requires a channel estimate. Unlike the downlink, where a common pilot signal is used, uplink transmissions originate from different locations. Therefore, a common pilot cannot be used and each user must have a separate pilot signal. This is carried on the *Dedicated Physical Control Channel* (DPCCH). The DPCCH also carries TPC bits, information about the transport format of the data transmitted on the DPDCH, and optionally a beamforming controlling bit. This information is required by the physical layer in the Node B for proper processing. User-specific scrambling is used in the WCDMA uplink, and the channelization codes are only used to separate different physical channels from the same UE. The same set of channelization codes can therefore be used by multiple UEs. As the transmissions from different UEs are not time synchronized, separation of different UEs by the use of OVSF codes is not possible. Hence, the uplink is said to be non-orthogonal, and different users' transmissions will interfere with each other.

The fact that the uplink is non-orthogonal implies that fast closed-loop power control is an essential feature of WCDMA. With fast closed-loop power control, the Node B measures the received *signal-to-interference ratio* (SIR) on the DPCCH received from each UE and once per slot commands the UEs to adjust their transmission power accordingly. The target of the power control is to ensure that the received SIR of the DPCCH is at an appropriate level for each user. This SIR target may very well be different for different users. If the SIR is below the target, that is too low for proper demodulation, the Node B commands the UE to increase its transmission power. Similarly, if the received SIR is above the target and therefore unnecessarily high, the UE is instructed to lower its transmission power. If power control would not be present, the inter-user interference could cause the transmissions from some users to be non-decodable. This is often referred to as the *near-far problem* – transmissions from users near the base station will be received at a significantly higher power level than transmissions from users far from the base station, making demodulation of far users impossible unless (closed-loop) power control is used.

The DPDCH power level is fixed relative to the DPCCH Tx power level. Thus, the power control also ensures that the DPDCH SIR meets a target. The ratio of DPDCH to DPCCH power may be set differently for each TFC, because each different data rate requires a different SIR for decoding. The ratio of DPDCH to DPCCH power can be signaled for each TFC, or computed for the current TFC based on signaled settings for reference TFC.

Fast closed-loop power control is used in the downlink as well, although not primarily motivated by the near-far problem as the downlink is orthogonal. However, it is still useful to combat the fast fading by varying the transmission power; when the channel conditions are favorable, less transmission power is used, and vice versa. This typically results in a lower average transmitted power than for the corresponding non-power-controlled case, thereby reducing the average (inter-cell) interference and improving the system capacity. *Soft handover* (SHO), or *macro diversity* as it is also called, is a key feature of WCDMA. It implies that a UE is communicating with multiple cells, in the general case with multiple Node Bs, simultaneously, and is mainly used for UEs close to the cell border to improve performance. The set of cells the UE is communicating with is known as the *active set*. The RNC determines, based on measurements from the UE, the set of cells forming the active set.

In the downlink, soft handover implies that data to a UE is transmitted simultaneously from multiple cells. This provides diversity against fast fading – the likelihood that the instantaneous channel conditions from all cells simultaneously are disadvantageous is lower the larger the number of cells in the active set.

Uplink soft handover implies that the transmission from the UE is received in multiple cells. The cells can often belong to different Node Bs. Reception of data at multiple locations is fundamentally beneficial as it provides diversity against fast fading. In case the cells receiving the uplink transmission are located in the same Node B (referred to as *softer handover*), combining of the received signals occurs in the physical-layer receiver. Typically, a RAKE receiver (see Chapter 3) is used for this purpose, although other receiver structures are also possible. However, if the cells belong to different Node Bs, combining cannot take place in the RAKE receiver. Instead, each Node B tries to decode the received signal and forwards correctly received data units to the RNC. As long as at least one Node B received the data correctly, the transmission is successful. The RNC can discard any duplicates in case multiple Node Bs correctly received the transmission and only forward one copy of each correctly received data unit.[1] Missing data units in case none of the Node Bs received a correct copy of the data unit can be detected by the RLC protocol and a retransmission can be requested. Soft handover is one of the main reasons why the RLC is located in the RNC and not in the Node B.

In addition to reception at multiple cells, uplink soft handover also implies power control from multiple cells. As it is sufficient if the transmission is received in one cell, the UE lowers its transmission power if the power control mechanism in at least one of the cells commands the UE to do so. Only if all cells in the active set request the UE to increase its transmission power is the power increased. This mechanism,

[1]Formally, the selection and duplicate removal mechanisms in case of inter-Node B macro diversity are part of the physical layer. Hence, the physical layer spans both the Node B and the RNC.

known as *or-of-the-downs*, ensures the average transmission power to be kept as low as possible. Because of the non-orthogonal property of the uplink, any reduction in average interference directly translates into an increased capacity.

In a mobile UE, it is of particular importance to be able to build an efficient *power amplifier* (PA). In order to operate the amplifier at its maximum efficiency point for the maximum possible proportion of the time, it is desirable that the ratio between the peak and the average of the uplink signal (*Peak to Average Power Ratio*, PAPR) is kept low. Low PAPR implies less need to back-off the average PA power from its peak power and thereby have an increased efficiency. PAPR can be improved by minimizing the number of OVSF codes in the uplink and by selecting good code combinations. Thus, the DPDCH in the uplink is in general configured to be transmitted on one code with a variable spreading factor, and both in release 99 and subsequent HSPA evolutions, care has been paid to minimize code usage and select codes carefully.

For reasons of mobility, it is necessary for the UE to make measurements of neighbor cells both on its own and other carriers and potentially also GSM/LTE cells on other carriers while transferring data. It is assumed that samples for neighbor cell reception on the serving carrier can be captured in parallel with data reception. For measuring on other carriers, though, if the UE only has one receiver chain it is necessary for the UE to stop receiving and retune its receiver to the other carrier. To facilitate this, a so-called compressed mode pattern of receiver gaps can be configured when receiving data.

In order to be able to initiate calls from the UE, it is necessary to have a means for the UE to access the network without prior notification. Such functionality is provided by a further uplink channel type, known as the *Physical Random Access Channel* (PRACH). Transmissions using the PRACH take two stages:

- In the first stage, short bursts of known sequences of chips, so-called preambles, are transmitted at pre-determined times by the UE. The transmission of preambles is intended to indicate to the network that a UE wishes to make a random access. Because the UE does not know its uplink pathloss conditions, it transmits preambles with increasing transmit power until it either receives a response from the network or has transmitted a maximum allowable number of preambles.
- If the network detects a preamble, it indicates a response to the UE by means of transmitting an *Acknowledgment Indicator* (AI) on AICH. The response can be either to block the UE from making its RACH access or to acknowledge the preamble and allow the UE to transmit. In this case, the UE may transmit a small amount of data in a 10 ms or 20 ms TTI.

## 6.5 RESOURCE HANDLING AND PACKET-DATA SESSION

With the above discussion in mind, a typical packet-data session in the first release of WCDMA can be briefly outlined. At call setup, when a connection is established between a UE and the radio-access network, the RNC checks the amount of resources the UE needs during the session.

In the downlink, the resources consist in principle of channelization codes and transmission power. Because the RNC is responsible for allocation of resources, it knows the fraction of the channelization code tree not reserved for any user. Measurements in the Node B provide the RNC with information about the average amount of available transmission power.

In the uplink, thanks to its non-orthogonal nature, there is no channelization code limitation. Instead, the resource consists of the amount of additional interference the cell can tolerate. To quantify this, the term *noise rise* or *rise-over-thermal* is often used when discussing uplink operation. Noise rise, defined as $(I_0 + N_0)/N_0$, where $I_0$ and $N_0$ are the power spectral densities due to uplink transmissions and background noise, respectively, is a measure of the increase in interference in the cell due to the transmission activity. For example, 0 dB noise rise indicates an unloaded system (no interference from other users) and 3 dB noise rise implies a power spectral density due to uplink transmission equal to the noise spectral density. Although noise rise as such is not of major interest, it has a close relation to coverage and uplink load. A large noise rise would result in loss of coverage of some channels—a UE may not have sufficient transmission power available to reach the required $E_b/N_0$ at the base station. Therefore, the radio resource control in the RNC must keep the noise rise within certain limits. The Node B provides uplink measurements that enable the RNC to estimate the uplink load.

Provided there are sufficient resources in both uplink and downlink, the RNC admits the UE into the cell and configures a dedicated (physical) channel in each direction. Several parameters are configured, one of them being the set of transport formats the MAC layer in the UE and the RNC are allowed to select from for uplink and downlink transmissions, respectively. Hence, the RNC reserves resources corresponding to the highest data rate the UE may transmit with during the packet call. In the downlink, a fraction of the code tree needs to be reserved corresponding to the smallest spreading factor that may be required during the session. Similarly, in the uplink, the RNC must ensure that the maximum interference in the cell is not exceeded even if the UE transmits at its highest data rate.

During the packet-data call, the data rate of the transmission may vary, depending on the traffic pattern. As a consequence, the transmission power will vary depending on the instantaneous data rate. However, the amount of the code tree allocated for a certain user in the downlink remains constant throughout the packet call (unless the parameters are reconfigured). Furthermore, as downlink transport format selection is located in the RNC, it is not aware of the instantaneous power consumption in the Node B. The RRC must therefore use a certain margin when admitting users into the system to ensure that sufficient transmission power is available for any transport format the MAC may select. Similarly, in the uplink, the RRC can only observe the average uplink interference and must therefore have sufficient margins to handle the case of all UEs suddenly selecting the transport format corresponding to the highest data rate and therefore the highest transmission power. In other words, the amount of resources reserved for a user is fairly static during the period of the packet call. This is suitable for services with a relatively constant data rate, for example, voice services or interactive video transmission, but as will be seen in subsequent chapters, significant enhancements are possible for packet-data services with rapidly varying resource requirements.

## 6.6 UE RRC STATES AND STATE TRANSITIONS

It is desirable to match UE and network resource usage to the amount of connectivity that the UE has with the network. During the long periods of time in which a release 99 user is not actively involved in voice or data calls, it is desirable that it does not consume network resources or drain its battery. To facilitate effective resource utilization, a WCDMA UE can be in several different states, as discussed below and illustrated in Figure 6.5.

- *Idle mode* is the state in which the UE current drain is lowest and network resource usage is lowest. In idle mode, the UE does not have an RRC connection and keeps its receiver switched off for most of the time, waking up periodically during paging occasions to receive paging messages. Mobility is UE driven based on the UE reselecting its serving cell based on measurements.
- *CELL_PCH* is a state in which the UE has an RRC connection but cannot send or receive data. The network knows the location of the last cell in which the UE was in cell FACH state. Mobility is determined by UE cell reselection; the UE needs to move to CELL_FACH to inform the network of a reselection. The UE is accessed via the paging channel.
- *URA_PCH* is similar to *CELL_PCH*, except that the UE's location is only known to within a paging area.

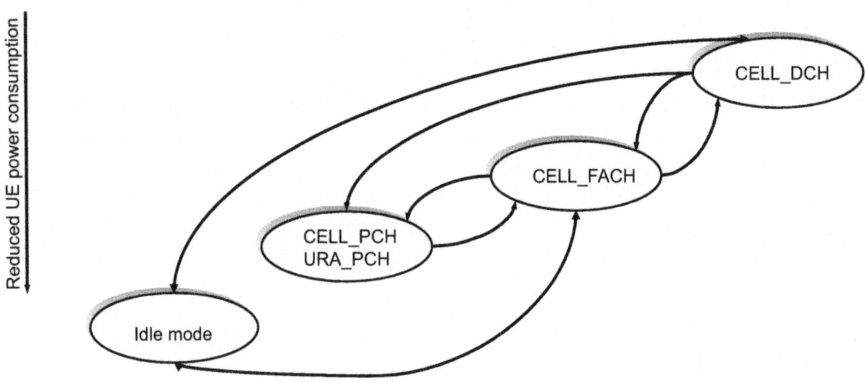


**FIGURE 6.5 RRC states and state transitions.**

- *CELL_FACH* is a higher connectivity state. The UE does not have dedicated channels allocated but can send and receive small amounts of data on FACH and RACH. Mobility is managed by the UE by means of making cell reselections. The UE receiver must be switched on continuously.
- *CELL_DCH* is the state used for all except the smallest amounts of data exchange. The UE is allocated a dedicated channel and must continuously transmit and receive at least the control parts. Thus, UE battery drain and network resource usage are at their highest. Mobility is managed by the network by means of handover and soft handover.

Moving between these states involves RRC signaling and is therefore associated with a certain latency. In later releases, enhancements have been added to improve the handling of the different UE states, as discussed in Chapter 14. Figure 6.5 outlines possible state transitions.

## REFERENCES


[1] H. Holma, A. Toskala, WCDMA for UMTS: Radio Access for Third Generation Mobile Communications, John Wiley & Sons, Chichester, UK, (2000).

[2] E. Dahlman, B. Gudmundsson, M. Nilsson, J. Sköld, UMTS/IMT-2000 based on wideband CDMA, IEEE Commun. Mag. (1998) 70–80.

[3] E. Dahlman, P. Beming, J. Knutsson, F. Ovesjö, M. Persson, C. Roobol, WCDMA – the radio interface for future mobile multimedia communications, IEEE Trans Vehicular Technol 47 (1998) 1105–1118.

[4] 3rd Generation Partnership Project; Technical Specification Group Radio Access Network; UTRAN Overall Description, 3GPP, 3GPP TS 25.401.

[5] 3rd Generation Partnership Project; Technical Specification Group Radio Access Network; Radio Interface Protocol Architecture, 3GPP, 3GPP TS 25.301.


## FURTHER READING


3rd Generation Partnership Project, In press 3rd Generation Partnership Project; Technical Specification Group Radio Access Network; Physical Channels and Mapping of Transport Channels onto Physical Channels (FDD), 3GPP, 3GP TS 25.211.

3rd Generation Partnership Project, In press 3rd Generation Partnership Project; Technical Specification Group Radio Access Network; Multiplexing and Channel Coding (FDD), 3GPP, 3GP TS 25.212.

3rd Generation Partnership Project, In press 3rd Generation Partnership Project; Technical Specification Group Radio Access Network; Spreading and Modulation (FDD), 3GPP, 3GP TS 25.213.

3rd Generation Partnership Project, In press 3rd Generation Partnership Project; Technical Specification Group Radio Access Network; Physical Layer Procedures (FDD), 3GPP, 3GP TS 25.214.

CHAPTER

# Introduction to HSPA

# 7

## CHAPTER OUTLINE



Following the standardization of WCDMA and the first 3G deployments in early 2000, it was soon recognized that the existing 3G technology would need to evolve to meet the challenges of the dawn of the era of mobile broadband. Shifts in technology, user behavior, traffic models, business models, and the learning curve of the early phases of 3G deployment drove a research effort that lead to the evolution from the release 99 specifications to modern HSPA. Today, HSPA stands alongside LTE as one of 3GPP's two market-leading air interfaces, capable of around 346 Mbps in downlink and 34 Mbps in uplink. With around two billion subscribers worldwide in 2014 [1], demand for traffic and technology evolution in HSPA continues to grow.

This chapter focuses on providing an overview of the evolution path over 3GPP releases 5 to 11 that has led to HSPA by describing each of the main building blocks, as illustrated in Figures 7.1 and 7.2. Companion chapters describe the technological functionality of each of the major building blocks in more detail, which complements

HSPA Evolution: The Fundamentals for Mobile Broadband. DOI: 10.1016/B978-0-08-099969-2.00007-7

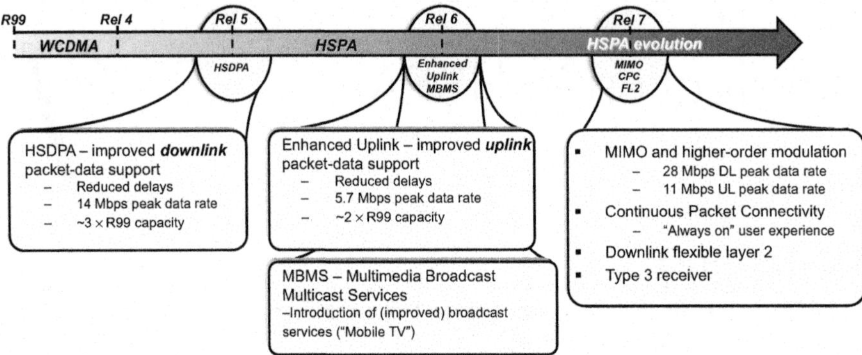


**FIGURE 7.1 HSPA features roadmap (release 99–release 7).**

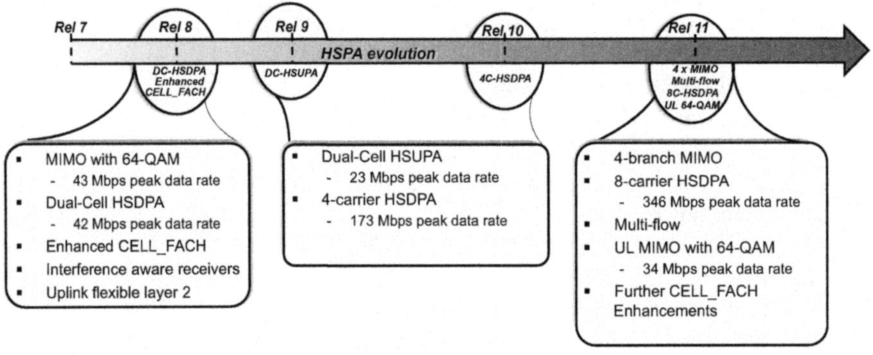


**FIGURE 7.2 HSPA features roadmap (release 8–release 11).**

the top-level view provided in Section 7.1. Section 7.2 provides an overview of some more minor features of HSPA.

## 7.1 HSPA EMERGES FROM WCDMA: FUNDAMENTAL BUILDING BLOCKS

The development of modern HSPA began with the completion of 3GPP release 5 in 2002 and with it the standardization of so-called *High Speed Downlink Packet Access* (HSDPA). Prior to HSDPA, the RAN network user- and control-plane was provided and managed centrally by the *Radio Network Controller* (RNC). Traffic channels could be configured and released by the RNC only, and management of each radio link needed to be performed at some distance from the user. For circuit-switched data, this approach was well motivated. However, the characteristics of packet data and the impact of the air interface on individual packets change with a significantly faster time scale than the latency of an Iub interface. The fundamental change in introducing HSDPA was to move a greater amount of responsibility for managing air interface resources and individual radio links out to the Node Bs, in order to improve responsiveness to changes in traffic and radio conditions. This was achieved through the introduction of sub-blocks such as a new shared channel, Node B scheduling, a reduced TTI length, and hybrid-ARQ. HSDPA is described in more detail in Chapter 8.

Following the successful introduction of HSDPA in release 5, attention turned to the uplink. Similarly to the downlink, uplink packet data and interference characteristics can change over a short time scale. This was addressed in 2005 in release 6 by the introduction of *High Speed Uplink Packet Access* (HSUPA), also referred to as *Enhanced Uplink* (EUL). HSUPA uses similar features such as Node B scheduling, reduced TTI lengths, and hybrid-ARQ, although there are some differences as well. HSUPA is described in more detail in Chapter 9.

HSDPA and HSUPA combine to create a downlink and uplink mobile broadband solution and are collectively known as HSPA. HSPA has been the most successful mobile broadband technology on the planet and during the building and deployment of networks and UEs, a large amount of improvements and optimizations have been implemented. Broadly speaking, these fall into a number of general categories:

- improved link-level performance, including increasing peak rates;
- improved handling of bursty traffic characteristics;
- improved handling of network and UE resources;
- improved utilization of frequency resources;
- improved utilization of multi-cell resources; and
- multi-antenna techniques support.

These improvements have been obtained by means of the introduction of a number of basic building blocks.

### 7.1.1 HIGHER-ORDER MODULATIONS

A variety of mechanisms can lead to higher SINRs becoming available. These include deployment aspects such as multi-antenna techniques, interference-mitigating receiver algorithms, and small indoor cells with large isolation, among other things. In order to take advantage of higher SINRs, higher modulation orders were introduced.

Release 5 downlink HSDPA supported a maximum modulation order of 16-QAM. This was extended in release 7 to 64-QAM, providing a peak rate of around 21 Mbps. In the uplink, release 6 HSUPA supported maximum QPSK, which was extended to 16-QAM in release 7 and even further to 64-QAM in release 11, providing a peak rate of around 17 Mbps per MIMO stream.

Higher-order modulation is described in Chapter 10.

### 7.1.2 ENHANCED RECEIVERS

When release 5 HSDPA was specified, RAKE receivers were commonly used. Technology improvements quickly lead, however, to the possibility of improved receivers, which directly increase link-level performance. In the downlink, support for improved receivers is captured by means of UE demodulation performance requirements in [2]. In release 7, performance requirements were introduced relating to Rx diversity receivers and for chip-level LMMSE equalizers (see Chapter 3). This was followed in release 8 by interference-aware receivers that are able to maximize the received signal from the serving cell while mitigating interference from the next strongest cell.

### 7.1.3 IMPROVED LAYER 2 FOR SUPPORTING HIGHER DATA RATES

The layer 2 MAC and RLC protocols designed for release 99 were based on the assumption that transport block sizes would not be selected dynamically and that the RNC would schedule user data. The introduction of HSDPA in release 5 differed from these assumptions, and with increasing data rates provided by features such as MIMO and higher-order modulation in subsequent releases, it became clear that the release 99 RLC and MAC protocols would become bottlenecks in throughput performance. Thus, in releases 6 and 7, an improved layer 2 processing was developed for downlink and uplink, respectively, which offers an increased flexibility in order to provide for differing and dynamically changing data rates. Layer 2 bottlenecks and improved layer 2 processing are described in Chapter 8.

### 7.1.4 MULTI-ANTENNA TECHNIQUES

MIMO techniques have a long history in HSPA; downlink MIMO was studied although never introduced during the standardization of downlink HSDPA in release 5. MIMO was eventually introduced in the form of so-called *dual TxAA* (D-TxAA) in release 7, enabling spatial multiplexing of up to two streams to each MIMO user. A further enhancement in release 9 was to introduce the possibility of 2×1[1] TxAA HS-PDSCH transmission in order that single-antenna UEs could also take advantage of beam-forming gains with HSDPA.

Early MIMO development was hindered by issues relating to the impact of the introduction of MIMO on legacy non-MIMO equalization performance, control channel overhead, and power balancing between *power amplifiers* (PAs). Fixes for these issues were retrospectively introduced into the specifications. In release 11, downlink MIMO evolved even further with the introduction of four-branch MIMO.

Uplink MIMO was somewhat later in its appearance in the 3GPP HSPA specifications than downlink MIMO because of the technological issues associated with building two transmit chains into a UE, the reduced need for highest peak rates in the uplink, and the late emergence of scenarios in which uplink SINR conditions could be such that MIMO would be useful. Uplink closed-loop Tx diversity (2×1) and dual-stream MIMO (2×2) were standardized during release 11.

Downlink and uplink multi-antenna transmission in the HSPA specifications is described in Chapter 11.

---

[1](2×1) refers to a transmission from two base station antennas to a single UE receive antenna.

### 7.1.5 CONTINUOUS PACKET CONNECTIVITY (CPC) AND RELATED RESOURCE-RELATED IMPROVEMENTS

Release 5 HSDPA and release 6 HSUPA operated in CELL_DCH, which requires continuous allocation of a low-rate dedicated channel in downlink for all HSDPA users and continuous transmission of DPCCH in uplink. Continuous allocation of dedicated channels consumes resources in both downlink and uplink. Downlink code resource usage was addressed in release 6 through the introduction of a so-called *Fractional DPCH* (F-DPCH), which is a downlink DPCH that carries TPC bits only, while all data is transmitted on HS-PDSCH. F-DPCH operation is described in Chapter 8.

Still, after release 6, it was necessary for HSPA users to keep the modem switched on continuously in downlink in order to be ready for scheduling via HS-PDSCH and to keep uplink transmission of DPCCH. For packet data, it is often the case that for short periods of time there is no activity in uplink and downlink. It is, however, undesirable to move users frequently to and from CELL_DCH. Continuous transmission and reception also drains the UE battery and, in the uplink case, causes additional interference that reduces data capacity. Thus, in release 7, so-called *Continuous Packet Connectivity* (CPC) was introduced, which introduced DRX in downlink, such that the UE does not need to keep its receiver on continuously, DTX in uplink, and a more efficient scheduling mechanism for streaming services such as VoIP.

In addition to the core efficiency improvements, another significant step during CPC was the introduction of physical layer orders. Physical layer orders are a re-use of the HSDPA signaling channel for layer 1 radio resource control signaling. The physical layer order format that was developed for CPC in release 7 has been extended to include layer 1 signaling for controlling other features in later releases and is now an essential part of the control plane functionality of HSPA. HSPA order functionality is described in more detail in Chapter 14. CPC is also described in more detail in Chapter 14.

### 7.1.6 CELL_FACH ENHANCEMENTS

The CELL_FACH RRC state is a state with intermediate power and resource consumption (lower than CELL_DCH but higher than idle mode) in which in release 6 the UE could receive small amounts of data on FACH via S-CCPCH and transmit small amounts of data on RACH. Increasing smartphone traffic lead to the recognition of a bursty traffic type consisting of frequent, small packets of data that are too resource intensive to handle in CELL_DCH but too large for efficient handling in release 99 CELL_FACH state. To improve the network's capability to efficiently deal with such packets and the impact on the UE's battery consumption, the CELL_FACH state was improved by, among other things, enabling limited HSDPA and HSUPA transmissions. The improved CELL_FACH state is known as *Enhanced CELL_FACH*. The improvements to the CELL_FACH state took place over several releases; the downlink operation was covered in release 7 and improved in release 8, whereas uplink E-DCH operation in CELL_FACH was introduced in release 8. In release 11, further improvements were introduced to the Enhanced CELL_FACH state based on further experience of smartphone traffic behavior and system bottlenecks. The Enhanced CELL_FACH state is described in more detail in Chapter 14.

### 7.1.7 MULTI-CARRIER HSDPA AND HSUPA

Many operators will typically own sufficient spectrum that they can operate on more than one HSPA carrier and thereby meet the traffic demand in their networks. The statistical nature of packet-based traffic is such that it is never the case that at each moment of time users are scheduled on every carrier in every cell, because such a situation would generally lead to very poor packet throughput performance. Statistically, while a user has traffic on one carrier other carriers may well not be utilized continuously. The fundamental idea behind multi-carrier HSPA is that by enabling scheduling of users on multiple carriers simultaneously, significant trunking gains can be obtained, leading to an increase in user experience and system capacity. An individual user can benefit from being scheduled on all carriers that are free and thus enjoying higher data rates, whereas the system benefits from the fact that capacity is used more effectively.

Multi-carrier was introduced into HSPA across multiple releases, with an evolution in the number of simultaneous carriers on which a UE can be configured to receive and the supportable multi-carrier functionality. In the downlink, so-called *Dual-Cell*[2] *HSDPA* (DC-HSDPA) was introduced in release 8, whereas four-cell HSDPA was introduced in release 10 and eight-cell HSDPA in release 11. Over these releases, increases in the functionality took place, too; in release 8, MIMO could not be supported with dual-cell operation and carriers needed to be in the same band. In release 9, dual band DC-HSDPA (that is, utilization of two carriers in different bands) and the possibility to configure MIMO was introduced. Release 9 also included support for aggregation of two contiguous uplink carriers (dual-cell HSUPA). Release 10 4C-HSDPA did not fully cover for non-contiguous carriers; this was enabled in release 11. It should be noted that the core multi-carrier HSDPA specifications are band agnostic, but supportable band combinations must be standardized in 3GPP TSG RAN4 according to operator needs.

Multi-carrier functionality is described in more detail in Chapter 12.

### 7.1.8 MULTI-FLOW

The same statistical considerations that lead to the trunking gains that are obtained through multi-carrier in principle apply spatially between different cell sites. Cell-edge UEs, who in general suffer from reduced data rates, may experience favorable carrier-to-interference ratio from more than one site, and thus trunking gains are obtainable if such UEs can be scheduled from more than one site simultaneously. Multi-flow enables independent and unsynchronized scheduling of a UE that is in soft handover from two geographically separated cells.

[2] In this context, $x$ cell HSDPA refers to simultaneous reception of HSDPA on $x$ carriers.

Scheduling a UE in two cells will, however, lead to interference from each cell when demodulating HS-DPSCH in the other cell. In order to obtain gains from multi-flow, it is in general necessary for a UE to use an interference mitigating receiver, such as a Type3i receiver.

In macro scenarios, multi-flow gains are somewhat limited by flow control considerations between the RNC and cell sites.

Multi-flow might well prove to be an enabling technology in heterogeneous network scenarios.

Multi-flow is described in more detail in Chapter 13.

### 7.1.9 MULTIMEDIA BROADCAST AND MULTICAST SERVICE (MBMS)

During release 6 standardization, a need was perceived to improve the ability of 3G networks to provide services to large groups of users by means of broadcast and multicast. Prior to release 6, such services needed to be supported via dedicated channels or HSDPA channels to each user, which could potentially be costly in terms of system resources. In release 6, *Multimedia Broadcast and Multicast Service* (MBMS) was introduced. MBMS included features such as a common traffic channel for broadcast services, soft combining between cells, counting of the number of users interested to receive a service, and subscription and charging mechanisms. In release 7, MBMS functionality was enhanced with functionality supporting synchronized simultaneous transmission from multiple cells on the same frequency (referred to as *Single-Frequency Network*, SFN) and a downlink-only MBMS carrier, and in release 8 with so-called *Integrated Mobile Broadcast* (IMB). Uptake of MBMS has never been significant.

MBMS is described in more detail in Chapter 15.

## 7.2 OTHER ENHANCEMENTS

The building blocks described above represent the most significant parts of the HSPA air interface. Apart from these, some other improvements increase the user experience and attractiveness of the technology.

### 7.2.1 INCREASED BAND COMBINATIONS AND MSR

The original core UMTS bands were enhanced with 1800 MHz and 900 MHz in release 5 and then with three further bands in release 6. Since then, as additional spectrum has become available and has become less technology focused in different regions, the number of bands has grown significantly. 3GPP TSG RAN4 has addressed not just new bands but also combinations of carriers and bands that can be supportable for, for example, multi-carriers.

From the base station perspective, a new generation of base stations supports multiple bands, carriers, and access technologies (such as GSM, HSPA, and LTE, and in the future potentially WiFi). To enable proper operation of such base stations, a new set of specifications known as *Multi Standard Radio* (MSR) was developed in release 9.

### 7.2.2 REMOTE ELECTRICAL TILT

The downlink of HSPA is in a macro scenario typically interference limited, whereas the uplink benefits from soft handover. This means that when doing cell planning, there is a contradiction between attempting to minimize cell overlap when considering downlink but including some cell overlap for uplink SHO. In release 6, 3GPP introduced the so-called IuAnt interface for enabling remote control of antenna tilt at cell sites. This would in principle allow for an increased possibility to reoptimize deployments given operating data.

### 7.2.3 SUPPORT FOR DIFFERENT TYPES OF BASE STATIONS

The development of the home Node B concept, in which operators provide their subscribers with home base stations whose user group may be limited to members of the household, lead to the introduction of the *Closed Subscriber Group* (CSG) base station concept in release 8. Supporting CSG cells included changes in the network architecture, mobility procedures to enable searching for, joining, and handing over to a CSG, and RF requirements to enable uncoordinated deployment.

The growth of so-called heterogeneous networks in recent years leads to base stations of varying size, with different transmit power levels, coverage areas, and cost levels. A single set of RF requirements that would be applicable both to wide-area macro base stations and indoor pico base stations is not an appropriate enabler of such technologies. Thus, 3GPP TSG RAN4 has developed the concept of base station classes, including wide-area (macro), medium-range (micro), local-area (pico), and home base stations such that RF requirements appropriate to each use-case can be developed (see Chapter 17).

### 7.2.4 MISCELLANEOUS MOBILITY IMPROVEMENTS

Other small mobility improvements have included improved mobility to GERAN, mobility to LTE, and a faster means of achieving HSPA serving cell change such that real-time services can be supported with HSDPA without interruptions at handover.

### 7.2.5 THE START OF THE SMARTPHONE ERA: FAST DORMANCY

Smartphones run apps that generate data packets. For the first generation of smartphones, the CELL_DCH state needs to be used for data transfer. In CELL_DCH, after the end of a packet transfer the UE will be held in CELL_DCH until a switch-down timer has expired. The length of the switch-down time is optimized from a network perspective depending on the traffic statistics and considering the latency, RNC, and signaling overhead of moving a UE between CELL_DCH and lower connectivity states. From a UE perspective, it is desirable not to spend an extended period of time in the CELL_DCH state since, even with the application of DRX and DTX in CPC, the UE battery consumption will be higher in CELL_DCH than in idle mode.

The application in the UE is aware of when a packet transfer has completed. Early smartphones took advantage of a signaling message known as *Signaling Connection Release Indicator* (SCRI) in order to leave CELL_DCH as soon as possible. A UE is able to send the SCRI to the network and then move to idle mode. The SCRI was originally intended for UEs to deal with a signaling connection failure; however, smartphones are able to use the message to move quickly to idle mode once a packet completes.

The problem with the use of this mechanism was that with a large amount of packet interactions, a considerable amount of processing was required in the network for setting up and tearing down RRC connections, which lead to an overloading of the control plane in some networks when smartphones were first deployed.

This problem was resolved in release 8 by means of introducing an extra signaling element to the SCRI message to indicate that the reason for the signaling connection release is the end of a packet session. A UE is not allowed to autonomously move to idle mode. Instead, the network can decide to move the UE to CELL_PCH for power saving. This avoids the need to set up and tear down connections and the large amount of network processing. If the UE is not moved to CELL_PCH, it can repeatedly send SCRI requests, considering a minimum time between such requests that must be maintained.

Fast dormancy assisted to solve early smartphone problems. In the future, deployment of the Enhanced CELL_FACH feature and in particular the release 11 further enhancements to CELL_FACH will enable the UE to use CELL_FACH for many more packet interactions and further reduce CELL_FACH power consumption and in this manner potentially cause the release 8 fast dormancy functionality to become less important.

### 7.2.6 CIRCUIT-SWITCHED VOICE OVER HSDPA

Release 7 efficiently supported transmission of *Voice over IP* (VoIP) using HSPA. However, it was not possible to map circuit-switched services onto HSDPA in release 7, and thus for supporting circuit-switched calls, DCH channels had to be configured. In release 8, it was recognized that it would be desirable to be able to also map circuit-switched calls to HSDPA, in order to improve capacity for circuit-switched calls in addition to VoIP and to avoid the need for configuration of DCHs, which take additional code resources.

### 7.2.7 EXTENSION OF WIDE-AREA DEPLOYMENT SCENARIOS

Prior to release 7, the maximum supportable cell size for WCDMA was restricted by the ability of the UTRAN to measure the PRACH propagation delay and report the round-trip time (RTT) across IuB. Because a delay of 1 chip corresponds to 78[3] m,

[3]This corresponds to the speed of light times the chip duration.

the maximum reportable RACH propagation delay of 765 chips allowed for a cell size of around 60 km.

It was recognized that in some scenarios, such as desert and ocean, propagation conditions could allow for a larger cell radius from a link budget point of view. Larger cells could be supported with GSM, and it was desirable to also be able to provide mobile broadband services into rural areas using UMTS.

The PRACH propagation delay and RTT reporting bottleneck was resolved in release 7 by introducing two new IEs, *extended propagation delay* and *extended round-trip time*. This additional signaling allows in principle for cell ranges up to 3069 chips, which corresponds to 240 km, although the measurement accuracy is only specified for round-trip times corresponding to distances of up to 180 km.

It should be noted that the changes were purely in Iub signaling; hence, the extended cell range remains compatible with legacy UEs.

## REFERENCES

[1] Ericsson Mobility report, November 2013 (www.ericsson.com/ericsson-mobility-report).
[2] User Equipment (UE) radio transmission and reception (FDD). 3GPP TS 25.101.

## FURTHER READING

J. Bergman, M. Ericson, D. Gerstenberger, B. Göransson, J. Peisa, S. Wager, HSPA evolution – boosting the performance of mobile broadband access, Ericsson Rev. (January 2008).

J. Bergman, D. Gerstenberger, F. Gunnarsson, S. Ström, Continued HSPA evolution of mobile broadband, Ericsson Rev. (January 2009).

H. Holma, A. Toskala, K. Ranta-aho, J. Pirskanen, J. Kaikkonen, High-speed packet access evolution in 3GPP release 7, IEEE Communications Magazine 45 (12) (2007) 29–35.

E. Jugl, M. Link, J. Mueckenheim, Impact of flexible RLC PDU size on HSUPA performance, Mobilkommunikation (2011).

J. Peisa, S. Wager, M. Sågfors, J. Torsner, B. Göransson, T. Fulghum, C. Cozzo, S. Grant, High speed packet access evolution – concept and technologies, IEEE Vehicular Technology Conference, 2007Maryland, MD, USA, September 30–October 3.

3rd Generation Partnership Project, In press 3rd Generation Partnership Project; Technical Specification Group Radio Access Network; UTRAN Iuant interface: Remote Electrical Tilting (RET) antennas Application Part (RETAP) signalling, 3GPP TR 25.463.

CHAPTER

# High-speed downlink packet access

# 8

## CHAPTER OUTLINE



The introduction of *High-Speed Downlink Packet Access* (HSDPA) in release 5 implied a major extension of WCDMA, enhancing the downlink packet-data performance and capabilities in terms of higher peak data rate, reduced latency, and increased capacity. This is achieved through the introduction of several of the techniques described in


**HSPA Evolution: The Fundamentals for Mobile Broadband. DOI: 10.1016/B978-0-08-099969-2.00008-9**

Part II, including *higher-order modulation, rate control, channel-dependent scheduling*, and *hybrid-ARQ with soft combining*. The HSDPA specifications are found in [1] and the references therein. HSDPA is the fundamental downlink building block of the subsequent HSPA evolution, and release 5 HSDPA is described in detail in this chapter. In addition, improvements to the L2 protocol made in release 7, which are needed in order to be able to fully benefit from the potential throughput gains offered by HSDPA, are described in this chapter. Further enhancements that have been introduced in later releases are then described in subsequent chapters.

## 8.1 OVERVIEW

### 8.1.1 SHARED-CHANNEL TRANSMISSION

A key characteristic of HSDPA is the use of *Shared-Channel Transmission*. Shared-channel transmission implies that a certain fraction of the total downlink radio resources available within a cell – channelization codes and transmission power in case of WCDMA – is seen as a common resource that is dynamically shared between users, primarily in the time domain. The use of shared-channel transmission in WCDMA, implemented through the *High-Speed Downlink Shared Channel* (HS-DSCH) as described below, enables the possibility to rapidly allocate a large fraction of the downlink resources for transmission of data to a specific user. This is suitable for packet-data applications that typically have bursty characteristics and thus rapidly varying resource requirements.

The basic HS-DSCH code and time structure is illustrated in Figure 8.1. The HS-DSCH code resource consists of a set of channelization codes of spreading factor 16 (see upper part of Figure 8.1), where the number of codes available for HS-DSCH transmission is configurable between 1 and 15. Codes not reserved for HS-DSCH transmission are used for other purposes, for example, related control signaling, MBMS services, or circuit-switched services.

The dynamic allocation of the HS-DSCH code resource for transmission to a specific user is done on a 2 ms TTI basis (see lower part of Figure 8.1). The use of such a short TTI for HSDPA reduces the overall delay and improves the tracking of fast channel variations exploited by the rate control and the channel-dependent scheduling, as discussed below.

In addition to being allocated a part of the overall code resource, a certain part of the total available cell power should also be allocated for HS-DSCH transmission. Note that the HS-DSCH is not power controlled but rate controlled, as discussed below. This allows the remaining power, after serving other channels, to be used for HS-DSCH transmission and enables efficient exploitation of the overall available power resource.

### 8.1.2 CHANNEL-DEPENDENT SCHEDULING

In each TTI, the scheduler decides to which user(s) the HS-DSCH should be transmitted and, in close cooperation with the rate-control mechanism, at what data rate.

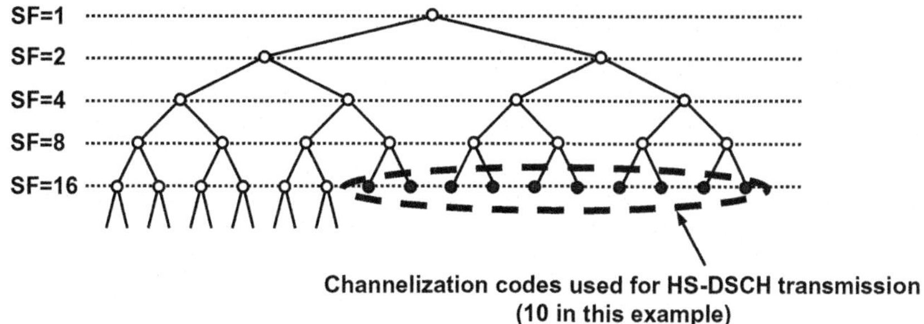


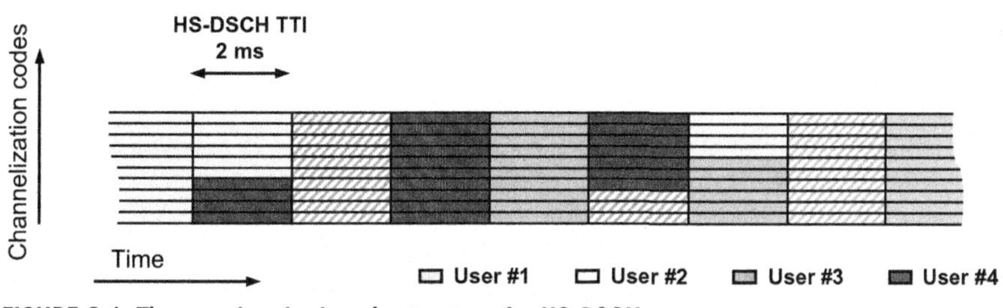


**FIGURE 8.1 Time- and code-domain structure for HS-DSCH.**

The scheduler is a key element and to a large extent determines the overall system performance, especially in a highly loaded network.

In Chapter 5, it was discussed how a significant increase in capacity can be obtained if the radio-channel conditions are taken into account in the scheduling decision, so-called *channel-dependent scheduling*. Because the radio conditions for the radio links to different UEs within a cell typically vary independently, at each point in time there is almost always a radio link whose channel quality is near its peak (see Figure 8.2). As this radio link is likely to have good channel quality, a high data

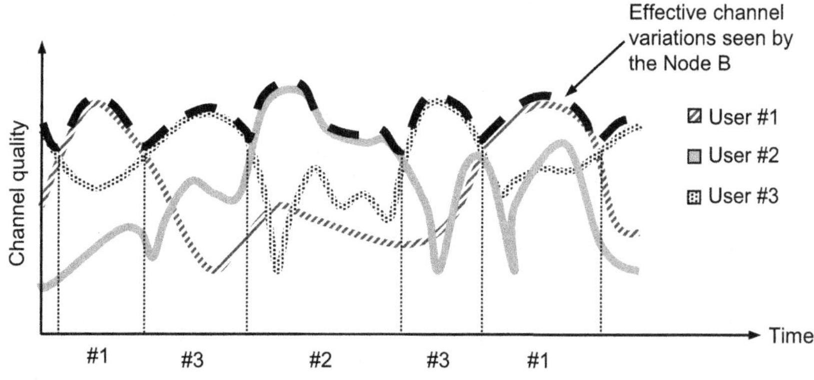


**FIGURE 8.2 Channel-dependent scheduling for HSDPA.**

rate can be used for this radio link. This translates into a high system capacity. The gain obtained by transmitting to users with favorable radio-link conditions is commonly known as *multi-user diversity* and the gains are larger, the larger the channel variations and the larger the number of users in a cell. Thus, in contrast to the traditional view that fast fading is an undesirable effect that has to be combated, with the possibility for channel-dependent scheduling, fading is potentially beneficial and should be exploited.

Several different scheduling strategies were discussed in Chapter 5. As discussed, a practical scheduler strategy exploits the short-term variations, for example, due to multipath fading and fast interference variations, while maintaining some degree of long-term fairness between the users. In principle, the larger the long-term unfairness, the higher the cell capacity. A trade-off between fairness and capacity is therefore required.

In addition to the channel conditions, traffic conditions are also taken into account by the scheduler. For example, there is obviously no purpose in scheduling a user with no data awaiting transmission, regardless of whether the channel conditions are beneficial or not. Furthermore, some services should preferably be given higher priority. As an example, streaming services should be ensured a relatively constant long-term data rate whereas background services, such as file download, have less stringent requirements on a constant long-term data rate.

### 8.1.3 RATE CONTROL AND HIGHER-ORDER MODULATION

In Chapter 5, rate control was discussed and, for packet-data services, shown to be a more efficient tool for link adaptation, compared to the fast power control typically used in CDMA-based systems, especially when used together with channel-dependent scheduling.

For HSDPA, rate control is implemented by dynamically adjusting the channel-coding rate as well as dynamically selecting between QPSK and 16-QAM modulation. In later releases, modulation was extended to also include 64-QAM modulation (see further Chapter 10). *Higher-order modulation*, such as 16-QAM, allows for higher bandwidth utilization than QPSK but requires higher received $E_b/N_0$, as described in Chapter 2. Consequently, 16-QAM is mainly useful in advantageous channel conditions. The data rate is selected independently for each 2 ms TTI by the Node B and the rate control mechanism can therefore track rapid channel variations.

### 8.1.4 HYBRID-ARQ WITH SOFT COMBINING

Fast *hybrid-ARQ with soft combining* allows the UE to request retransmission of erroneously received transport blocks, effectively fine-tuning the effective code rate and compensating for errors made by the link-adaptation mechanism. The UE attempts to decode each transport block it receives and reports to the Node B its success or failure five ms after the reception of the transport block. This allows for rapid retransmissions of unsuccessfully received data and significantly reduces the delays associated with retransmissions compared to release 99.

*Soft combining* implies that the UE does not discard soft information in case it cannot decode a transport block as in traditional hybrid-ARQ protocols, but combines soft information from previous transmission attempts with the current retransmission to increase the probability of successful decoding. *Incremental redundancy*, (IR), is used as the basis for soft combining in HSDPA; that is, the retransmissions may contain parity bits not included in the original transmission. From Chapter 5, it is known that IR can provide significant gains when the code rate for the initial transmission attempts is high as the additional parity bits in the retransmission results in a lower overall code rate. Thus, IR is mainly useful in bandwidth-limited situations, for example, when the UE is close to the base station and the amount of channelization codes, but not the transmission power, limits the achievable data rate. The set of coded bits to use for the retransmission is controlled by the Node B, taking the available UE memory into account.

### 8.1.5 ARCHITECTURE

From the previous discussion, it is clear that the basic HSDPA techniques rely on fast adaptation to rapid variations in the radio conditions. Therefore, these techniques need to be placed close to the radio interface on the network side, that is, in the Node B. At the same time, an important design objective of HSDPA was to retain the release 99 functional split between layers and nodes as far as possible. Minimization of the architectural changes is desirable as it simplifies introduction of HSDPA in already deployed networks and also secures operation in environments where not all cells have been upgraded with HSDPA functionality. Therefore, HSDPA introduces a new MAC sublayer in the Node B, the *MAC-hs*, responsible for scheduling, rate control, and hybrid-ARQ protocol operation. The MAC-hs was enhanced as MAC-ehs in release 8, as described in Section 8.4. Hence, apart from the necessary enhancements to the RNC, such as admission control of HSDPA users, the introduction of HSDPA mainly affects the Node B (Figure 8.3).

Each UE using HSDPA will receive HS-DSCH transmission from one cell, the *serving cell*[1]. The serving cell is responsible for scheduling, rate control, hybrid-ARQ, and all other MAC-hs functions used by HSDPA. Uplink soft handover is supported, in which case the uplink data transmission will be received in multiple cells and the UE will receive power control commands from multiple cells.

Mobility from a cell supporting HSDPA to a cell that is not supporting HSDPA is easily handled. Uninterrupted service to the user can be provided, albeit at a lower data rate, by using channel switching in the RNC and switching the user to a dedicated channel in the non-HSDPA cell. Similarly, a user equipped with an HSDPA-capable UE may be switched from a dedicated channel to HSDPA when the user enters a cell with HSDPA support.

---

[1]The introduction of multi-flow in release 11 facilitates scheduling of HS-DSCH from more than one cell; see Chapter 13.

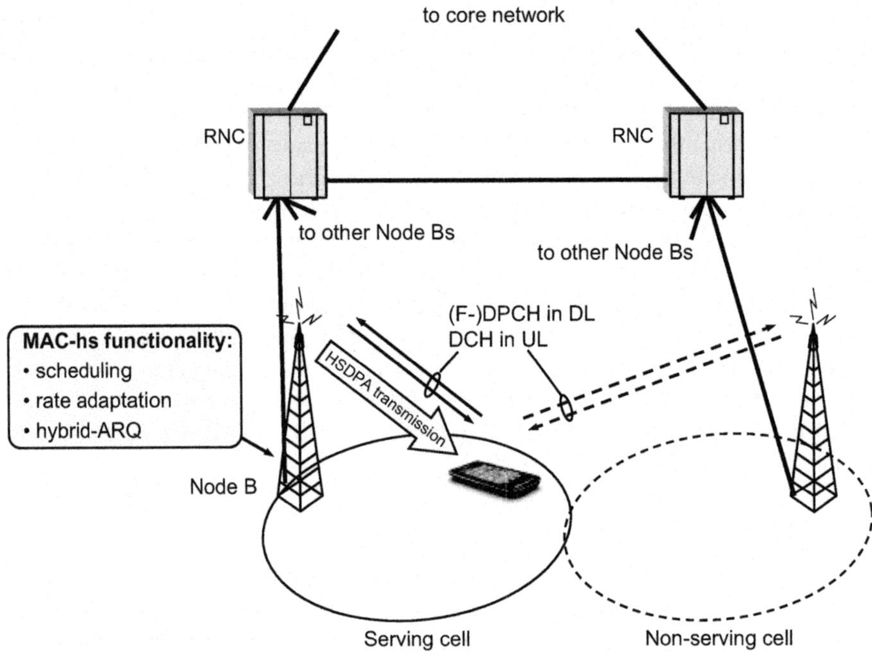


**FIGURE 8.3 Illustration of the HSDPA architecture.**

## 8.2 DETAILS OF HSDPA

### 8.2.1 HS-DSCH: INCLUSION OF FEATURES IN WCDMA RELEASE 5

The *High-Speed Downlink Shared Channel* (HS-DSCH) is the transport channel used to support shared-channel transmission and the other basic technologies in HSDPA, namely channel-dependent scheduling, rate control (including higher-order modulation), and hybrid-ARQ with soft combining. As discussed in the introduction and illustrated in Figure 8.1, the HS-DSCH corresponds to a set of channelization codes, each with spreading factor 16. Each such channelization code is also known as an *High-Speed Physical Downlink Shared Channel* (HS-PDSCH).

In addition to HS-DSCH, there is a need for other channels as well, for example, for circuit-switched services and for control signaling. To allow for a trade-off between the amount of code resources set aside for HS-DSCH and the amount of code resource used for other purposes, the number of channelization codes available for HS-DSCH can be configured, ranging from 1 to 15 codes. Codes not reserved for HS-DSCH transmission are used for other purposes, for example, related control signaling and circuit-switched services. The first node in the code tree can never be used for HS-DSCH transmission as this node includes mandatory physical channels such as the common pilot.

Sharing of the HS-DSCH code resource should primarily take place in the time domain. The reason is to fully exploit the advantages of channel-dependent scheduling and rate control, because the quality at the UE varies in the time domain but is (almost) independent of the set of codes (physical channels) used for transmission. However, sharing of the HS-DSCH code resource in the code domain is also supported, as illustrated in Figure 8.1. With code-domain sharing, two or more UEs are scheduled simultaneously by using different parts of the common code resource (different sets of physical channels). The reasons for code-domain sharing are twofold: support of UEs that are, for complexity reasons, not able to despread the full set of codes and efficient support of small payloads when the transmitted data does not require the full set of allocated HS-DSCH codes. In either of these cases, it is obviously a waste of resources to assign the full code resource to a single UE.

In addition to being allocated a part of the overall code resource, a certain part of the total available cell power should also be used for HS-DSCH transmission. To maximize the utilization of the power resource in the base station, the remaining power after serving other power-controlled channels should preferably be used for HS-DSCH transmission, as illustrated in Figure 8.4. In principle, this results in a (more or less) constant transmission power in a cell. Because the HS-DSCH is rate controlled, as discussed below, the HS-DSCH data rate can be selected to match the radio conditions and the amount of power instantaneously available for HS-DSCH transmission.

To obtain rapid allocation of the shared resources, and to obtain a small end-user delay, the TTI should be selected as small as possible. At the same time, a too small TTI would result in excessive overhead as control signaling is required for each transmission. For HSDPA, this trade-off resulted in the selection of a 2 ms TTI.

Downlink control signaling is necessary for the operation of HS-DSCH in each TTI. Obviously, the identity of the UE(s) currently being scheduled must be signaled as well as the physical resource (the channelization codes) used for transmission to this UE. The UE also needs to be informed about the transport format used for the transmission as well as hybrid-ARQ-related information. The resource and transport-format information consists of the part of the code tree used for data transmission, the modulation scheme used, and the transport-block size. The downlink control signaling is carried on the *High-Speed Shared Control Channel* (HS-SCCH), transmitted in parallel to the HS-DSCH using a separate channelization code. The HS-SCCH is a shared channel received by all UEs for which an HS-DSCH is configured to find out whether the UE has been scheduled or not.

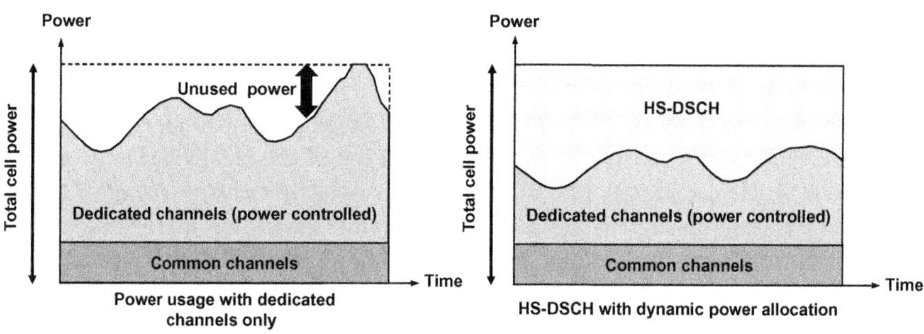


**FIGURE 8.4 Dynamic power usage with HS-DSCH.**

Several HS-SCCHs can be configured in a cell, but as the HS-DSCH is shared mainly in the time domain and only the currently scheduled UE needs to receive the HS-SCCH, there is typically only one or, if code-domain sharing is supported in the cell, a few HS-SCCHs configured in each cell. However, each HS-DSCH-capable UE is required to be able to monitor up to four HS-SCCHs. Four-HS-SCCH has been found to provide sufficient flexibility in the scheduling of multiple UEs; if the number was significantly smaller, the scheduler would have been restricted in which UEs to schedule simultaneously in case of code-domain sharing.

HSDPA transmission also requires uplink control signaling as the hybrid-ARQ mechanism must be able to inform the Node B whether the downlink transmission was successfully received or not. For each downlink TTI in which the UE has been scheduled, an ACK or NACK will be sent on the uplink to indicate the result of the HS-DSCH decoding. This information is carried on the uplink *High-Speed Dedicated Physical Control Channel* (HS-DPCCH). One HS-DPCCH is set up for each UE with an HS-DSCH configured. In addition, the Node B needs information about the instantaneous downlink channel conditions at the UE for the purpose of channel-dependent scheduling and rate control. Therefore, each UE also measures the instantaneous downlink channel conditions and transmits a *Channel-Quality Indicator* (CQI) on the HS-DPCCH.

In addition to HS-DSCH and HS-SCCH, an HSDPA UE needs to receive power control commands for support of fast closed-loop power control of the uplink in the same way as any WCDMA UE. This can be achieved by a downlink dedicated physical channel, DPCH, for each UE. In addition to power control commands, this channel can also be used for user data not carried on the HS-DSCH, for example circuit-switched services.

In release 6, support for *fractional DPCH* (F-DPCH) was added to reduce the consumption of downlink channelization codes. In principle, the only use for a dedicated channel in the downlink is to carry power control commands to the UE in order to adjust the uplink transmission. If all data transmissions, including higher-layer signaling radio bearers, are mapped to the HS-DSCH, it is a waste of scarce code resources to use a dedicated channel with spreading factor 256 per UE for power control only. The F-DPCH resolves this by allowing multiple UEs to share a single downlink channelization code (which can also be seen as using a substantially higher spreading factor than 256).

To summarize, the overall channel structure with HSDPA is illustrated in Figure 8.5.

Neither the HS-PDSCH, nor the HS-SCCH, is subject to downlink macro-diversity or soft handover. The basic reason is the location of the HS-DSCH scheduling in the Node B. Hence, it is not possible to simultaneously transmit the HS-DSCH to a single UE from multiple Node Bs, which prohibits the use of inter–Node B soft handover. Furthermore, it should be noted that within each cell, multiuser diversity is exploited by the channel-dependent scheduler. Basically, the scheduler only transmits to a user when the instantaneous radio conditions are favorable and thus the additional gain from macro-diversity is reduced.

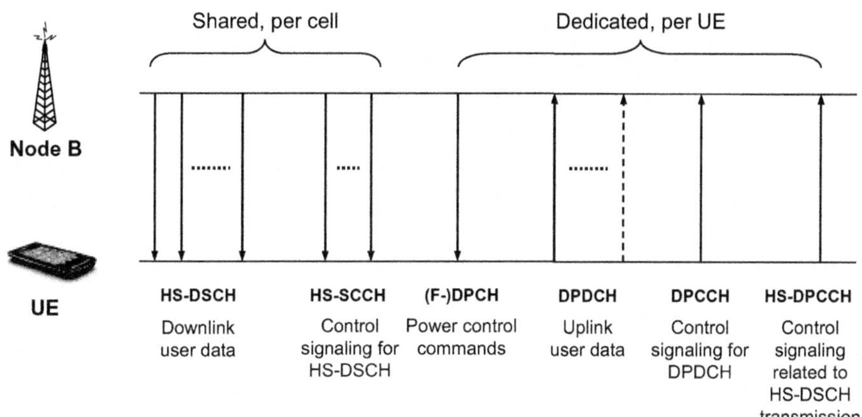


**FIGURE 8.5 Channel structure with HSDPA.**

However, the uplink channels, as well as any dedicated downlink channels, can be in soft handover. As these channels are not subject to channel-dependent scheduling, macro-diversity provides a direct coverage benefit.

### 8.2.2 MAC-HS AND PHYSICAL-LAYER PROCESSING

As mentioned in the introduction, the MAC-hs is a new sublayer located in the Node B and responsible for the HS-DSCH scheduling, rate control, and hybrid-ARQ protocol operation. In release 7, MAC-hs was improved, as described in Section 8.3.10, to better support even higher data rates. To support these features, the physical layer has also been enhanced with the appropriate functionality, for example, support for soft combining in the hybrid-ARQ. In Figure 8.6, the MAC-hs and physical-layer processing is illustrated.

The MAC-hs consists of scheduling, priority handling, transport-format selection (rate control), and the protocol parts of the hybrid-ARQ mechanism. Data, in the form of a single transport block with dynamic size, passes from the MAC-hs via the HS-DSCH transport channel to the HS-DSCH physical-layer processing.

The HS-DSCH physical-layer processing is straightforward. A 24-bit CRC is attached to each transport block. The CRC is used by the UE to detect errors in the received transport block.

Demodulation of 16-QAM, which is one of the modulation schemes supported by the HS-DSCH, requires amplitude knowledge at the receiver in order to correctly form the soft values prior to Turbo decoding. This is different from QPSK, where no such knowledge is required as all information is contained in the phase of the received signal. To ease the estimation of the amplitude reference in the receiver, the bits after CRC attachment are scrambled. This results in a sufficiently random sequence out from the Turbo coder to cause both inner and outer signal points in the 16-QAM constellation to be used, thereby aiding the UE in the estimation of the amplitude

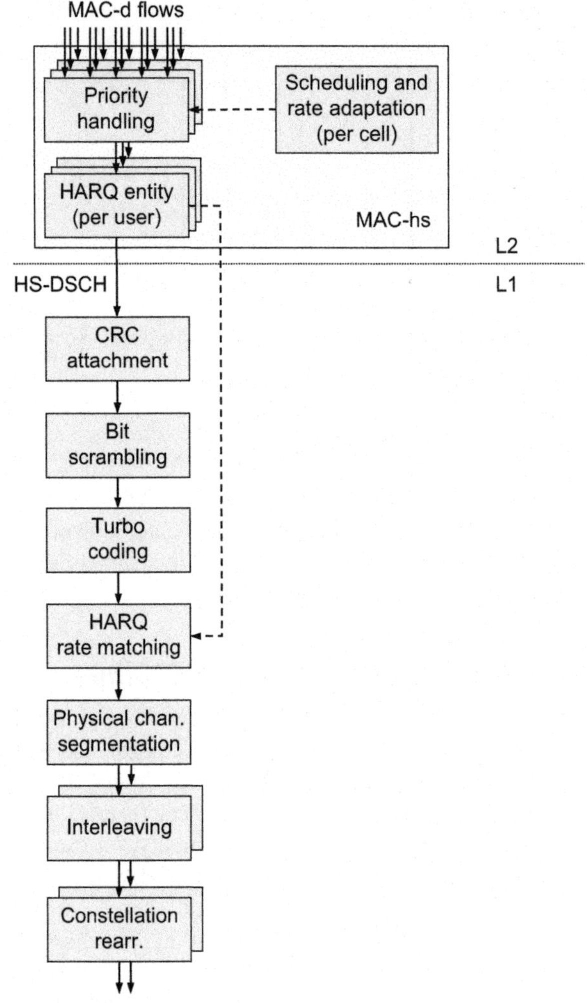


**FIGURE 8.6 MAC-hs and physical-layer processing.**

reference. Note that bit scrambling is done regardless of the modulation scheme used, even if it is strictly speaking only needed in case of 16-QAM modulation.

The fundamental coding scheme in HSDPA is rate-1/3 Turbo coding. To obtain the code rate selected by the rate-control mechanism in the MAC-hs, rate matching, that is, puncturing or repetition, is used to match the number of coded bits to the number of physical-channel bits available. The rate-matching mechanism is also part of the physical-layer hybrid-ARQ and is used to generate different redundancy versions for incremental redundancy. This is done through the use of different puncturing patterns; different bits are punctured for initial transmissions and retransmission.

Physical-channel segmentation distributes the bits to the channelization codes used for transmission, followed by channel interleaving.

Constellation rearrangement is used only for 16-QAM. If Chase combining is used in combination with 16-QAM, a gain in performance can be obtained if the signal constellation is changed between retransmissions. This is further elaborated upon below.

### 8.2.3 SCHEDULING

One of the basic principles for HSDPA is the use of channel-dependent scheduling. The scheduler in the MAC-hs controls what part of the shared code and power resource is assigned to which user in a certain TTI. It is a key component and to a large extent determines the overall HSDPA system performance, especially in a loaded network. At lower loads, only one or a few users are available for scheduling and the differences between different scheduling strategies are less pronounced.

Although the scheduler is implementation specific and not specified by 3GPP, the overall goal of most schedulers is to take advantage of the channel variations between users and preferably schedule transmissions to a user when the channel conditions are advantageous. As discussed in Chapter 5, several scheduling strategies are possible. However, efficient scheduling strategies require at least

- information about the instantaneous channel conditions at the UE; and
- information about the buffer status and priorities of the data flows.

Information about the instantaneous channel quality at the UE is typically obtained through a 5-bit CQI, which each UE feeds back to the Node B at regular intervals. The CQI is calculated at the UE based on the signal-to-noise ratio of the received common pilot. Instead of expressing the CQI as a received signal quality, the CQI is expressed as a recommended transport-block size, taking into account also the receiver performance. This is appropriate as the quantity of relevance is the instantaneous data rate a UE can support rather than the channel quality alone. Hence, a UE with a more advanced receiver, being able to receive data at a higher rate at the same channel quality, will report a larger CQI than a UE with a less advanced receiver, all other conditions being identical.

In addition to the instantaneous channel quality, the scheduler should typically also take buffer status and priority levels into account. Obviously, for UEs for which there is no data awaiting, transmission should not be scheduled. There could also be data that is important to transmit within a certain maximum delay, regardless of the channel conditions. One important example hereof is RRC signaling, for example, related to cell change in order to support mobility, which should be delivered to the UE as soon as possible. Another example, although not as time critical as RRC signaling, is streaming services, which have an upper limit on the acceptable delay of a packet to ensure a constant average data rate. To support priority handling in the scheduling decision, a set of priority queues is defined into which the data is inserted

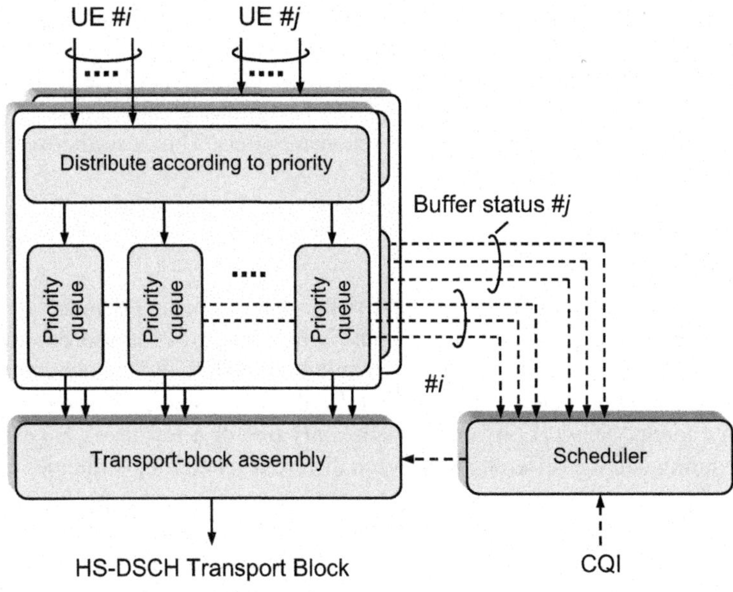


**FIGURE 8.7 Priority handling in the scheduler.**

according to the priority of the data, as illustrated in Figure 8.7. The scheduler selects data from these priority queues for transmission based on the channel conditions, the priority of the queue, and any other relevant information. To efficiently support streaming applications, which require a minimum average data rate, there is a possibility for the RNC to "guarantee" this data rate by providing information about the average data rate to the scheduler in the Node B. The scheduler can take this constraint into account in the scheduling process.

### 8.2.4 RATE CONTROL

As described in Chapter 5, rate control denotes the process of adjusting the data rate to match the instantaneous radio conditions. The data rate is adjusted by changing the modulation scheme and the channel-coding rate. For each TTI, the rate-control mechanism in the scheduler selects, for the scheduled user(s), the transport format(s) and channelization-code resources to use. The transport format consists of the modulation scheme (QPSK or 16-QAM) and the transport-block size.

The resulting code rate after Turbo coding and rate matching is given implicitly by the modulation scheme, the transport-block size, and the channelization-code set allocated to the UE for the given TTI. The number of coded bits after rate matching is given by the modulation scheme and the number of channelization codes, while the number of information bits prior to coding is given by the transport-block size. Hence, by adjusting some or all of these parameters, the overall code rate can be adjusted.

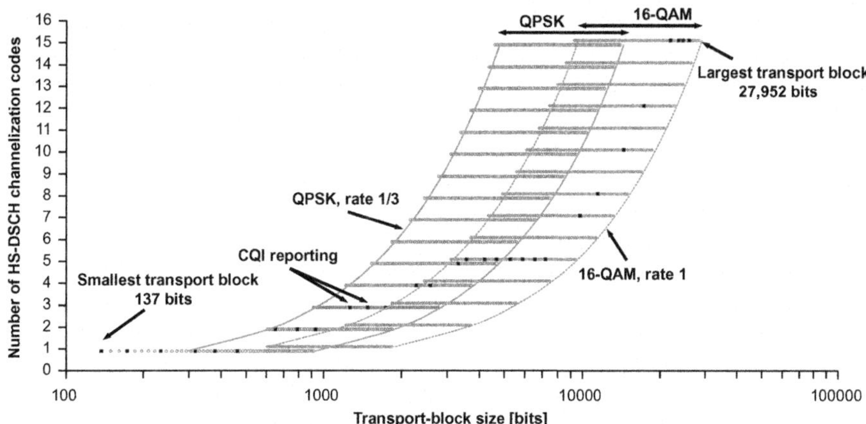


**FIGURE 8.8 Transport-block sizes vs. the number of channelization codes for QPSK and 16-QAM modulation. The transport-block sizes used for CQI reporting are also illustrated.**

Rate control is implemented by allowing the MAC-hs to set the transport format independently for each 2 ms HS-DSCH TTI. Hence, both the modulation scheme and the instantaneous code rate can be adjusted to obtain a data rate suitable for the current radio conditions. The relatively short TTI of 2 ms allows the rate control to track reasonable rapid variations in the instantaneous channel quality.

The HS-DSCH transport-block size can take one of 254 values. These values, illustrated in Figure 8.8, are listed in the specifications and therefore known to both the UE and the Node B. Thus, there is no need for (re)configuration of transport-block sizes at channel setup or when switching serving cell, which reduces the amount of overhead associated with mobility. Each combination of HS-DSCH channelization codes and modulation scheme defines a subset containing 63 of the 254 different transport-block sizes, and the 6-bit "HS-DSCH transport-block size information" indicates one of the 63 transport-block sizes possible for this subset. With this scheme, transport-block sizes in the range of 137–27,952 bits can be signaled, with channel-coding rates ranging from 1/3 up to 1.

For retransmissions, there is a possibility for a code rate >1. This is achieved by exploiting the fact that the transport-block size cannot change between transmission and retransmission. Hence, instead of signaling the transport-block size for the retransmission, a reserved value can be used, indicating that no transport-block-size information is provided by the HS-SCCH and the value from the original transmission should be used. This is useful for additional scheduling flexibility, for example, to retransmit only a small amount of parity bits in case the latest CQI report indicates that the UE was "almost" able to decode the original transmission.

As stated in the introduction, the primary way of adapting to rapid variations in the instantaneous channel quality is rate control as no fast closed-loop power control is specified for HS-DSCH. This does not imply that the HS-DSCH transmission power cannot change for other reasons, for example, because of variations in the power required by other downlink channels. In Figure 8.4, an example of a dynamic HS-DSCH power allocation scheme is illustrated, where the HS-DSCH uses all power not used by other channels. Of course, the overall interference created in the cell must be taken into account when allocating the amount of HS-DSCH power. This is the responsibility of the radio-resource control in the RNC, which can set an upper limit on the power used by the Node B for the HS-DSCHs and all HS-SCCHs.[2] As long as the Node B stays within this limit, the power allocation for HSDPA is up to the Node B implementation. Corresponding measurements, used by the Node B to report the current power usage to the RNC, are also defined. Knowledge about the amount of power used for non-HSDPA channels is useful to the admission control functionality in the RNC. Without this knowledge, the RNC would not be able to determine whether there are resources left for non-HSDPA users trying to enter the cell.

Unlike QPSK, the demodulation of 16-QAM requires an amplitude reference at the UE. How this is achieved is implementation specific. One possibility is to use a channel estimate formed from the common pilot and obtain the ratio between the HS-DSCH and common pilot received powers through averaging over 2 ms. The instantaneous amplitude estimate necessary for 16-QAM demodulation can then be obtained from the common pilot and the estimated offset. This is the reason for the bit level scrambling prior to Turbo coding in Figure 8.6; with scrambling, both the inner and outer signal points in the 16-QAM constellation will be used with a high probability, and an accurate estimate of the received HS-DSCH power can be formed.

What criteria to use for the rate control, that is, the transport-format selection process in the MAC-hs, is implementation specific and not defined in the standard. Principally, the target for the rate control is to select a transport format resulting in transmitting as large a transport block as possible with a reasonable error probability, given the instantaneous channel conditions. Naturally, selecting a transport-block size larger than the amount of data to be transmitted in a TTI is not useful, regardless of whether the instantaneous radio conditions allow for a larger transport block to be transmitted. Hence, the transport-format selection does not only depend on the instantaneous radio conditions but also on the instantaneous source traffic situation.

Because the rate control typically depends on the instantaneous channel conditions, rate control relies on the same estimate of the instantaneous radio quality at the UE as the scheduler. As discussed above, this knowledge is typically obtained from the CQI although other quantities may also be useful. This is further elaborated upon in Section 8.3.6.

### 8.2.5 HYBRID-ARQ WITH SOFT COMBINING

The hybrid-ARQ functionality spans both the MAC-hs and the physical layer. As the MAC-hs is located in the Node B, erroneous transport blocks can be rapidly retransmitted. Hybrid-ARQ retransmissions are therefore significantly less costly in

[2]If the cell is configured to support E-DCH as well, this limit also covers the power used for the related E-DCH downlink control signaling. See Chapter 9.

terms of delay compared to RLC-based retransmissions. There are two fundamental reasons for this difference:

1. There is no need for signaling between the Node B and the RNC for the hybrid-ARQ retransmission. Consequently, any Iub/Iur delays are avoided for retransmissions. Handling retransmission in the Node B is also beneficial from a pure Iub/Iur capacity perspective; hybrid-ARQ retransmissions come at no cost in terms of transport-network capacity.
2. The RLC protocol is typically configured with relatively infrequent status reports of erroneous data blocks (once per several TTIs) to reduce the signaling load, whereas the HSDPA hybrid-ARQ protocol allows for frequent status reports (once per TTI), thus reducing the round-trip time.

In HSDPA, the hybrid-ARQ operates per transport block or, equivalently, per TTI. That is, whenever the HS-DSCH CRC indicates an error, a retransmission representing the same information as the original transport block is requested. As there is a single transport block per TTI, the content of the whole TTI is retransmitted in case of an error. This reduces the amount of uplink signaling, as a single ACK/NACK bit per TTI is sufficient. Furthermore, studies during the HSDPA design phase indicated that the benefits of having multiple transport blocks per TTI with the possibility for individual retransmissions were quite small. A major source of transmission errors are sudden interference variations on the channel and errors in the link-adaptation mechanism. Thanks to the short TTI, the channel is relatively static during the transmission of a transport block and in most cases errors are evenly distributed over the TTI. This limits the potential benefits of individual retransmissions.

Incremental redundancy is the basic scheme for soft combining, that is, retransmissions may consist of a different set of coded bits than the original transmission. Different redundancy versions, that is, different sets of coded bits, are generated as part of the rate-matching mechanism. The rate matcher uses puncturing (or repetition) to match the number of code bits to the number of physical channel bits available. By using different puncturing patterns, different sets of coded bits, that is, different redundancy versions, result. This is illustrated in Figure 8.9. Note that Chase combining is a special case of incremental redundancy; the Node B decides whether to use incremental redundancy or Chase combining by selecting the appropriate puncturing pattern for the retransmission.

The UE receives the coded bits and attempts to decode them. In case the decoding attempts fails, the UE buffers the received soft bits and requests a retransmission by sending a NACK. Once the retransmission occurs, the UE combines the buffered soft bits with the received soft bits from the retransmission and tries to decode the combination.

For soft combining to operate properly, the UE needs to know whether the transmission is a retransmission of previously transmitted data or whether it is transmission of new data. For this purpose, the downlink control signaling includes a new-data indicator, used by the UE to control whether the soft buffer should be cleared (the

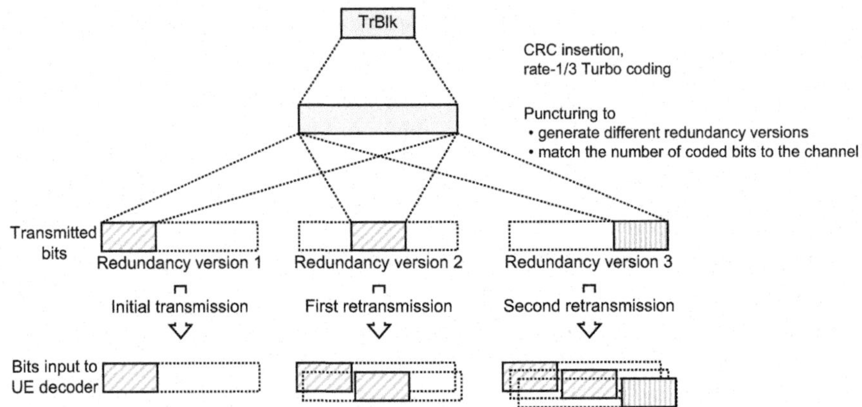


**FIGURE 8.9 Generation of redundancy versions.**

current transmission is new data) or whether soft combining of the soft buffer and the received soft bits should take place (retransmission).

To minimize the delay associated with a retransmission, the outcome of the decoding in the UE should be reported to the Node B as soon as possible. At the same time, the amount of overhead from the feedback signaling should be minimized. This lead to the choice of a stop-and-wait structure for HSDPA, where a single bit is transmitted from the UE to the Node B at a well-specified time, approximately five ms, after the reception of a transport block. To allow for continuous transmission to a single UE, multiple stop-and-wait structures, or *hybrid-ARQ processes*, are operated in parallel, as illustrated in Figure 8.10. Hence, for each user there is one *hybrid-ARQ entity*, each consisting of *multiple* hybrid-ARQ processes.

The number of hybrid-ARQ processes should match the round-trip time between the UE and Node B, including their respective processing time, to allow for

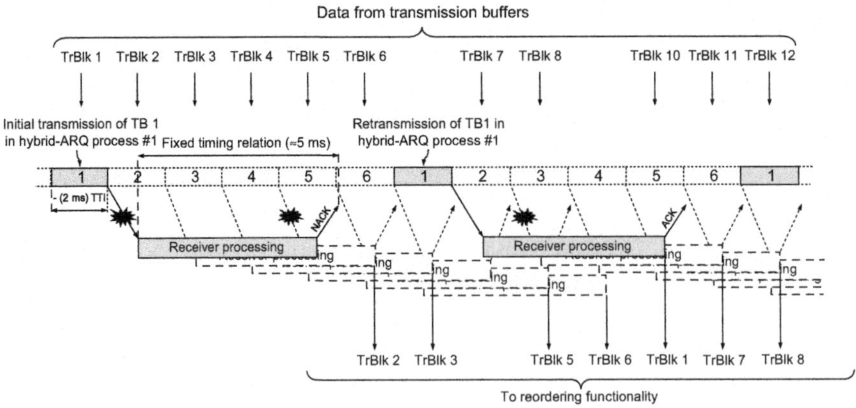


**FIGURE 8.10 Multiple hybrid-ARQ process (six in this example).**

continuous transmission to a UE. Using a larger number of processes than motivated by the round-trip time does not provide any gains but introduces unnecessary delays between retransmissions.

Because the Node B processing time may differ between different implementations, the number of hybrid-ARQ processes is configurable. Up to eight processes can be set up for a user, although a typical number of processes is six. This provides approximately 2.8 ms of processing time in the Node B from the reception of the ACK/NACK until the Node B can schedule a (re)transmission to the UE in the same hybrid-ARQ process.

Downlink control signaling is used to inform the UE which of the hybrid-ARQ processes is used for the current TTI. This is important information to the UE as it is needed to do soft combining with the correct soft buffer; each hybrid-ARQ process has its own soft buffer.

One result of having multiple independent hybrid-ARQ processes operated in parallel is that decoded transport blocks may appear out of sequence. For example, a retransmission may be needed in hybrid-ARQ process number one, whereas process number two did successfully receive the data after the first transmission attempt. Therefore, the transport block transmitted in process number two will be available for forwarding to higher layers at the receiver side before the transport block transmitted in process number one, although the transport blocks were originally transmitted in a different order. This is illustrated in Figure 8.10. As the RLC protocol assumes data to appear in the correct order, a reordering mechanism is used between the outputs from the multiple hybrid-ARQ processes and the RLC protocol. The reordering mechanism is described in more detail in Section 8.3.4.

### 8.2.6 DATA FLOW

To illustrate the flow of user data through the different layers, an example radio-interface protocol configuration is shown in Figure 8.11. For the UE in this example, an IP-based service is assumed, where the user data is mapped to the HS-DSCH.

For signaling purposes in the radio network, several signaling radio bearers are configured in the control plane. In release 5, signaling radio bearers cannot be mapped to the HS-DSCH, and consequently dedicated transport channels must be used, while this restriction is removed in release 6 to allow for operation completely without dedicated transport channels in the downlink.

Figure 8.12 illustrates the data flow at the reference points shown in Figure 8.11. In this example, an IP-based service is assumed. The PDCP performs (optional) IP header compression. The output from the PDCP is fed to the RLC protocol entity. After possible concatenation, the RLC SDUs are segmented into smaller blocks of typically 40 bytes. An RLC PDU is composed of a data segment and the RLC header. If logical-channel multiplexing is performed in MAC-d, a 4-bit header is added to form a MAC-d PDU. In MAC-hs, a number of MAC-d PDUs, possibly of varying sizes, are assembled and a MAC-hs header is attached to form one transport block, subsequently coded and transmitted by the physical layer.

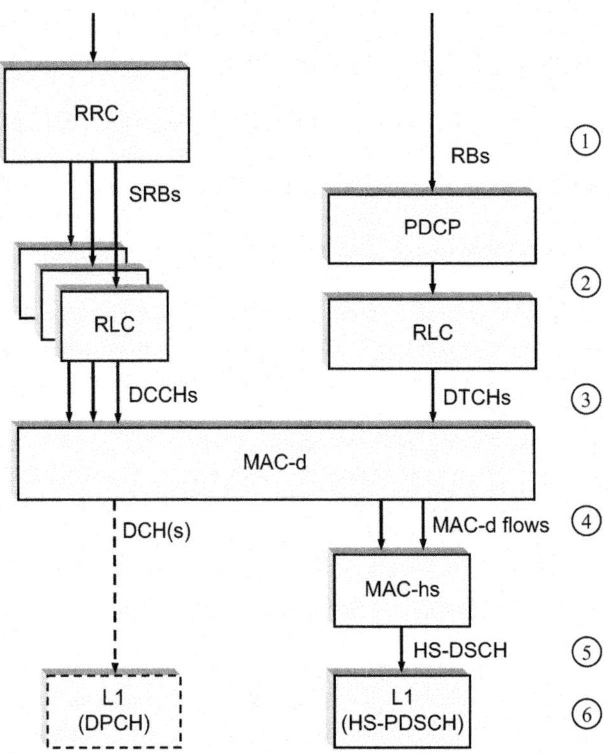


**FIGURE 8.11 Protocol configuration when HS-DSCH is assigned. The numbers in the rightmost part of the figure correspond to the numbers to the right in Figure 8.12.**

### 8.2.7 RESOURCE CONTROL FOR HS-DSCH

With the introduction of HSDPA, parts of the radio resource management are handled by the Node B instead of the RNC. This is a result of introducing channel-dependent scheduling and rate control in the Node B in order to exploit rapid channel variations. However, the RNC still has the overall responsibility for radio-resource management, including admission control and handling of inter-cell interference. Therefore, new measurement reports from the Node B to the RNC have been introduced, as well as mechanisms for the RNC to set the limits within which the Node B is allowed to handle the HSDPA resources[3] in the cell.

To limit the transmission power used for HSDPA, the RNC can set the maximum amount of power the Node B is allowed to use for HSDPA-related downlink transmissions. This ensures that the RNC has control of the maximum amount of interference a cell may generate to neighboring cells. Within the limitation set by the RNC, the Node B is free to manage the power spent on the HSDPA downlink channels. If

[3]Note that many of these measurements were extended in release 6 to include Enhanced Uplink Downlink control channels in addition to the HSDPA-related channels.

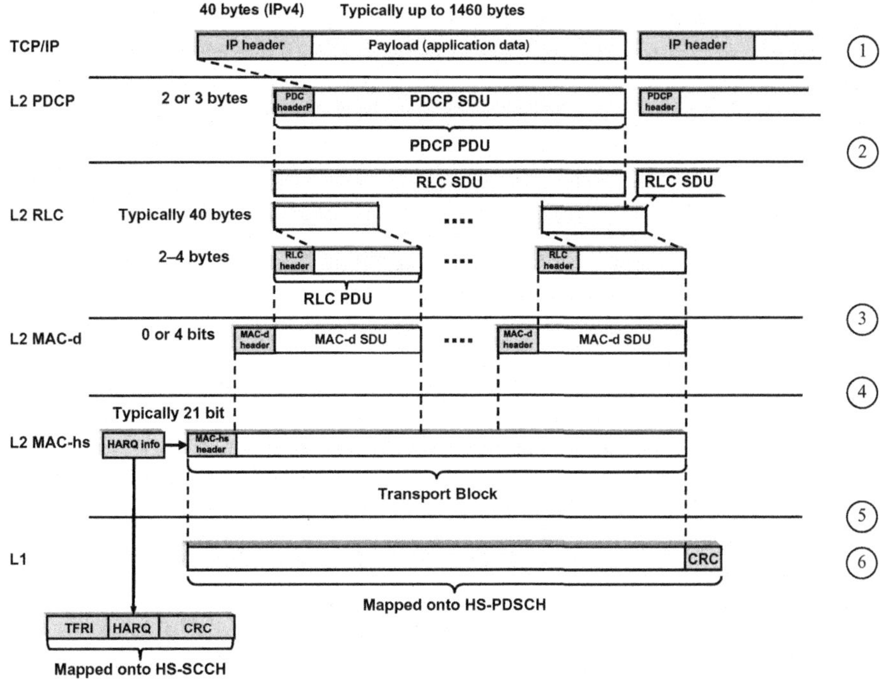


**FIGURE 8.12 Data flow at UTRAN side.**

the quantity is absent (or larger than the total Node B power), the Node B may use all available power for downlink transmissions on the HS-DSCH and HS-SCCH.

Admission control in the RNC needs to take into account the amount of power available in the Node B. Only if there is a sufficient amount of transmission power available in the Node B can a new user be admitted into the cell. The *transmitted carrier power* measurement is available for this purpose. However, with the introduction of HSDPA, the Node B can transmit at full power, even with a single user in the cell, to maximize the data rates. To the admission control in the RNC, it would appear as if the cell is fully loaded and no more users would be admitted. Therefore, a new measurement, *transmitted carrier power of all codes not used for HS-PDSCH or HS-SCCH*, is introduced, which can be used in admission control to determine whether new users can be admitted into the cell or not (Figure 8.13).

In addition to the power-related signaling discussed above, there is also signaling useful to support streaming services. To efficiently support streaming, where a certain minimum data rate needs to be provided on average, the RNC can signal the *MAC-hs Guaranteed Bit Rate*. The scheduler can use this information to ensure that, averaged over a longer period of time, a sufficiently high data rate is provided for a certain MAC-d priority queue. To monitor the fulfillment of this, and to be able to observe the load in the cell due to these restrictions, the Node B can report the required transmission power for each priority class configured by the RNC in order to

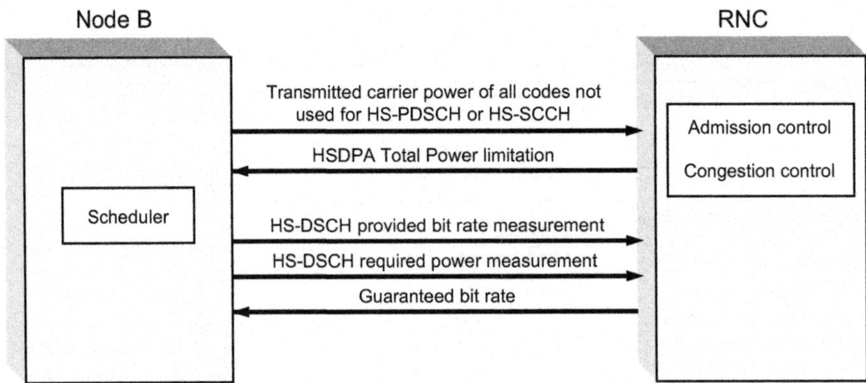


**FIGURE 8.13 Measurements and resource limitations for HSDPA.**

identify "costly" UEs. The Node B can also report the data rate (averaged over 100 ms) it actually provides for each priority class.

## 8.2.8 MOBILITY

Mobility for HSDPA, that is, change of serving cell, is handled through RRC signaling using similar procedures as for dedicated channels. The basics for mobility are network-controlled handover and UE measurement reporting. Measurements are reported from the UE to the RNC, which, based on the measurements, reconfigures the UE and involved Node Bs, resulting in a change of serving cell.

Several measurement mechanisms are already specified in the first release of WCDMA and used for, for example, active set update, hard handover, and intra-frequency measurements. One example is Measurement Event 1D, "change of best cell," which is reported by the UE whenever the common-pilot strength from a neighboring cell (taking any measurement offsets into account) becomes stronger than for the current best cell. This can be used to determine when to switch the HS-DSCH serving cell, as illustrated in Figure 8.14. Updates of the active set are not included in this example; it is assumed that both the source serving cell and the target serving cell are part of the active set.

The reconfiguration of the UE and involved Node Bs can be either synchronous or asynchronous. With synchronous reconfiguration, an activation time is defined in the reconfiguration message, ensuring that all involved parties change their reconfiguration at the same time. Because of unknown delays between the Node B and the RNC, as well as processing and protocol delays, a suitable margin may need to be taken into account in the choice of activation time. Asynchronous reconfiguration implies that the involved nodes obey the reconfiguration message as soon it is received. However, in this case, data transmission from the new cell may start before the UE has been switched from the old cell, which would result in some data loss that has to be retransmitted by the RLC protocol. Hence, synchronous reconfigurations are typically used for HS-DSCH serving cell change. The MAC-hs protocol is reset

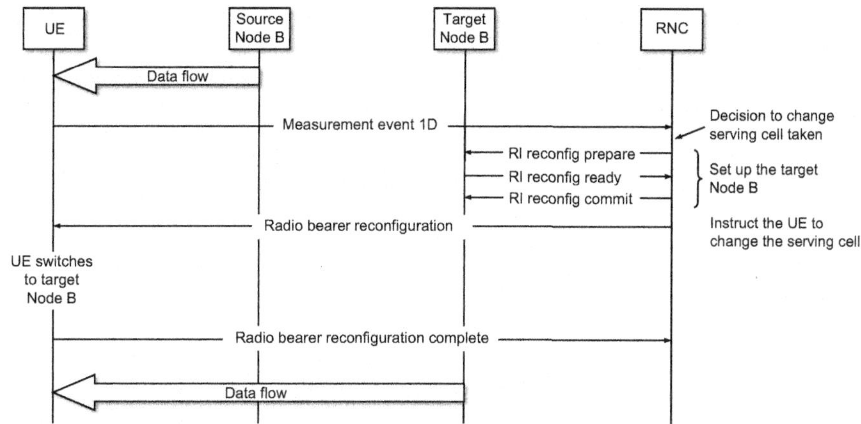


**FIGURE 8.14 Change of serving cell for HSPA. It is assumed that both the source and target Node B are part of the active set.**

when moving from one Node B to another. Thus the hybrid-ARQ protocol state is not transferred between the two Node Bs. Any packet losses at the time of cell change are instead handled by the RLC protocol.

Related to mobility is the flow control between the Node B and the RNC, used to control the amount of data buffered in the MAC-hs in the Node B and avoid overflow in the buffers. The requirements on the flow control are, to some extent, conflicting, as they shall ensure that MAC-hs buffers should be large enough to contain a sufficient amount of data to fully utilize the physical channel resources (in case of advantageous channel conditions), while at the same time MAC-hs buffers should be kept as small as possible to minimize the amount of packets that need to be resent to a new Node B in case of inter–Node B handover.

## 8.2.9 UE CATEGORIES

To allow for a range of UE implementations, different UE capabilities are specified. The UE capabilities are divided into a number of parameters, which are sent from the UE at the establishment of a connection and if/when the UE capabilities are changed during an ongoing connection. The UE capabilities may then be used by the network to select a configuration that is supported by the UE. Several of the UE capabilities applicable to other channels are valid for HS-DSCH as well, but there are also some HS-DSCH-specific capabilities.

Basically, the physical-layer UE capabilities are used to limit the requirements for three different UE resources: the despreading resource, the soft-buffer memory used by the hybrid-ARQ functionality, and the Turbo decoder. The despreading resource is limited in terms of the maximum number of HS-PDSCH codes the UE simultaneously needs to despread. In release 5, three different capabilities exist in terms of despreading resources, corresponding to the capability to despread a maximum of 5, 10, or 15 physical channels (HS-PDSCH).

The amount of soft-buffer memory is in the range of 14,400–172,800 soft bits for release 5, depending on the UE category. Note that this is the total amount of buffer memory for all hybrid-ARQ processes, not the value per process. The memory is divided among the multiple hybrid-ARQ processes, typically with an equal amount of memory per process, although nonequal allocation is also possible.

The requirements on the Turbo-decoding resource are defined through two parameters: the maximum number of transport-channel bits that can be received within an HS-DSCH TTI and the minimum inter-TTI interval, that is, the distance in time between subsequent transport blocks. The decoding time in a Turbo decoder is roughly proportional to the number of information bits, which thus provides a limit on the required processing speed. In addition, for low-end UEs, there is a possibility to avoid continuous data transmission by specifying an inter-TTI interval larger than one.

In order to limit the number of possible combinations of UE capabilities and to avoid parameter combinations that do not make sense, the UE capability parameters relevant for the physical layer are lumped into 12 different *categories* in release 5, as illustrated in Table 8.1. New features introduced into subsequent releases, such as MIMO, have added further categories.

**Table 8.1** HSDPA UE Categories [3]

| HS-DSCH Category | Maximum Number of HS-DSCH Codes Received | Minimum Inter-TTI Interval | Maximum Transport-Block Size (Mbit/s) | Maximum Number of Soft Bits | Supported Modulation Schemes |
|---|---|---|---|---|---|
| 1 | 5 | 3 | 7298 (3.6) | 19,200 | 16-QAM, QPSK |
| 2 | 5 | 3 | 7298 (3.6) | 28,800 | 16-QAM, QPSK |
| 3 | 5 | 2 | 7298 (3.6) | 28,800 | 16-QAM, QPSK |
| 4 | 5 | 2 | 7298 (3.6) | 38,400 | 16-QAM, QPSK |
| 5 | 5 | 1 | 7298 (3.6) | 57,600 | 16-QAM, QPSK |
| 6 | 5 | 1 | 7298 (3.6) | 67,200 | 16-QAM, QPSK |
| 7 | 10 | 1 | 14,411 (7.2) | 115,200 | 16-QAM, QPSK |
| 8 | 10 | 1 | 14,411 (7.2) | 134,400 | 16-QAM, QPSK |
| 9 | 15 | 1 | 20,251 (10.1) | 172,800 | 16-QAM, QPSK |
| 10 | 15 | 1 | 27,952 (14) | 172,800 | 16-QAM, QPSK |
| 11 | 5 | 2 | 3630 (1.8) | 14,400 | QPSK |
| 12 | 5 | 1 | 3630 (1.8) | 28,800 | QPSK |
| 13 | 15 | 1 | 35,280 (17.6) | 259,200 | QPSK, 16QAM, 64QAM |
| 14 | 15 | 1 | 42,192 (21) | 259,200 | QPSK, 16QAM, 64QAM |
| 15 | 15 | 1 | 23,370[1] (23) | 345,600 | QPSK, 16QAM |
| 16 | 15 | 1 | 27,952[1] (28) | 345,600 | QPSK, 16QAM |

*(Continued)*

**Table 8.1** HSDPA UE Categories [3] *(cont.)*

| HS-DSCH Category | Maximum Number of HS-DSCH Codes Received | Minimum Inter-TTI Interval | Maximum Transport-Block Size (Mbit/s) | Maximum Number of Soft Bits | Supported Modulation Schemes |
|---|---|---|---|---|---|
| 17a | 15 | 1 | 35,280[2] (17.6) | 259,200 | QPSK, 16QAM, 64QAM |
| 17b | 15 | 1 | 23,370[2] (23) | 345,600 | QPSK, 16QAM |
| 18a | 15 | 1 | 42,192[2] (21) | 259,200 | QPSK, 16QAM, 64QAM |
| 18b | 15 | 1 | 27,952[2] (28) | 345,600 | QPSK, 16QAM |
| 19 | 15 | 1 | 35,280[1] (35) | 518,400 | QPSK, 16QAM, 64QAM |
| 20 | 15 | 1 | 42,192[1] (42) | 518,400 | QPSK, 16QAM, 64QAM |
| 21 | 15 | 1 | 23,370[3] (23) | 345,600 | QPSK, 16QAM |
| 22 | 15 | 1 | 27,952[3] (28) | 345,600 | QPSK, 16QAM |
| 23 | 15 | 1 | 35,280[3] (35) | 518,400 | QPSK, 16QAM, 64QAM |
| 24 | 15 | 1 | 42,192[3] (42) | 518,400 | QPSK, 16QAM, 64QAM |
| 25 | 15 | 1 | 23,370[1,3] (46) | 691,200 | QPSK, 16QAM |
| 26 | 15 | 1 | 27,952[1,3] (56) | 691,200 | QPSK, 16QAM |
| 27 | 15 | 1 | 35,280[1,3] (70) | 1,036,800 | QPSK, 16QAM, 64QAM |
| 28 | 15 | 1 | 42,192[1,3] (84) | 1,036,800 QPSK, 16QAM, 64QAM | |
| 29 | 15 | 1 | 42,192[4] (63) | 777,600 | 3 QPSK, 16QAM, 64QAM |
| 30 | 15 | 1 | 42,192[1,4] (126) | 1,555,200 QPSK, 16QAM, 64QAM | |
| 31 | 15 | 1 | 42,192[5] (84) | 1,036,800 | QPSK, 16QAM, 64QAM |
| 32 | 15 | 1 | 42,192[1,5] (168) | 2,073,600 | QPSK, 16QAM, 64QAM |
| 33 | 15 | 1 | 42,192[6] (126) | 1,555,200 | QPSK, 16QAM, 64QAM |
| 34 | 15 | 1 | 42,192[1,6] (253) | 3,110,400 | QPSK, 16QAM, 64QAM |

*(Continued)*

**Table 8.1** HSDPA UE Categories [3] *(cont.)*

| HS-DSCH Category | Maximum Number of HS-DSCH Codes Received | Minimum Inter-TTI Interval | Maximum Transport-Block Size (Mbit/s) | Maximum Number of Soft Bits | Supported Modulation Schemes |
|---|---|---|---|---|---|
| 35 | 15 | 1 | 42,192[7] (169) | 2,073,600 | QPSK, 16QAM, 64QAM |
| 36 | 15 | 1 | 42,192[1,7] (338) | 4,147,200 | QPSK, 16QAM, 64QAM |
| 37 | 15 | 1 | 42,192[2,8] (169) | 2,073,600- | QPSK, 16QAM, 64QAM |
| 38 | 15 | 1 | 42,192[5,8] (338) | 4,147,200 | QPSK, 16QAM, 64QAM |

[1]*Supports 2 × 2 MIMO.*
[2]*Categories 17 and 18 support either 64-QAM with no MIMO or MIMO with max 16-QAM.*
[3]*Supports DC-HSDPA.*
[4]*Supports 3-carrier MC-HSDPA.*
[5]*Supports 4-carrier MC-HSDPA.*
[6]*Supports 6-carrier MC-HSDPA.*
[7]*Supports 8-carrier MC-HSDPA.*
[8]*Supports 4 × 4 MIMO.*

# 8.3 FINER DETAILS OF HSDPA

## 8.3.1 HYBRID-ARQ REVISITED: PHYSICAL-LAYER PROCESSING

Hybrid-ARQ with soft combining has been described above, although some details of the physical-layer and protocol operation were omitted in order to simplify the description. This section provides a more detailed description of the processing, aiming at filling in the missing gaps.

As already mentioned, the hybrid-ARQ operates on a single transport block, that is, whenever the HS-DSCH CRC indicates an error, a retransmission representing the same information as the original transport block is requested. Because there is a single transport block per TTI, this implies that it is not possible to mix transmissions and retransmissions within the same TTI.

Because incremental redundancy is the basic hybrid-ARQ soft-combining scheme, retransmissions generally consist of a different set of coded bits. Furthermore, the modulation scheme, the channelization-code set, and the transmission power can be different compared to the original transmission. Incremental redundancy generally has better performance, especially for high initial code rates, but poses higher requirements on the soft buffering in the UE because soft bits from all transmission attempts must be buffered prior to decoding. Therefore, the Node B needs to have knowledge about the soft-buffer size in the UE (for each active hybrid-ARQ process). Coded bits that do not fit within the soft buffer shall not be transmitted. For

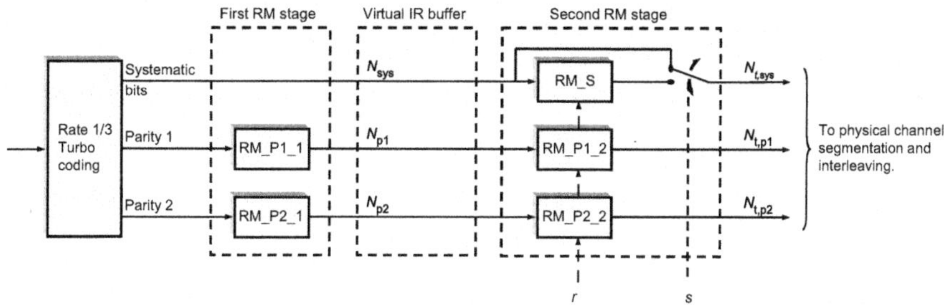


**FIGURE 8.15 The principle of two-stage rate matching.**

HSDPA, this problem is solved through the use of *two-stage rate matching*. The first rate-matching stage limits the number of coded bits to what is possible to fit in the soft buffer, whereas the second rate-matching stage generates the different redundancy versions.

Each rate-matching stage uses several identical rate-matching blocks, denoted RM in Figure 8.15. An RM block can be configured to puncture or repeat every $n$th bit.

The first rate-matching stage is used to limit the number of coded bits to the available UE soft buffer for the hybrid-ARQ process currently being addressed. A sufficient number of coded bits are punctured to ensure that all coded bits at the output of the first rate-matching stage will fit in the soft buffer (known as *virtual IR buffer* at the transmitter side). Hence, depending on the soft-buffer size in the UE, the lowest code rate may be higher than the rate-1/3 mother code rate in the Turbo coder. Note that if the number of bits from the channel coding does not exceed the UE soft-buffering capability, the first rate-matching stage is transparent and no puncturing is performed.

The second rate-matching stage serves two purposes:

- Matching the number of bits in the virtual IR buffer to the number of available channel bits. The number of available channel bits is given by the size of the channelization-code set and the modulation scheme selected for the TTI.
- Generating different sets of coded bits as controlled by the two redundancy-version parameters $r$ and $s$, described below.

Equal repetition for all three streams is applied if the number of available channel bits is larger than the number of bits in the virtual IR buffer, otherwise puncturing is applied.

To support full incremental redundancy, that is, to have the possibility to transmit only/mainly parity bits in a retransmission, puncturing of systematic bits is possible, as controlled by the parameter $s$. Setting $s = 1$ implies that the systematic bits are prioritized and puncturing is primarily applied with an equal amount to the two parity-bit streams. On the other hand, for a transmission prioritizing the parity bits, $s = 0$ and primarily the systematic bits are punctured. If, for a transmission prioritizing the systematic bits, the number of coded bits is larger than the number of physical channel bits, despite all the parity bits having been punctured, further puncturing is applied to the systematic bits. Similarly, if puncturing the systematic bits is not sufficient for a transmission prioritizing the parity bits, puncturing is applied to the parity bits as well.

For good performance, all systematic bits should be transmitted in the initial transmission, corresponding to $s = 1$, and the code rate should be set to less than one. For the retransmission (assuming the initial transmission did not succeed), different strategies can be applied. If the Node B received neither ACK nor NACK in response to the initial transmission attempt, the UE may have missed the initial transmission. Setting $s = 1$ also for the retransmission is therefore appropriate. This is also the case if NACK is received and Chase combining is used for retransmissions. However, if a NACK is received and incremental redundancy is used, that is, the parity bits should be prioritized, setting $s = 0$ is appropriate.

The parameter $r$ controls the puncturing pattern in each rate-matching block in Figure 8.15 and determines which bits to puncture. Typically, $r = 0$ is used for the initial transmission attempt. For retransmissions, the value of $r$ is typically increased, effectively leading to a different puncturing pattern. Thus, by varying $r$, multiple, possibly partially overlapping, sets of coded bits representing the same set of information bits can be generated. It should be noted that changing the number of channel bits by changing the modulation scheme or the number of channelization codes also affects which coded bits are transmitted even if the $r$ and $s$ parameters are unchanged between the transmission attempts.

With the two-stage rate-matching scheme, both incremental redundancy and Chase combining can easily be supported. By setting $s = 1$ and $r = 0$ for all transmission attempts, the same set of bits is used for the retransmissions as for the original transmission, that is, Chase combining. Incremental redundancy is easily achieved by setting $s = 1$ and using $r = 0$ for the initial transmission, while retransmissions use $s = 0$ and $r > 0$. Partial IR, that is, incremental redundancy with the systematic bits included in each transmission, results if $s = 1$ for all the retransmissions as well as the initial transmission.

In Figure 8.16, a simple numerical example is shown to further illustrate the operation of the physical-layer hybrid-ARQ processing of data. Assume that, as an example, a transport block of 2404 bits is to be transmitted using one of the hybrid-ARQ processes. Furthermore, assume the hybrid-ARQ process in question is capable of buffering at the most 7000 soft values because of memory limitations in the UE and the soft memory configuration set by higher layers. Finally, the channel can carry 3840 coded bits in this example (QPSK modulation, four channelization codes).

A 24-bit CRC is appended to the transport block, rate-1/3 Turbo coding is applied, and a 12-bit tail appended, resulting in 7296 coded bits. The coded bits are fed to the first stage rate matching, which punctures parity bits such that 2432 systematic bits and $2 \times 2284$ parity bits, in total 7000 bits, are fed to the second-stage rate-matching block. Because at most 7000 coded bits can be transmitted, the lowest possible code rate is 2432/7000 = 0.35, which is slightly higher than the mother code's rate of 1/3 because of the limited soft buffer in the UE.

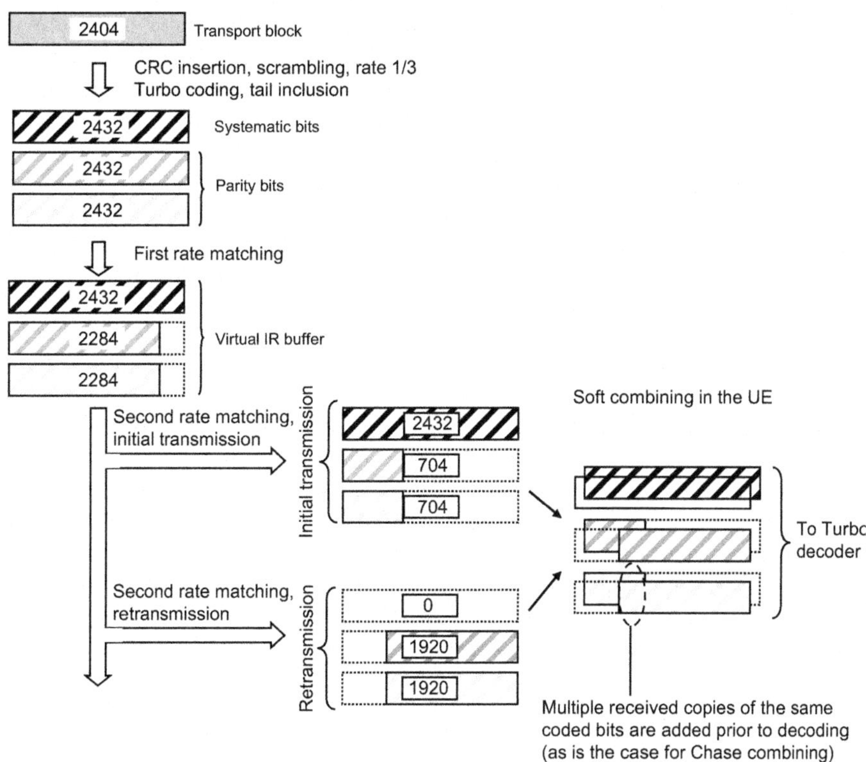


**FIGURE 8.16 An example of the generation of different redundancy versions in the case of IR. The numbers indicate the number of bits after the different stages using the example case in the text.**

For the initial transmission, the second-stage rate matching matches the 7000 coded bits to the 3840 channel bits by puncturing the parity bits only. This is achieved by using $r = 0$ and $s = 1$, that is, a self-decodable transmission, and the resulting code rate is 2432/3840 = 0.63.

For retransmissions, either Chase combining or incremental redundancy can be used, as chosen by the Node B. If Chase combining is used by setting $s = 1$ and $r = 0$, the same 3840 bits as used for the initial transmission are retransmitted (assuming unchanged modulation scheme and channelization-code set). The resulting effective code rate remains 0.63 as no additional parity has been transmitted, but an energy gain has been obtained as, in total, twice the amount of energy has been transmitted for each bit. Note that this example assumed identical transport formats for the initial transmission and the retransmission.

If incremental redundancy is used for the retransmission, for example, by using $s = 0$ and $r = 1$, the systematic bits are punctured and only parity bits are retransmitted, of which 3840 (out of 4568 parity bits available after the first stage rate matching) fit into the physical channel. Note that some of these parity bits were already included in the original transmission as the number of unique parity bits is not large enough to fill both the original transmission and the retransmissions. After the retransmission, the resulting code rate is 2432/7000 = 0.35. Hence, contrary to Chase combining, there is a coding gain in addition to the energy gain.

### 8.3.2 INTERLEAVING AND CONSTELLATION REARRANGEMENT

For 16-QAM, two of the four bits carried by each modulation symbol will be more reliable at the receiver because of the difference in the number of nearest neighbors in the constellation. This is in contrast to QPSK, where both bits are of equal reliability. Furthermore, for Turbo codes, systematic bits are of greater importance in the decoding process, compared to parity bits. Hence, it is desirable to map as many of the systematic bits as possible to the more reliable positions in a 16-QAM symbol. A dual interleaver scheme, illustrated in Figure 8.17, has been adopted for HS-DSCH in order to control the mapping of systematic and parity bits onto the 16-QAM modulation symbols.

For QPSK, only the upper interleaver in Figure 8.17 is used, whereas for 16-QAM, two identical interleavers are used in parallel. Systematic bits are primarily fed into the upper interleaver, whereas parity bits are primarily fed into the lower interleaver. The 16-QAM constellation is defined such that the output from the upper interleaver is mapped onto the reliable bit positions and the output from the lower interleaver onto the less reliable positions.

If 16-QAM is used in conjunction with hybrid-ARQ using Chase combining, there is a performance gain by rearranging the 16-QAM symbol constellations between multiple transmission attempts as this provides an averaging effect among the reliability of the bits. However, note that this gain is only available for retransmissions and not for the initial transmission. Furthermore, the gains with constellation rearrangement in combination with incremental redundancy are minor. Hence, its use is mainly applicable when Chase combining is used.

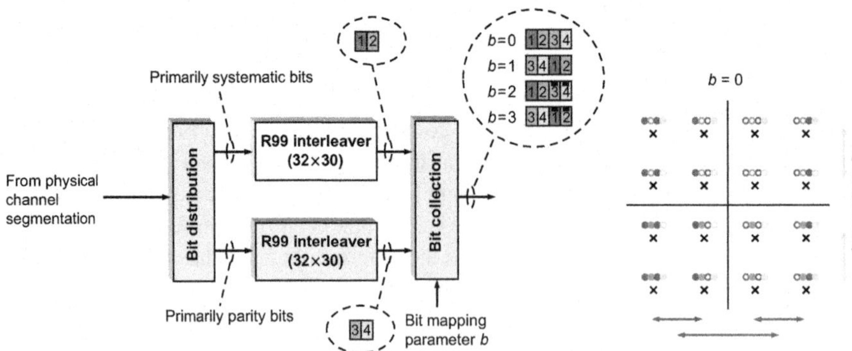


**FIGURE 8.17 The channel interleaver for the HS-DSCH. The shaded parts are only used for 16-QAM. Colors illustrate the mapping order for a sequence of four bits, where a bar on top of the figure denotes bit inversion.**

Constellation rearrangement is obtained through bit manipulations in the bit collector block and is controlled by a four-state bit mapping parameter, controlling two independent operations. First, the output of the two interleavers can be swapped. Second, the output of the lower interleaver (or the upper interleaver if swapping is used) can be inverted. In essence, this results in the selection of one out of four different signal constellations for 16-QAM.

### 8.3.3 HYBRID-ARQ REVISITED: PROTOCOL OPERATION

As stated earlier, each hybrid-ARQ entity is capable of supporting multiple (up to eight) stop-and-wait hybrid-ARQ processes. The motivation behind this is to allow for continuous transmission to a single UE, which cannot be achieved by a single stop-and-wait scheme. The number of hybrid-ARQ processes is configurable by higher-layer signaling. Preferably, the number of hybrid-ARQ processes is chosen to match the round-trip time, consisting of the TTI itself, any radio-interface delay in downlink and uplink, the processing time in the UE, and the processing time in the Node B.

The protocol design assumes a well-defined time between the end of the received transport block and the transmission of the ACK/NACK, as discussed in Section 8.2.5. In essence, this time is the time the UE has available for decoding of the received data. From a delay perspective, this time should be as small as possible, but a too small value would put unrealistic requirements on the UE processing speed. Although in principle the time could be made a UE capability, this was not felt necessary and a value of 5 ms was agreed as a good trade-off between performance and complexity. This value affects the number of hybrid-ARQ processes necessary. Typically, a total of six processes are configured, which leaves around 2.8 ms for processing of retransmissions in the Node B.

Which of the hybrid-ARQ processes is used for the current transmission is controlled by the scheduler and explicitly signaled to the UE. Note that the hybrid-ARQ processes can be addressed in any order. The amount of soft-buffering memory available in the UE is semi-statically split between the different hybrid-ARQ processes. Thus, the larger the number of hybrid-ARQ processes, the smaller the amount of soft-buffer memory available to a hybrid-ARQ process for incremental redundancy. The split of the total soft-buffer memory between the hybrid-ARQ processes is controlled by the RNC and does not necessarily have to be such that the soft-buffer memory per hybrid-ARQ process is the same. Some hybrid-ARQ processes can be configured to use more soft-buffer memory than others, although the typical case is to split the available memory equally among the processes.

Whenever the current transmission is not a retransmission, the Node B MAC-hs increments the single-bit new-data indicator. Hence, for each new transport block, the bit is toggled. The indicator is used by the UE to clear the soft buffer for initial transmissions because, by definition, no soft combining should be done for an initial transmission. The indicator is also used to detect error cases in the status signaling, for example, if the "new-data" indicator is not toggled despite the fact that the previous data for the hybrid-ARQ process in question was correctly decoded and acknowledged, an error in the uplink signaling has most likely occurred. Similarly, if the indicator is toggled but the previous data for the hybrid-ARQ process was not correctly decoded, the UE will replace the data previously in the soft buffers with the new received data.

Errors in the status (ACK/NACK) signaling will impact the overall performance. If an ACK is misinterpreted as a NACK, an unnecessary hybrid-ARQ retransmission will take place, leading to a (small) reduction in the throughput. On the other hand, misinterpreting a NACK as an ACK will lead to loss of data as the Node B will not perform a hybrid-ARQ retransmission even if the UE was not able to successfully decode the data. Instead, the missing data has to be retransmitted by the RLC protocol, a more time-consuming procedure than hybrid-ARQ retransmissions. Therefore, the requirements on the ACK/NACK errors are typically asymmetric with $\Pr\{NACK|ACK\} = 10^{-2}$ and $\Pr\{ACK|NACK\} = 10^{-3}$ (or $10^{-4}$) as typical values. With these error probabilities, the impact on the end-user TCP performance due to hybrid-ARQ signaling errors is small [2].

### 8.3.4 IN-SEQUENCE DELIVERY

The multiple hybrid-ARQ processes cannot themselves ensure in-sequence delivery as there is no interaction between the processes. Hence, in-sequence delivery must be implemented on top of the hybrid-ARQ processes and a reordering queue in the UE MAC-hs is used for this purpose. Related to the reordering queues in the UE are the priority queues in the Node B, used for handling priorities in the scheduling process.

The Node B MAC-hs receives MAC-d PDUs in one or several MAC-d flows. Each such MAC-d PDU has a priority assigned to it and MAC-d PDUs with different priorities can be mixed in the same MAC-d flow. The MAC-d flows are split if necessary and the MAC-d PDUs are sorted into priority queues, as illustrated in Figure 8.18. Each priority queue corresponds to a certain MAC-d flow and a certain MAC-d priority, where RRC signaling is used to set up the mapping between the priority queues and the MAC-d flows. Hence, the scheduler in the MAC-hs can, if desired, take the priorities into account when making the scheduling decision. One or several MAC-d PDUs from one of the priority queues are assembled into a data block, where the number of MAC-d PDUs and the priority queue selection are controlled by the scheduler. A MAC-hs header containing, among others, queue identity and a transmission sequence number, is added to form a transport block. The transport block is forwarded to the physical layer for further processing. As there is only a single transmission sequence number and queue identity in the transport block, all MAC-d PDUs within the same transport block come from the same priority queue. Thus, mixing MAC-d PDUs from different priority queues within the same TTI is not possible.

In the UE, the reordering-queue identity is used to place the received data block, containing received MAC-d PDUs, into the correct reordering queue, as illustrated in Figure 8.18. Each reordering queue corresponds to a priority queue in the Node B,

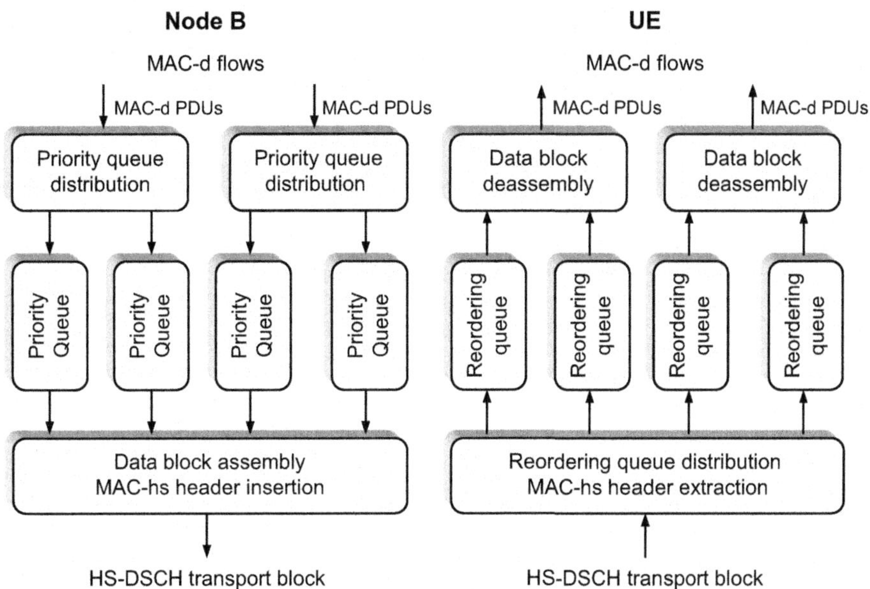


**FIGURE 8.18 The priority queues in the Node B MAC-hs (left) and the reordering queues in the UE MAC-hs (right).**

although the priority queues buffer MAC-d PDUs, while the reordering queues buffer data blocks. Within each reordering queue, the transmission sequence number sent in the MAC-hs header is used to ensure in-sequence delivery of the MAC-d PDUs. The transmission sequence number is unique within the reordering queue but not between different reordering queues.

The basic idea behind reordering, illustrated in Figure 8.19, is to store data blocks in the reordering queue until all data blocks with lower sequence numbers have been delivered. As an example, at time $t_0$ in Figure 8.19, the Node B has transmitted data blocks with sequence numbers 0 through 3. However, the data block with sequence number 1 has not yet reached the MAC-hs reordering queue in the UE, possibly due

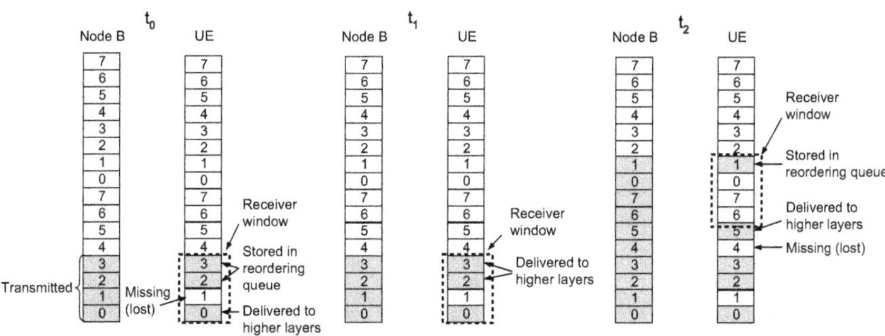


**FIGURE 8.19 Illustration of the principles behind reordering queues.**

to hybrid-ARQ retransmissions or errors in the hybrid-ARQ uplink signaling. Data block 0 has been disassembled into MAC-d PDUs and delivered to upper layers by the UE MAC-hs, whereas data blocks 2 and 3 are buffered in the reordering queue because data block 1 is missing.

Evidently, there is a risk of stalling the reordering queue if missing data blocks (data block 1 in this example) are not successfully received within a finite time. Therefore, a timer-based stall avoidance mechanism is defined for the MAC-hs. Whenever a data block is successfully received but cannot be delivered to higher layers, a timer is started. In Figure 8.19, this occurs when data block 2 is received because data block 1 is missing in the reordering buffer. Note that there is at maximum one stall avoidance timer active. Therefore, no timer is started upon reception of data block 3 as there is already one active timer started for data block 2. Upon expiration of the timer, which occurs at time $t_1$ in Figure 8.19, data block 1 is considered to be lost. Any subsequent data blocks up to the first missing data block are to be disassembled into MAC-d PDUs and delivered to higher layers. In Figure 8.19, data blocks 2 and 3 are delivered to higher layers.

Relying on the timer-based mechanism alone would limit the possible values of the timer and limit the performance if the sequence numbers are to be kept unique. Hence, a window-based stall avoidance mechanism is defined in addition to the timer-based mechanism to ensure a consistent UE behavior. If a data block with a sequence number higher than the end of the window is received by the reordering function, the data block is inserted into the reordering buffer at the position indicated by the sequence number. The receiver window is advanced such that the received data block forms the last data block within the window. Any data blocks not within the window after the window advancement are delivered to higher layers. In the example in Figure 8.19, the window size of 4 is used, but the MAC-hs window size is configurable by RRC. In Figure 8.19, a data block with sequence number 1 is received at time $t_2$, which causes the receiver window to be advanced to cover sequence numbers 6 through 1. Data block 4 is considered to be lost, because it is now outside the window, whereas data block 5 is disassembled and delivered to higher layers. In order for the reordering functionality in the UE to operate properly, the Node B should not retransmit MAC-hs PDUs with sequence numbers lower than the highest transmitted sequence number minus the UE receiver window size.

### 8.3.5 MAC-HS HEADER

To support reordering and de-multiplexing of MAC-d PDUs in the UE, as discussed above, the necessary information needs to be signaled to the UE. As this information is required only after successful decoding of a transport block, in-band signaling in the form of a MAC-hs header can be used.

The MAC-hs header contains

- reordering-queue identity;
- *Transmission Sequence Number* (TSN); and
- number and size of the MAC-d PDUs.

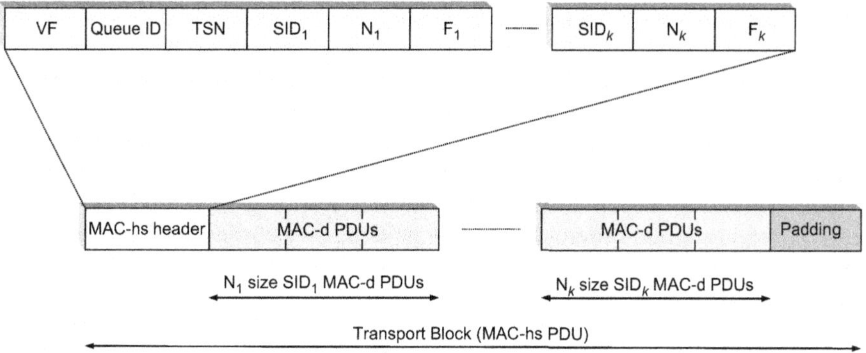


**FIGURE 8.20 The structure of the MAC-hs header.**

The structure of the MAC-hs header is illustrated in Figure 8.20. The *Version Flag* (VF) is identical to zero and reserved for future extensions of the MAC-hs header. The 3-bit Queue ID identifies the reordering queue to be used in the receiver. All MAC-d PDUs in one MAC-hs PDU belong to the same reordering queue. The 6-bit TSN field identifies the transmission sequence number of the MAC-hs data block. The TSN is unique within a reordering buffer but not between different reordering buffers. Together with the Queue ID, the TSN provides support for in-sequence delivery, as described in the previous section.

The MAC-hs payload consists of one or several MAC-d PDUs. The 3-bit *SID*, size index identifier, provides the MAC-d PDU size and the 7-bit *N* field identifies the number of MAC-d PDUs. The flag *F* is used to indicate the end of the MAC-hs header. One set of *SID*, *N*, and *F* is used for each set of consecutive MAC-d PDUs, and multiple MAC-d PDU sizes are supported by forming groups of MAC-d PDUs of equal size. Note that all the MAC-d PDUs within a data block must be in consecutive order because the sequence numbering is per data block. Hence, if a sequence of MAC-d PDUs with sizes given by $SID_1$, $SID_2$, $SID_1$ is to be transmitted, three groups have to be formed despite that there are only two MAC-d PDU sizes. Finally, the MAC-hs PDU is padded (if necessary) such that the MAC-hs PDU size equals a suitable block size. It should be noted that, in most cases, there is only a single MAC-d PDU size and, consequently, only a single set of *SID*, *N*, and *F*.

It should be noted that the original MAC-hs in release 5 was found to present a significant protocol bottleneck that could potentially restrict achievable throughputs, and thus an improved MAC was introduced in release 7, as described in Section 8.3.10.

### 8.3.6 CQI AND OTHER MEANS TO ASSESS THE DOWNLINK QUALITY

Obviously, some of the key HSDPA functions, primarily scheduling and rate control, rely on rapid adaptation of the transmission parameters to the instantaneous channel conditions as experienced by the UE. The Node B is free to form an estimate of the channel conditions using any available information, but, as already discussed, uplink control signaling from the UEs in the form of a CQI is typically used.

The CQI does not explicitly indicate the channel quality but rather the data rate supported by the UE given the current channel conditions. More specifically, the CQI is a recommended transport-block size (which is equivalent to a recommended data rate).

The reason for not reporting an explicit channel-quality measure is that different UEs might support different data rates in identical environments, depending on the exact receiver implementation. By reporting the data rate rather than an explicit channel-quality measure, the fact that a UE has a relatively better receiver can be utilized to provide better service (higher data rates) to such a UE. It is interesting to note that this provides a benefit with advanced receiver structures for the end user. For a power-controlled channel, the gain from an advanced receiver is seen as a lower transmit power at the Node B, thus providing a benefit for the network but not the end user. This is in contrast to the HS-DSCH using rate control, where a UE with an advanced receiver can receive the HS-DSCH with a higher data rate compared to a standard receiver.

Each 5-bit CQI value corresponds to a given transport-block size, modulation scheme, and number of channelization codes. These values are shown in Figure 8.8 (assuming a high-end UE, capable of receiving 15 codes). Different tables are used for different UE categories as a UE shall not report a CQI exceeding its capabilities. For example, a UE only supporting five codes shall not report a CQI corresponding to 15 codes, while a 15-code UE may do so. Therefore, power offsets are used for channel qualities exceeding the UE capabilities. A power offset of $x$ dB indicates that the UE can receive a certain transport-block size but at $x$ dB lower transmission power than the CQI report was based upon. This is illustrated in Table 8.2 for some different UE categories. UEs belonging to categories 1–6 can only receive up to five HS-DSCH channelization codes and therefore must use a power offset for the highest CQI values, whereas category 10 UEs are able to receive up to 15 codes.

The CQI values listed are sorted in ascending order, and the UE shall report the highest CQI for which transmission with parameters corresponding to the CQI results in a block error probability not exceeding 10%. The CQI values are chosen such that an increase in CQI by one step corresponds to approximately 1 dB increase in the instantaneous carrier-to-interference ratio on an AWGN channel. Measurements on the common pilot form the basis for the CQI. The CQI represents the instantaneous channel conditions in a predefined three-slot interval ending one slot prior to the CQI transmission. Specifying which interval the CQI relates to allows the Node B to track changes in the channel quality between the CQI reports by using the power control commands for the associated downlink (F-) DPCH, as described below. The timing of the CQI reports and the earliest possible time the report can be used for scheduling purposes is illustrated in Figure 8.21.

The rate of the channel-quality reporting is configurable in the range of one report per 2–160 ms. The CQI reporting can also be switched off completely.

**Table 8.2** Example of CQI Reporting for Two Different UE Categories [4]

| CQI Value | Transport-Block Size | | Modulation Scheme | Number of HS-DSCH Channelization Codes | | Power Offset (dB) | |
|---|---|---|---|---|---|---|---|
| | Categories 1–6 | Category 10 | | Categories 1–6 | Category 10 | Categories 1–6 | Category 10 |
| 0 | N/A | | | Out of range | | | |
| 1 | 137 | | QPSK | 1 | | 0 | |
| 2 | 173 | | QPSK | 1 | | 0 | |
| 3 | 233 | | QPSK | 1 | | 0 | |
| 4 | 317 | | QPSK | 1 | | 0 | |
| 5 | 377 | | QPSK | 1 | | 0 | |
| 6 | 461 | | QPSK | 1 | | 0 | |
| 7 | 650 | | QPSK | 2 | | 0 | |
| 8 | 792 | | QPSK | 2 | | 0 | |
| 9 | 931 | | QPSK | 2 | | 0 | |
| 10 | 1262 | | QPSK | 3 | | 0 | |
| 11 | 1483 | | QPSK | 3 | | 0 | |
| 12 | 1742 | | QPSK | 3 | | 0 | |
| 13 | 2279 | | QPSK | 4 | | 0 | |
| 14 | 2583 | | QPSK | 4 | | 0 | |
| 15 | 3319 | | QPSK | 5 | | 0 | |
| 16 | 3565 | | 16-QAM | 5 | | 0 | |
| 17 | 4189 | | 16-QAM | 5 | | 0 | |
| 18 | 4664 | | 16-QAM | 5 | | 0 | |
| 19 | 5287 | | 16-QAM | 5 | | 0 | |
| 20 | 5887 | | 16-QAM | 5 | | 0 | |
| 21 | 6554 | | 16-QAM | 5 | | 0 | |
| 22 | 7168 | | 16-QAM | 5 | | 0 | |
| 23 | 7168 | 9719 | 16-QAM | 5 | 7 | −1 | 0 |
| 24 | 7168 | 11,418 | 16-QAM | 5 | 8 | −2 | 0 |
| 25 | 7168 | 14,411 | 16-QAM | 5 | 10 | −3 | 0 |
| 26 | 7168 | 17,237 | 16-QAM | 5 | 12 | −4 | 0 |
| 27 | 7168 | 21,754 | 16-QAM | 5 | 15 | −5 | 0 |
| 28 | 7168 | 23,370 | 16-QAM | 5 | 15 | −6 | 0 |
| 29 | 7168 | 24,222 | 16-QAM | 5 | 15 | −7 | 0 |
| 30 | 7168 | 25,558 | 16-QAM | 5 | 15 | −8 | 0 |

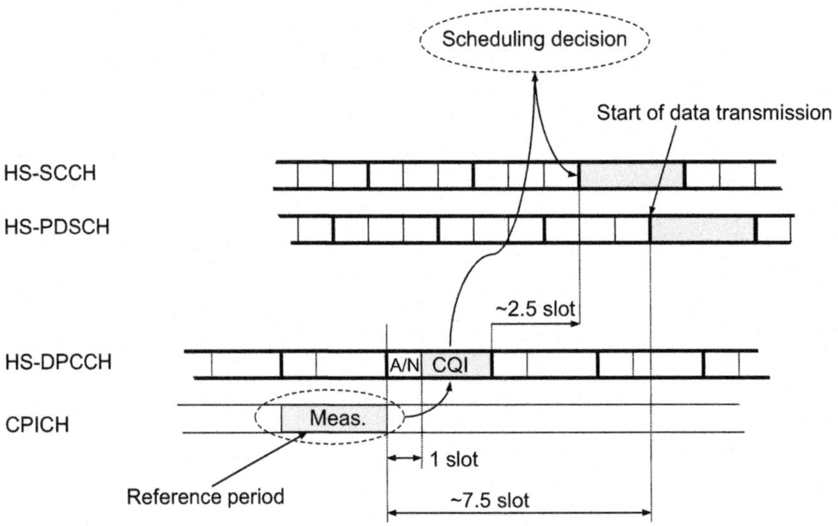


**FIGURE 8.21 Timing relation for the CQI reports.**

As the scheduling and rate-adaptation algorithms are vendor specific, it is possible to perform rate control based on other criteria than the UE reports as well, either alone or in combination. Using the transmit power level of the associated DPCH is one such possibility, where a high transmit power indicates unfavorable channel conditions and a low DPCH transmit power indicates favorable conditions. Because the power level is a relative measure of the channel quality and does not reflect an absolute subjective channel quality, this technique is advantageously combined with infrequent UE quality reports. The UE reports provide an absolute quality status, and the transmission power of the power-controlled DPCH can be used to update this quality report between the reporting instances. This combined scheme works quite well and can significantly reduce the frequency of the UE CQI reports as long as the DPCH is not in soft handover. In soft handover, the transmit power of the different radio links involved in the soft handover are power controlled such that the combined received signal is of sufficient quality. Consequently, the DPCH transmit power at the serving HS-DSCH cell does not necessarily reflect the perceived UE channel quality. Hence, more frequent UE quality reports are typically required in soft handover scenarios.

### 8.3.7 DOWNLINK CONTROL SIGNALING: HS-SCCH

The HS-SCCH, sometimes referred to as the *shared control channel*, is a shared downlink physical channel that carries control signaling information needed for a UE to be able to properly despread, demodulate, and decode the HS-DSCH.

In each 2 ms interval corresponding to one HS-DSCH TTI, one HS-SCCH carries physical-layer signaling to a single UE. As HSDPA supports HS-DSCH transmission to multiple users in parallel by means of code multiplexing (see Section 8.1.1), multiple HS-SCCH may be needed in a cell. According to the specification, a UE should be able to decode four HS-SCCHs in parallel. However, more than four HS-SCCHs can be configured within a cell, although the need for this is rare.

HS-SCCH uses a spreading factor of 128 and has a time structure based on a subframe of length 2 ms that is the same length as the HS-DSCH TTI. The following information is carried on the HS-SCCH:

- The HS-DSCH transport format, consisting of:
  - HS-DSCH channelization-code set [7 bits];
  - HS-DSCH modulation scheme, QPSK/16-QAM [1 bit]; and
  - HS-DSCH transport-block size information [6 bits].
- Hybrid-ARQ-related information, consisting of:
  - hybrid-ARQ process number [3 bits];
  - redundancy version [3 bits]; and
  - new-data indicator [1 bit].
- A UE ID that identifies the UE for which the HS-SCCH information is intended [16 bits]. As will be described below, the UE ID is not explicitly transmitted but implicitly included in the CRC calculation and HS-SCCH channel coding.

As described in Section 8.2.4, the HS-DSCH transport block can take one of 254 different sizes. Each combination of channelization-code-set size and modulation scheme corresponds to a subset of these transport-block sizes, where each subset consists of 63 possible transport-block sizes. The 6-bit "HS-DSCH transport-block size information" indicates which one of the 63 possible transport-block sizes is actually used for the HS-DSCH transmission in the corresponding TTI. The transport-block sizes have been defined to make full use of code rates ranging from 1/3 to 1 for initial transmissions. For retransmissions, instantaneous code rates larger than one can be achieved by indicating "the transport-block size is identical to the previous transmission in this hybrid-ARQ process." This is indicated setting the "HS-DSCH transport-block size information" field to 111111. This is useful for additional scheduling flexibility, for example, to retransmit only a small amount of parity bit in case the latest CQI report indicates the UE "almost" was able to decode the original transmission.

Requirements on when different parts of the HS-SCCH information need to be available to the UE have affected the detailed structure of the HS-SCCH channel coding and physical-channel mapping. For UE complexity reasons, it is beneficial if the channelization-code set is known to the UE prior to the start of the HS-DSCH transmission. Otherwise, the UE would have to buffer the received signal on a sub-chip level prior to despreading or, alternatively, despread all potential HS-DSCH codes up to the maximum of 15 codes. Knowing the modulation scheme prior to the HS-DSCH subframe is also preferred as it allows for "on-the-fly" demodulation. On the other hand, the transport-block size and the hybrid-ARQ-related information are only needed at HS-DSCH decoding/soft combining, which can not start until the end of the HS-DSCH TTI. Thus, the HS-SCCH information is split into two parts:

- part 1 consisting of channelization-code set and modulation scheme [total of 8 bits]; and
- part 2 consisting of transport-block size and hybrid-ARQ-related parameters [total of 13 bits].

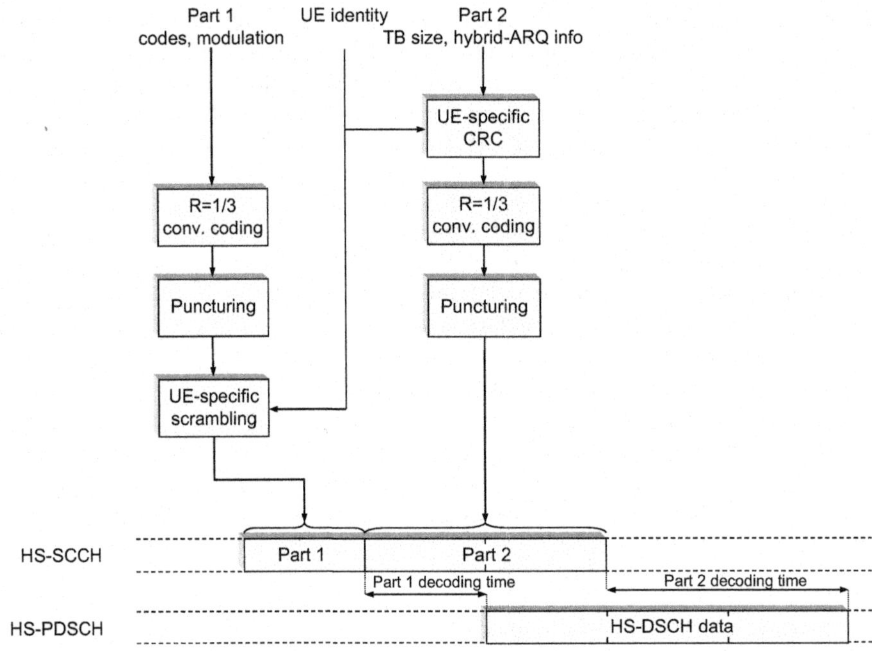


**FIGURE 8.22 HS-SCCH channel coding.**

The HS-SCCH coding, physical-channel mapping, and timing relation to the HS-DSCH transmission is illustrated in Figure 8.22. The HS-DSCH channel coding is based on rate-1/3 convolutional coding, carried out separately for part 1 and part 2. Part 1 is coded and rate matched to 40 bits to fit into the first slot of the HS-SCCH subframe. Before mapping to the physical channel, the coded part 1 is scrambled by a 40-bits UE-specific bit sequence. The sequence is derived from the 16-bits UE ID using rate-1/2 convolutional coding followed by puncturing. With the scheme of Figure 8.22, the part 1 information can be decoded after one slot of the HS-SCCH subframe. Furthermore, in case of more than one HS-SCCH, the UE can find the correct HS-SCCH from the soft metric of the channel decoder already after the first slot. One possible way for the UE to utilize the soft metric for determining which (if any) of the multiple HS-SCCHs carries control information for the UE is to form the log-likelihood ratio between the most likely code word and the second most likely code word for each HS-SCCH. The HS-SCCH with the largest ratio is likely to be intended for the UE and can be selected for further decoding of the part-2 information.

Part 2 is coded and rate matched to 80 bits to fit into the second and third slot of the HS-SCCH. Part 2 includes a UE-specific CRC for error detection. The CRC is calculated over all the information bits, both part 1 and part 2, as well as the UE identity. The identity is not explicitly transmitted, but by including its ID when calculating the CRC at the receiver, the UE can decide whether it was the intended recipient or not. If the transmission is intended for another UE, the CRC will not check.

In case of HS-DSCH transmission to a single UE in consecutive TTIs, the UE must despread the HS-SCCH in parallel to the HS-DSCH channelization codes. To reduce the number of required despreaders, the same HS-SCCH shall be used when HS-DSCH transmission is carried out in consecutive TTI. This implies that, when simultaneously receiving HS-DSCH, the UE only needs to despread a single HS-SCCH.

In order to avoid waste of capacity, the HS-SCCH transmit power should be adjusted to what is needed to reach the intended UE. Similar information used for rate control of the HS-DSCH, for example the CQI reports, can be used to power control the HS-SCCH.

#### 8.3.7.1 HS-SCCH types in releases 7–11

The set of HSPA features has expanded in releases 7–11, and several of the new features require additional signaling to be conveyed on the HS-SCCH. To keep backward compatibility with release 5 HSDPA, the HS-SCCH format standardized in release 5, as described in Section 8.3.7, was retained but renamed to "HS-SCCH type 1." Several other HS-SCCH types were then defined. The set of HS-SCCH types defined for release 11 is as follows:

- HS-SCCH type 1 is the original HS-SCCH type, as defined for release 5.
- HS-SCCH type 2 is intended for retransmissions as part of HS-SCCH less transmission operation. HS-SCCH less transmission was introduced as part of CPC and is described in more detail in Chapter 14.
- HS-SCCH type 3 is intended for release 7 dual branch MIMO, as described in Chapter 11.
- HS-SCCH type 4 is intended for release 11 four branch MIMO operation, as described in Chapter 11

All of the HS-SCCH types retain the basic three slot structure of the HS-SCCH. The difference between the HS-SCCH types is the information mapping and coding. HS-SCCH types 1, 2, and 4 can carry HS-SCCH orders (see Chapter 14).

### 8.3.8 DOWNLINK CONTROL SIGNALING: F-DPCH

As described in Section 8.2.1, for each UE for which HS-DSCH can be transmitted, there is also an associated downlink DPCH. In principle, if all data transmission, including RRC signaling, is mapped to the HS-DSCH, there is no need to carry any user data on the DPCH. Consequently, there is no need for downlink *Transport-Format Combination Indicator* (TFCI) or dedicated pilots on such a DPCH. In this case, the only use for the downlink DPCH in case of HS-DSCH transmission is to carry power control commands to the UE in order to adjust the uplink transmission power. This fact is exploited by the F-DPCH or *fractional DPCH*, introduced in release 6 as a means to reduce the amount of downlink channelization codes used for dedicated channels. Instead of allocating one DPCH with spreading factor 256 for the sole purpose of transmitting one power control command per slot, the

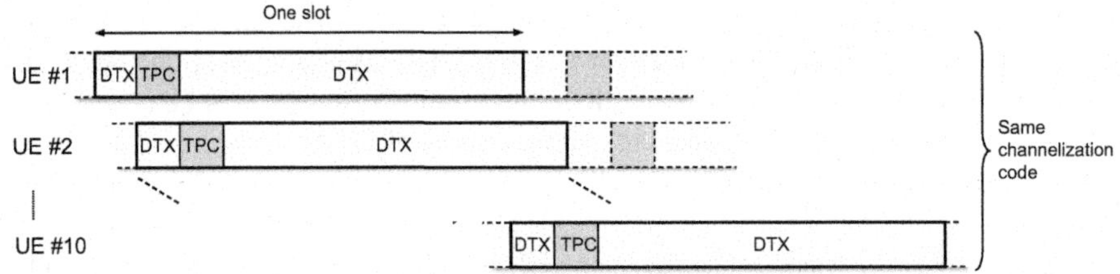


**FIGURE 8.23 Fractional DPCH (F-DPCH), introduced in release 6.**

F-DPCH allows up to 10 UEs to share a single channelization code for this purpose. In essence, the F-DPCH is a slot format supporting TPC bits only. Two TPC bits (one QPSK symbol) are transmitted in 1/10th of a slot, using a spreading factor 256, and the rest of the slot remains unused. By setting the downlink timing of multiple UEs appropriately, as illustrated in Figure 8.23, up to 10 UEs can then share a single channelization code. This can also be seen as time-multiplexing power control commands to several users on one channelization code.

## 8.3.9 UPLINK CONTROL SIGNALING: HS-DPCCH

For operation of the hybrid-ARQ protocol and to provide the Node B with knowledge about the instantaneous downlink channel conditions, uplink control signaling is required. This signaling is carried on an additional new uplink physical channel, the HS-DPCCH, using a channelization code separate from the conventional uplink DPCCH. The use of a separate channelization code for the HS-DPCCH makes the HS-DPCCH "invisible" to non-HSDPA-capable base stations and allows for the uplink being in soft handover even if not all Node Bs in the active set support HSDPA.

The HS-DPCCH uses a spreading factor of 256 and is transmitted in parallel with the other uplink channels, as illustrated Figure 8.24. To reduce the uplink peak-to-average ratio, the channelization code used for HS-DPCCH and if the HS-DPCCH is mapped to the I or Q branch of this code depends on the maximum number of

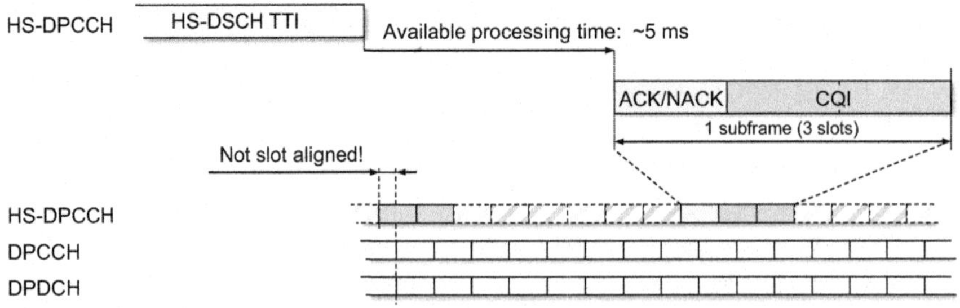


**FIGURE 8.24 Basic structure of uplink signaling with IQ/code-multiplexed HS-DPCCH.**

**Table 8.3** Channelization Code Used for HS-DPCCH in Release 5

| Nmax-dpdch | Channelization Code $C_{ch}$ | I/Q Mapping |
|---|---|---|
| 1 | $C_{ch,256,64}$ | I-branch |
| 2,4,6 | $C_{ch,256,1}$ | Q-branch |
| 3,5 | $C_{ch,256,32}$ | I-branch |

DPDCHs used by the transport-format combination set configured for the UE. The channelization code configuration for release 5 is shown in Table 8.3.

As the HS-DPCCH spreading factor is 256, the HS-DPCCH allows for a total of 30[4] channel bits per 2 ms subframe (three slots). The HS-DPCCH information is structured in such a way that the hybrid-ARQ acknowledgement is transmitted in the first slot of the subframe whereas the channel-quality indication is transmitted in the second and third slots; see Figure 8.24.

In order to minimize the hybrid-ARQ round-trip time, the HS-DPCCH transmission timing is not slot aligned to the other uplink channels. Instead, the HS-DPCCH timing is defined relative to the end of the subframe carrying the corresponding HS-DSCH data, as illustrated in Figure 8.24. The timing is such that there are approximately 7.5 slots (19,200 chips) of UE processing time available, from the end of the HS-DSCH TTI to the transmission of the corresponding uplink hybrid-ARQ acknowledgement. If the HS-DPCCH had been slot aligned to the uplink DPCCH, there would have been an uncertainty of one slot in the HS-DSCH/HS-DPCCH timing. This uncertainty would have reduced the processing time available for the UE/ Node B by one slot.

Because of the alignment between the uplink HS-DPCCH and the downlink HS-DSCH, the HS-DPCCH will not necessarily be slot aligned with the uplink DPDCH/DPCCH. However, note that the HS-DPCCH is always aligned to the uplink DPCCH/DPDCH on a 256-chip basis in order to keep uplink orthogonality. As a consequence, the HS-DPCCH cannot have a completely fixed transmit timing relative to the received HS-DSCH. Instead, the HS-DPCCH transmit timing varies in an interval 19,200 chips to 19,200 + 255 chips. Note that CQI and ACK/NACK are transmitted independently of each other. In subframes where no ACK/NACK or CQI are to be transmitted, nothing is transmitted in the corresponding HS-DPCCH field.

The hybrid-ARQ acknowledgement consists of a single information bit, ACK or NACK, indicating whether the HS-DSCH was correctly decoded (the CRC checked) or not. ACK or NACK is only transmitted in case the UE correctly received the HS-SCCH control signaling. If no HS-SCCH control signaling intended for the UE was detected, nothing is transmitted in the ACK/NACK field (DTX). This reduces the uplink load as only the UEs to which HS-DSCH data was actually sent in a TTI transmit an ACK/NACK on the uplink. The single-bit ACK is repetition coded into 10 bits to fit into the first slot of an HS-DPCCH subframe.

[4]Three slots × 2560 chips/slot/SF256 and BPSK modulation yield 30-channel bits.

Reliable reception of the uplink ACK/NACK requires a sufficient amount of energy. In some situations where the UE is power limited, it may not be possible to collect enough energy by transmitting the ACK/NACK over a single slot. Therefore, there is a possibility to configure the UE to repeat the ACK/NACK in $N$ subsequent ACK/NACK slots. Naturally, when the UE is configured to transmit repeated acknowledgements, it cannot receive HS-DSCH data in consecutive TTIs, as the UE would then not be able to acknowledge all HS-DSCH data. Instead there must be at least $N-1$ idle 2 ms subframes between each HS-DSCH TTI in which data is to be received. Examples when repetition of the acknowledgements can be useful are very large cells or in some soft handover situations. In soft handover, the uplink can be power controlled by multiple Node Bs. If any of the non-serving Node Bs has the best uplink, the received HS-DPCCH quality at the serving Node B may not be sufficient and repetition may therefore be necessary. The number of consecutive ACK/NACK slots used for a single hybrid-ARQ message, the so-called *N_acknack_transmit*, is configured via RRC signaling and supported values in release 5 are {1, 2, 3, 4}. The ACK/NACK repetition factor is consequently given by *N_acknack_transmit* – 1.

As mentioned earlier, the impact of ACK-to-NACK and NACK-to-ACK errors is different, leading to different requirements. In addition, the DTX-to-ACK error case also has to be handled. If the UE misses the scheduling information and the Node B misinterprets the DTX as ACK, data loss in the hybrid-ARQ will occur. An asymmetric decision threshold in the ACK/NACK detector should therefore preferably be used, as illustrated in Figure 8.25. Based on the noise variance at the ACK/NACK detector, the threshold can be computed to meet a certain DTX-to-ACK error probability, for example, $10^{-2}$, after which the transmission power of the ACK and NACK can be set to meet the remaining error requirements (ACK-to-NACK and NACK-to-ACK).

In release 6 of the WCDMA specifications, an enhancement to the ACK/NACK signaling has been introduced. In addition to the ACK and NACK, the UE may also transmit two additional code words, PRE and POST, on the HS-DPCCH. A UE configured to use the enhancement will transmit PRE and POST in the subframes preceding and succeeding, respectively, the ACK/NACK (unless these subframes were used by the ACK/NACK for other transport blocks). Thus, an ACK will cause a

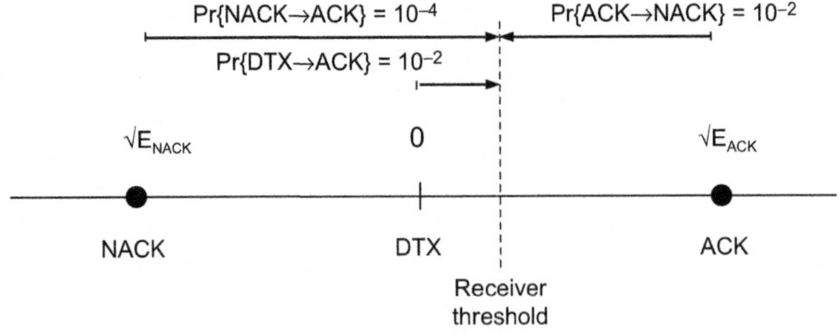


**FIGURE 8.25 Detection threshold for the ACK/NACK field of HS-DPCCH.**

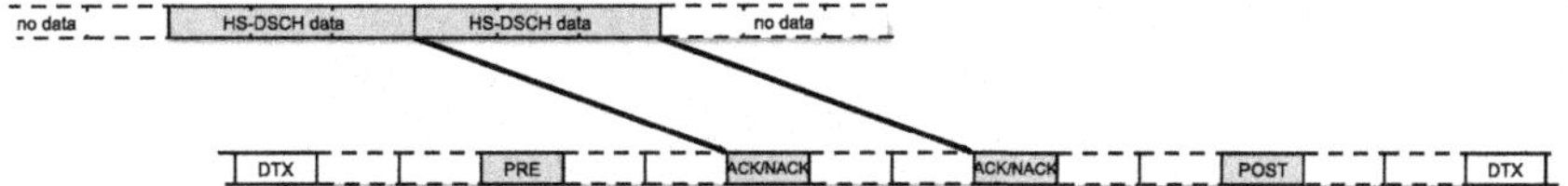


**FIGURE 8.26 Enhanced ACK/NACK using PRE and POST.**

transmission spanning multiple subframes and the power can therefore be reduced while maintaining the same ACK-to-NACK error rate (Figure 8.26).

The CQI consists of five information bits. A (20, 5) block code is used to code this information to 20 bits, which corresponds to two slots on the HS-DPCCH. Similarly to the ACK/NACK, repetition of the CQI field over multiple 2 ms subframes is possible and can be used to provide improved coverage. The number of consecutive subframes used for a single CQI message, referred to as *N_cqi_transmit*, is configured via RRC signaling, and supported values in release 5 are {1, 2, 3, 4}. How frequently the UE should transmit a new CQI is controlled by the so-called *CQI feedback cycle* parameter signaled via RLC. Supported values in release 5 are {0, 2, 4, 8, 10, 20, 40, 80, 160} ms, where a value zero indicates that no CQI reports are transmitted.

The power setting of the HS-DPCCH depends on RRC signaled parameters and what information is conveyed on the HS-DPCCH. Separate power settings for the ACK/NACK slot and the CQI slots can be used, and different power settings can be used for an ACK and a NACK. The HS-DPCCH gain factor, $\beta_{hs}$, is determined according to

$$\beta_{hs} = \beta_c \cdot A_{hs}, \tag{8.1}$$

where $\beta_c$ is the gain factor for DPCCH and the quantized amplitude value $A_{hs}$ is determined from an index $\Delta$ according to Table 8.4, where the index $\Delta$ depends on a number of RRC signaled parameters ($\Delta_{ACK}$, $\Delta_{NACK}$, and $\Delta_{CQI}$) according to Table 8.5. An increase in $\Delta$ by +1 results in 2 dB extra power allocated to HS-DPCCH. The power setting of HS-DPCCH is illustrated in Figure 8.27.

**Table 8.4** Quantized Amplitude Ratios $A_{hs}$ Supported in Release 5

| Index $\Delta$ | Quantized Amplitude Ratio $A_{hs} = \beta_{hs}/\beta_c$ | Quantized Power Offsets (dB) |
|---|---|---|
| 8 | 30/15 | 6.02 |
| 7 | 24/15 | 4.08 |
| 6 | 19/15 | 2.05 |
| 5 | 15/15 | 0 |
| 4 | 12/15 | –1.94 |
| 3 | 9/15 | –4.43 |
| 2 | 8/15 | –5.46 |
| 1 | 6/15 | –7.96 |
| 0 | 5/15 | –9.54 |

Table 8.5 Mapping From Information Element to Δ

| | HARQ message | Δ |
|---|---|---|
| HARQ Slot | ACK | $\Delta_{ACK}$ |
| | NACK | $\Delta_{NACK}$ |
| | POST or PRE | $\max(\Delta_{ACK}, \Delta_{NACK})$ |
| CQI slot | | $\Delta_{CQI}$ |

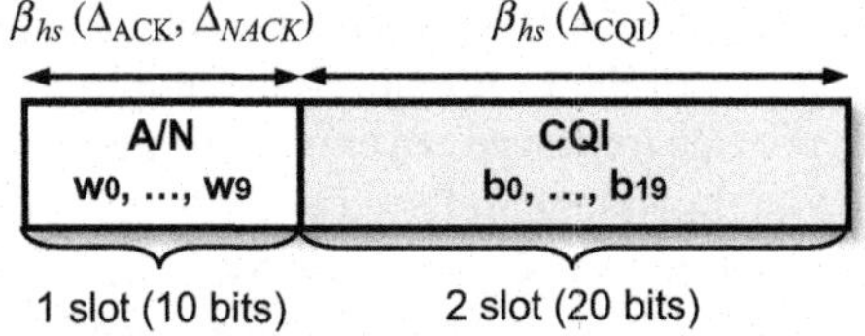


FIGURE 8.27 Illustration of HS-DPCCH power setting.

## 8.4 RELEASE 7 IMPROVED LAYER 2

Release 5 HSDPA, as well as release 6 HSUPA, reused the same RLC and MAC protocols as release 99. The release 99 RLC protocol was designed considering that the RNC controlled the transport-block sizes on the DCH and that there was no physical layer hybrid-ARQ. Data integrity was ensured by means of L2 ACKs of data and retransmission of data for which no ACK was received.

The principle of *acknowledged mode* (AM) operation is illustrated in Figure 8.28. The RLC in the RNC generates a PDU and passes it over Iub to the MAC-hs in an HSDPA Node B. The Node B schedules the MAC PDU and transfers it to the UE. The UE passes the PDU to the UE RLC, which generates an L2 ACK. The L2 ACK is transmitted via Uu to the Node B and via Iub to the RNC. The time that elapses between the RNC generating the RLC PDU and receiving the L2 ACK is twice the round-trip time between the RNC and UE.

When an RLC PDU fails, the UE RLC needs to buffer RLC PDUs until the missing PDU is retransmitted and received; for this purpose, memory is required in the UE. In order to avoid a so-called buffer overrun in the UE caused by receiving too many new RLC PDUs before a retransmission, the size of RLC PDUs is fixed and the transmit side is limited to transmitting a limited number of new RLC PDUs after the last PDU for which it received an ACK. The number of PDUs that can be transmitted since the last ACK is called the *RLC window size*. If the data rate at which RLC PDUs can be transmitted is such that the full RLC window can be transmitted in less time than is required for an ACK to reach the Node B, then so-called stalling occurs in which the transmitter has to stop transmitting new data and wait for each ACK to arrive. This situation is illustrated in Figure 8.29. In the figure, the round-trip time between the RNC and UE is, for the sake of illustration, five times the length of time

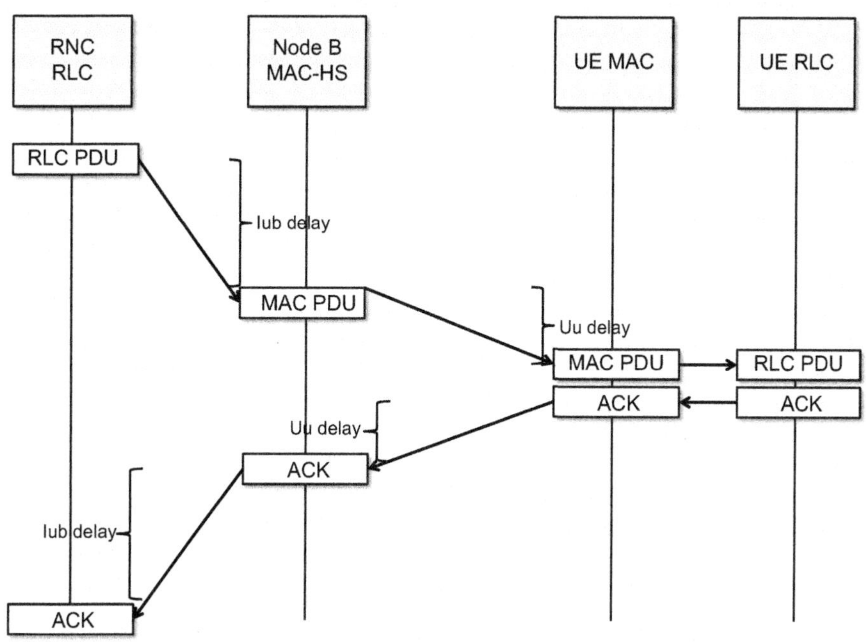


**FIGURE 8.28 Acknowledged mode RLC operation.**

taken to transmit an RLC PDU to the UE, and the UE's RLC buffer size is 3 PDUs. After transmitting 3 PDUs, the RLC still has not received an L2 ACK for the first PDU. Because it knows that the UE cannot buffer more than 3 PDUs but does not know if the first PDU has been successfully received, the RLC has to pause transmission until an ACK is received for the first PDU. Because of this protocol operation, the air interface may be capable of transmitting more PDUs than the *Acknowledged Mode* (AM) RLC will provide.

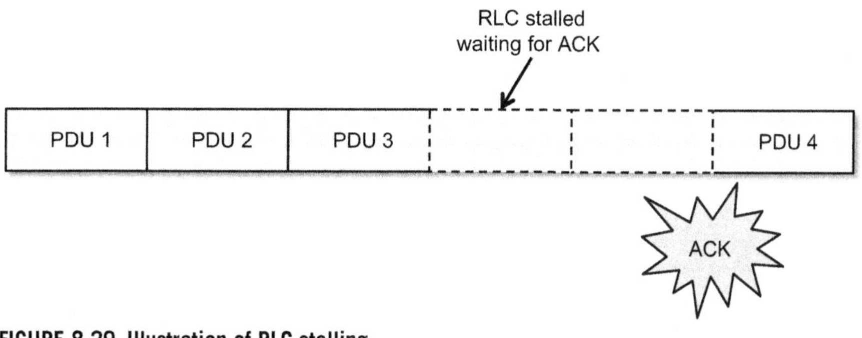


**FIGURE 8.29 Illustration of RLC stalling.**

With increasing data rates for HSDPA, the release 99 RLC protocol design leads to problems for a number of reasons. The fixed, one-size-fits-all RLC PDU size needs to be kept relatively small, because otherwise large RLC PDU sizes would lead to problems for small data packets that were significantly smaller than the RLC PDU size, as to build a full RLC PDU, small payloads would require padding in RLC, which would represent a significant overhead. For larger files though, the RLC PDU should ideally be large to avoid segmentation of the data into lots of small PDUs, incurring a protocol overhead. More seriously, because the RLC window size is a fixed number of PDUs, with high over the air bit rates but limited opportunities for reducing Iub latency for sending ACKs, RLC window stalling is likely to limit the achievable user throughput that would otherwise be possible with the more advanced air interface. As an example, if the RLC PDU size would be 40 bytes, the RLC window size 2047, and the RTT between UE and RNC 150 ms, then the RLC protocol would limit the throughput to less than 5 Mbit/s regardless of the air interface capability.

To exacerbate this problem, in release 5 the MAC is not able to segment RLC PDUs but only to concatenate them. Thus, the granularity that can be applied by the MAC scheduler in selecting transport-block sizes is limited by the RLC PDU size, and padding of MAC PDUs to account for the difference in transport-block size length and RLC PDU size was often needed, further increasing overhead. This is illustrated in Figure 8.30a. In this part of the figure, six fixed-size RLC PDUs are generated, and two MAC PDUs are shown. The size of the MAC PDUs varies according to the air interface conditions. The MAC can only fit whole RLC PDUs into a transport block. In each MAC PDU, there are some additional bits available that are

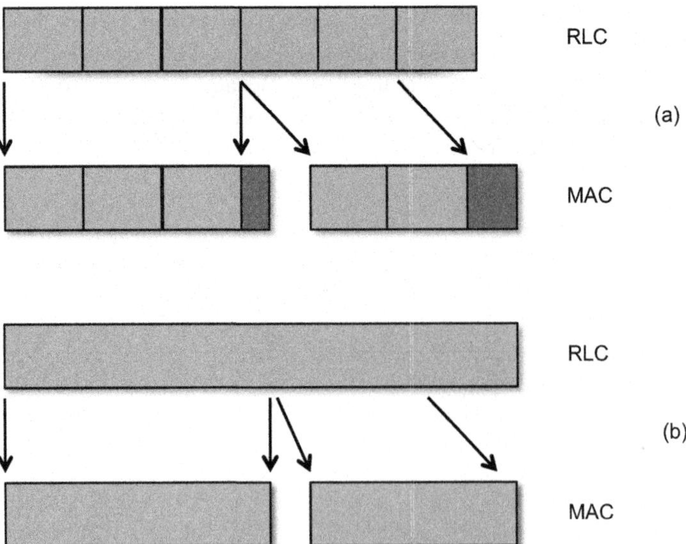


**FIGURE 8.30 Release 6 RLC and MAC operation, and improved L2 operation.**

less than the size of an RLC PDU, which have to be filled with padding bits. Thus, the L2 became the limiting factor in HSDPA systems.

The limitations caused by the L2 arose from the disconnect caused by moving some radio resource management and retransmission, such as scheduling and hybrid-ARQ, to the MAC scheduler while keeping the L2 RLC functionality in the RNC, and thus the solution was effectively to also move segmentation and reordering functionality to the Node B MAC. The L2 improvements consisted of two principal steps: flexible RLC PDU sizes and MAC segmentation. Flexible RLC sizes enable the RLC PDU length to be set appropriately for the RLC SDU payload. (An SDU is a unit of data that is input to a protocol; a PDU is a data block that is output. Thus, a protocol receives SDUs, carries out processing, and delivers PDUs, which are in turn SDUs for the next layer.) Small RLC packets can thereby be created for small SDUs, whereas for large SDUs, large RLC packets can be created. For example, a single RLC PDU can be created for an entire TCP segment of, for example, 1500 bytes.

Because the Node B MAC is responsible for adaptive modulation and coding as well as transport-block size selection, it makes sense for the Node B MAC to be able to segment RLC PDUs, and thus MAC segmentation and reassembly of RLC PDUs was also implemented.

MAC-level hybrid-ARQ already existed prior to the L2 enhancements and reduces the probability of an RLC retransmission of an entire RLC packet to become very small. Thus, there is no real need for RLC to limit the packet size it sends to MAC.

The flexible layer 2 is illustrated in Figure 8.30b, in which the RLC can be seen to send a single PDU of the required size, while the MAC segments the PDU according to the transport-block sizes available at the air interface and fills the transport blocks exactly, eliminating the need for padding.

This so-called flexible layer 2 structure in release 7, known as MAC-ehs in the downlink and MAC-i/is in the uplink, removed the RLC protocol bottleneck and enabled proper utilization of higher data rates, MIMO, and higher-order modulation.

With increasing uplink rates, a similar approach to the L2 was adopted in release 8 for uplink.

The improved layer 2 is an essential implementation for realizing HSPA user experience in CELL_DCH. For the CELL_FACH enhancements, use of the improved L2 RLC and MAC is mandatory.

## REFERENCES


[1] S. Parkvall, E. Englund, P. Malm, T. Hedberg, M. Persson, J. Peisa, WCDMA evolved-high-speed packet-data services, Ericsson Rev., No. 02, 2003, pp. 56–65. Telefonaktiebolaget LM Ericsson, Stockholm, Sweden.

[2] J. Sköld, M. Lundevall, S. Parkvall, M. Sundelin, Broadband data performance of third-generation mobile systems, Ericsson Rev., No. 01, 2005, pp. 1–10. Telefonaktiebolaget LM Ericsson, Stockholm, Sweden.

[3] 3rd Generation Partnership Project; Technical Specification Group Radio Access Network; UE Radio Access capabilities, 3GPP, 3GPP TS 25.306.

[4] 3rd Generation Partnership Project; Technical Specification Group Radio Access Network; Physical layer procedures (FDD), 3GPP, 3GPP TS 25.214.

## FURTHER READING

3rd Generation Partnership Project; Technical Specification Group Radio Access Network; High Speed Downlink Packet Access: Physical Layer Aspects (Release 5), 3GPP, 3GPP TR 25.858.

3rd Generation Partnership Project; Technical Specification Group Radio Access Network; High Speed Downlink Packet Access (HSDPA); Overall Description; Stage 2, 3GPP, 3GPP TS 25.308.

P. Frenger, S. Parkvall, E. Dahlman, Performance comparison of HARQ with chase combining and incremental redundancy for HSDPA, Proceedings of the IEEE Vehicular Technology Conference, Atlantic City, NJ, USA, October 2001, pp. 1829–1833.

Minutes of HSDPA Simulation Ad-hoc, Document R4-040770, 3GPP TSG-RAN WG4 meeting 33, Shin-Yokohama, Japan, November 2004.

X. Zhaoji, B. Sébire, Impact of ACK/NACK signalling errors on high speed uplink packet access (HSUPA), IEEE International Conference on Communications, Vol. 4, May 2005, pp. 2223–2227, 16–20.

CHAPTER

# High-speed uplink packet access

## CHAPTER OUTLINE



**HSPA Evolution: The Fundamentals for Mobile Broadband. DOI: 10.1016/B978-0-08-099969-2.00009-0**

*Enhanced Uplink,* also known as *High-Speed Uplink Packet Access* (HSUPA), was introduced in WCDMA release 6.[1] It provides uplink improvements in uplink in terms of higher uplink data rates, reduced latency, and improved system capacity and is therefore a natural complement to the downlink enhancements introduced with HSDPA (see Chapter 8). Together, HSDPA and HSUPA are commonly referred to as *High-Speed Packet Access* (HSPA). This chapter describes release 6 HSUPA features. Subsequent chapters then elaborate on additional uplink enhancements introduced in later releases.

## 9.1 OVERVIEW

At the core of HSUPA are two basic technologies also used for HSDPA: fast scheduling and fast hybrid-ARQ with soft combining. For similar reasons as for HSDPA, HSUPA also introduces a short 2 ms uplink TTI. These enhancements are implemented in WCDMA through a new transport channel, the *Enhanced Dedicated Channel* (E-DCH).

Although the same technologies are used both for HSDPA and HSUPA, there are fundamental differences between downlink and uplink transmission directions, which have affected the details of the features:

- In the downlink, the shared resource is transmission power and the code space, both of which are located in *one central* node, the Node B. In the uplink, the shared resource is the amount of allowed uplink interference, which depends on the transmission power of *multiple distributed* nodes, the UEs.
- The scheduler and the transmission buffers are located in the same node in the downlink. In contrast, in the uplink the scheduler is located in the Node B while the data buffers are distributed among the UEs. Hence, the UEs need to signal buffer status information to the scheduler.
- The WCDMA uplink, also with HSUPA, is inherently non-orthogonal and subject to interference between uplink transmissions within the same cell. This is in contrast to the downlink, where different channels within the same cell are *orthogonal* at least at the transmitter side (the radio channel will partly destroy the orthogonality). Fast power control is therefore essential for the

[1]In this book, we use the term HSUPA.

uplink to handle the near–far problem.[2] The E-DCH is transmitted with a power offset relative to the power-controlled uplink control channel (DPCCH), and by adjusting the maximum allowed power offset, the scheduler can control the E-DCH data rate. This is in contrast to HSDPA, where a (more or less) constant transmission power with rate adaptation is used.

- Soft handover is supported by the E-DCH. *Receiving* data from a UE in multiple cells is fundamentally beneficial as it provides diversity, while transmission from multiple cells in case of HSDPA is cumbersome and with questionable benefits, as discussed in the previous chapter. Soft handover also implies *power control by multiple cells*, which is necessary to limit the amount of interference generated in neighboring cells and to maintain backward compatibility and coexistence with UEs not using the E-DCH for data transmission.
- In the downlink, higher-order modulation, which trades power efficiency for bandwidth efficiency, is useful to provide high data rates in some situations, for example, when the scheduler has assigned a small number of channelization codes for a transmission but the amount of available transmission power is relatively high. The situation in the uplink is different; there is no need to share channelization codes between users, and the channel coding rates are therefore typically lower than for the downlink. Hence, unlike the downlink, higher-order modulation was seen less useful in the release 6 time frame uplink deployment scenarios and was therefore not part of the first release of HSUPA.[3]

With these differences in mind, the basic principles behind HSUPA can be discussed.

### 9.1.1 SCHEDULING

For HSUPA, the scheduler is a key element, controlling *when* and *at what data rate* the UE is allowed to transmit. The higher the data rate a UE is using, the higher the UE's received power at the Node B must be to maintain the $E_b/N_0$ required for successful demodulation. By increasing the transmission power, the UE can transmit at a higher data rate. However, because of the non-orthogonal uplink, the received power from one UE represents interference for other UEs. Hence, the shared resource for HSUPA is the amount of tolerable interference in the cell. If the interference level is too high, some transmissions in the cell, control channels, and non-scheduled uplink transmissions may not be received properly. On the other hand, a too low interference level may indicate that UEs are artificially throttled and the full system capacity not exploited. Therefore, HSUPA relies on a scheduler to give users with data transmit permission to use an as high data rate as possible without exceeding the maximum tolerable interference level in the cell.

[2]The near–far problem describes the problem of detecting a weak user, located far from the transmitter, when a user close to the transmitter is active. Power control ensures the signals are received at a similar strength, thereby enabling detection of both users' transmissions.

[3]Uplink higher-order modulation is introduced in release 7 and is relevant for some small cell scenarios; see Chapter 10 for further details.

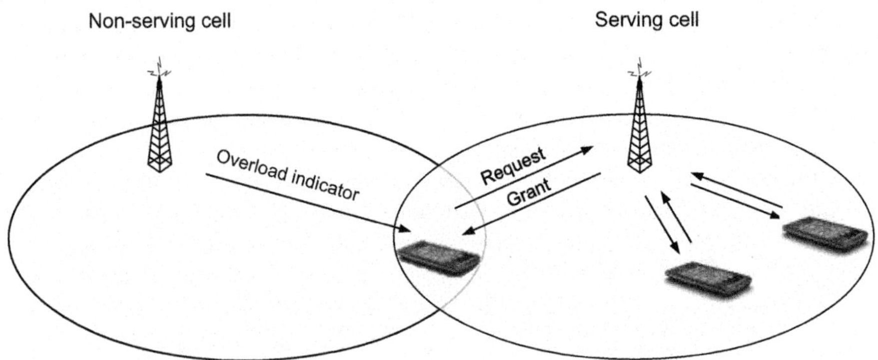


**FIGURE 9.1 HSUPA scheduling framework.**

Unlike HSDPA, where the scheduler and the transmission buffers both are located in the Node B, the data to be transmitted resides in the UEs for the uplink case. At the same time, the scheduler is located in the Node B to coordinate different UEs' transmission activities in the cell. Hence, a mechanism for communicating the scheduling decisions to the UEs and to provide buffer information from the UEs to the scheduler is required. The scheduling framework for HSUPA is based on *scheduling grants* sent by the Node B scheduler to control the UE transmission activity and *scheduling requests* sent by the UEs to request resources. The scheduling grants control the maximum allowed E-DCH-to-pilot power ratio the UE may use; a larger grant implies the UE may use a higher data rate but also contributes more to the interference level in the cell. Based on measurements of the (instantaneous) interference level, the scheduler controls the scheduling grant in each UE to maintain the interference level in the cell at a desired target (Figure 9.1).

In HSDPA, typically a single user is addressed in each TTI. For HSUPA, the implementation-specific uplink scheduling strategy in most cases schedules multiple users in parallel. The reason is the significantly smaller transmit power of a UE compared to a Node B: a single UE typically cannot utilize the full cell capacity on its own.

Inter-cell interference also needs to be controlled. Even if the scheduler has allowed a UE to transmit at a high data rate based on an acceptable inter-cell interference level, this may cause non-acceptable interference to neighboring cells. Therefore, in soft handover, the *serving cell* has the main responsibility for the scheduling operation, but the UE monitors scheduling information from all cells with which the UE is in soft handover. The non-serving cells can request all their non-served users to lower their E-DCH data rate by transmitting an *overload indicator* in the downlink. This mechanism ensures a stable network operation.

Fast scheduling allows for a more relaxed connection admission strategy. A larger number of bursty high-rate packet-data users can be admitted to the system as the scheduling mechanism can handle the situation when multiple users need to transmit in parallel. If this creates an unacceptably high interference level in the system, the scheduler can rapidly react and restrict the data rates they may use. Without fast scheduling, the admission control would have to be more conservative and reserve a margin in the system in case of multiple users transmitting simultaneously.

## 9.1.2 HYBRID-ARQ WITH SOFT COMBINING

Fast hybrid-ARQ with soft combining is used by HSUPA for basically the same reason as HSDPA – to provide robustness against occasional transmission errors. A similar scheme as for HSDPA is used. For each transport block received in the uplink, a single bit is transmitted from the Node B to the UE to indicate successful decoding (ACK) or to request a retransmission of the erroneously received transport block (NACK).

One main difference compared to HSDPA stems from the use of soft handover in the uplink. When the UE is in soft handover, this implies that the hybrid-ARQ protocol is *terminated in multiple cells*. Consequently, in many cases, the transmitted data may be successfully received in some Node Bs but not in others. From a UE perspective, it is sufficient if at least one Node B successfully receives the data. Therefore, in soft handover, all involved Node Bs attempt to decode the data and transmit an ACK or a NACK. If the UE receives an ACK from at least one of the Node Bs, the UE considers the data to be successfully received.

Hybrid-ARQ with soft combining can be exploited not only to provide robustness against unpredictable interference but also to improve the link efficiency to increase capacity and/or coverage. One possibility to provide a data rate of $x$ Mbit/s is to transmit at $x$ Mbit/s and set the transmission power to target a low error probability (in the order of a few percent) in the first transmission attempt. Alternatively, the same resulting data rate can be provided by transmitting using $n$ times higher data rate at an unchanged transmission power and using multiple hybrid-ARQ retransmissions. From the discussion in Chapter 5, this approach on average results in a lower cost per bit (a lower $E_b/N_0$) than the first approach. The reason is that, on average, less than $n$ transmissions will be used. This is sometimes known as *early termination gain* and can be seen as implicit rate adaptation. Additional coded bits are only transmitted when necessary. Thus, the code rate after retransmissions is determined by what was needed by the instantaneous channel conditions. This is exactly what rate adaptation also tries to achieve, the main difference being that rate adaptation tries to find the correct code rate prior to transmission. The same principle of implicit rate adaptation can also be used for HS-DSCH in the downlink to improve the link efficiency.

## 9.1.3 ARCHITECTURE

For efficient operation, the scheduler should be able to exploit rapid variations in the interference level and the channel conditions. Hybrid-ARQ with soft combining also benefits from rapid retransmissions as this reduces the cost of retransmissions. These two functions should therefore reside close to the radio interface. As a result, and for similar reasons as for HSDPA, the scheduling and hybrid-ARQ functionalities of HSUPA are located in the Node B. Furthermore, also similar to the HSDPA design, it is preferable to keep all radio-interface layers above MAC intact. Hence, ciphering, admission control, etc. is still under the control of the RNC. This also allows for a smooth introduction of HSUPA in selected areas; in cells not supporting E-DCH transmissions, channel switching can be used to map the user's data flow onto the DCH instead.

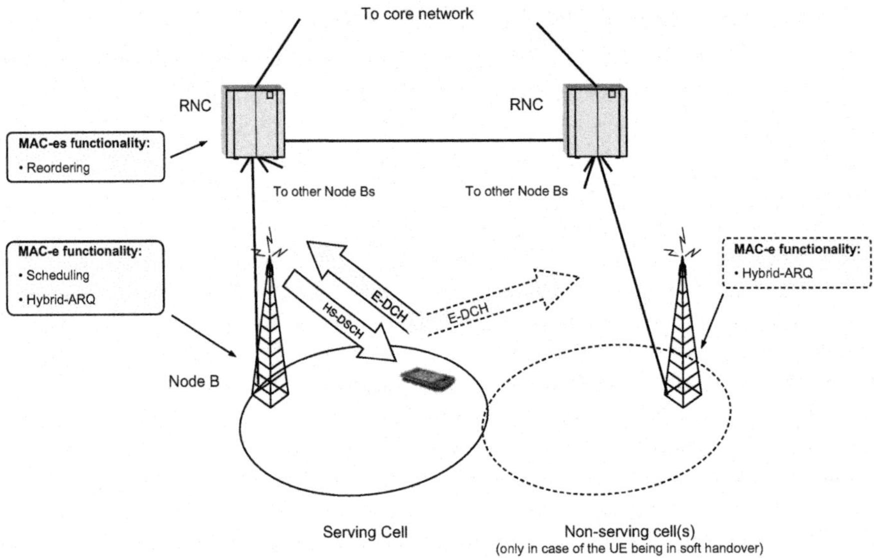


**FIGURE 9.2 The architecture with E-DCH (and HS-DSCH) configured.**

Following the HSDPA design philosophy, a new MAC entity, the *MAC-e*, is introduced in the UE and Node B. In the Node B, the MAC-e is responsible for support of fast hybrid-ARQ retransmissions and scheduling, while in the UE, the MAC-e is responsible for selecting the data rate within the limits set by the scheduler in the Node B MAC-e. In release 8, the MAC-e was enhanced to better support even higher data rates; see Chapter 8.

When the UE is in soft handover with multiple Node Bs, different transport blocks may be successfully decoded in different Node Bs. Consequently, one transport block may be successfully received in one Node B while another Node B is still involved in retransmissions of an earlier transport block. Therefore, to ensure in-sequence delivery of data blocks to the RLC protocol, a reordering functionality is required in the RNC in the form of a new MAC entity, the MAC-es. In soft handover, multiple MAC-e entities are used per UE as the data is received in multiple cells. However, the MAC-e in the serving cell has the main responsibility for the scheduling; the MAC-e in a non-serving cell is mainly handling the hybrid-ARQ protocol (Figure 9.2).

## 9.2 DETAILS OF HSUPA

To support uplink scheduling and hybrid-ARQ with soft combining in WCDMA, a new transport-channel type, the *Enhanced Dedicated Channel* (E-DCH) has been introduced in release 6. The E-DCH can be configured simultaneously with one or several DCHs. Thus, high-speed packet-data transmission on the E-DCH can occur at the same time as services using the DCH from the same UE.

A low delay is one of the key characteristics of HSUPA and is required for efficient packet-data support. Therefore, a short TTI of 2 ms is supported by the E-DCH to allow for rapid adaptation of transmission parameters and reduction of the end-user delays associated with packet-data transmission. Not only does this reduce the cost of a retransmission, but the transmission time for the initial transmission is also reduced. Physical-layer processing delay is typically proportional to the amount of data to process, and the shorter the TTI, the smaller the amount of data to process in each TTI for a given data rate. At the same time, in deployments with relatively modest data rates, for example, in large cells, a longer TTI may be beneficial as the payload in a 2 ms TTI can become unnecessarily small and the associated relative overhead too large. Hence, the E-DCH supports two TTI lengths, 2 and 10 ms, and the network can configure the appropriate value. In principle, different UEs can be configured with different TTIs.

The E-DCH is mapped to a set of uplink channelization codes known as *E-DCH Dedicated Physical Data Channels* (E-DPDCHs). Depending on the instantaneous data rate, the number of E-DPDCHs and their spreading factors are both varied.

Simultaneous transmission of E-DCH and DCH is possible, as discussed above. Backward compatibility requires the E-DCH processing to be invisible to a Node B not supporting HSUPA. This has been solved by separate processing of the DCH and E-DCH and mapping to different channelization code sets, as illustrated in Figure 9.3.

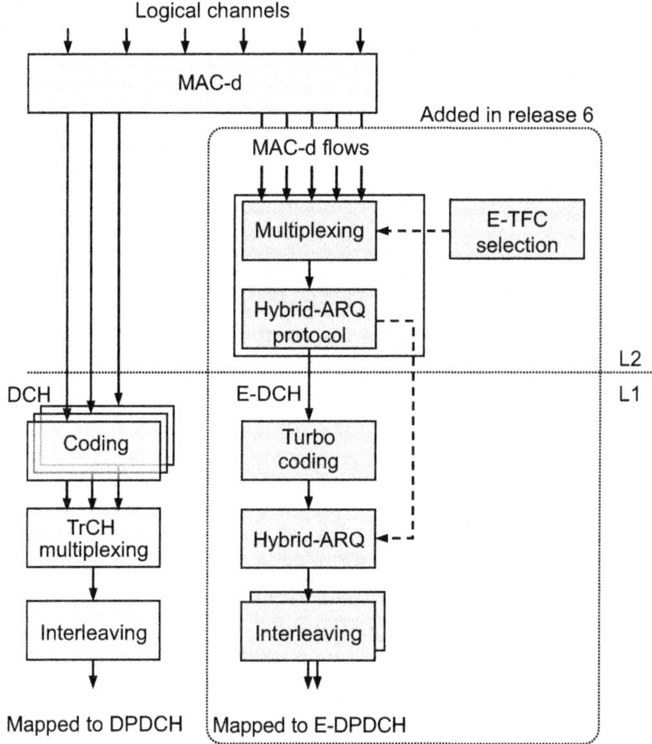


**FIGURE 9.3 Separate processing of E-DCH and DCH.**

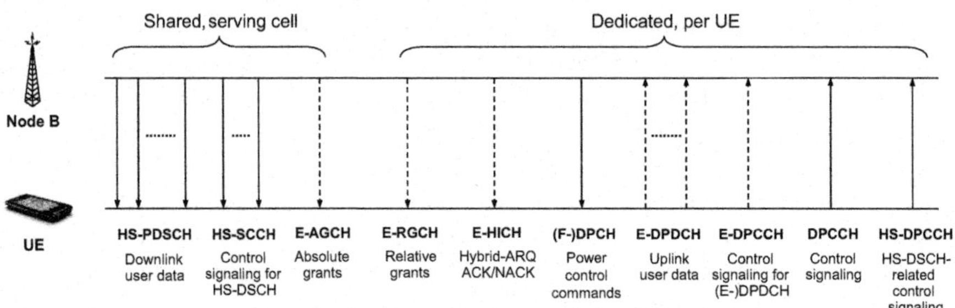


**FIGURE 9.4 Overall channel structure with HSDPA and HSUPA. The new channels introduced as part of HSUPA are shown with dashed lines.**

If the UE is in soft handover with multiple cells, of which some do not support HSUPA, the E-DCH transmission is invisible to these cells. This allows for a gradual introduction of HSUPA in an existing network, an aspect that was important at the time of design with a large number of WCDMA networks already deployed. An additional benefit with the structure is that it simplifies the introduction of the 2 ms TTI and also provides greater freedom in the selection of hybrid-ARQ processing.

Downlink control signaling is necessary for the operation of the E-DCH. The downlink, as well as uplink, control channels used for E-DCH support are illustrated in Figure 9.4, together with the channels used for HSDPA.

Obviously, the Node B needs to be able to request retransmissions from the UE as part of the hybrid-ARQ mechanism. This information, the ACK/NACK, is sent on a new downlink dedicated physical channel, the *E-DCH Hybrid-ARQ Indicator Channel* (E-HICH). Each UE with E-DCH configured receives one E-HICH of its own from each of the cells that the UE is in soft handover with.

Scheduling grants, sent from the scheduler to the UE to control when and at what data rate the UE is transmitting, can be sent to the UE using the shared *E-DCH Absolute Grant Channel* (E-AGCH). The E-AGCH is sent from the serving cell only as this is the cell having the main responsibility for the scheduling operation and is received by all UEs with an E-DCH configured. In addition, scheduling grant information can also be conveyed to the UE through an *E-DCH Relative Grant Channel* (E-RGCH). The E-AGCH is typically used for large changes in the data rate, while the E-RGCH is used for smaller adjustments during an ongoing data transmission. This is further elaborated upon in the discussion on scheduling operation below.

Since the uplink by design is non-orthogonal, fast closed-loop power control is necessary to address the near–far problem. The E-DCH is no different from any other uplink channel and is therefore power controlled in the same way as other uplink channels. The Node B measures the received signal-to-interference ratio and sends power control commands in the downlink to the UE to adjust the DPCCH transmission power. Power control commands can be transmitted using DPCH or, to save channelization codes, the *fractional DPCH*, (F-DPCH).

In the uplink, control signaling is required to provide the Node B with the necessary information to be able to demodulate and decode the data transmission. Even though, in principle, the serving cell could have this knowledge as it has issued the scheduling grants, the non-serving cells in soft handover clearly do not have this information. Furthermore, as discussed below, the E-DCH also supports non-scheduled transmissions. Hence, there is a need for out-band control signaling in the uplink, and the *E-DCH Dedicated Physical Control Channel* (E-DPCCH) is used for this purpose.

## 9.2.1 MAC-E AND PHYSICAL LAYER PROCESSING

Similar to HSDPA, short delays and rapid adaptation are important aspects of the HSUPA. This is implemented by introducing the MAC-e, a new entity in the Node B responsible for scheduling and hybrid-ARQ protocol operation. The physical layer is also enhanced to provide the necessary support for a short TTI and for soft combining in the hybrid-ARQ mechanism.

In soft handover, uplink data can be received in multiple Node Bs. Consequently, there is a need for a MAC-e entity in each of the involved Node Bs to handle the hybrid-ARQ protocol. The MAC-e in the serving cell is, in addition, responsible for handling the scheduling operation.

To handle the HSUPA processing in the UE, there is also a MAC-e entity in the UE. This can be seen in Figure 9.5, where the HSUPA processing in the UE is illustrated. The MAC-e in the UE consists of MAC-e multiplexing, transport format selection, and the protocol parts of the hybrid-ARQ mechanism.

Mixed services, for example, simultaneous file upload and VoIP, are supported. Hence, as there is only a single E-DCH transport channel, data from multiple MAC-d flows can be multiplexed through MAC-e multiplexing. The different services are in this case typically transmitted on different MAC-d flows as they may have different quality-of-service requirements.

Only the UE has accurate knowledge about the buffer situation and power situation in the UE at the time of transmission of a transport block in the uplink. Hence, the UE is allowed to autonomously select the data rate or, strictly speaking, the *E-DCH Transport Format Combination* (E-TFC). Naturally, the UE needs to take the scheduling decisions into account in the transport format selection; the scheduling decision represents an upper limit of the data rate the UE is not allowed to exceed. However, it may well use a lower data rate, for example, if the transmit power does not support the scheduled data rate. E-TFC selection, including MAC-e multiplexing, is discussed further in conjunction with scheduling.

The hybrid-ARQ protocol is similar to the one used for HSDPA, that is, multiple stop-and-wait hybrid-ARQ processes operate in parallel. There is one major difference though – when the UE is in soft handover with several Node Bs, the hybrid-ARQ protocol is terminated in multiple nodes.

Physical layer processing is straightforward and has several similarities with the HS-DSCH physical layer processing. From the MAC-e in the UE, data is passed to

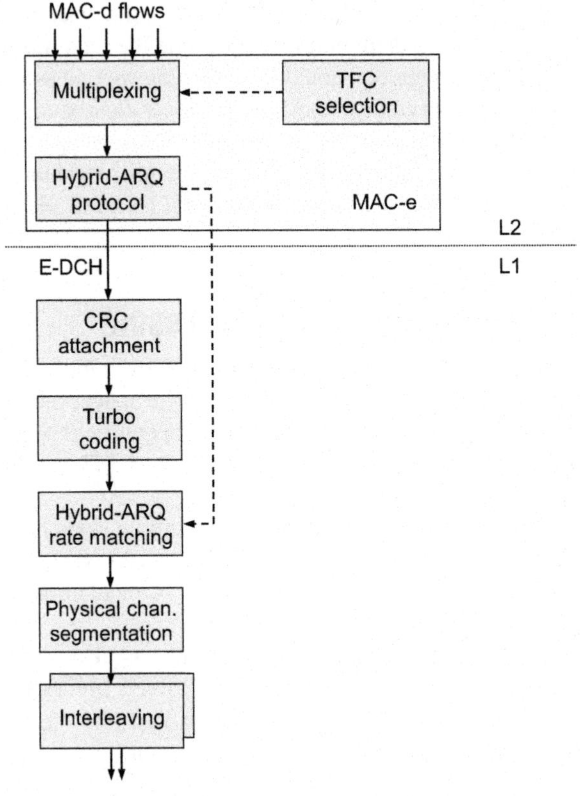


**FIGURE 9.5 MAC-e and physical-layer processing.**

the physical layer in the form of one transport block per TTI on the E-DCH. Compared to the DCH coding and multiplexing chain, the overall structure of the E-DCH physical layer processing is simpler as there is only a single E-DCH and hence no transport channel multiplexing.

A 24-bit CRC is attached to the single E-DCH transport block to allow the hybrid-ARQ mechanism in the Node B to detect any errors in the received transport block. Coding is done using the same rate 1/3 Turbo coder as used for HSDPA.

The physical layer hybrid-ARQ functionality is implemented in a similar way as for HSDPA. Repetition or puncturing of the bits from the Turbo coder is used to adjust the number of coded bits to the number of channel bits. By adjusting the puncturing pattern, different redundancy versions can be generated.

Physical channel segmentation distributes the coded bits to the different channelization codes used, followed by interleaving and modulation.

### 9.2.2 SCHEDULING

Scheduling is one of the fundamental technologies behind HSUPA. In principle, scheduling is possible already in the first version of WCDMA, but HSUPA supports a significantly faster scheduling operation thanks to the location of the scheduler in the Node B.

The responsibility of the scheduler is to control *when* and *at what data rate* a UE is allowed to transmit, thereby controlling the amount of interference affecting other users at the Node B. This can be seen as controlling each UE's consumption of common resources, which in case of HSUPA is the amount of tolerable interference, that is, the total received power at the base station. The amount of common uplink resources a UE is using depends on the data rate used. Generally, the higher the data rate, the larger the required transmission power and thus the higher the resource consumption.

The term *noise rise* or *rise-over-thermal* is often used when discussing uplink operation. Noise rise, defined as $(I_0 + N_0)/N_0$ where $N_0$ and $I_0$ are the noise and interference power spectral densities, respectively, is a measure of the increase in interference in the cell due to the transmission activity. For example, 0 dB noise rise indicates an unloaded system and 3 dB noise rise implies a power spectral density due to uplink transmission equal to the noise spectral density. Although noise rise as such is not of major interest, it has a close relation to coverage and uplink load. A too large noise rise would result in loss of coverage for some channels – a UE may not have sufficient transmission power available to reach the required $E_b/N_0$ at the base station. Hence, the uplink scheduler must keep the noise rise within acceptable limits.

*Channel-dependent scheduling*, which typically is used in HSDPA, is possible for the uplink as well, but it should be noted that the benefits are different. As fast power control is used for the uplink, a UE transmitting when the channel conditions are favorable will generate the same amount of interference in the cell as a UE transmitting in unfavorable channel conditions, given the same data rate for the two. This is in contrast to HSDPA, where in principle a constant transmission power is used and the data rates are adapted to the channel conditions, resulting in a higher data rate for users with favorable radio conditions. However, for the uplink the transmission power will be different for the two UEs. Hence, the amount of interference generated in *neighboring* cells will differ. Channel-dependent scheduling will therefore result in a lower noise rise in the system, thereby improving capacity and/or coverage.

In practical cases, the transmission power of a UE is limited by several factors, both regulatory restrictions and power amplifier implementation restrictions. For WCDMA, different power classes are specified, limiting the maximum power the UE can use, where 21 dBm is a common value of the maximum power. This affects the discussion on uplink scheduling, making channel-dependent scheduling beneficial also from an inter-cell perspective. A UE scheduled when the channel conditions are beneficial encounters a reduced risk of hitting its transmission power limitation. This implies that the UE is likely to be able to transmit at a higher data rate if scheduled to transmit at favorable channel conditions. Therefore, taking channel conditions into account in the uplink scheduling decisions will improve the capacity, although the difference between non-channel-dependent and channel-dependent scheduling in most cases is not as large as in the downlink case.

Round-robin scheduling is one simple example of an uplink scheduling strategy, where UEs take turns in transmitting in the uplink. Similar to round-robin scheduling in HSDPA, this results in TDMA-like operation and avoids inter-cell interference due to the non-orthogonal uplink. However, as the maximum transmission power of the UEs is limited, a single UE may not be able to fully utilize the uplink capacity when transmitting and thus reducing the uplink capacity in the cell. The larger the cells, the higher the probability that the UE does not have sufficient transmit power available.

To overcome this, an alternative is to assign the same data rate to all users having data to transmit and to select this data rate such that the maximum cell load is respected. This results in maximum fairness in terms of the same data rate for all users but does not maximize the capacity of the cell. One of the benefits is the simple scheduling operation – there is no need to estimate the uplink channel quality and the transmission power status for each UE. Only the buffer status of each UE and the total interference level in the cell are required.

With *greedy filling*, the UE with the best radio conditions is assigned an as high data rate as possible. If the interference level at the receiver is smaller than the maximum tolerable level, the UE with the second best channel conditions is allowed to transmit as well, continuing with more and more UEs until the maximum tolerable interference level at the receiver is reached. This strategy maximizes the radio-interface utilization but is achieved at the cost of potentially large differences in data rates between users. In the extreme case, a user at the cell border with poor channel conditions may not be allowed to transmit at all.

Strategies between these two can also be considered such as different proportional fair strategies. This can be achieved by including a weighting factor for each user, proportional to the ratio between the instantaneous and average data rates, into the greedy filling algorithm. In a practical scenario, it is also necessary to take the transport network capacity and the processing resources in the base station into account in the scheduling decision, as well as the priorities for different data flows.

The above discussion of different scheduling strategies assumed all UEs having an infinite amount of data to transmit (full buffers). Similarly as the discussion for HSDPA, the traffic behavior is important to take into account when comparing different scheduling strategies. Packet-data applications are typically bursty in nature with large and rapid variations in their resource requirements. Hence, the overall target of the scheduler is to allocate a large fraction of the shared resource to users momentarily requiring high data rates while ensuring stable system operation by keeping the noise rise within limits.

A particular benefit of fast scheduling is the fact that it allows for a more relaxed connection admission strategy. For the DCH, admission control typically has to reserve resources relative to the peak data rate as there are limited means to recover from an event when many or all users transmit simultaneously with their maximum rate. Admission relative to the peak rate results in a rather conservative admission strategy for bursty packet-data applications. With fast scheduling, a larger number of packet-data users can be admitted since fast scheduling provides means to control the load in case many users request for transmission simultaneously.

#### 9.2.2.1 Scheduling framework for HSUPA

The scheduling framework for HSUPA is generic in the sense that the control signaling allows for several different scheduling implementations. One major difference between uplink and downlink scheduling is the location of the scheduler and the information necessary for the scheduling decisions.

In HSDPA, the scheduler and the buffer status are located at the same node, the Node B. Hence, the scheduling strategy is completely implementation dependent and there is no need to standardize any buffer status signaling to support the scheduling decisions.

In HSUPA, the scheduler is still located in the Node B to control the transmission activity of different UEs, while the buffer status information is distributed among the UEs. In addition to the buffer status, the scheduler also needs information about the available transmission power in the UE: if the UE is close to its maximum transmission power there is no use in scheduling a (significantly) higher data rate. Hence, there is a need to specify signaling to convey buffer status and power availability information from the UE to the Node B.

The basis for the scheduling framework is *scheduling grants* sent by the Node B to the UE and limiting the E-DCH data rate and *scheduling requests* sent from the UE to the Node B to request permission to transmit (at a higher rate than currently allowed). Scheduling decisions are taken by the *serving cell*, which has the main responsibility for scheduling, as illustrated in Figure 9.6 (in case of simultaneous

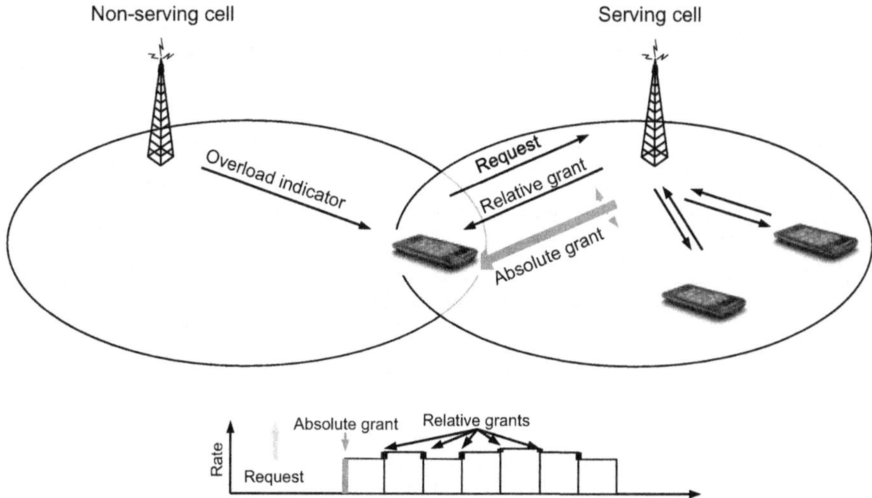


**FIGURE 9.6 Overview of the scheduling operation.**

HSDPA and HSUPA, the same cell is the serving cell for both). However, when in soft handover, the non-serving cells have a possibility to influence the UE behavior to control the inter-cell interference.

Providing the scheduler with the necessary information about the UE situation, taking the scheduling decision based on this information, and communicating the decision back to the UE takes a non-zero amount of time. The situation at the UE in terms of buffer status and power availability may therefore be different at the time of transmission compared to the time of providing the information to the Node B UE buffer situation. For example, the UE may have less data to transmit than assumed by the scheduler, high-priority data may have entered the transmission buffer, or the channel conditions may have worsened such that the UE has less power available for data transmission. To handle such situations and to exploit any interference reductions due to a lower data rate, the scheduling grant does not set the E-DCH data rate but rather an *upper limit* of the resource usage. The UE selects the data rate or, more precisely, the *E-DCH Transport Format Combination* (E-TFC) within the restrictions set by the scheduler.

The *serving grant* is an internal variable in each UE, used to track the maximum amount of resource the UE is allowed to use. It is expressed as a maximum E-DPDCH-to-DPCCH power ratio and the UE is allowed to transmit from any MAC-d flow and using any transport-block size as long as it does not exceed the serving grant. Hence, the scheduler is responsible for scheduling between UEs, while the UEs themselves are responsible for scheduling between MAC-d flows according to rules in the specifications. Basically, a high-priority flow should be served before a low-priority flow.

Expressing the serving grant as a maximum power ratio is motivated by the fact that the fundamental quantity the scheduler is trying to control is uplink interference, which is directly proportional to transmission power. The E-DPDCH transmission power is defined relative to the DPCCH to ensure the E-DPDCH is affected by the power control commands. As the E-DPDCH transmission power typically is significantly larger than the DPCCH transmission power, the E-DPDCH-to-DPCCH power ratio is roughly proportional to the total transmission power, $(P_{\text{E-DPDCH}} + P_{\text{DPCCH}})/P_{\text{DPCCH}} \approx P_{\text{E-DPDCH}}/P_{\text{DPCCH}}$, and thus setting a limit on the maximum E-DPCCH-to-DPCCH power ratio corresponds to control of the maximum transmission power of the UE.

The Node B can update the serving grant in the UE by sending an *absolute grant* or a *relative grant* to the UE (Figure 9.7). Absolute grants are transmitted on the shared E-AGCH and are used for absolute changes of the serving grant. Typically, these changes are relatively large, for example, to assign the UE a high data rate for an upcoming packet transmission.

Relative grants are transmitted on the E-RGCH and are used for relative changes of the serving grant. Unlike the absolute grants, these changes are small; the change in transmission power due to a relative grant is typically in the order of 1 dB. Relative grants can be sent from both serving and, in case of the UE being in soft handover, also from the non-serving cells. However, there is a significant difference between the two, and the two cases deserve to be treated separately.

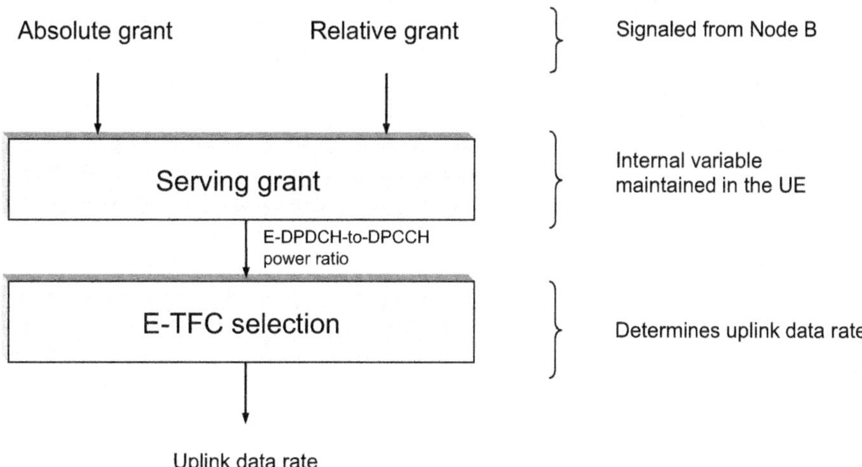


**FIGURE 9.7 The relation between absolute grant, relative grant, and serving grant.**

Relative grants from the serving cell are dedicated for a single UE; that is, each UE receives its own relative grant to allow for individual adjustments of the serving grants in different UEs. The relative grant is typically used for small, possibly frequent, updates of the data rate during an ongoing packet transmission. A relative grant from the serving cell can take one of the three values: "UP," "HOLD," or "DOWN." The "up" ("down") command instructs the UE to increase (decrease) the serving grant, that is, to increase (decrease) the maximum allowed E-DPDCH-to-DPCCH power ratio compared to the last used power ratio, where the last used power ratio refers to the previous TTI in the same hybrid-ARQ process. The "hold" command instructs the UE not to change the serving grant. An illustration of the operation is found in Figure 9.8.

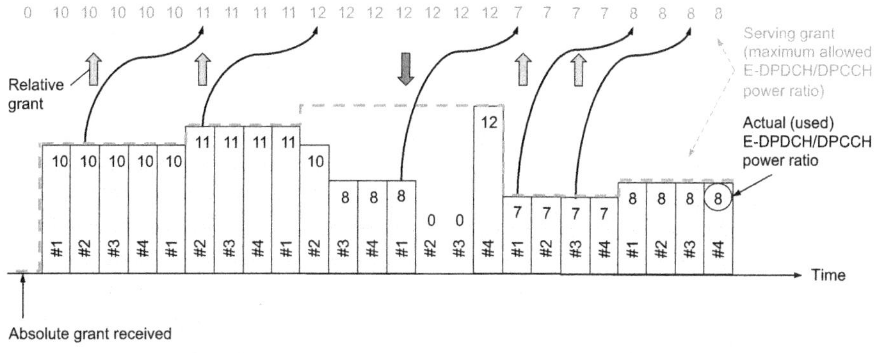


**FIGURE 9.8 Illustration of relative grant usage.**

Relative grants from non-serving cells are used to control inter-cell interference. The scheduler in the serving cell has no knowledge about the interference to neighboring cells due to the scheduling decisions taken. For example, the load in the serving cell may be low and from that perspective, it may be perfectly fine to schedule a high-rate transmission. However, the neighboring cell may not be able to cope with the additional interference caused by the high-rate transmission. Hence, there must be a possibility for the non-serving cell to influence the data rates used. In essence, this can be seen as an "emergency brake" or an "overload indicator," commanding non-served UEs to lower their data rate.

Although the name "relative grant" is used also for the overload indicator, the operation is quite different from the relative grant from the serving cell. First, the overload indicator is a common signal received by all UEs. Since the non-serving cell is only concerned about the total interference level from the neighboring cell, and not which UE is causing the interference, a common signal is sufficient. Furthermore, as the non-serving cell is not aware of the traffic priorities, etc. of the UEs it is not serving, there would be no use in having dedicated signaling from the nonserving cell.

Second, the overload indicator only takes two, not three, values, "DTX" and "down," where the former does not affect the UE operation. All UEs receiving "down" from any of the non-serving cells decrease their respective serving grant relative to the previous TTI in the same hybrid-ARQ process.

#### 9.2.2.2 Scheduling information

For efficient scheduling, the scheduler obviously needs information about the UE situation, both in terms of buffer status and in terms of the available transmission power. Naturally, the more detailed the information is, the better the possibilities for the scheduler to take accurate and efficient decisions. However, at the same time, the amount of information sent in the uplink should be kept low so as not to consume excessive uplink capacity. These requirements are, to some extent, contradicting and are in HSUPA addressed by providing two mechanisms complementing each other: the out-band "happy bit" transmitted on the E-DPCCH and in-band scheduling information transmitted on the E-DCH.

Out-band signaling is done through a single bit on the E-DPCCH, the "happy bit." Whenever the UE has available power for the E-DCH to transmit at a higher data rate compared to what is allowed by the serving grant, and the number of bits in the buffer would require more than a certain number of TTIs, the UE shall set the bit to "not happy" to indicate that it would benefit from a higher serving grant. Otherwise, the UE shall declare "happy." Note that the happy bit is only transmitted in conjunction with an ongoing data transmission as the E-DPCCH is only transmitted together with the E-DPDCH.

In-band scheduling information provides detailed information about the buffer occupancy, including priority information and the transmission power available for the E-DCH. The in-band information is transmitted in the same way as user data, either alone or as part of a user data transmission. Consequently, this information benefits from hybrid-ARQ with soft combining. As in-band scheduling information is the only mechanism for the unscheduled UE to request resources, the scheduling information can be sent non-scheduled and can therefore be transmitted regardless of the serving grant. Non-scheduled transmissions are not restricted to scheduling information only; the network can configure non-scheduled transmissions also for other data.

### 9.2.3 E-TFC SELECTION

The E-TFC selection is responsible for selecting the transport format of the E-DCH, thereby determining the data rate to be used for uplink transmission, and to control MAC-e multiplexing. Clearly, the selection needs to take the scheduling decisions taken by the Node B into account, which is done through the serving grant, as previously discussed. MAC-e multiplexing, on the other hand, is handled autonomously by the UE. Hence, while the scheduler handles resource allocation *between* UEs, the E-TFC selection controls resource allocation between flows *within* the UE. The rules for multiplexing of the flows are given by the specification; in principle, high-priority data shall be transmitted before any data of lower priority.

Introduction of the E-DCH needs to take coexistence with DCHs into account. If this is not done, services mapped onto DCHs could be affected. This would be a non-desirable situation as it may require reconfiguration of parameters set for DCH transmission. Therefore, a basic requirement is to serve DCH traffic first and only spend otherwise unused power resources on the E-DCH. Comparisons can be made with HSDPA, where any dedicated channels are served first and the HS-DSCH may use the otherwise unused transmission power. Therefore, TFC selection is performed in two steps. First, the normal DCH TFC selection is performed as in previous releases. The UE then estimates the remaining power and a second TFC selection step is performed where E-DCH can use the remaining power. The overall E-TFC selection procedure is illustrated in Figure 9.9.

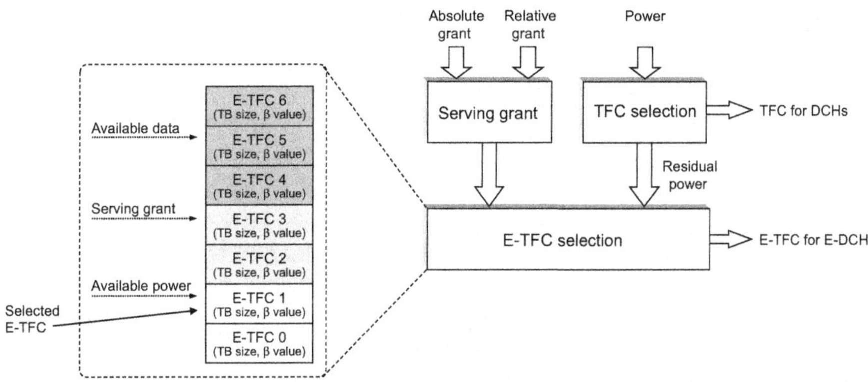


**FIGURE 9.9 Illustration of the E-TFC selection process.**

Each E-TFC has an associated E-DPDCH-to-DPCCH power offset. Clearly, the higher the data rate, the higher the power offset. When the required transmitter power for different E-TFCs has been calculated, the UE can calculate the possible E-TFCs to use from a power perspective. The UE then selects the E-TFC by maximizing the amount of data that can be transmitted given the power constraint and the scheduling grant.

The possible transport-block sizes being part of the E-TFCs are predefined in the specifications, similar to HS-DSCH. This reduces the amount of signaling, for example at handover between cells, as there is no need to configure a new set of E-TFCs at each cell change. Generally, conformance tests to ensure the UE obeys the specifications are also simpler the smaller the amount of configurability in the UE.

To allow for some flexibility in the transport-block sizes, there are four tables of E-TFCs specified in release 6; for each of the two TTIs specified there is one table optimized for common RLC PDU sizes and one general table with constant maximum relative overhead. Which one of the predefined tables the UE shall use is determined by the TTI and RRC signaling. Enhancements in subsequent releases have led to the inclusion of further tables.

### 9.2.4 HYBRID-ARQ WITH SOFT COMBINING

Hybrid-ARQ with soft combining for HSUPA serves a similar purpose as the hybrid-ARQ mechanism for HSDPA – to provide robustness against transmission errors. However, hybrid-ARQ with soft combining is not only a tool for providing robustness against occasional errors, but it can also be used for enhanced capacity, as discussed in the introduction. As hybrid-ARQ retransmissions are fast, many services allow for a retransmission or two. Combined with incremental redundancy, this forms an implicit rate control mechanism. Thus, hybrid-ARQ with soft combining can be used in several (related) ways:

- to provide robustness against variations in the received signal quality; and
- to increase the link efficiency by targeting multiple transmission attempts, for example by imposing a maximum number of transmission attempts and operating the outer loop power control on the residual error event after soft combining.

To a large extent, the requirements on hybrid-ARQ are similar to HSDPA and, consequently, the hybrid-ARQ design for HSUPA is fairly similar to the design used for HSDPA. There are some differences as well, mainly originating from the support of soft handover in the uplink.

Similar to HSDPA, HSUPA hybrid-ARQ spans both the MAC layer and the physical layer. The use of multiple parallel stop-and-wait processes for the hybrid-ARQ protocol has proven efficient for HSDPA and is used for HSUPA for the same reasons – fast retransmission and high throughput combined with low overhead of the ACK/NACK signaling. Upon reception of the single transport block transmitted in a certain TTI and intended for a certain hybrid-ARQ process, the Node B attempts to decode the set of bits and the outcome of the decoding attempt, ACK or NACK, is signaled to the UE. To minimize the cost of the ACK/NACK, a single bit is used.

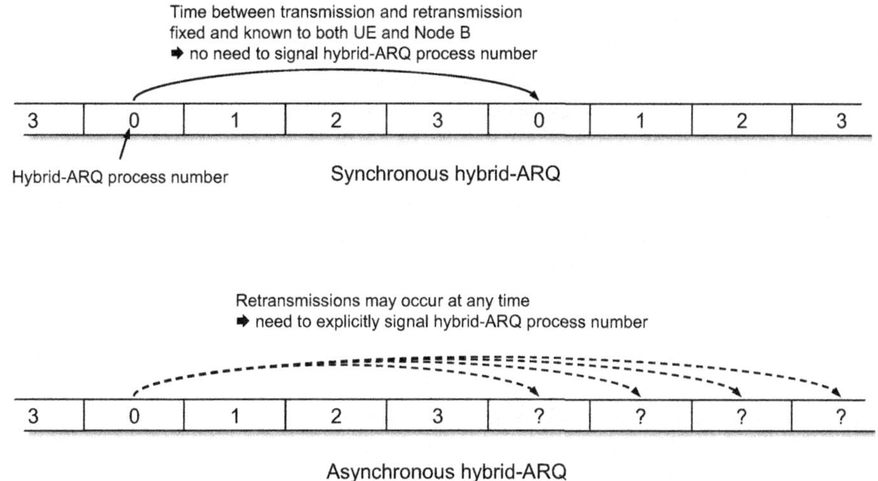


**FIGURE 9.10 Synchronous versus asynchronous hybrid-ARQ.**

Clearly, the UE must know which hybrid-ARQ process a received ACK/NACK bit is associated with. Again, this is solved using the same approach as in HSDPA, that is, the timing of the ACK/NACK is used to associate the ACK/NACK with a certain hybrid-ARQ process. A well-defined time after reception of the uplink transport block on the E-DCH, the Node B generates an ACK/NACK. Upon reception of a NACK, the UE performs a retransmission and the Node B performs soft combining using incremental redundancy.

The handling of retransmissions, more specifically when to perform a retransmission, is one of the major differences between the hybrid-ARQ operation in the uplink and the downlink (Figure 9.10). For HSDPA, retransmissions are scheduled as any other data and the Node B is free to schedule the retransmission to the UE at any time instant and using a redundancy version of its choice. It may also address the hybrid-ARQ processes in any order, that is, it may decide to perform retransmissions for one hybrid-ARQ process but not for another process in the same UE. This type of operation is often referred to as adaptive asynchronous hybrid-ARQ. Adaptive since the Node B may change the transmission format and asynchronous since retransmissions may occur at any time after receiving the ACK/NACK.

For the uplink, on the other hand, a synchronous, non-adaptive hybrid-ARQ operation is used. Hence, thanks to the synchronous operation, retransmissions occur a predefined time after the initial transmission; that is, they are not explicitly scheduled. Likewise, the non-adaptive operation implies that the transport format and redundancy version to be used for each of the retransmissions is also known from the time of the original transmission. Therefore, there is neither a need for explicitly scheduling the retransmissions nor is there a need for signaling the redundancy version the UE shall use. This is the main benefit of synchronous operation of the hybrid-ARQ – minimizing the control signaling overhead. Naturally, the possibility to adapt the transmission format of the retransmissions to any changes in the channel conditions is lost, but as the uplink scheduler in the Node B has less knowledge of the transmitter status – this information is located in the UE and provided to the Node B using in-band signaling not available until the hybrid-ARQ has successfully decoded the received data – than the downlink scheduler, this loss is by far outweighed by the gain in reduced control signaling overhead.

Apart from the synchronous versus asynchronous operation of the hybrid-ARQ protocol, the other main difference between uplink and downlink hybrid-ARQ is the use of soft handover in the former case. In soft handover between different Node Bs, the hybrid-ARQ protocol is terminated in multiple nodes, namely, all the involved Node Bs. For HSDPA, on the other hand, there is only a single termination point for the hybrid-ARQ protocol – the UE. In HSUPA, the UE therefore needs to receive ACK/NACK from all involved Node Bs. As it, from the UE perspective, is sufficient if at least one of the involved Node Bs receives the transmitted transport block correctly, it considers the data to be successfully delivered to the network if at least one of the Node Bs signals an ACK. This rule is sometimes called "or-of-the-ACKs." A retransmission occurs only if all involved Node Bs signal a NACK, indicating that none of them has been able to decode the transmitted data.

As known from the HSDPA description, the use of multiple parallel hybrid-ARQ processes cannot itself provide in-sequence delivery and a reordering mechanism is required (Figure 9.11). For HSDPA, reordering is obviously located in the UE. The same aspect with out-of-sequence delivery is also valid for the uplink, which calls for a reordering mechanism also in this case. However, because of the support of soft handover, reordering cannot be located in the Node B. Data transmitted in one hybrid-ARQ process may be successfully decoded in one Node B, while data transmitted in the next hybrid-ARQ process may happen to be correctly decoded in another Node B. Furthermore, in some situations, several involved Node Bs may succeed in decoding the same transport block. For these reasons, the reordering mechanism needs to have access to the transport blocks delivered from all involved Node Bs, and therefore the reordering is located in the RNC. Reordering also removes any duplicates of transport blocks detected in multiple Node Bs.

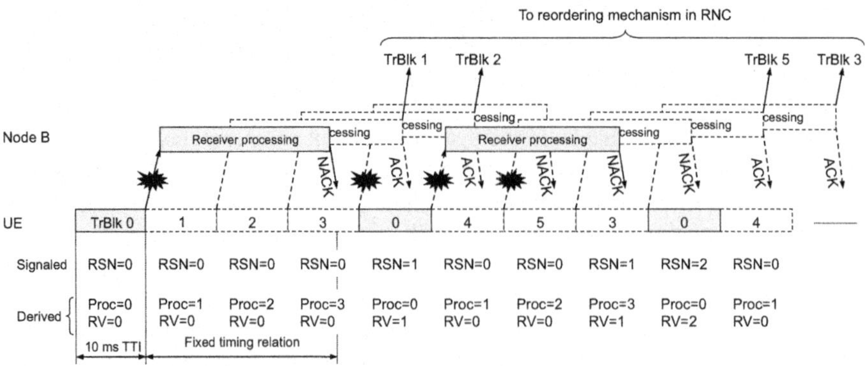


**FIGURE 9.11 Multiple hybrid-ARQ processes for HSUPA.**

The presence of soft handover in the uplink has also impacted the design of the control signaling. Similar to HSDPA, there is a need to indicate to the receiving end whether the soft buffer should be cleared, that is, the transmission is an initial transmission, or if soft combining with the soft information stored from previous transmissions in this hybrid-ARQ process should take place. HSDPA relies on a single-bit new-data indicator for this purpose. If the Node B misinterpreted an uplink NACK as an ACK and continues with the next packet, the UE can capture this error event by observing the single-bit "new-data indicator" that is incremented for each new packet transmission. If the new-data indicator is incremented, the UE will clear the soft buffer, despite that its contents were not successfully decoded, and try to decode the new transmission. Although a transport block is lost and has to be retransmitted by the RLC protocol, the UE does not attempt to soft combine coded bits originating from different transport blocks and therefore the soft buffer is not corrupted. Only if the uplink NACK *and* the downlink new-data indicator are *both* misinterpreted, which is a rare event, will the soft buffer be corrupted.

For HSUPA, a single-bit new-data indicator would work in absence of soft handover. Only if the downlink NACK *and* the uplink control signaling both are misinterpreted will the Node B soft buffer be corrupted. However, in the presence of soft handover, this simple method is not sufficient. Instead, a 2-bit *Retransmission Sequence Number* (RSN) is used for HSUPA. The initial transmission sets RSN to zero and for each subsequent transmission the RSN is incremented by one. Even if the RSN can only take values in the range of 0 to 3, any number of retransmissions is possible; the RSN simply remains at 3 for the third and later retransmissions. Together with the synchronous protocol operation, the Node B knows when a retransmission is supposed to occur and with what RSN. The simple example in Figure 9.12 illustrates the operation. As the first Node B acknowledged packet A, the UE continues with packet B, despite that the second Node B did not correctly decode the packet. At the point of transmission of packet B, the second Node B expects a retransmission of packet A, but because of the uplink channel conditions at this point in time, the second Node B does not even detect the presence of a transmission. Again, the first Node B acknowledged the transmission and the UE continues with

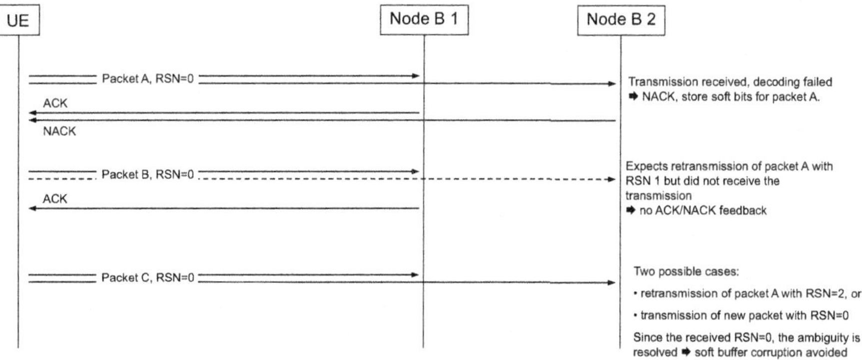


**FIGURE 9.12 Retransmissions in soft handover.**

packet C. This time, the second Node B does receive the transmissions and, thanks to the synchronous hybrid-ARQ operation, can immediately conclude that it must be a transmission of a new packet. If it was a retransmission of packet A, the RSN would have been equal to two. This example illustrates the improved robustness from a 2-bit RSN together with a synchronous hybrid-ARQ operation. A scheme with a single-bit "new-data indicator," which can be seen as a 1-bit RSN, would not have been able to handle the fairly common case of a missed transmission in the second Node B. The new-data indicator would in this case be equal to zero, both in the case of a retransmission of packet A and in the case of an initial transmission of packet C, thereby leading to soft buffer corruption.

Soft combining in the hybrid-ARQ mechanism for HSUPA is based on incremental redundancy. Generation of redundancy versions is done in a similar way as for HSDPA by using different puncturing patterns for the different redundancy versions. The redundancy version is controlled by the RSN according to a rule in the specifications (see further Section 9.3.2).

For Turbo codes, the systematic bits are of higher importance than the parity bits, as discussed in Chapter 5. Therefore, the systematic bits should be included in the initial transmission to allow for decodability already after the first transmission attempts. Furthermore, for the best gain with incremental redundancy, the retransmissions should contain additional parity. This leads to a design where the initial transmission is self-decodable and includes all systematic bits as well as some parity bits, while the retransmission mainly contains additional parity bits not previously transmitted.

However, in soft handover, not all involved Node Bs may have received all transmissions. There is a risk that a Node B did not receive the first transmission with the systematic bits but only the parity bits in the retransmission. As this would lead to degraded performance, it is preferable if all redundancy versions used when in soft handover are self-decodable and contain the systematic bits. The above-mentioned rule used to map RSN into redundancy versions does this by making all redundancy versions self-decodable for lower data rates, which typically are used in the soft handover region at the cell-edge, while using full incremental redundancy for the highest data rates, unlikely to be used in soft handover.

### 9.2.5 PHYSICAL CHANNEL ALLOCATION

The E-DPDCH physical channel layout is shown in Figure 9.13. Up to four simultaneous E-DPDCHs can be transmitted and the only supported modulation order in release 6 is BPSK (later releases include both 16-QAM and 64-QAM; see Chapter 10). Modulation in the uplink is performed before I/Q-mapping. Hence, it can be seen that for multi-code E-DPDCH transmission, each code is independently BPSK modulated and mapped to the I- or Q-branch, effectively creating QPSK modulation. The spreading code configuration and I/Q mapping depend on the configuration according to Table 9.1. The setting of gain factors, $\beta_{ed,k}$, are discussed further in Section 9.3.3.

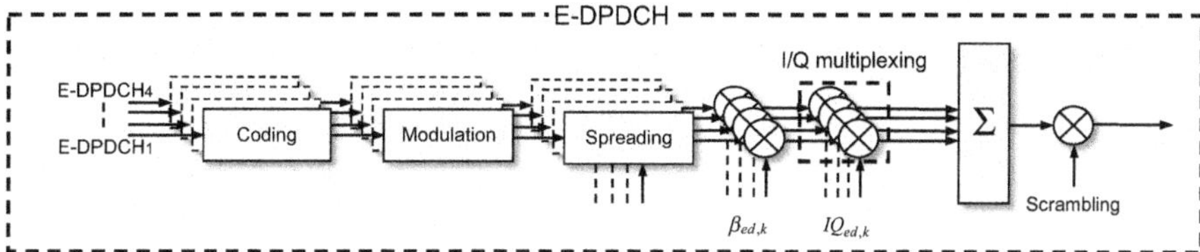


**FIGURE 9.13 Physical channel layout for E-DPDCH.**

The mapping of the coded E-DCH onto the physical channels is straightforward. As illustrated in Figure 9.13, the E-DCH is mapped to one or several E-DPDCHs, separate from the DPDCH. Depending on the E-TFC selected, a different number of E-DPDCHs is used. For the lowest data rates, a single E-DPDCH with a spreading factor inversely proportional to the data rate is sufficient. When more than one E-DPDCH is transmitted, pairs of E-DPDCHs are I/Q multiplexed using the same channelization code. Hence, there is no configuration supporting, for example, three simultaneous E-DPDCHs.

To maintain backward compatibility, the mapping of the DPCCH, DPDCH, and HS-DPCCH remains unchanged compared to previous releases.

The order in which the E-DPDCHs are allocated is chosen to minimize the *Peak-to-Average Power Ratio* (PAPR) in the UE, and it also depends on whether the HS-DPCCH and the DPDCH are present or not. The higher the PAPR, the larger the back-off required in the UE power amplifier, which impacts the uplink coverage. Hence, a low PAPR is a highly desirable property. PAPR is also the reason why SF2 is introduced, as it can be shown that two codes of SF2 have a lower PAPR than four codes of SF4. For the highest data rates, a mixture of spreading factors,

**Table 9.1** Channelization Code and I/Q Mapping for E-DPDCH

| $N_{max\text{-}DPDCH}$ | HS-DSCH Configured | E-DPDCH$_k$ | Channelization Code $C_{ed,k}$ | $IQ_{ed,k}$ |
|---|---|---|---|---|
| 0 | No/Yes | E-DPDCH$_1$ | $C_{ch,SF,SF/4}$ if SF $\geq$ 4<br>$C_{ch,2,1}$ if SF = 2 | 1 |
| | | E-DPDCH$_2$ | $C_{ch,4,1}$ if SF = 4<br>$C_{ch,2,1}$ if SF = 2 | $j$ |
| | | E-DPDCH$_3$ | $C_{ch,4,1}$ | 1 |
| | | E-DPDCH$_4$ | $C_{ch,4,1}$ | $j$ |
| 1 | No | E-DPDCH$_1$ | $C_{ch,SF,SF/2}$ | $j$ |
| | | E-DPDCH$_2$ | $C_{ch,4,2}$ if SF = 4<br>$C_{ch,2,1}$ if SF = 2 | 1 |
| | Yes | E-DPDCH$_1$ | $C_{ch,SF,SF/2}$ | 1 |
| | | E-DPDCH$_2$ | $C_{ch,4,2}$ if SF = 4<br>$C_{ch,2,1}$ if SF = 2 | $j$ |

**Table 9.2** Possible Physical Channel Configurations

| #DPCCH | #DPDCH | #HS-DPCCH | #E-DPCCH | #E-DPDCH | Comment |
|---|---|---|---|---|---|
| 1 | 1–6 | 0 or 1 | – | – | Rel 5 configurations |
| 1 | 0 or 1 | 0 or 1 | 1 | 1 × SF ≥ 4 | 0.96 Mbit/s E-DPDCH raw data rate |
| 1 | 0 or 1 | 0 or 1 | 1 | 2 × SF4 | 1.92 Mbit/s E-DPDCH raw data rate |
| 1 | 0 or 1 | 0 or 1 | 1 | 2 × SF2 | 3.84 Mbit/s E-DPDCH raw data rate |
| 1 | 0 | 0 or 1 | 1 | 2 × SF2 + 2 × SF4 | 5.76 Mbit/s E-DPDCH raw data rate |

*Note: The E-DPDCH data rates are raw data rates: the maximum E-DCH data rate will be lower because of coding and limitations set by the UE categories.*

2 × SF2 + 2 × SF4, is used. The physical channel configurations possible are listed in Table 9.2, and in Figure 9.14 the physical channel allocation with a simultaneous HS-DPCCH is illustrated.

### 9.2.6 POWER CONTROL

The E-DCH power control works in a similar manner as for the DCH and there is no change in the overall power control architecture with the introduction of the E-DCH. A single inner power control loop adjusts the transmission power of the DPCCH. The E-DPDCH transmission power is set by the E-TFC selection relative to the DPCCH power in a similar way as the DPDCH transmission power is set by the TFC selection. The inner loop power control located in the Node B bases its decision on the SIR target set by the outer loop power control located in the RNC.

The outer loop in earlier releases is primarily driven by the DCH BLER visible to the RNC. If a DCH is configured, the outer loop, which is an implementation-specific algorithm, may continue to operate on the DCH only. This approach works well as long as there are sufficiently frequent transmissions on the DCH, but the performance is degraded if DCH transmissions are infrequent.

If no DCH is configured, or if only infrequent transmissions occur at the DCH, information on the E-DCH transmissions needs to be taken into account. However, because of the introduction of hybrid-ARQ for the E-DCH, the residual E-DCH BLER may not be an adequate input for the outer loop power control. In most cases, the residual E-DCH BLER visible to the RNC is close to zero, which would cause the outer loop to lower the SIR target and potentially cause a loss of the uplink DPCCH if the residual E-DCH BLER alone is used as input to the outer loop mechanism. Therefore, to assist the outer loop power control, the number of retransmissions

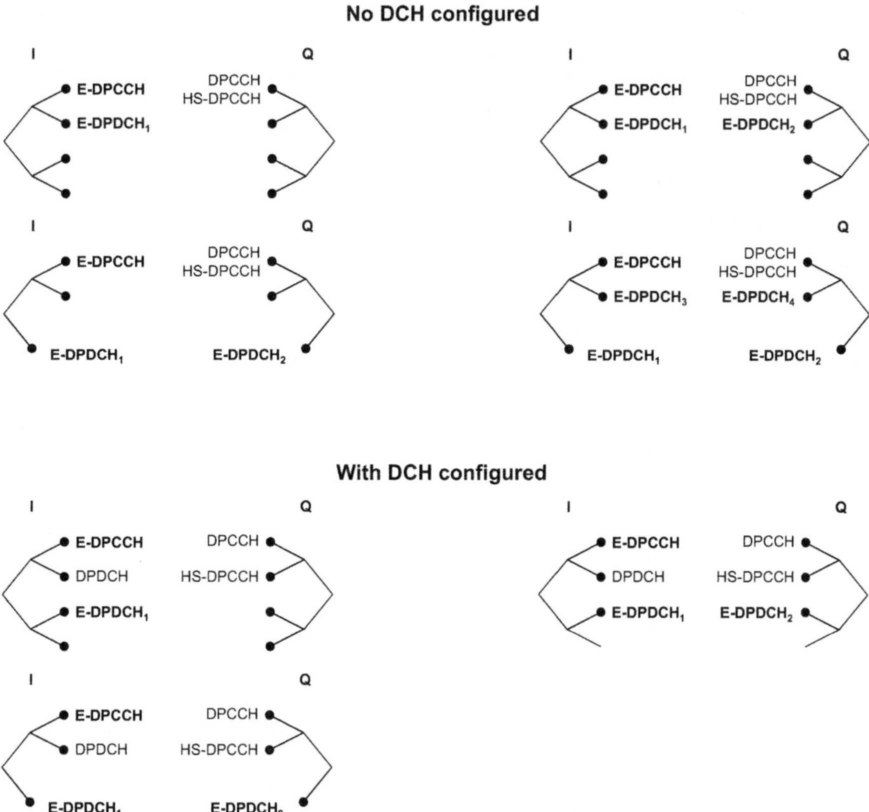


FIGURE 9.14 Code allocation in case of simultaneous E-DCH and HS-DSCH operation (note that the code allocation is slightly different when no HS-DPCCH is configured). Channels with SF >4 are shown on the corresponding SF4 branch for illustrative purposes.

actually used for transmission of a transport block is signaled from the Node B to the RNC. The RNC can use this information as part of the outer loop to set the SIR target in the inner loop.

## 9.2.7 DATA FLOW

In Figure 9.15, the data flow from the application through all the protocol layers is illustrated in a similar way as for HSDPA. In this example, an IP service is assumed. The PDCP optionally performs IP header compression. The output from the PDCP is fed to the RLC. After possible concatenation, the RLC SDUs are segmented into smaller blocks of typically 40 bytes and an RLC header is attached. The RLC PDU is passed via the MAC-d, which is transparent for HSUPA, to the MAC-e. The MAC-e concatenates one or several MAC-d PDUs from one or several MAC-d flows and inserts MAC-es and MAC-e headers to form a transport block, which is forwarded on the E-DCH to the physical layer for further processing and transmission.

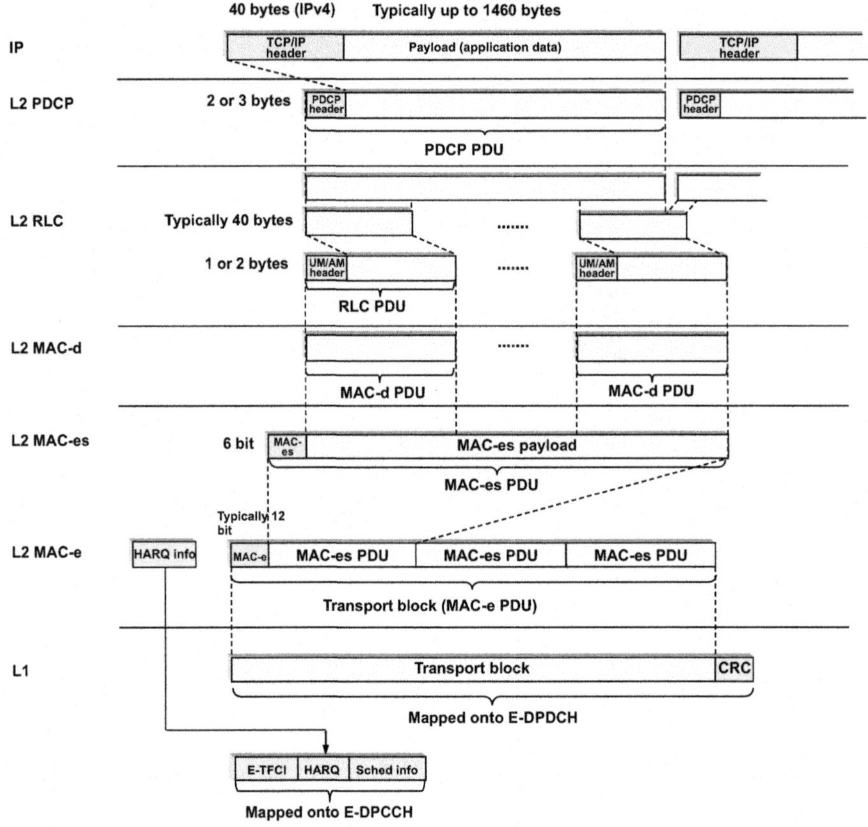


**FIGURE 9.15 Data flow.**

## 9.2.8 RESOURCE CONTROL FOR E-DCH

Similar to HSDPA, the introduction of HSUPA implies that a part of the radio resource management is handled by the Node B instead of the RNC. However, the RNC still has the overall responsibility for radio resource management, including admission control and handling of inter-cell interference. Thus, there is a need to monitor and control the resource usage of E-DCH channels to achieve a good balance between E-DCH and non-E-DCH users. This is illustrated in Figure 9.16.

For admission control purposes, the RNC relies on the *Received Total Wideband Power* (RTWP) measurement, which indicates the total uplink resource usage in the cell. Admission control may also exploit the *E-DCH provided bit rate*, which is a Node B measurement reporting the aggregated served E-DCH bit rate per priority class. Together with the RTWP measurement, it is possible to design an admission algorithm evaluating the E-DCH scheduler headroom for a particular priority class.

To control the load in the cell, the RNC may signal an RTWP target to the Node B in which case the Node B should schedule E-DCH transmissions such that the RTWP

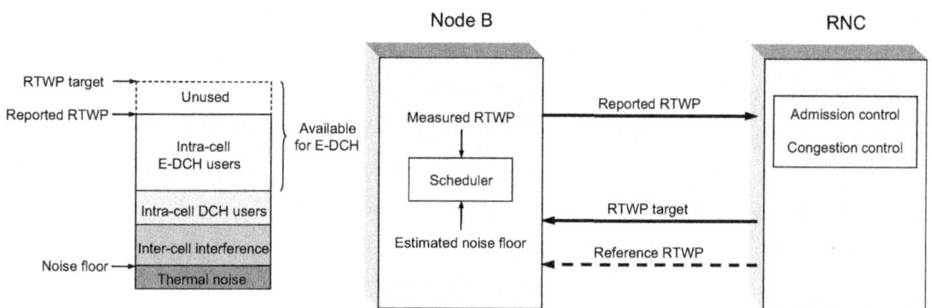


**FIGURE 9.16 Illustration of the resource sharing between E-DCH and DCH channels.**

is within this limit. The RNC may also signal a reference RTWP, which the Node B may use to improve its estimate of the uplink load in the cell. Note that whether the scheduler uses an absolute measure such as the RTWP or a relative measure such as noise rise is not specified. Internally, the Node B performs any measurements useful to a particular scheduler design.

To provide the RNC with a possibility to control the ratio between inter-cell and inter-cell interference, the RNC may signal a *Target Non-serving E-DCH to Total E-DCH Power Ratio* to the Node B. The scheduler must obey this limitation when setting the overload indicator and is not allowed to suppress non-serving E-DCH UEs unless the target is exceeded. This prevents a cell from starve users in neighboring cells. If this was not the case, a scheduler could in principle permanently set the overload indicator to "steal" resources from neighboring cells – a situation that definitely is not desirable.

Finally, the measurement *Transmitted carrier power of all codes not used for HS-PDSCH, HS-SCCH, E-AGCH, E-RGCH, or E-HICH transmission* also includes the E-DCH-related downlink control signaling.

### 9.2.9 MOBILITY

Active set management for the E-DCH uses the same mechanisms as for release 99 DCH, that is, the UE measures the signal quality from neighboring cells and informs the RNC. The RNC may then take a decision to update the active set. Note that the E-DCH active set is a subset of the DCH active set. In most cases, the two sets are identical, but in situations where only part of the network supports E-DCH, the E-DCH active set may be smaller than the DCH active set as the former only includes cells capable of E-DCH reception.

Changing serving cell is performed in the same way as for HSDPA (see Chapter 8) as the same cell is the serving cell for both E-DCH and HS-DSCH.

### 9.2.10 UE CATEGORIES

Similar to HSDPA, in release 6 the physical layer UE capabilities were grouped into six categories. Fundamentally, two major physical layer aspects, the number of

**Table 9.3** E-DCH UE Categories [2]

| E-DCH Category | Max # E-DPDCHs | Supports 2 ms TTI | Maximum Transport-Block Size (Mbit/s) | |
|---|---|---|---|---|
| | | | 10 ms TTI | 2 ms TTI |
| 1 | 1 × SF4 | – | 7110 (0.7) | – |
| 2 | 2 × SF4 | Yes | 14,484 (1.4) | 2798 (1.4) |
| 3 | 2 × SF4 | – | 14,484 (1.4) | – |
| 4 | 2 × SF2 | Yes | 20,000 (2) | 5772 (2.9) |
| 5 | 2 × SF2 | – | 20,000 (2) | – |
| 6 | 2 × SF4 + 2 × SF2 | Yes | 20,000 (2) | 11,484 (5.74) |
| 7[1] | 2 × SF4 + 2 × SF2 | Yes | 20,000 (2) | 22,996 (11.4)[2] |
| 8[1] | 2 × SF4 + 2 × SF2 | Yes | – | 11,484 (11.4)[3] |
| 9[1] | 2 × SF4 + 2 × SF2 | Yes | – | 22,996 (22.8)[2,3] |
| 10[1] | 2 × SF4 + 2 × SF2 | Yes | – | 34,507 (17.2)[4] |
| 11[1] | 2 × SF4 + 2 × SF2 | Yes | – | 22,996 (22.8)[5] |
| 12[1] | 2 × SF4 + 2 × SF2 | Yes | – | 34,507 (34.4)[4,5] |

[1]*Introduced after release 6.*
[2]*Supports 16-QAM.*
[3]*Supports dual-cell HSUPA with up to two transport blocks.*
[4]*Supports 64-QAM.*
[5]*Supports UL MIMO.*

supported channelization codes and the supported TTI values, are determined by the category number. The releases 6-11 E-DCH UE categories are listed in Table 9.3. Support for 10 ms E-DCH TTI is mandatory for all release 6 UE categories, while only a subset of the categories supports a 2 ms TTI. Furthermore, note that the highest data rate supported with 10 ms TTI is 2 Mbit/s. The reason for this is to limit the amount of buffer memory for soft combining in the Node B; a larger transport-block size translates into a larger soft buffer memory in case of retransmissions. A UE supporting E-DCH must also be able to support HS-DSCH.

## 9.3 FINER DETAILS OF HSUPA

### 9.3.1 SCHEDULING – THE SMALL PRINT

The use of a serving grant as a means to control the E-TFC selection has already been discussed, as has the use of absolute and relative grants to update the serving grant. Absolute grants are transmitted on the shared E-AGCH physical channel and are used for absolute changes of the serving grant, as already stated. In addition to conveying the maximum E-DPDCH-to-DPCCH power ratio, the E-AGCH also contains an activation flag, whose usage will be discussed below. Obviously, the E-AGCH is also carrying the identity of the UE for which the E-AGCH information is intended. However, although the UE receives only *one* E-AGCH, it is assigned *two* identities,

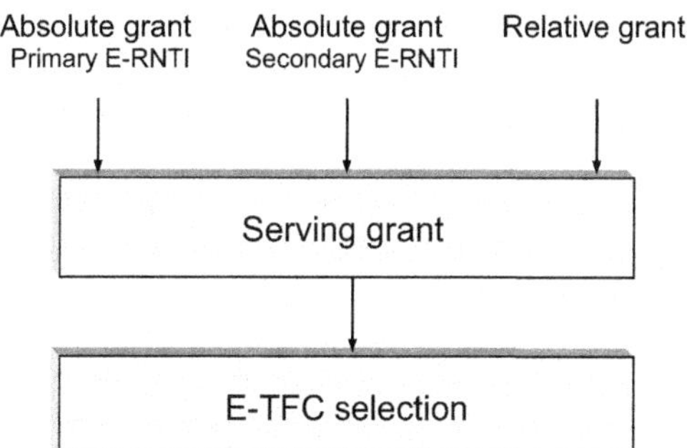


**FIGURE 9.17 The relation between absolute grant, relative grant, and serving grant.**

one primary and one secondary. The primary identity is UE specific and unique for each UE in the cell, while the secondary identity is a group identity shared by a group of UEs. The reason for having two identities is to allow for scheduling strategies based on both common, or groupwise, scheduling, where multiple UEs are addressed with a single identity and individual per-UE scheduling (Figure 9.17).

Common scheduling means that multiple UEs are assigned the same identity; the secondary identity is common to multiple UEs. A grant sent with the secondary identity is therefore valid for multiple UEs, and each of these UEs may transmit up to the limitation set by the grant. Hence, this approach is suitable for scheduling strategies not taking the uplink radio conditions into account, for example, CDMA-like strategies where the scheduler mainly strives to control the total cell interference level. A low-load condition is one such example. At low cell load, there is no need to optimize for capacity. Optimization can instead focus on minimizing the delays by assigning the same grant level to multiple UEs using the secondary identity. As soon as a UE has data to transmit, the UE can directly transmit up to the common grant level. There is no need to go through a request phase first, as the UE already has been assigned a non-zero serving grant. Note that multiple UEs may, in this case, transmit simultaneously, which must be taken into account when setting the serving grant level.

Individual per-UE scheduling provides tighter control of the uplink load and is useful to maximize the capacity at high loads. The scheduler determines which user is allowed to transmit and set the serving grant of the intended user by using the primary identity, unique for a specific UE. In this case, the UEs' resource utilization is individually controlled, for example, to exploit advantageous uplink channel conditions. The greedy filling strategy discussed earlier is one example of a strategy requiring individual grants.

Which of the two identities, the primary or the secondary, a UE is obeying can be described by a state diagram, illustrated in Figure 9.18. Depending on the state the UE is in, it follows either grants sent with the primary or the secondary identity. Addressing the UE with its unique primary identity causes the UE to stop obeying grants sent using the secondary common identity. There is also a mechanism to force

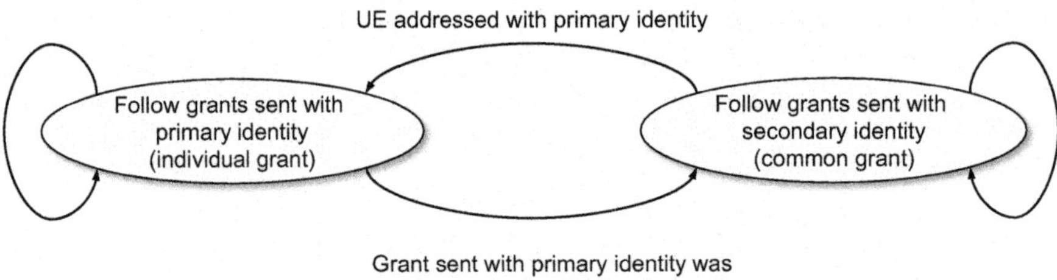


**FIGURE 9.18 Illustration of UE monitoring of the two identities.**

the UE back to follow the secondary, common grant level. The usefulness of this is best illustrated with the example below.

Consider the example in Figure 9.19, illustrating the usage of common and dedicated scheduling. The UEs are all initialized to follow the secondary identity and a suitable common grant level is set using the secondary identity. Any UE that has been assigned a grant level using the secondary identity may transmit using a data rate up to the common grant level; a level that is adjusted as the load in the system varies, for example, due to non-scheduled transmissions. As time evolves, UE #1 is in need of a high data rate to upload a huge file. Note that UE #1 may start the upload using the common grant level while waiting for the scheduler to grant a higher level. The scheduler decides to lower the common grant level using the secondary, common identity to reduce the load from other UEs. A large grant is sent to UE #1 using UE #1's primary and unique identity to grant UE #1 a high data rate (or, more accurately, a higher E-DPDCH-to-DPCCH power ratio). This operation also causes UE #1 to enter the "primary" state in Figure 9.18. At a later point in time, the scheduler decides to send a zero grant to UE #1 with the activation flag set to *all*, which forces UE #1 back to follow the secondary identity (back to common scheduling).

From this example, it is seen that the two identities each UE is assigned – one primary, UE-specific identity and one secondary, common identity – facilitates a flexible scheduling strategy.

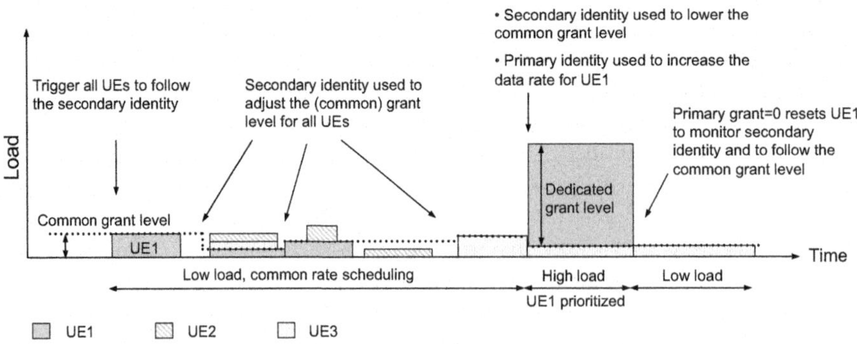


**FIGURE 9.19 Example of common and dedicated scheduling.**

#### 9.3.1.1 Relative grants

Relative grants from the serving cell can take one of the values "up," "down," and "no change." This is used to fine-tune an individual UE's resource utilization, as already discussed. To implement the increase (decrease) of the serving grant, the UE maintains a table of possible E-DPDCH-to-DPCCH power ratios, as illustrated in Figure 9.20. The up/down commands correspond to an increase/decrease of power ratio in the table by one step compared to the power ratio used in the previous TTI in the same hybrid-ARQ process. There is also a possibility to have a larger increase (but not decrease) for small values of the serving grant. This is achieved by (through RRC signaling) configuring two thresholds in the E-DPDCH-to-DPCCH power ratio table, below which the UE may increase the serving grant by three and two steps, respectively, instead of only a single step. The use of the table and the two thresholds allows the network to increase the serving grant efficiently without extensive repetition of relative grants for small data rates (small serving grants) and at the same time avoiding large changes in the power offset for large serving grants.

The "overload indicator" (relative grant from non-serving cells) is used to control the inter-cell interference (in contrast to the grants from the serving cell, which provide the possibility to control the inter-cell interference). As previously described, the overload indicator can take one of two values: "down" or "DTX," where the latter does not affect the UE operation. If the UE receives "down" from any of the non-serving cells, the serving grant is decreased relative to the previous TTI in the same hybrid-ARQ process.

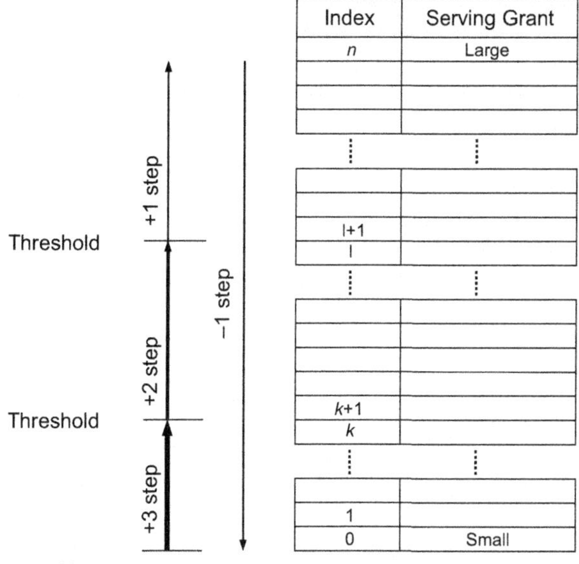


**FIGURE 9.20 Grant table.**

"Ping-pong effects" describes a situation where the serving grant level in a UE starts to oscillate. One example when this could happen is if a non-serving cell requests the UEs to lower their transmission power (and hence data rate) due to a too high interference level in the non-served cell. When the UE has reacted to this, the serving cell will experience a lower interference level and may decide to increase the grant level in the UE. The UE utilizes this increased grant level to transmit at a higher power, which again triggers the overload indicator from the non-serving cell, and the process repeats.

To avoid "ping-pong effects," the UE ignores any "up" commands from the serving cell for one hybrid-ARQ round-trip time after receiving an "overload indicator." During this time, the UE shall not allow the serving grant to increase beyond the limit resulting from the "overload indicator." This avoids situations where the non-serving cell reduces the data rate to avoid an overload situation in the non-serving cell, followed by the serving cell increasing the data rate to utilize the interference headroom suddenly available, thus causing the overload situation to reappear. The serving cell may also want to be careful with immediately increasing the serving grant to its previous level as the exact serving grant in the UE is not known to the serving cell in this case (although it may partly derive it by observing the happy bit). Furthermore, to reduce the impact from erroneous relative grants, the UE shall ignore relative grants from the serving cell during one hybrid-ARQ round-trip time after having received an absolute grant with the primary identity.

#### 9.3.1.2 Per-process scheduling

Individual hybrid-ARQ processes can be (de)activated, implying that not all processes in a UE are available for data transmission, as illustrated in Figure 9.21. Process (de)activation is only possible for 2 ms TTI; for 10 ms TTI all processes are permanently enabled. The reason for process deactivation is mainly to be able to limit the data rate in case of 2 ms E-DCH TTI (with 320-bit RLC PDUs and 2 ms TTI, the minimum data rate is 160 kbit/s unless certain processes are disabled), but it can also be used to enable TDMA-like operation in case of all UEs' uplink transmissions being synchronized. Activation of individual processes can either be done through RRC signaling or by using the activation flag being part of the absolute grant. The activation flag indicates whether only the current process is activated or whether all processes not disabled by RRC signaling are activated.

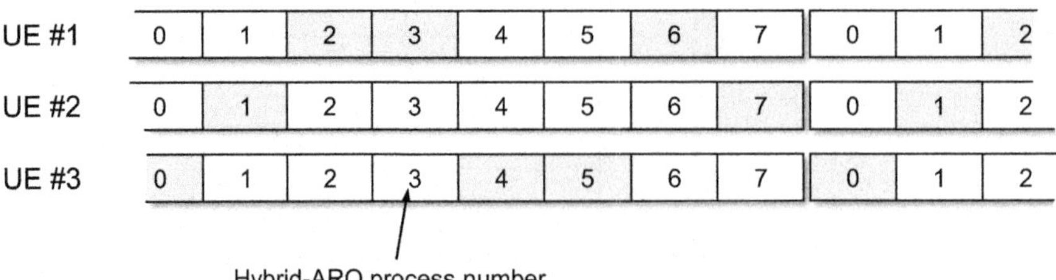


**FIGURE 9.21 Example of activation of individual hybrid-ARQ processes.**

Non-scheduled transmission can be restricted to certain hybrid-ARQ processes. The decision is taken by the serving Node B and informed to the other Node Bs in the active set through the RNC. Normally, the scheduler needs to operate with a certain margin to be able to handle any non-scheduled transmissions that may occur, and restricting non-scheduled transmissions to certain processes can therefore allow the scheduler to operate with smaller margins in the remaining processes.

#### 9.3.1.3 Scheduling requests

The scheduler needs information about the UE status and, as already discussed, two mechanisms are defined in HSUPA to provide this information: the in-band *scheduling information* and the out-band *happy bit*.

In-band scheduling information can be transmitted either alone or in conjunction with uplink data transmission. From a baseband perspective, scheduling information is no different from uplink user data. Hence, the same baseband processing and hybrid-ARQ operation is used.

In case of a standalone scheduling information, the E-DPDCH-to-DPCCH power offset to be used is configured by RRC signaling. To ensure that the scheduling information reaches the scheduler, the transmission is repeated until an ACK is received from the *serving* cell (or the maximum number of transmission attempts is reached). This is different from a normal data transmission, where an ACK from *any* cell in the active set is sufficient.

In case of simultaneous data transmission, the scheduling information is transmitted using the same hybrid-ARQ profile as the highest-priority MAC-d flow in the transmission (see Section 9.3.1.5 for a discussion on hybrid-ARQ profiles). In this case, periodic triggering will be relied upon for reliability. Scheduling information can be transmitted using any hybrid-ARQ process, including processes deactivated for data transmission. This is useful to minimize the delay in the scheduling mechanism.

The in-band scheduling information consists of 18 bits, containing information about

- identity of the highest-priority logical channel with data awaiting transmission, four bits;
- buffer occupancy, five bits indicating the total number of bytes in the buffer and four bits indicating the fraction of the buffer occupied with data from the highest-priority logical channel; and
- available transmission power relative to DPCCH, five bits.

Since the scheduling information contains information about both the total number of bits and the number of bits in the highest-priority buffer, the scheduler can ensure that UEs with high-priority data are served before UEs with low-priority data, a useful feature at high loads. The network can configure for which flows scheduling information should be transmitted.

Several rules for when to transmit scheduling information are defined. These are as follows:

- If padding allows transmission of scheduling information. Clearly, it makes sense to fill up the transport block with useful scheduling information rather than dummy bits.

- If the serving grant is zero or all hybrid-ARQ processes are deactivated and data enters the UE buffer. Obviously, if data enters the UE but the UE has no valid grant for data transmission, a grant should be requested.
- If the UE does have a grant, but incoming data has higher priority than the data currently in the buffer. The presence of higher-priority data should be conveyed to the Node B as it may affect its decision to scheduler the UE in question.
- Periodically as configured by RRC signaling (although scheduling information is not transmitted if the UE buffer size equals zero).
- At serving cell, change to provide the new cell with information about the UE status.

#### 9.3.1.4 Node B hardware resource handling in soft handover

From a Node B internal hardware allocation point of view, there is a significant difference between the serving cell and the non-serving cells in soft handover: the serving cell has information about the scheduling grant sent to the UE and, therefore, knowledge about the maximum amount of hardware resources needed for processing transmissions from this particular UE, information that is missing in the non-serving cells. Internal resource management in the non-serving cells therefore requires some attention when designing the scheduler. One possibility is to allocate sufficient resources for the highest possible data rate the UE is capable of. Obviously, this does not imply any restrictions to the data rates the serving cell may schedule, but may, depending on the implementation, come at a cost of less efficient usage of processing resources in the non-serving Node Bs. To reduce this cost, the highest data rates the scheduler is allowed to assign could be limited by the scheduler design. Alternatively, the non-serving Node B may under-allocate processing resources, knowing that it may not be able to decode the first few TTIs of a UE transmission. Once the UE starts to transmit at a high data rate, the non-serving Node B can reallocate resources to this UE, assuming that it will continue to transmit for some time. Non-serving cells may also try to listen to the scheduling requests from the UE to the serving cell to get some information about the amount of resources the UE may need.

#### 9.3.1.5 Quality-of-service support

The scheduler operates per UE, that is it determines which UE is allowed to transmit and at what maximum resource consumption. However, each UE may have several different flows of different priority. For example, VoIP and RRC signaling typically have a higher priority than a background file upload. Since the scheduler operates per UE, the control of different flows within a UE is *not* directly controlled by the scheduler. In principle, this could be possible, but it would increase the amount of downlink control signaling. For HSUPA, an E-TFC-based mechanism for quality-of-service support has been selected. Hence, as described earlier, the scheduler handles resource allocation *between UEs*, while the E-TFC selection controls resource allocation between flows *within the UE.*

The basis for QoS support is so-called hybrid-ARQ profiles, one per MAC-d flow in the UE. A hybrid-ARQ profile consists of a power offset attribute and a maximum number of transmissions allowed for a MAC-d flow.

The power offset value is used to determine the hybrid-ARQ operating point, which is directly related to the number of retransmissions. In many cases, several retransmissions may fit within the allowed delay budget. Exploiting multiple transmission attempts together with soft combining is useful to reduce the cost of transmitting at a certain data rate, as discussed in conjunction with hybrid-ARQ.

However, for certain high-priority MAC-d flows, the delays associated with multiple hybrid-ARQ retransmissions may not be acceptable. This could, for example, be the case for RRC signaling such as handover messages for mobility. Therefore, for these flows, it is desirable to increase the E-DPDCH transmission power, thereby increasing the probability for the data to be correctly received at the first transmission attempt. This is achieved by configuring a higher power offset for hybrid-ARQ profiles associated with high-priority flows. Of course, the transmission must be within the limits set by the serving grant. Therefore, the payload is smaller when transmitting high-priority data with a larger power offset than when transmitting low-priority data with a smaller power offset.

The power needed for the transmission of an E-DCH transport block is calculated including the power offset obtained from the hybrid-ARQ profile for flow to be transmitted. The required transmit power for each possible transport-block size can then be calculated by adding (in dB) the E-DPDCH-to-DPCCH power offset given by the transport block-size and the power offset associated with the hybrid-ARQ profile. The UE then selects the largest possible payload, taking these power offsets into account, which can be transmitted within the power available for the E-DCH (Figure 9.22).

Absolute priorities for logical channels are used; that is, the UE maximizes the data rate for high-priority data and only transmits data from a low priority in a TTI if

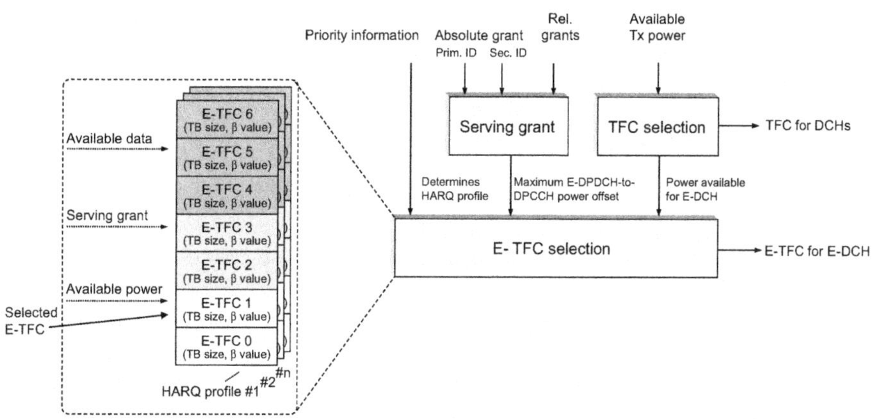


**FIGURE 9.22 E-TFC selection and hybrid-ARQ profiles.**

all data with higher priority has been transmitted. This ensures that any high-priority data in the UE is served before any low-priority data.

If data from more than one MAC-d flow is included in a TTI, the power offset associated with the MAC-d flow with the logical channel with the highest priority shall be used in the calculation. Therefore, if multiple MAC-d flows are multiplexed within a given transport block, the low-priority flows will get a "free ride" in this TTI when multiplexed with high-priority data.

There are two ways of supporting guaranteed bit rate services: scheduled and nonscheduled transmissions. With scheduled transmission, the Node B schedules the UE sufficiently frequently and with a sufficiently high bit rate to support the guaranteed bit rate. With nonscheduled transmission, a flow using non-scheduled transmission is defined by the RNC and configured in the UE through RRC signaling. The UE can transmit data belonging to such a flow without first receiving any scheduling grant. An advantage with the scheduled approach is that the network has more control of the interference situation and the power required for downlink ACK/NACK signaling and may, for example, allocate a high bit rate during a fraction of the time while still maintaining the guaranteed bit rate in average. Non-scheduled transmissions, on the other hand, are clearly needed at least for transmitting the scheduling information in case the UE does not have a valid scheduling grant.

### 9.3.2 FURTHER DETAILS ON HYBRID-ARQ OPERATION

Hybrid-ARQ for HSUPA serves a similar purpose as for the HSDPA – to provide robustness against occasional transmissions errors. It can also, as already discussed, be used to increase the link efficiency by targeting multiple hybrid-ARQ retransmission attempts.

The hybrid-ARQ for the E-DCH operates on a single transport block; that is, whenever the E-DCH CRC indicates an error, the MAC-e in the Node B can request a retransmission representing the same information as the original transport block. Note that there is a single transport block per TTI. Thus, it is not possible to mix initial transmission and retransmissions within the same TTI.

Incremental redundancy is used as the basic soft combining mechanism; that is, retransmissions typically consist of a different set of coded bits than the initial transmission. Note that, per definition, the set of information bits must be identical between the initial transmission and the retransmissions. For HSUPA, this implies that the E-DCH transport format, which is defined by the transport-block size and includes the number of physical channels and their spreading factors, remains unchanged between transmission and retransmission. Thus, the number of channel bits is identical for the initial transmission and the retransmissions. However, the rate matching pattern will change in order to implement incremental redundancy. The transmission power may also be different for different transmission attempts, for example, due to DCH activity.

The physical layer processing supporting the hybrid-ARQ operation is similar to the one used for HS-DSCH, although only a single rate-matching stage is used. The

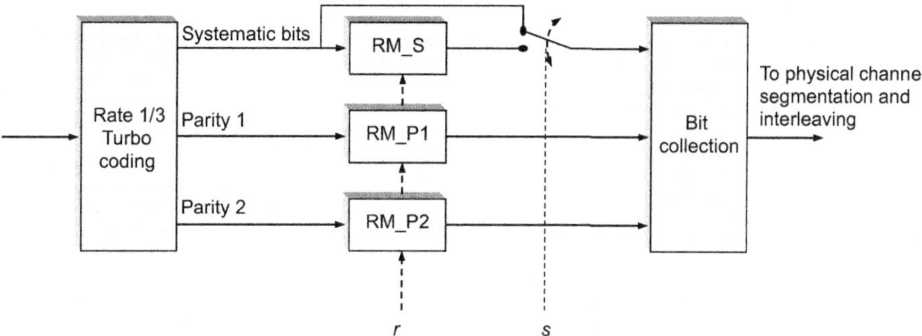


**FIGURE 9.23 E-DCH rate matching and the *r* and *s* parameters. The bit collection procedure is identical to the QPSK bit collection for HS-DSCH.**

reason for two-stage rate matching for HS-DSCH was to handle memory limitations in the UE, but for the E-DCH, any Node B memory limitations can be handled by proper network configuration. For example, the network could restrict the number of E-TFCs in the UE such that the UE cannot transmit more bits than the Node B can buffer.

The purpose of the E-DCH rate matching, illustrated in Figure 9.23, is twofold:

- to match the number of coded bits to the number of physical channel bits on the E-DPDCH available for the selected E-DCH transport format; and
- to generate different sets of coded bits for incremental redundancy as controlled by the two parameters *r* and *s*, as described below.

The number of physical channel bits depends on the spreading factor and the number of E-DPDCHs allocated for a particular E-DCH transport format. In other words, part of the E-TFC selection is to determine the number of E-DPDCHs and their respective spreading factors. From a performance perspective, coding is always better than spreading and, preferably, the number of channelization codes should be as high as possible and their spreading factor as small as possible. This would avoid puncturing and result in full utilization of the rate 1/3 mother Turbo code. At the same time, there is no point in using a lower spreading factor than necessary to reach rate 1/3 as this would only lead to excessive repetition in the rate-matching block. Furthermore, from an implementation perspective, the number of E-DPDCHs should be kept as low as possible to minimize the processing cost in the Node B receiver as each E-DPDCH requires one set of despreaders.

To fulfill these partially contradicting requirements, the concept of *Puncturing Limit* (PL) is used to control the maximum amount of puncturing the UE is allowed to perform. The UE will select an as small number of channelization codes and as high spreading factor as possible without exceeding the puncturing limits, that is, not puncture more than a fraction of $(1 - PL)$ of the coded bits. This is illustrated in Figure 9.24, where it is seen that puncturing is allowed up to a limit before additional E-DPDCHs are used. Two puncturing limits, $PL_{\text{max}}$ and $PL_{\text{non-max}}$, are defined.

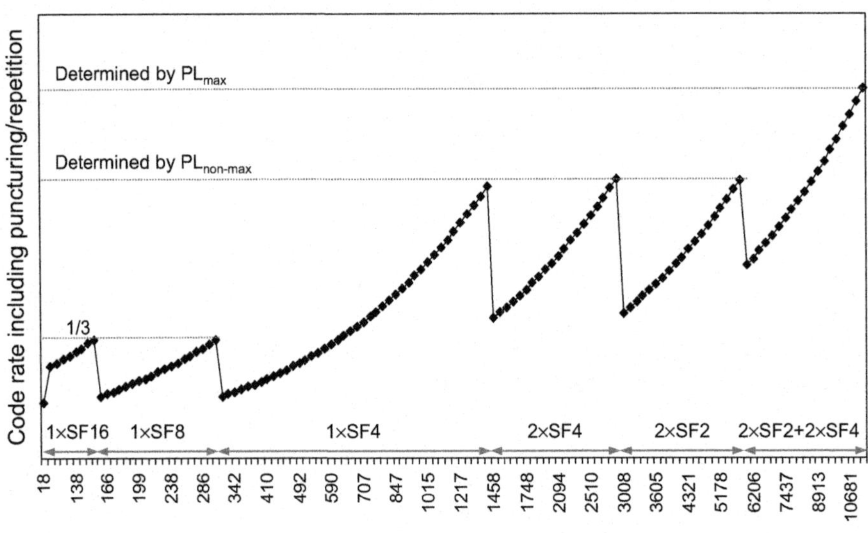


**FIGURE 9.24 Amount of puncturing as a function of the transport-block size.**

The limit $PL_{max}$ is determined by the UE category and is used if the number of E-DPDCHs and their spreading factors are equal to the UE capability, and the UE therefore cannot increase the number of E-DPDCHs. Otherwise, $PL_{non\text{-}max}$, which is signaled to the UE at the setup of the connection, is used. The use of two different puncturing limits, instead of a single one as for the DCH, allows for a higher maximum data rate as more puncturing can be applied for the highest data rates. Typically, additional E-DPDCHs are used when the code rate is larger than approximately 0.5. For the highest data rates, on the other hand, a significantly larger amount of puncturing is necessary as it is not possible to further increase the number of codes.

The puncturing (or repetition) is controlled by the two parameters *r* and *s* in the same way as for the second HS-DSCH rate matching stage (Figure 9.23). If $s = 1$, systematic bits are prioritized and an equal amount of puncturing is applied to the two streams of parity bits, while if $s = 0$, puncturing is primarily applied to the systematic bits. The puncturing pattern is controlled by the parameter *r*. For the initial transmission attempt *r* is set to zero and is increased for the retransmissions. Thus, by varying *r*, multiple, possibly partially overlapping, sets of coded bits representing the same set of information bits can be generated. Note that a change in *r* also affects the puncturing pattern, even if *r* is unchanged, as different amounts of systematic and parity bits will be punctured for the two possible values of *s*.

Equal repetition for all three streams is applied if the number of available channel bits is larger than the number of bits from the Turbo coder; otherwise, puncturing is applied. Unlike the DCH, but in line with the HS-DSCH, the E-DCH rate matching may puncture the systematic bits as well and not only the parity bits. This is used for incremental redundancy, where some retransmissions contain mainly parity bits.

The values of $s$ and $r$ are determined from the *redundancy version* (RV), which in turn is linked to the RSN. The RSN is set to zero for the initial transmission and incremented by one for each retransmission, as described earlier.

Compared to the HS-DSCH, one major difference is the support for soft handover on the E-DCH. As briefly mentioned above, not all involved cells may receive all transmission attempts in soft handover. Self-decodable transmissions, $s = 1$, are therefore beneficial in these situations as the systematic bits are more important than the parity bits for successful decoding. If full incremental redundancy is used in soft handover, there is a possibility that the first transmission attempt, containing the systematic bits ($s = 1$), is not reliably received in a cell, while the second transmission attempt, containing mostly parity bits ($s = 0$), is received. This could result in degraded performance. However, the data rates in soft handover are typically somewhat lower (the code rate is lower) as the UE in most cases is far from the base station when entering soft handover. Therefore, the redundancy versions are defined such that all transmissions are self-decodable ($s = 1$) for transport formats where the initial code rate is less than 0.5, while the remaining transport formats include retransmissions that are not self-decodable. Thus, thanks to this design, self-decodability "comes for free" when in soft handover. The design is also well matched to the fact that incremental redundancy (i.e., $s = 0$ for some of the retransmissions) provides most of the gain when the initial code rate is high.

The mapping from RSN via RV to the $r$ and $s$ parameters, illustrated in Figure 9.25, is mandated in the specification and is not configurable, with the exception that higher layer signaling can be used to mandate the UE to always use RV = 0, regardless of the RSN. This implies that the retransmission consists of exactly the same coded bits as the initial transmission (Chase combining) and can be used if the memory capabilities of the Node B are limited. Note that, for RSN = 3, the RV is linked to the (sub)frame number. The reason is to allow for variations in the puncturing pattern even for situations when more than three retransmissions are used.

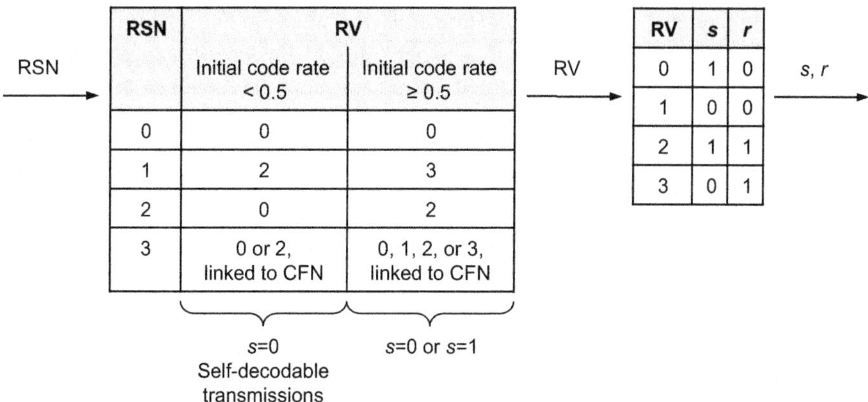


**FIGURE 9.25 Mapping from RSN via RV to *s* and *r*.**

#### 9.3.2.1 Protocol operation

The hybrid-ARQ protocol uses multiple stop-and-wait hybrid-ARQ processes similar to HS-DSCH. The motivation is to allow continuous transmission, which cannot be achieved with a single stop-and-wait scheme, and at the same time have some of the simplicity of a stop-and-wait protocol.

As already touched upon several times, the support for soft handover is one of the major differences between the uplink and the downlink. This has also impacted the number of hybrid-ARQ processes. For HSDPA, this number is configurable to allow for different Node B implementations. Although the same approach could be taken for HSUPA, soft handover between Node Bs from different vendors would be complicated. In soft handover, all involved Node Bs need to use the same number of hybrid-ARQ processes, which partially removes the flexibility of having a configurable number as all Node Bs must support at least one common configuration. To simplify the overall structure, a fixed number of hybrid-ARQ processes to be used by all Node Bs are defined. The number of processes is strongly related to the timing of the ACK/NACK transmission in the downlink (see the discussion on control signaling timing for details). For the two TTIs of 10 and 2 ms, the number of processes, $N_{HARQ}$, is 4 and 8, respectively. This results in a total hybrid-ARQ round-trip time of $4 \times 10 = 40$ ms and $8 \times 2 = 16$ ms, respectively.

The use of a synchronous hybrid-ARQ protocol is a distinguishing feature compared to HSDPA. In a synchronous scheme, the hybrid-ARQ process number is derived from the (sub)frame number and is not explicitly signaled. This implies that the transmissions in a given hybrid-ARQ process can only be made once every $N_{HARQ}$ TTI. This also implies that a retransmission (when necessary) always occurs $N_{HARQ}$ TTIs after the previous (re)transmission. Note that this does not affect the delay until a first transmission can be made since a data transmission can be started in any available process. Once the transmission of data in a process has started, retransmissions will be made until either an ACK is received or the maximum number of retransmissions has been reached (the maximum number of retransmissions is configurable by the RNC via RRC signaling). The retransmissions are done without the need for scheduling grants; only the initial transmission needs to be scheduled. As the scheduler in the Node B is aware of whether a retransmission is expected or not, the interference from the (non-scheduled) retransmissions can be taken into account when forming the scheduling decision for other users.

#### 9.3.2.2 In-sequence delivery

Similar to the case for HS-DSCH, the multiple hybrid-ARQ processes of E-DCH cannot, in themselves, ensure in-sequence delivery, as there is no interaction between the processes. Also, in soft handover situations, data is received independently in several Node Bs and can therefore be received in the RNC in a different order than transmitted. In addition, differences in Iub/Iur transport delay can cause out-of sequence delivery to RLC. Hence, in-sequence delivery must be implemented on top of the MAC-e entity, and a reordering entity in the RNC has been defined for this purpose in a separate MAC entity, the MAC-es. In E-DCH, the reordering is always performed per logical channel such that all data for a logical channel is delivered in-sequence to the corresponding RLC entity. This can be compared to HS-DSCH where the reordering is performed in configurable reordering queues.

The actual mechanism to perform reordering in the RNC is implementation specific and not standardized, but typically similar principles as specified for the HS-DSCH are used. Therefore, each MAC-es PDU transmitted from the UE includes a *Transmission Sequence Number* (TSN), which is incremented for each transmission on a logical channel. By ordering the MAC-es PDUs based on TSN, in-sequence delivery to the RLC entities is possible.

To illustrate the reordering mechanism, consider the situation shown in Figure 9.26. The MAC-es PDUs 0, 2, 3, and 5 have been received in the RNC while MAC-es PDUs 1 and 4 have not yet been received. The RNC can in this situation not know why PDUs 1 and 4 are missing and needs to store PDUs 2, 3, and 5 in the reordering buffer. As soon as PDU 1 arrives, PDU 1, 2, and 3 can be delivered to RLC.

The reordering mechanism also needs to handle the situation where PDUs are permanently lost due to, for example, loss over Iub, errors in the hybrid-ARQ signaling, or in case the maximum number of retransmissions was reached without successful decoding. In those situations, a stall avoidance mechanism is needed, that is, a mechanism to prevent the reordering scheme that waits for PDUs that never will arrive. Otherwise, PDU 5 in Figure 9.26 would never be forwarded to RLC.

Stall avoidance can be achieved with a timer similar to what is specified for the UE in HS-DSCH. The stall avoidance timer delivers packets to the RLC entity if a PDU has been missing for a certain time. If the stall avoidance mechanism delivers PDUs to the RLC entity too early, it may result in unnecessary RLC retransmissions when the PDU is only delayed, for example, because of too many hybrid-ARQ retransmissions. If, on the other hand, the PDUs are kept too long in the reordering buffer, it will also degrade the performance since the delay will increase.

To improve the stall avoidance mechanism, the Node B signals the time (frame and sub-frame number) when each PDU was correctly decoded to the RNC, as well as how many retransmissions were needed before the PDU was successfully received. The RNC can use this information to optimize the reordering functionality. Consider the example in Figure 9.26. If PDU 5 in the example above needed four retransmissions and the maximum number of retransmission attempts configured equals five,

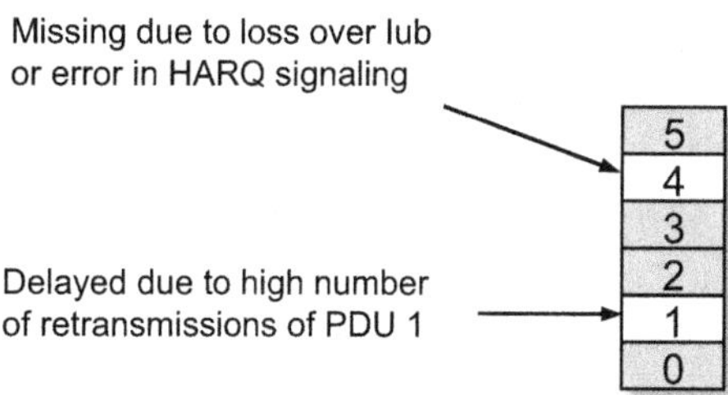


**FIGURE 9.26 Reordering mechanism.**

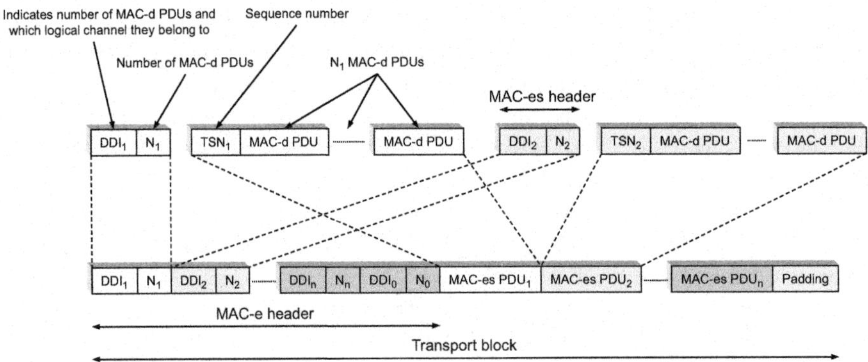


**FIGURE 9.27 Structure and format of the MAC-e/es PDU.**

the RNC knows that if PDU 4 has not arrived within one hybrid-ARQ round-trip time (plus some margin to consider variations in Iub delay) after PDU 5, it is permanently lost. In this case, the RNC only has to wait one round-trip time before delivering PDU 5 to RLC.

To support reordering and demultiplexing of the PDUs from different MAC-d flows, the appropriate information is signaled in-band in the form of MAC-es and MAC-e headers. The structures of the MAC-e/es headers are illustrated in Figure 9.27.

Several MAC-d PDUs of the same size and from the same logical channel are concatenated. The *Data Description Indicator* (DDI) provides information about the logical channel from which the MAC-d PDUs belong, as well as the size of the MAC-d PDUs. The number of MAC-d PDUs is indicated by *N*. The TSN, used to support reordering, as described in the previous section, is also attached to the set of MAC-d PDUs.

The MAC-e header consists of a number of *DDI* and *N* pairs. A mapping is provided by RRC from the *DDI* field to a MAC-d PDU size, logical channel ID, and MAC-d flow ID. The logical channel also uniquely identifies the reordering queue since reordering in E-DCH is performed per logical channel.

The sequence of *DDI* and *N* fields is completed with a predefined value of *DDI* to indicate the end of the MAC-e header. After the MAC-e header follows a number of MAC-es PDUs, where the number of MAC-es PDUs is the same as the number of *DDI* and *N* pairs in the MAC-e header (not counting the predefined DDI value indicating the end of the MAC-e header). After the last MAC-es PDU, the MAC-e PDU may contain padding to fit the current transport-block size.

When appropriate, the MAC-e header also includes 18 bits of scheduling information using a special DDI value.

### 9.3.3 POWER SETTING FOR E-DPDCH

As discussed in Chapter 2, achieving larger data rates in a given bandwidth requires more power. The power of E-DPDCH is set relative to the power of DPCCH by means of gain factors ($\beta_{ed}$). Strictly speaking, since the gain factors are expressed as amplitude values, the power ratio between E-DPDCH and DPCCH is given by the square of the gain factor. To increase the E-DPDCH power, and thereby the supported rate, either the DPCCH power or the gain factors can be increased. In general, it is beneficial to decouple DPCCH power and data rate. For system stability and predictability reasons, it is desirable to operate the system at a rather fixed DPCCH power level. Also, a low DPCCH power means less overhead and thereby decreased interference, especially for inactive users only transmitting DPCCH. Clearly, it is unfortunate if an inactive user creates excessive interference because it needs to support high rates when scheduled. Thus, a better solution is to adapt the gain factors as a function of the rate. Consequently, there is a close interaction between the E-TFC selection procedure and the gain factor setting procedure.

The E-TFC selection procedure is discussed in detail in Sections 9.2.3 and 9.3.1. The mapping from serving grant (or available data power) to transport-block size is essentially determined by two factors: the hybrid-ARQ profile and two sets of reference values. As discussed in 9.3.1.5, different hybrid-ARQ profiles can be used as QoS support. A high-priority profile corresponds to less data given a fixed serving grant (power) compared to a low-priority profile, but on the other hand, a high-priority profile increases the likelihood that the transmission can be decoded successfully with fewer hybrid-ARQ transmissions. The second and most important input for determining the mapping from power to data rate is provided by two sets of network configured reference values. One set contains reference amplitude values (or rather indexes that are mapped to amplitude values) and the other set the corresponding E-TFCI (transport-block size) values. These two sets provide flexibility in how the network can be operated. In general, configuring a more aggressive reference setting means that a fixed serving grant corresponds to a larger transport-block size compared to configuring a more moderate reference setting. It is possible to have several reference values in order to allow different mappings depending on the operating point, for example, one mapping in the power-limited region and another in the bandwidth-limited region.

There are two different schemes for determining the gain factors for E-DPDCH: *extrapolation* and *interpolation*. Extrapolation was introduced in release 6, whereas interpolation was standardized in release 7. Release 7 contained a number of enhancements related to the E-DCH power setting because of the introduction of 16-QAM, which required new tools to facilitate more flexible power settings and improved channel estimation. Most notably, the interpolation scheme and E-DPCCH boosting were introduced. The intention with E-DPCCH boosting is to provide means for enhanced channel estimation by ensuring that a fixed data-to-total-pilot power ratio is maintained irrespectively of the E-DPDCH power. E-DPCCH boosting is described in detail in Chapter 10.

#### 9.3.3.1 Extrapolation formula

The extrapolation formula, as the name suggests, uses rudimentary extrapolation to set the gain factor as a function of the transport-block size. In fact, the technique corresponds to linear extrapolation through (0, 0) corresponding to no power and no data and a reference point. In other words, the prediction follows a straight line through the origin and the values in the reference sets (see Figure 9.28 for an illustration).

The unquantized amplitude factor is given from the predicted value corresponding to the current transport-block size. The unquantized amplitude value ($\beta_{ed,uq}/\beta_c$) is then essentially rounded to the nearest existing quantized value in a set of in total 30 possible quantized values ranging between –4.8 dB and 10.5 dB. Special considerations apply during compressed mode and power-limited scenarios.

The reference values need to be chosen such that the resulting mapping between gain factor and transport-block size (and vice versa) is unambiguous. This means that the gain factor as a function of transport-block size needs to follow a strictly increasing function in order to have a 1-to-1 mapping between power (serving grant) and transport-block size. The reason is that given a power offset (serving grant) there should not be an ambiguity about what transport-block size to use, and similar, given a transport-block size there should be only one corresponding power offset. The process of mapping power to transport-block size and vice versa using the extrapolation scheme is illustrated in Figure 9.28.

As discussed earlier, the maximum allowed power offset for E-DPDCH is determined by the serving grant. In fact, in scenarios where the transport-block size is fully determined by the serving grant, the extrapolation formula yields amplitude values such that the serving grant is distributed evenly between the E-DPDCHs.

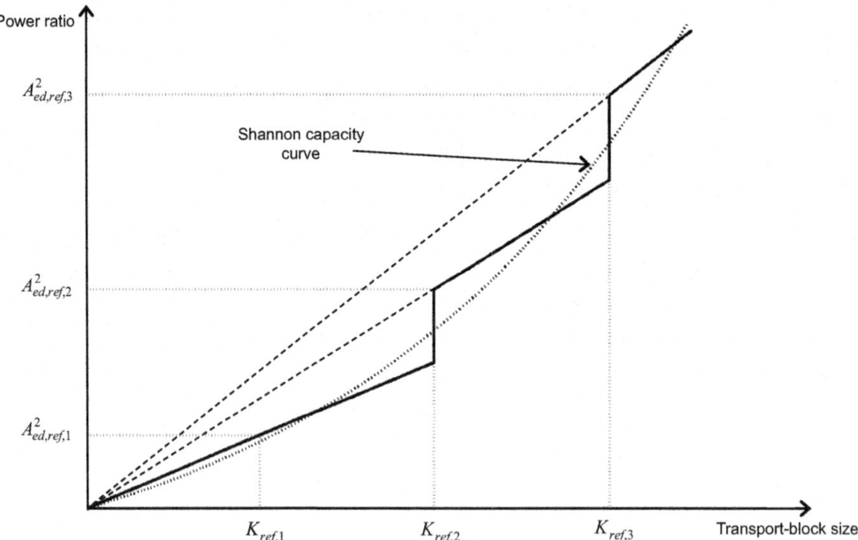


**FIGURE 9.28 Illustration of power to transport-block size mapping using the extrapolation formula. $A_{ed,ref},k$ and $K_{ref},k$ correspond to amplitude reference values and corresponding transport-block sizes.**

#### 9.3.3.2 Interpolation formula

The interpolation formula was introduced in release 7 since it was observed that the extrapolation formula was not flexible enough to efficiently set the power over a wide range of transport-block sizes. The reason is that while the relation between SNR and spectral efficiency is more or less linear in the power limited region (low to moderate rates), it is close to exponential in the bandwidth limited region (high rates). The interpolation formula employs linear interpolation to set the gain factor as a function of the transport-block size given two reference points, which provides more flexibility compared to the extrapolation approach, which essentially uses one reference point.

The process of mapping power to transport-block size and vice versa using the interpolation scheme is illustrated in Figure 9.29. More specifically, it is possible to get a mapping representing, for example, an exponential or logarithmic shape. The former would correspond to the ideal Shannon capacity curve, but the latter can be useful in practical scenarios where a higher rate may require a larger DPCCH SIR target. In practice, the optimal gain factor setting and corresponding reference values will depend on a number of factors, such as radio conditions and receiver algorithms. Irrespective of how the gain factor mapping is configured, the outer loop will adjust the SIR target to ensure that the appropriate receive power is achieved. Nevertheless, the outer loop converges rather slowly and in general it is undesirable to have a close connection between rate and SIR target. One factor that affects the power setting in

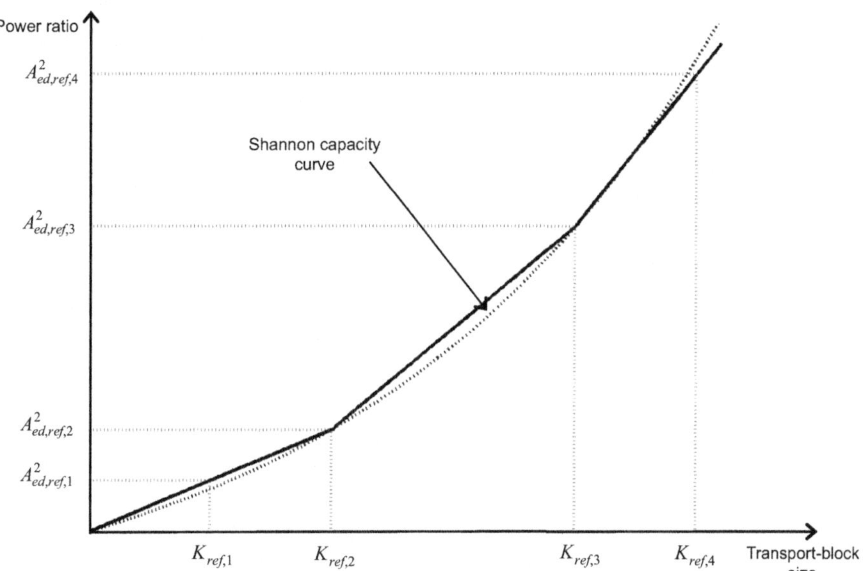


**FIGURE 9.29 Illustration of power to transport-block size mapping using the interpolation formula. $A_{ed,ref},k$ and $K_{ref},k$ correspond to amplitude reference values and corresponding transport-block sizes.**

practice is channel estimation. Higher rates require better channel estimation, especially when higher-order modulation is enabled. E-DPCCH boosting is one alternative for improving the channel estimate quality; see Chapter 10.

#### 9.3.3.3 Power scaling

The power of DPCCH is adjusted each slot based on the inner-loop power control TPC commands, and the powers of other uplink channels are set relative to DPCCH via the gain factors. The gain factor procedure for E-DPDCH is described in the previous section. If the total transmit power after DPCCH power adjustment and gain factor calculations would exceed the maximum allowed value, additional power scaling needs to be applied. In general, the E-TFC selection procedure ensures that the total transmit power satisfies the maximum power constraint. However, there are situations when additional power scaling is required, for example

- The E-TFC selection procedure is performed once every TTI, whereas the DPCCH power is updated every slot. Hence, there is a possibility that the total transmit power increases beyond the maximum allowed value, especially if the E-TFC selection procedure results in a power allocation close to the maximum allowed value.
- Because of a retransmission, a transport block with an E-TFCI corresponding to a power setting exceeding the allowed maximum value may need to be transmitted.
- In a coverage-limited scenario, the UE is transmitting close to its maximum power even though the E-TFC selection procedure allocates a potentially very small transport-block size. Also, in this case it is likely that the 10 ms TTI is configured, meaning that the inner-loop power control mechanism adjusts the DPCCH power up to 15 times per E-TFC selection procedure cycle, further emphasizing the number of power changes between E-TFC decisions.

The power scaling procedure depends on whether E-DCH and/or DCH are configured. The details of the power scaling procedure can be found in [1]. An illustration of the power scaling procedure is found in Figure 9.30.

For simplicity, consider a scenario where E-DCH is configured without an associated DCH. In this case, the power scaling procedure is illustrated in Figure 9.31. The first step is to scale down all E-DPDCH gain factors by an equal scaling factor to respective values $\beta_{ed,k,\text{reduced}}$ such that the maximum transmit power is satisfied. Needless to say, if only E-DCH is configured and the first power scaling step results in an E-DPDCH power too small to sustain a reasonable data rate, it can be questioned why send anything at all. Hence, in a second step it is ensured that the E-DPDCH power remains reasonable by imposing a minimum value for $\beta_{ed,k,\text{reduced}}$. The minimum value, $\beta_{ed,k,\text{reduced,min}}$, was fixed to 8/15 in release 6 but made RRC configurable in release 7. If the total UE transmit power is still greater than the maximum allowed value after step 2, a third step is performed where the DPCCH power is scaled down while keeping all gain factors fixed. The DPCCH power is scaled down such that the maximum allowed power is fulfilled. The power scaling procedure has been updated in later releases due to, for example, the introduction of dual-cell HSUPA, closed-loop transmit diversity, MIMO, and DTX/DRX [1].

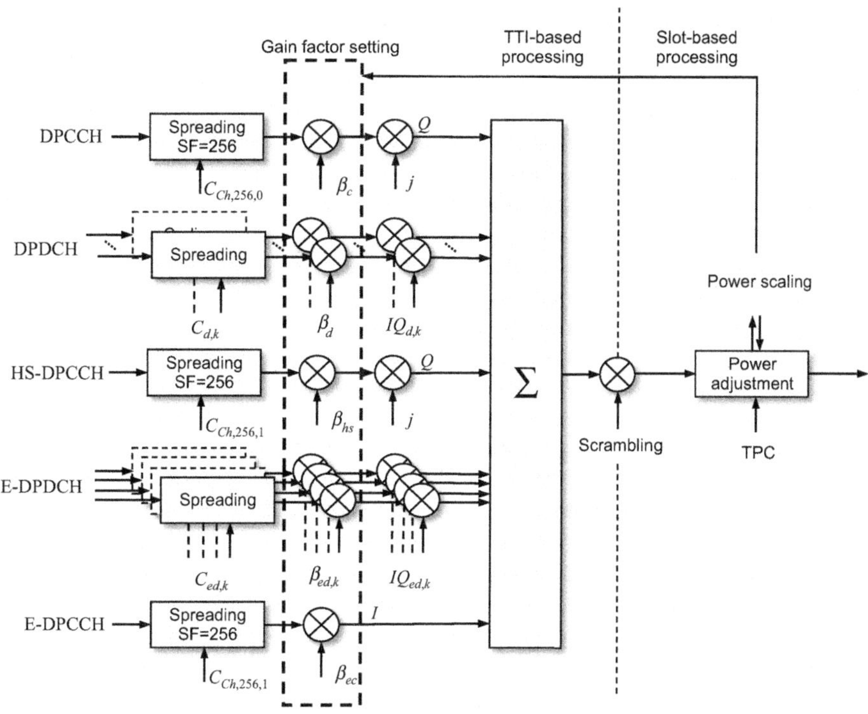


**FIGURE 9.30 Illustration of power scaling for the uplink.**

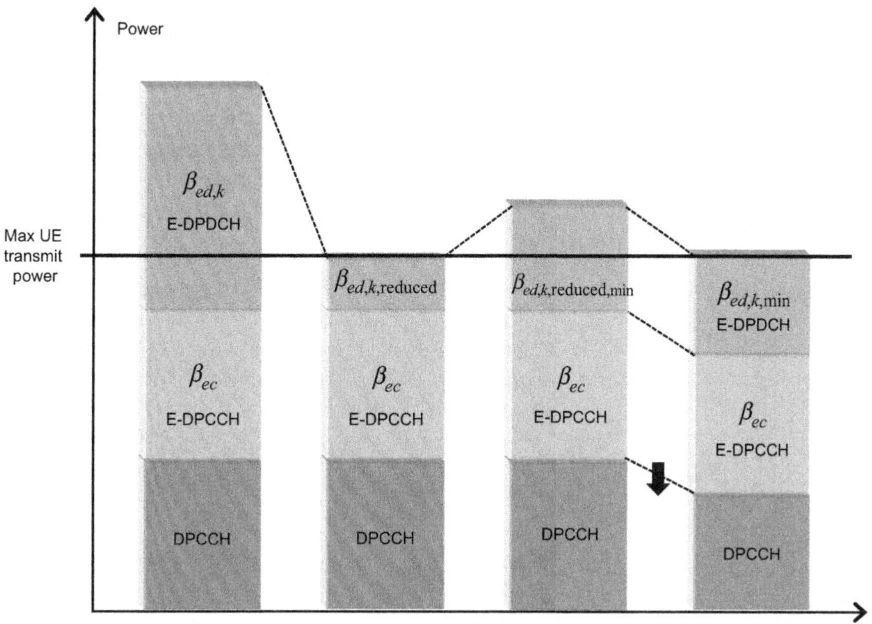


**FIGURE 9.31 Illustration of power scaling for E-DCH.**

### 9.3.4 CONTROL SIGNALING

To support E-DCH transmissions in the uplink, three downlink channels carrying out-band control signaling are defined:

1. The E-HICH is a dedicated physical channel transmitted from each cell in the active set and used to carry the hybrid-ARQ acknowledgments.
2. The E-AGCH is a shared physical channel transmitted from the serving cell only and used to carry the absolute grants.
3. The E-RGCH carries relative grants. From the serving cell, the E-RGCH is a dedicated physical channel, carrying the relative grants. From non-serving cells, the E-RGCH is a common physical channel, carrying the overload indicator.

Thus, a single UE will receive multiple downlink physical control channels. From the serving cell, the UE receives the E-HICH, E-AGCH, and E-RGCH. From each of the non-serving cells, the UE receives the E-HICH and the E-RGCH.

Out-band uplink control signaling is also required to indicate the E-TFC the UE selected, the RSN, and the happy bit. This information is carried on the uplink E-DPCCH.

In addition to the E-DCH-related out-band control signaling, downlink control signaling for transmission of power control bits is required. This is no different from WCDMA in general and is carried on the (F-)DPCH. Similarly, the DPCCH is present in the uplink to provide a phase reference for coherent demodulation as well. The overall E-DCH-related out-band control signaling is illustrated in Figure 9.32.

#### *9.3.4.1 E-HICH*

The E-HICH is a downlink dedicated physical channel, carrying the binary hybrid-ARQ acknowledgments to inform the UE about the outcome of the E-DCH detection at the Node B. The Node B transmits either ACK or NACK, depending on whether the decoding of the corresponding E-DCH transport block was successful or a retransmission is requested. To not unnecessarily waste downlink transmission power, nothing is transmitted on the E-HICH if the Node B did not detect a transmission attempt, that is, no energy was detected on the E-DPCCH or the E-DPDCH.

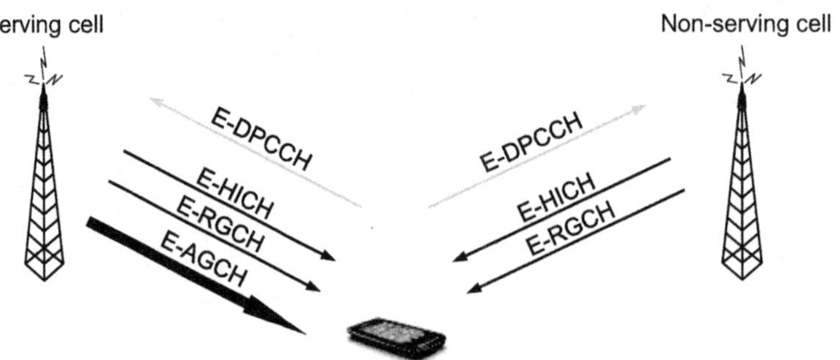


**FIGURE 9.32 E-DCH-related out-band control signaling.**

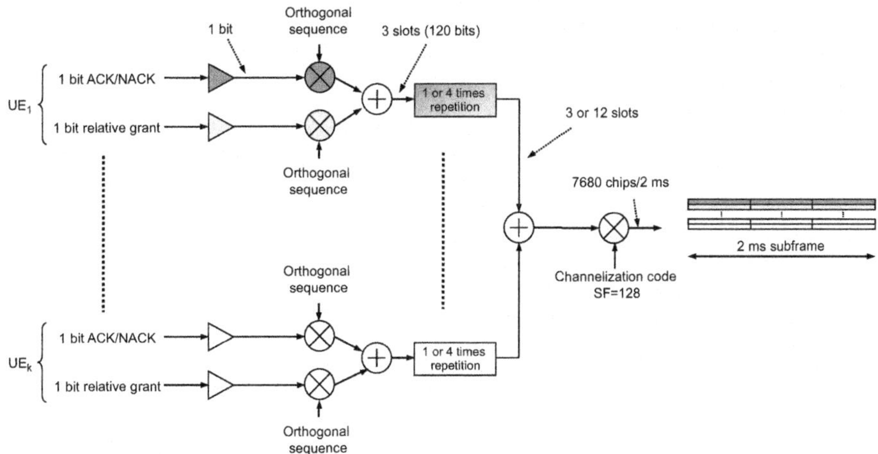


**FIGURE 9.33 E-HICH and E-RGCH structures (from the serving cell).**

Despite the fact that the ACK/NACK is a single bit of information, the ACK/NACK is transmitted with a duration of 2 or 8 ms, depending on the TTI configured.[4] This ensures that a sufficient amount of energy can be obtained to satisfy the relatively stringent error requirements of the ACK/NACK signaling, without requiring a too high peak power for the E-HICH.

To save channelization codes in the downlink, multiple ACK/NACKs are transmitted on a single channelization code of spreading factor 128. Each user is assigned an orthogonal signature sequence to generate a signal spanning 2 or 8 ms, as illustrated in Figure 9.33. The single-bit ACK/NACK is multiplied with a signature sequence of length 40 bits,[5] which equals one slot of bits at the specified spreading factor of 128. The same procedure is used for three or 12 slots, depending on the E-DCH TTI, to obtain the desired signaling interval of 2 or 8 ms. This allows multiple UEs to share a single channelization code and significantly reduces the amount of channelization codes that needs to be assigned for E-HICH.

As the mutual correlation between different signature sequences varies with the sequence index, signature sequence hopping is used to average out these differences. With hopping, the signature sequence of a certain UE changes from slot to slot using a hopping pattern,[6] as illustrated in Figure 9.34.

Both the E-HICH and the E-RGCH use the same structure and to simplify the UE implementation, the E-RGCH and E-HICH for a certain UE shall be allocated the same channelization code and scrambling code. Thus, with length 40 signature sequences, 20 users, each with one E-RGCH and one E-HICH, can share a single channelization

[4]The reason for 8 ms and not 10 ms is to provide some additional processing time in the Node B. See the timing discussion for further details.

[5]In essence, this is identical to defining a spreading factor of $40 \times 128 = 5120$.

[6]The use of hopping could also be expressed as a corresponding three-slot-long signature sequence.

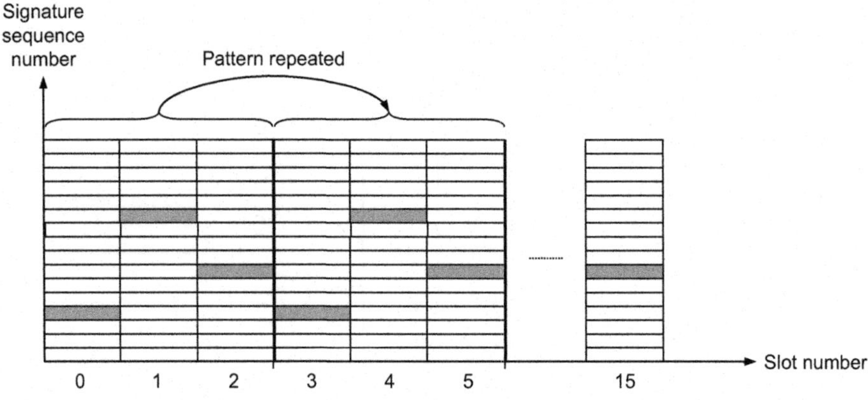


**FIGURE 9.34 Illustration of signature sequence hopping.**

code. Note that the power for different users' E-HICH and E-RGCH can be set differently, despite the fact that they share the same code.

When a single Node B is handling multiple cells and a UE is connected to several of those cells, that is, the UE is in softer handover between these cells, it is reasonable to assume that the Node B will transmit the same ACK/NACK information to the UE in all these cells. Hence, the UE shall perform soft combining of the E-HICH in this case, and the received signal on each of the E-HICH-es being received from the same Node B shall be coherently added prior to decoding. This is the same approach as used for combining of power control bits already from the first release of WCDMA.

The modulation scheme used for the E-HICH is different for the serving and the non-serving cells. In the serving radio link set, BPSK is used, while for non-serving radio link sets, *On–Off Keying* (OOK) is used such that NACK is mapped to DTX (no energy transmitted). The reason for having different mappings is to minimize downlink power consumption. Generally, BPSK is preferable if ACK is transmitted for most of the transmissions, while the average power consumption is lower for OOK when NACK is transmitted more than 75% of the time as no energy is transmitted for the NACK. When the UE is not in soft handover, there is only the serving cell in the active set and this cell will detect the presence of an uplink transmission most of the time. Thus, BPSK is preferred for the serving cells. In soft handover, on the other hand, at most one cell is typically able to decode the transmission, implying that most of the cells will transmit a NACK, making OOK attractive. However, note that the Node B will only transmit an ACK or a NACK in case it detected the presence of an uplink transmission attempt. If even the presence of data transmission is not detected in the Node B, nothing will be transmitted, as described above. Hence, the E-HICH receiver in the UE must be able to handle the DTX case as well, although from a protocol point of view only the values ACK and NACK exist.

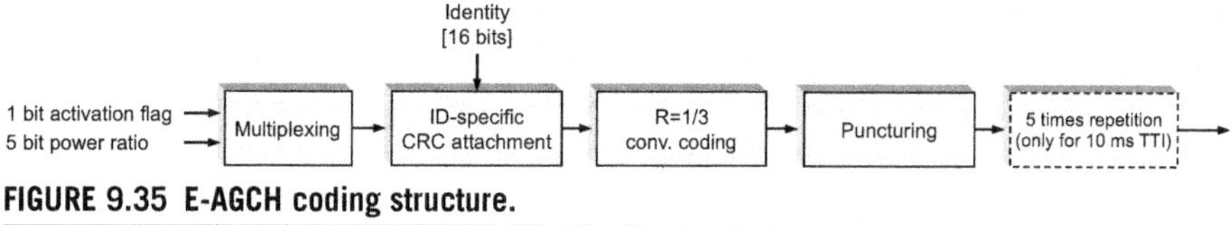


**FIGURE 9.35 E-AGCH coding structure.**

### *9.3.4.2 E-AGCH*

The E-AGCH is a shared channel, carrying absolute scheduling grants consisting of

- The maximum E-DPDCH/DPCCH power ratio the UE is allowed to use for the E-DCH (five bits).
- An activation flag (one bit), used for (de)activating individual hybrid-ARQ processes.
- An identity that identifies the UE (or group of UEs) for which the E-AGCH information is intended (16 bits). The identity is not explicitly transmitted but implicitly included in the CRC calculation.

Rate 1/3 convolutional coding is used for the E-AGCH and the coded bits are rate matched to 30 bits, corresponding to 2 ms duration at the E-AGCH spreading factor of 256 (Figure 9.35). In case of a 10 ms E-DCH TTI, the 2 ms structure is repeated five times. Note that a single channelization code can handle a cell with both TTIs and therefore it is not necessary to reserve two channelization codes in a cell with mixed TTIs. UEs with 2 ms TTI will attempt to decode each sub-frame of a 10 ms E-AGCH without finding its identity. Similarly, a 10 ms UE will combine five sub-frames before decoding and the CRC check will fail unless the grant was 10 ms long. For groupwise scheduling, it is unlikely that both 2 and 10 ms UEs will be given the same grant (although the above behavior might be exploited) and the absolute grants for these two groups of UEs can be sent separated in time on the same channelization code.

Each E-DCH-enabled UE receives one E-AGCH (although there may be one or several E-AGCH configured in a cell) from the serving cell. Although the UE is required to monitor the E-AGCH for valid information every TTI, a typical scheduling algorithm may only address a particular UE using the E-AGCH occasionally. The UE can discover whether the information is valid or not by checking the ID-specific CRC.

### *9.3.4.3 E-RGCH*

Relative grants are transmitted on the E-RGCH and the transmission structure used for the E-RGCH is identical to that of the E-HICH. The UE is expected to receive one relative grant per TTI from each of the cells in its active set. Thus, relative grants can be transmitted from both the serving and the non-serving cells.

From the serving cell, the E-RGCH is a dedicated physical channel and the signaled value can be one of +1, DTX, and −1, corresponding to UP, HOLD, and DOWN, respectively. Similar to the E-HICH, the duration of the E-RGCH equals 2 or 8 ms, depending on the E-DCH TTI configured.

From the non-serving cells, the E-RGCH is a common physical channel, in essence a common "overload indicator" used to limit the amount of inter-cell interference. The value on the E-RGCH from the non-serving cells (overload indicator) can only take the values DTX and −1, corresponding to "no overload" and DOWN, respectively. E-RGCH from the non-serving cells span 10 ms, regardless of the E-DCH TTI configured. Note that Figure 9.33 is representative for the serving cell as each UE is assigned a separate relative grant (from the non-serving cell, the E-RGCH is common to multiple UEs).

#### *9.3.4.4 Timing*

The timing structure for the E-DCH downlink control channels (E-AGCH, E-RGCH, E-HICH) is designed to fulfill a number of requirements. Additional timing bases in the UE are not desirable from a complexity perspective, and hence the timing relation should either be based on the common pilot or the downlink DPCH as the timing of those channels needs to be handled by the UE.

Common channels, the E-RGCH from the non-serving cell and the E-AGCH, are monitored by multiple UEs and must have a common timing. Therefore, the timing relation of these channels is defined as an offset relative to the common pilot. The duration of the E-AGCH is equal to the E-DCH TTI for which the UE is configured. For the E-RGCH from the non-serving cell, the duration is always 10 ms, regardless of the TTI. This simplifies mixing UEs with different TTIs in a single cell while providing sufficiently rapid inter-cell interference control.

Dedicated channels, the E-RGCH from the serving cell and the E-HICH, are unique for each UE. To maintain a similar processing delay in the UE and Node B, regardless of the UE timing offset to the common pilot, their timing is defined relative to the downlink DPCH.

The structure of the E-HICH, where multiple E-HICHs share a common channelization code, has influenced the design of the timing relations. To preserve orthogonality between users sharing the same channelization code, the (sub)frame structure of the E-HICHs must be aligned. Therefore, the E-HICH timing is derived from the downlink DPCH timing, adjusted to the closest 2 ms sub-frame not violating the smallest UE processing requirement.

The number of hybrid-ARQ processes directly affects the delay budget in the UE and Node B. The smaller the number of hybrid-ARQ processes, the better from a round-trip time perspective but also the tighter the implementation requirements. The number of hybrid-ARQ processes for E-DCH is fixed to four processes in case of a 10 ms TTI and 10 processes in case of a 2 ms TTI. The total delay budget is split between the UE and the Node B as given by the expressions relating the downlink DPCH timing to the corresponding E-HICH sub-frame. To allow for 2 ms extra Node B processing delays, without tightening the UE requirements, the E-HICH duration is 8 ms, rather than 10 ms in case of a 10 ms E-DCH TTI. Note that the acceptable UE and Node B processing delays vary in a 2 ms interval depending on the downlink DPCH timing configuration. For the UE, this effect is hard to exploit as it has no control over the network configuration, and the UE design therefore

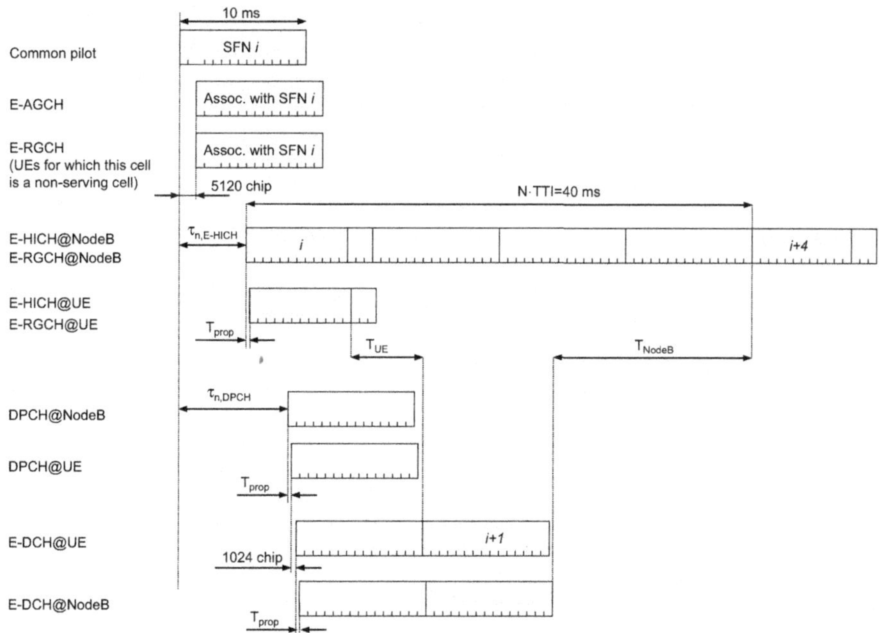


**FIGURE 9.36 Timing relation for downlink control channels, 10 ms TTI.**

must account for the worst case. The Node B, on the other hand, may, at least in principle, exploit this fact if the network is configured to obtain the maximum Node B processing time.

For simplicity, the timing of the E-RGCH from the serving cell is identical to that of the E-HICH. This also matches the interpretation of the relative grant in the UE as it is specified relative to the previous TTI in the same hybrid-ARQ process, that is, the same relation that is valid for the ACK/NACK.

The downlink timing relations are illustrated in Figure 9.36 for 10 ms E-DCH TTI and in Figure 9.37 for 2 ms TTI. An overview of the approximate processing delays in the UE and Node B can be found in Table 9.4.

## 9.3.5 UPLINK CONTROL SIGNALING: E-DPCCH

The uplink E-DCH-related out-band control signaling, transmitted on the E-DPCCH physical channel, consists of

- 2-bit RSN,
- 7-bit E-TFCI, and
- 1-bit rate request ("happy bit").

The E-DPCCH is transmitted in parallel to the uplink DPCCH on a separate channelization code with spreading factor 256. In this way, backward

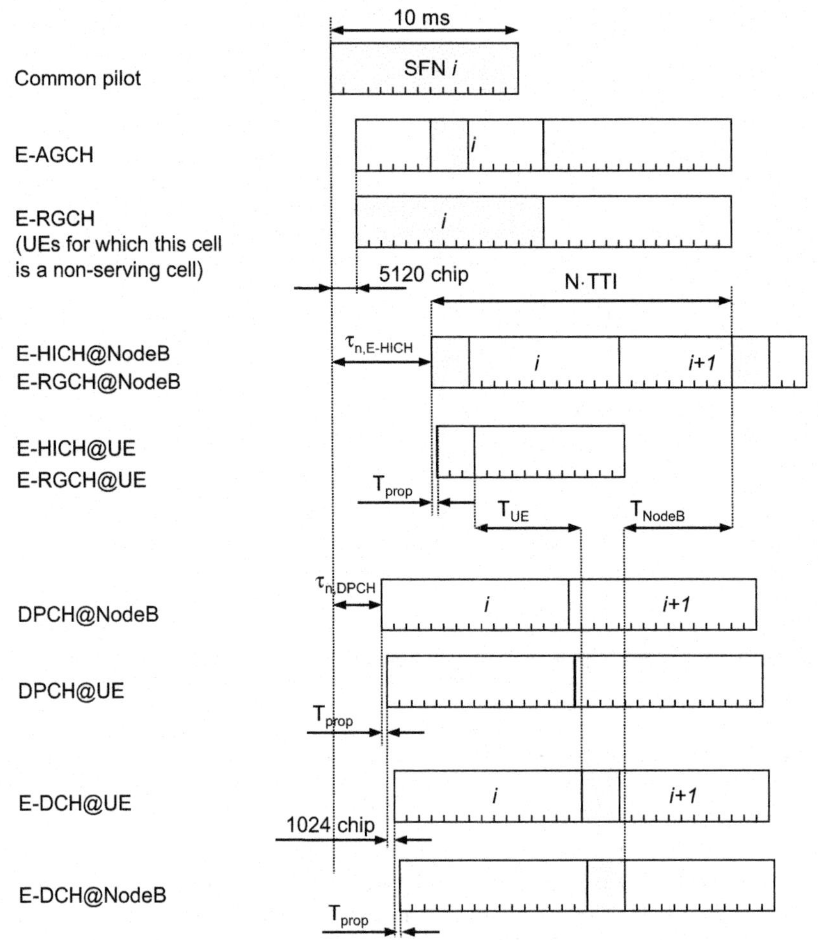


**FIGURE 9.37 Timing relation for downlink control channels, 2 ms TTI.**

**Table 9.4** Minimum UE and Node B Processing Time

| | 10 ms E-DCH TTI | 2 ms E-DCH TTI |
|---|---|---|
| Number of hybrid-ARQ processes | 4 | 8 |
| Minimum UE processing time | 5.56 ms | 3.56 ms |
| Minimum Node B processing time | 14.1 ms | 6.1 ms |

*Note that the propagation delay has to be included in the Node B timing budget.*

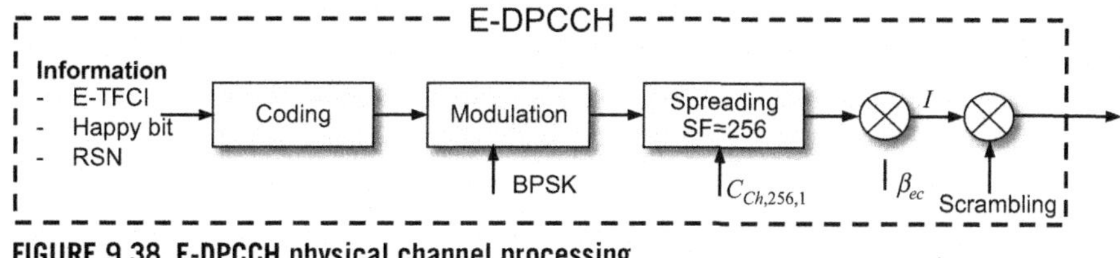


**FIGURE 9.38 E-DPCCH physical channel processing.**

compatibility is ensured in the sense that the uplink DPCCH has retained exactly the same structure as in earlier WCDMA releases. An additional benefit of transmitting the DPCCH and the E-DPCCH in parallel, instead of time multiplexed on the same channelization code, is that it allows for independent power-level setting for the two channels. This is useful as the Node B performance may differ between implementations. Figure 9.38 illustrates the physical layout of E-DPCCH. In release 6, the power of E-DPCCH is set relative to the DPCCH via the gain factor $\beta_{ec}$ according to

$$\beta_{ec} = \beta_c \cdot A_{ec}, \tag{9.1}$$

where $\beta_c$ denotes the gain factor of the DPCCH and the quantized amplitude value $A_{ec}$ is determined from a corresponding RRC signaled index. In total, nine quantized amplitude values raging between –4.8 dB and 3.0 dB exist in release 6. In release 7, E-DPCCH boosting was introduced, which provides an alternative way of setting the power of the E-DPCCH (see Chapter 10).

The complete set of 10 E-DPCCH information bits are encoded into 30 bits using a second-order Reed–Müller code (the same block code as used for coding of control information on the DPCCH). The 30 bits are transmitted over three E-DPCCH slots for the case of 2 ms E-DCH TTI (Figure 9.39). In case of 10 ms E-DCH TTI, the 2 ms structure is repeated five times. The E-DPCCH timing is aligned with the DPCCH (and consequently the DPDCH and the E-DPDCH).

To minimize the interference generated in the cell, the E-DPCCH is only transmitted when the E-DPDCH is transmitted. Consequently, the Node B has to detect whether the E-DPCCH is present or not in a certain sub-frame (DTX detection) and, if present, decode the E-DPCCH information. Several algorithms are possible for DTX detection, for example, comparing the E-DPCCH energy against a threshold depending on the noise variance.

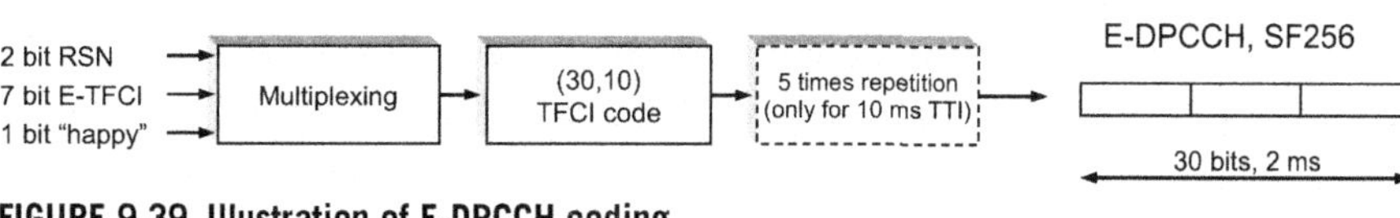


**FIGURE 9.39 Illustration of E-DPCCH coding.**

## REFERENCES

[1] 3rd Generation Partnership Project; Technical Specification Group Radio Access Network; Physical Layer Procedures (FDD), 3GPP, 3GPP TS 25.214.

[2] 3rd Generation Partnership Project; Technical Specification Group Radio Access Network; UE Radio Access Capabilities, 3GPP, 3GPP TS 25.306.

## FURTHER READING

3rd Generation Partnership Project; Technical Specification Group Radio Access Network; FDD Enhanced Uplink; Overall Description; Stage 2, 3GPP, 3GPP TS 25.309.

3rd Generation Partnership Project; Technical Specification Group Radio Access Network; Spreading and Modulation (FDD), 3GPP, 3GPP TS 25.213.

3rd Generation Partnership Project; Technical Specification Group Radio Access Network; Medium Access Control (MAC) protocol specification, 3GPP, 3GPP TS 25.321.

D. Zhang, S. Sambhwani, B. Mohanty, HSUPA scheduling algorithms utilizing RoT measurements and interference cancellations, IEEE Int. Conf. Commun. (2008) 5033–5037.

X. Zhaoji, B. Sébire, Impact of ACK/NACK Signalling Errors on High Speed Uplink Packet Access (HSUPA), IEEE International Conference on Communications, Vol. 4, May 2005, pp. 2223–2227, 16–20.

CHAPTER

# Higher-order modulation 10

## CHAPTER OUTLINE



The digital modulation scheme determines how bits are mapped to the phase and amplitude of transmitted signals. Higher-order modulation, meaning that the modulation alphabet is extended to include additional signaling alternatives, allows more information bits to be conveyed per modulation symbol, thereby increasing the spectral efficiency and peak data rate. The WCDMA/HSPA standard supports a variety of modulation schemes, both in uplink and downlink, facilitating a flexible and large dynamic range of possible transport block sizes and offering a good complement to spatial-multiplexing multiantenna techniques for increasing the peak rate.

## 10.1 OVERVIEW

A general introduction to higher-order modulation is given in Chapter 2 and a description of modulation functionality in release 5 (HSDPA) and release 6 (HSUPA) is given in Chapters 8 and 9, respectively. The modulation schemes provided by release 99 WCDMA are QPSK for the downlink and BPSK per I/Q-branch for the uplink. Release 5 HSDPA offers support for QPSK and 16-QAM and release 6 HSUPA supports multicode transmission, where each code is independently BPSK modulated and mapped to the I- or Q-branch, thereby effectively creating QPSK modulation.

In release 7, the modulation order was extended with 64-QAM for the downlink and 16-QAM, or rather 4-PAM per code, for the uplink. The next evolution

HSPA Evolution: The Fundamentals for Mobile Broadband. DOI: 10.1016/B978-0-08-099969-2.00010-7

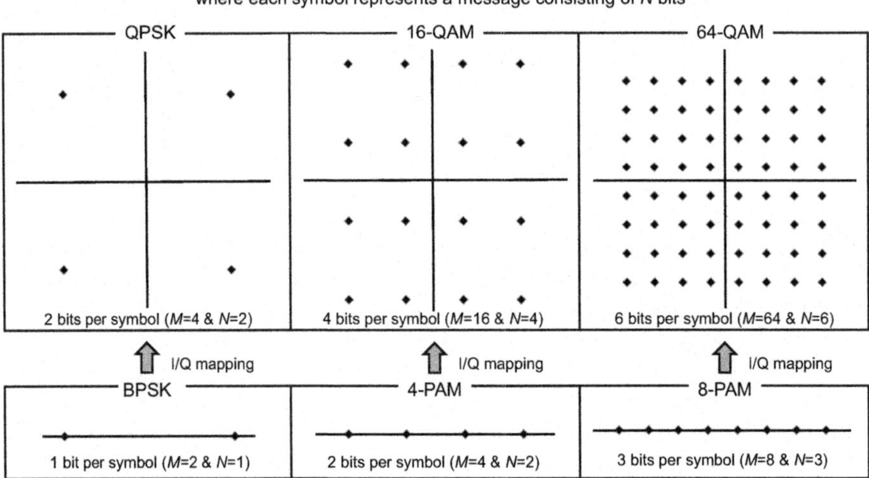


**FIGURE 10.1 A number of modulation schemes available in the WCDMA/HSPA standard. The modulation alphabet is represented in a constellation diagram, showing the amplitude of the I-branch at the *x*-axis and the amplitude of the Q-branch at the *y*-axis for each symbol.**

step came in release 11 where 64-QAM (8-PAM per code) in the uplink was introduced. Figure 10.1 gives an illustration of modulation schemes that are relevant for WCDMA/HSPA. In the figure, the modulation alphabet is represented in a constellation diagram, showing the amplitude of the I-branch at the *x*-axis and the amplitude of the Q-branch at the *y*-axis for each symbol. Table 10.1 shows how HSPA peak rates have evolved in different releases.

Higher-order modulation is typically more utilized in the downlink compared to the uplink due to reasons such as larger downlink transmit power creating more favorable SINR conditions and the non-orthogonality between uplink transmissions, which makes it more challenging to enable good coverage and high rates simultaneously, especially in multiuser scenarios. Small cell deployments are examples of scenarios where SINRs facilitating higher-order modulation are expected to be seen more frequently. Also, the use of advanced receivers expands the applicability of higher-order modulation.

The introduction of higher-order modulation into the specifications is relatively straightforward and affects mainly some isolated parts of the physical-layer specifications. There are no updates to the RLC or MAC layers from a conceptual point of view. Larger data rates need, however, to be handled, and if the RLC PDU size is too small or the RLC window size is too small, the rate will be limited by the ACK feedback rate (RLC protocol stall). The so-called flexible layer-2 structure is used to alleviate this problem (see Section 8.4). More specifically, configuration of downlink 64-QAM requires MAC-ehs to handle the increased data amount more efficiently. Similarly, MAC-i/is is required for uplink 64-QAM but not for uplink 16-QAM.

**Table 10.1** Illustration of Theoretical Peak Rate Statistics for Higher-Order Modulation

| | Features | Peak Spectral Efficiency (Bits/s/Hz) | Peak Data Rate (Mbps) | Gain Compared to Baseline |
|---|---|---|---|---|
| HSDPA | Release 5 with QPSK (baseline) | 1.4 | 7.2 | – |
| | Release 5 with 16-QAM | 2.9 | 14.4 | ×2 |
| | Release 7 with 64-QAM | 4.3 | 21.6 | ×3 |
| | Release 8 with 2×2 MIMO and 64-QAM | 8.6 | 43.2 | ×6 |
| | Release 11 with 4 carrier, 4×4 MIMO, and 64-QAM | 17.3 | 346 | ×12/×48 |
| HSUPA | Release 6 with BPSK (baseline) | 1.2 | 5.8 | – |
| | Release 7 with 16-QAM | 2.3 | 11.5 | ×2 |
| | Release 11 with 64-QAM | 3.5 | 17.3 | ×3 |
| | Release 11 with 2×2 MIMO and 64-QAM | 6.9 | 34.6 | ×6 |

## 10.2 DOWNLINK 64-QAM

The introduction of downlink 64-QAM is only applicable to the HS-PDSCH and offers a peak data rate increase of 50% compared to 16-QAM. The main specification impacts due to the introduction of 64-QAM are found in the coding chain, the modulation mapper, the associated L1/L2 control signaling, and in terms of increased transport-block sizes with associated CQI tables. The introduced functionality is, in general, a straightforward extension of the corresponding functionality introduced to support 16-QAM in release 5 (see Section 8.3.2).

### 10.2.1 CODING

The reliability of the channel bits that are Gray-mapped onto the modulated symbols vary significantly for 64-QAM depending on the minimum distance between signaling points corresponding to different values (0 or 1) for a given bit. Similar to that for 16-QAM, there is a desire to map systematic bits to the most reliable positions and to randomize the mapping between different transmission attempts to enhance the Turbo decoder performance (see Section 8.3.2 for further details).

The number of bits in one TTI for one physical channel increases to 2880[1] for 64-QAM operation. As discussed in Section 8.3.2, a second interleaver was introduced for 16-QAM in order to control the mapping of systematic and parity bits onto the 16-QAM modulation symbols. The same design philosophy is adopted for

[1]2560 chips per slot, three slots per TTI, spreading factor 16, and 64-QAM yield $2560 \times 3/16 \times 6 = 2880$ channel bits per physical channel in one TTI.

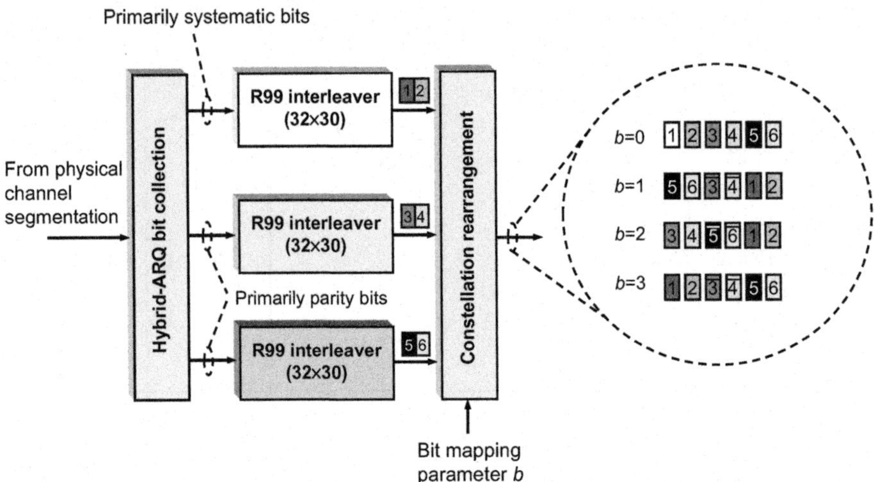


**FIGURE 10.2 Hybrid-ARQ interleaving and constellation rearrangement for 64-QAM. A bar over a number represents bit inversion.**

64-QAM, and a third (32 × 30) interleaver is consequently added. Systematic bits are primarily fed into the first R99 interleaver, while parity bits serve primarily as input to the second and third R99 interleavers. The constellation rearrangement is also updated to handle six consecutive bits, and depending on the bit mapping parameter, *b*, different output sequences are generated rendering a randomization of the bit reliability of the modulated symbols. An illustration can be found in Figure 10.2 (see also Figure 8.17 in Section 8.3.2 for the corresponding figure for 16-QAM).

## 10.2.2 MODULATION

A new modulation mapper is introduced for 64-QAM. Six consecutive binary symbols $n_k, n_{k+1}, n_{k+2}, n_{k+3}, n_{k+4}, n_{k+5}$ (where $k \bmod 6 = 0$) are serial-to-parallel converted to three consecutive binary symbols ($i_1 = n_k$, $i_2 = n_{k+2}$, $i_3 = n_{k+4}$) on the I-branch and three consecutive binary symbols ($q_1 = n_{k+1}$, $q_2 = n_{k+3}$, $q_3 = n_{k+5}$) on the Q-branch and then Gray-mapped to 64-QAM symbols using the mapping shown in Figure 10.3. A new HS-PDSCH slot format is defined for 64-QAM, supporting spreading factor SF = 16 and 960 bits/slot.

## 10.2.3 CONTROL SIGNALING

Downlink control signaling is carried on the HS-SCCH. As described in Section 8.3.7, the release 5 HS-SCCH is divided into two parts, where Part I consists of one information bit (Mod) indicating whether the modulation order is QPSK or 16-QAM and seven bits indicating the *channelization code set* (CCS). Part II carries information related to the *transport block size* (TBS), *hybrid-ARQ process number* (HAP), *redundancy version* (RV), *new data indicator* (NDI), and *UE identity* (UE Id).

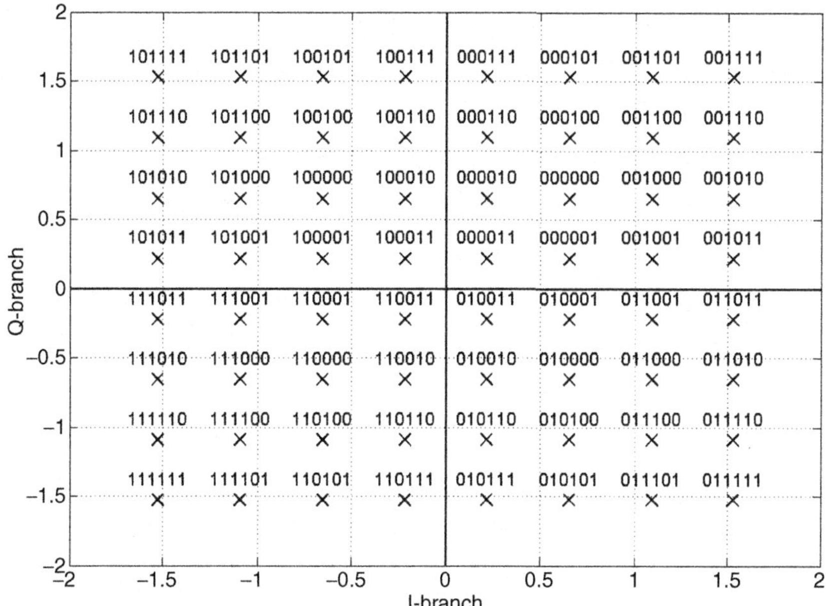


**FIGURE 10.3 Illustration of modulation mapper for 64-QAM.**

The HS-SCCH format is updated to support the introduction of 64-QAM. More specifically, Part II of the HS-SCCH is not impacted by the introduction of 64-QAM, while Part I is updated to support signaling of 64-QAM. The total number of information bits in Part I is not impacted but the meaning of some of the bits has changed, and a trick utilizing whether the HS-SCCH number is even or odd is employed (see Figure 10.4 for an overview of the type-1 HS-SCCH).

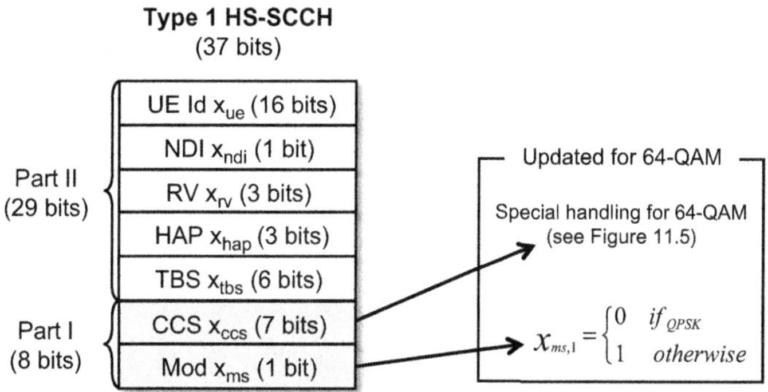


**FIGURE 10.4 Illustration of type-1 HS-SCCH. Part I consists of the modulation scheme mapping (mod) and the channelization code set (CCS), while Part II carries information related to the transport-block size (TBS), hybrid-ARQ process number (HAP), redundancy version (RV), new data indicator (NDI), and UE identity (UE Id).**

The modulation scheme mapping bit $x_{ms}$ is given by

$$x_{ms} = \begin{cases} 0, & \text{QPSK} \\ 1, & \text{Otherwise} \end{cases}, \tag{10.1}$$

and the channelization code set consists of seven bits $x_{ccs,1}$, $x_{ccs,2}$, …, $x_{ccs,7}$. The first three bits (*code group indicator*) are determined as the unsigned binary representation of the following expression

$$x_{ccs,1}, x_{ccs,2}, x_{ccs,3} = \min(P-1, 15-P), \tag{10.2}$$

where $P$ denotes the number of multicodes, and the last four bits (*code-offset indicator*) are determined as the unsigned binary representation of the following expression

$$x_{ccs,4}, x_{ccs,5}, x_{ccs,6}, x_{ccs,7} = \left| O - 1 - \lfloor P/8 \rfloor x 15 \right|, \tag{10.3}$$

where $O$ represents the starting position of the multicode block. If 64-QAM is not configured or if 64-QAM is configured and the modulation order is QPSK, then the bit mapping related to modulation scheme and channelization code set is finished. Otherwise (if 64-QAM is configured and the modulation order is not QPSK) some special considerations apply. First of all, the HS-SCCH number must be chosen such that the following relation is satisfied

$$(\text{HS-SCCH number}) \bmod 2 = x_{ccs,7}, \tag{10.4}$$

where $x_{ccs,7}$ is defined from (10.3), meaning that the HS-SCCH number must be even if $x_{ccs,7} = 0$ and odd if $x_{ccs,7} = 1$. Second, the $x_{ccs,7}$ bit is reset according to

$$x_{ccs,7} = \begin{cases} 0, & \text{16-QAM} \\ 1, & \text{64-QAM} \end{cases}. \tag{10.5}$$

Consequently, for 64-QAM modulation, the $x_{ccs,7}$ bit in (10.3) is indicated by the HS-SCCH code number (even or odd) and the $x_{ccs,7}$ bit in the channelization code set represents 16-QAM or 64-QAM instead. Alternatively phrased, there is a constraint added for which P/O combinations can be signaled for even and odd HS-SCCH. This is shown in Figure 10.5, where shadowed and non-shadowed combinations require odd and even HS-SCCH, respectively. For example, to signal 64-QAM with 10 multicodes ($P$ = 10) starting at position 1 ($O$ = 1), the $x_{ccs}$ sequence should be 1011111 and the HS-SCCH needs to be odd. If the HS-SCCH would be even and $x_{ccs}$ = 1011111, the UE would interpret this as 64-QAM with 10 multicodes ($P$ = 10) starting at position 2 ($O$ = 2).

The mapping from *transport format resource indicator* (TFRI) to transport block size and the associated transport block size table are also updated for 64-QAM. A total of 295 different transport block sizes can be signaled, compared to 254 in release 5 (see Section 8.2.4). Each combination of HS-DSCH channelization code set and modulation scheme defines a subset containing 63 out of the total

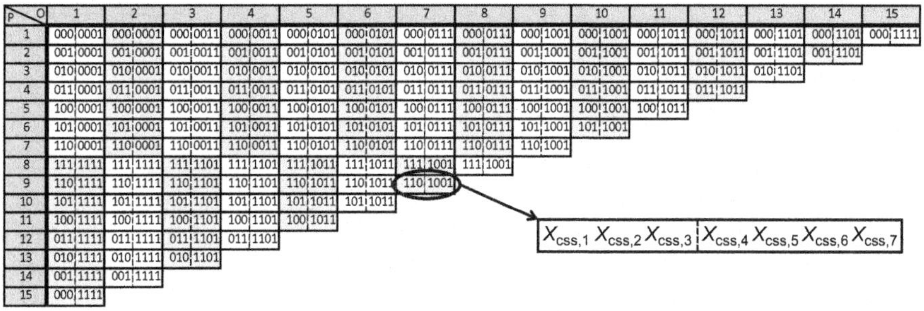

| P \ O | 1 | 2 | 3 | 4 | 5 | 6 | 7 | 8 | 9 | 10 | 11 | 12 | 13 | 14 | 15 |
|---|---|---|---|---|---|---|---|---|---|---|---|---|---|---|---|
| 1 | 000 0001 | 000 0001 | 000 0011 | 000 0011 | 000 0101 | 000 0101 | 000 0111 | 000 0111 | 000 1001 | 000 1001 | 000 1011 | 000 1011 | 000 1101 | 000 1101 | 000 1111 |
| 2 | 001 0001 | 001 0001 | 001 0011 | 001 0011 | 001 0101 | 001 0101 | 001 0111 | 001 0111 | 001 1001 | 001 1001 | 001 1011 | 001 1011 | 001 1101 | 001 1101 | |
| 3 | 010 0001 | 010 0001 | 010 0011 | 010 0011 | 010 0101 | 010 0101 | 010 0111 | 010 0111 | 010 1001 | 010 1001 | 010 1011 | 010 1011 | 010 1101 | | |
| 4 | 011 0001 | 011 0001 | 011 0011 | 011 0011 | 011 0101 | 011 0101 | 011 0111 | 011 0111 | 011 1001 | 011 1001 | 011 1011 | 011 1011 | | | |
| 5 | 100 0001 | 100 0001 | 100 0011 | 100 0011 | 100 0101 | 100 0101 | 100 0111 | 100 0111 | 100 1001 | 100 1001 | 100 1011 | | | | |
| 6 | 101 0001 | 101 0001 | 101 0011 | 101 0011 | 101 0101 | 101 0101 | 101 0111 | 101 0111 | 101 1001 | 101 1001 | | | | | |
| 7 | 110 0001 | 110 0001 | 110 0011 | 110 0011 | 110 0101 | 110 0101 | 110 0111 | 110 0111 | 110 1001 | | | | | | |
| 8 | 111 1111 | 111 1111 | 111 1101 | 111 1101 | 111 1011 | 111 1011 | 111 1001 | 111 1001 | | | | | | | |
| 9 | 110 1111 | 110 1111 | 110 1101 | 110 1101 | 110 1011 | 110 1011 | 110 1001 | | | | | | | | |
| 10 | 101 1111 | 101 1111 | 101 1101 | 101 1101 | 101 1011 | 101 1011 | | | | | | | | | |
| 11 | 100 1111 | 100 1111 | 100 1101 | 100 1101 | 100 1011 | | | | | | | | | | |
| 12 | 011 1111 | 011 1111 | 011 1101 | 011 1101 | | | | | | | | | | | |
| 13 | 010 1111 | 010 1111 | 010 1101 | | | | | | | | | | | | |
| 14 | 001 1111 | 001 1111 | | | | | | | | | | | | | |
| 15 | 000 1111 | | | | | | | | | | | | | | |



**FIGURE 10.5 Illustration of mapping from P/O combinations and the HS-SCCH bits $x_{ccs,1}$, $x_{ccs,2}$, ..., $x_{ccs,7}$ when 64-QAM is configured and the instantaneous modulation order is 16-QAM or 64-QAM. Signaling of shadowed P/O combinations requires an odd HS-SCCH, whereas signaling of non-shadowed P/O combinations requires an even HS-SCCH.**

295 different transport-block sizes and the 6-bit "HS-DSCH transport-block size information" signaled on the HS-SCCH indicates which one of the 63 possibilities is scheduled. This framework, where a total of $15\times3\times63=2835$ different combinations of channelization code set, modulation order, and "HS-DSCH transport-block size information" can be scheduled while the total number of different transport block sizes is 295, provides flexibility in the sense that a wide range of transport block sizes can be scheduled with code rates ranging from 1/3 to 1 (see Figure 10.6 for an illustration).

Uplink control information is carried on the HS-DPCCH and includes hybrid-ARQ ACK/NACKs and CQI/PCI information. The introduction of 64-QAM does not require a change to the HS-DPCCH format. Two new CQI tables are, however, introduced to facilitate 64-QAM and match the larger transport block sizes. These tables are referred to as Tables F and G [1] and correspond to HS-DSCH UE categories 13 and 14, respectively. The increased transport block sizes offered by 64-QAM require larger SINR operation points, something that is reflected by these new CQI tables. More specifically, to cover the increased dynamic SINR range for 64-QAM given the same number of total CQI values (30), the SINR step size for the largest CQI values is increased to roughly 2 dB. This is illustrated in Figure 10.7, which shows the SINR step size for CQI Table G given an AWGN simulation.

## 10.3 UPLINK HIGHER-ORDER MODULATION

The introduction of 16-QAM and 64-QAM in the uplink is only applicable to E-DPDCH and offers a peak data rate increase of 100% and 150%, respectively, compared to QPSK. As discussed in Section 9.2.5, modulation in the uplink is performed before I/Q-mapping. Strictly speaking, the modulation schemes introduced for higher-order modulation are therefore 4-PAM and 8-PAM per code. The standard

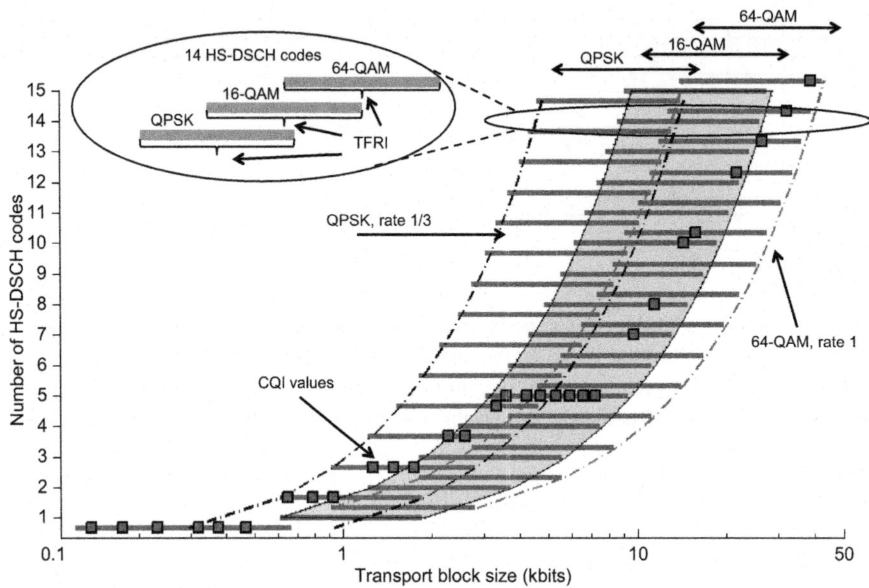


**FIGURE 10.6 Mapping from number of HS-DSCH codes, modulation order, and TFRI to transport block size. The transport-block sizes for CQI reporting using Table G (HS-DSCH category 14) are also shown.**

impact for introducing higher-order modulation in the uplink is very similar to the introduction of higher-order modulation in the downlink, and the main specification impacts are found in the coding chain, the modulation mapper, the associated L1/L2 control signaling, and in terms of increased transport-block sizes.

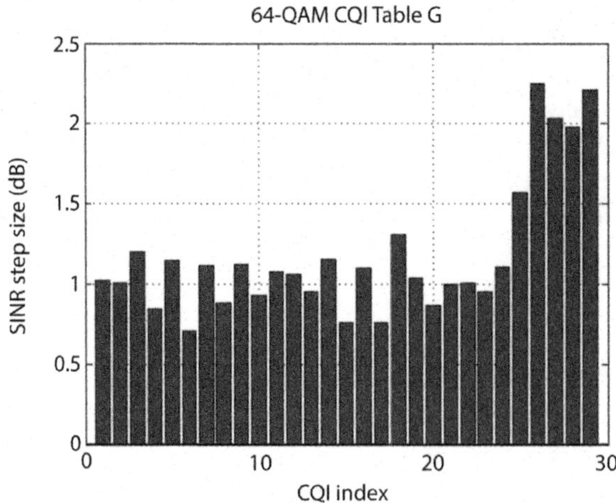


**FIGURE 10.7 SINR step size for CQI Table G given an ideal Shannon based mapping between transport-block size and SINR.**

A key area discussed during the standardization phase of uplink higher-order modulation was pilot reference enhancements. As opposed to BPSK, 4-PAM and 8-PAM require amplitude knowledge at the receiver in order to correctly form the soft values prior to Turbo decoding. Also, higher-order modulation is more sensitive to channel-estimation errors. Three main alternatives for improved channel estimation were discussed (see [2]): power boosting of DPCCH, power boosting of E-DPCCH, and introducing a secondary pilot channel. Among these, power boosting of E-DPCCH was selected and is described in Section 10.3.3.

Since some bits in the 4-PAM and 8-PAM constellations are more sensitive to estimation errors than others, there were also discussions regarding enhancements to the hybrid-ARQ bit mapping, constellation re-arrangement, and the interleaver. It was, however, shown (see [3]) that at the operating points where uplink higher-order modulation is typically used, that is, high rates with few retransmissions, this would offer only limited performance improvements, so in the end none of these changes were adopted. This is one difference in the design between the uplink and the downlink.

## 10.3.1 CODING

Compared to the downlink, no special hybrid-ARQ bit mapping or constellation rearrangement functionality is added when introducing higher-order modulation for the uplink. The process of determining spreading factor, modulation scheme, and number of E-DPDCH physical channels is, however, updated. Higher-order modulation is only supported for the 2×SF2 + 2×SF4 OVSF code configuration, and four new E-DPDCH slot formats, two per modulation order, are introduced to support 4-PAM and 8-PAM (see Table 10.2).

**Table 10.2** E-DPDCH Slot Formats. Shadowed Formats Correspond to Higher-order Modulation.

| Slot Format #i | Bits/ Symbol M | SF | Bits/ Sub-frame | Bits/Slot $N_{data}$ | $r$ (TTI = 2 ms) |
|---|---|---|---|---|---|
| 0 | 1 | 256 | 30 | 10 | 1 |
| 1 | 1 | 128 | 60 | 20 | 2 |
| 2 | 1 | 64 | 120 | 40 | 4 |
| 3 | 1 | 32 | 240 | 80 | 8 |
| 4 | 1 | 16 | 480 | 160 | 16 |
| 5 | 1 | 8 | 960 | 320 | 32 |
| 6 | 1 | 4 | 1920 | 640 | 64 |
| 7 | 1 | 2 | 3840 | 1280 | 128 |
| 8 | 2 | 4 | 3840 | 1280 | 64 |
| 9 | 2 | 2 | 7680 | 2560 | 128 |
| 10 | 3 | 4 | 5760 | 1920 | 64 |
| 11 | 3 | 2 | 11520 | 3840 | 128 |

As discussed in Section 9.3.2, for HSUPA in release 6, two parameters were introduced to control the maximum amount of allowed puncturing before additional E-DPDCHs are employed. More specifically, for multicode transmissions, the parameter $PL_{non\_max}$ controls when to switch from one code configuration to the next and the parameter $PL_{max}$ determines the maximum allowed code rate when operating at maximum allowed code configuration. $PL_{non\_max}$ is signaled via higher layers, while $PL_{max}$ is set by the specification to 1/3 if the maximum code configuration (2×SF2 + 2×SF4) is allowed and 0.44 otherwise.

To control when to change modulation order, two additional parameters are introduced. When operating at maximum code configuration (2×SF2 + 2×SF4), the parameters $PL_{mod\_switch}$ and $PL_{mod\_switch\_2}$ control when to switch from BPSK to 4-PAM and from 4-PAM to 8-PAM, respectively. These parameters are set by the specification to $PL_{mod\_switch} = 0.468$ and $PL_{mod\_switch\_2} = 0.4$ and were determined based on link simulations. All these puncturing limit parameters are set relative to the mother code of $R_b = 1/3$. Figure 10.8 illustrates the process of determining spreading factor, modulation, and code configuration.

The E-DCH interleaving, which operates on the output bits from the physical channel segmentation block, is updated for higher-order modulation. A second

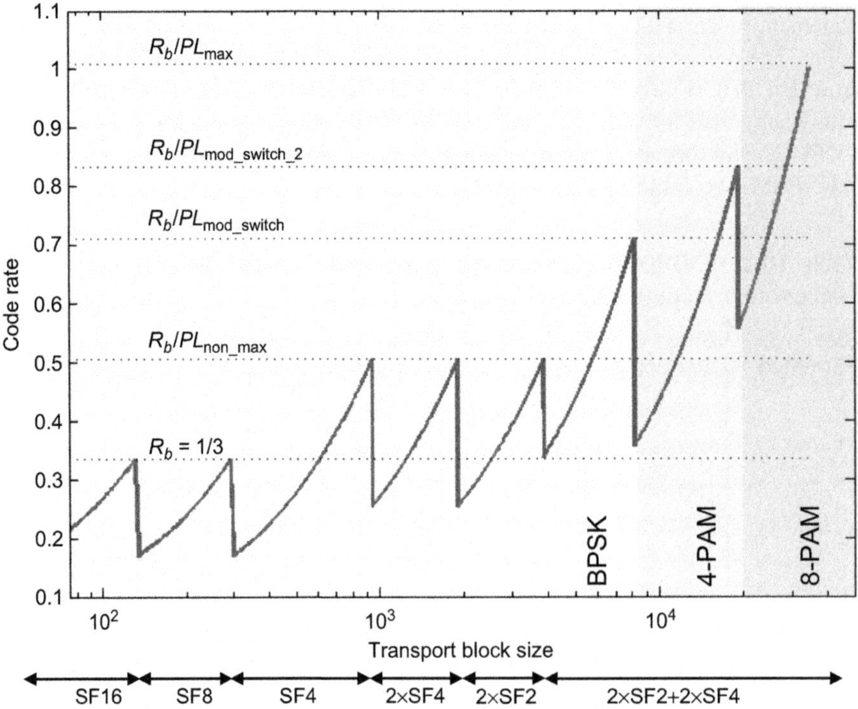


**FIGURE 10.8 Illustrations of the process of determining spreading factor (SF), modulation, and code configuration. $PL_{max}$ = 1/3 and $PL_{non\_max}$ = 2/3.**

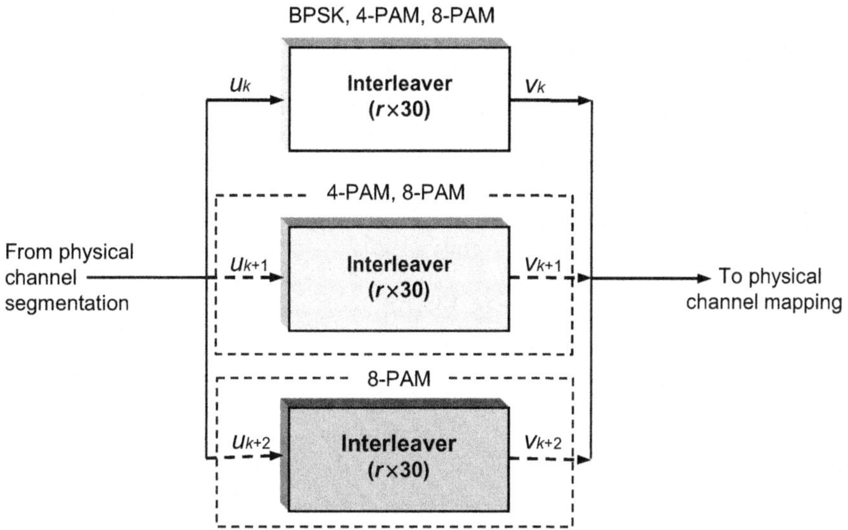


**FIGURE 10.9 Interleaving for E-DPDCH depending on modulation order.**

$r \times 30$ interleaver is added for 4-PAM and a third $r \times 30$ interleaver is introduced for 8-PAM, where the number of rows $r$ is set to match the number of coded bits in one TTI per physical channel and is given in Table 10.2. These two interleavers are transparent for BPSK modulation. The operation is illustrated in Figure 10.9.

### 10.3.2 MODULATION

New modulation mappers are introduced for 4-PAM and 8-PAM. In case of 4-PAM modulation, a set of two consecutive binary symbols $n_k$, $n_{k+1}$ (where $k$ mod 2 = 0) is converted to a real valued sequence, while for 8-PAM modulation, a set of three consecutive binary symbols $n_k$, $n_{k+1}$, $n_{k+2}$ (where $k$ mod 3 = 0) is converted to a real valued sequence. The mapping is illustrated in Figure 10.10. This procedure is done for all E-DPDCHs.

### 10.3.3 E-DPDCH POWER SETTING AND E-DPCCH BOOSTING

Demodulation of the increased transport block sizes facilitated by higher-order modulation requires increased E-DPDCH power. Higher-order modulation is also more sensitive to various impairments, meaning that the quality of the channel estimate needs to be improved. If the decoding quality becomes too poor, the outer-loop power control (OLPC) will trigger an increase of the SINR target leading to better channel estimates and increased E-DPDCH power but potentially also to a decreased grant and thereby lower data rate. In general, it is undesirable to have a strong dependency between rate and SINR target since the OLPC is a slow process that leads to inefficiencies and potential oscillations of power and grant and in the long term a risk of system instability. One possibility to enhance the quality of the channel

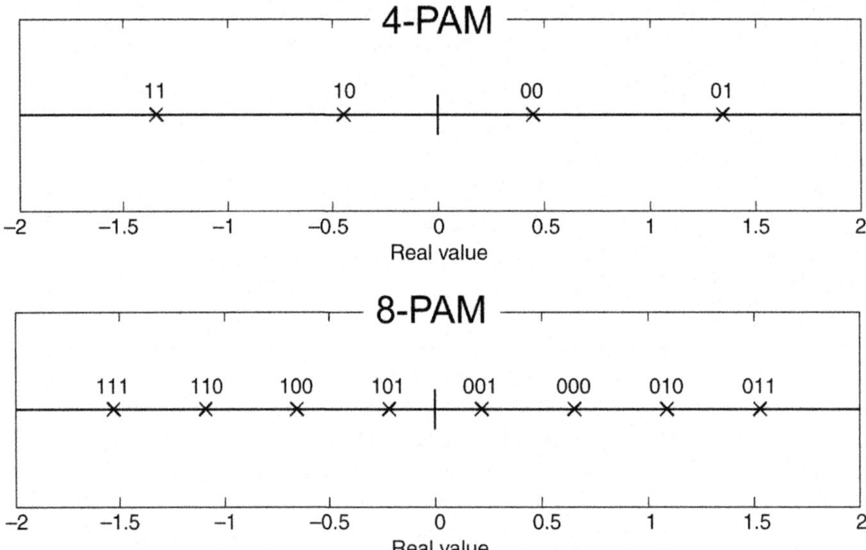


**FIGURE 10.10 Illustrations of modulation mappers for 4-PAM and 8-PAM.**

estimate would be to power boost the DPCCH whenever higher-order modulation is used, but this is non-trivial because of a number of reasons. For example, the power of all other uplink channels is set relative to the DPCCH, so boosting the DPCCH will either boost all channels or, if boosting is applied once the power of all other channels is set, the resulting power of other channels will be unknown unless the exact boosting value is known in all relevant nodes (serving and non-serving). A better approach would be to boost a specific uplink channel in a deterministic manner.

To facilitate better channel estimation quality for higher-order modulation, E-DPCCH power boosting is introduced. The idea is to increase (boost) the power of the E-DPCCH as a function of the transport block size and to treat detected E-DPCCH symbols as known symbols that can be used as additional pilots in order to demodulate E-DPDCH. Increasing the power of the E-DPCCH improves the probability of correctly detecting and decoding the E-DPCCH and implies that detected E-DPCCH symbols can serve as additional high-quality reference symbols. Needless to say, there is always a risk that the E-DPCCH is not being correctly decoded, which would reduce the quality of the E-DPCCH aided channel estimate. However, if the E-DPCCH is not successfully decoded, then it is very likely that E-DPDCH cannot be correctly decoded either since E-DPCCH carries control information required to decode E-DPDCH.

The E-DPCCH boosting procedure aims at keeping a fixed *traffic-to-total-pilot-power-ratio* ($\Delta_{T2TP}$) for E-TFCIs larger than a higher-layer configured E-TFCI$_{ec,boost}$, where the $\Delta_{T2TP}$ ratio is defined as

$$\Delta_{\mathrm{T2TP}} \triangleq 10\log_{10}\left(\frac{P_{\mathrm{E\text{-}DPDCH}}}{P_{\mathrm{E\text{-}DPCCH}}+P_{\mathrm{DPCCH}}}\right), \tag{10.6}$$

with $P_x$ denoting the power (in linear domain) of the physical channel $x$. This is achieved by letting the unquantized E-DPCCH gain factor be set according to (see [1])

$$\beta_{ec,uq} = \beta_c \cdot \sqrt{\max\left( A_{ec}^2, \frac{\sum_{k=1}^{k_{\max}} \left(\frac{\beta_{ed,k}}{\beta_c}\right)^2}{10^{\frac{\Delta_{\mathrm{T2TP}}}{10}}} - 1 \right)}, \tag{10.7}$$

whenever the E-TFCI is larger than E-TFCI$_{\mathrm{ec,boost}}$ and according to legacy procedures otherwise (see Section 9.3.4.5). In (10.7) $\beta_c$ and $\beta_{ed,k}$ denote the gain factors used for DPCCH and E-DPDCHs and $A_{ec}$ is the E-DPCCH amplitude ratio. Needless to say, power boosting the E-DPCCH as a function of the transport block size will require more and potentially larger quantized E-DPCCH gain factors. If E-TFCI is larger than E-TFCI$_{\mathrm{ec,boost}}$, then the unquantized amplitude value ($\beta_{ec,uq}/\beta_c$) is rounded to the nearest existing quantized value in a set of 18 possible quantized values ranging between –4.8 and 12 dB, otherwise the unquantized gain value is rounded to the nearest value in the legacy set described in Chapter 9. The power offset value $\Delta_{\mathrm{T2TP}}$ (ranging from 10 to 16 dB in steps of 1 dB) is signaled via higher layers. Special considerations apply for compressed frames. A closer look at the $\Delta_{\mathrm{T2TP}}$ ratio (10.6) reveals that (let $\alpha$ denote $\Delta_{\mathrm{T2TP}}$ in linear scale and assume $\beta_c = 1$)

$$\alpha = \frac{P_{\text{E-DPDCH}}}{P_{\text{E-DPCCH}} + P_{\text{DPCCH}}} = \frac{\sum_k \beta_{ed,k}^2}{\beta_{ec}^2 + 1}, \tag{10.8}$$

which means that the total power $P_{\mathrm{tot}}$ is given by

$$\begin{aligned} P_{\mathrm{tot}} &\triangleq P_{\text{E-DPDCH}} + P_{\text{E-DPCCH}} + P_{\text{DPCCH}} \\ &= \left(1 + \frac{\beta_{ec}^2 + 1}{\sum_k \beta_{ed,k}^2}\right) \sum_k \beta_{ed,k}^2 P_{\text{DPCCH}} \\ &= \frac{\alpha + 1}{\alpha} \sum_k \beta_{ed,k}^2 P_{\text{DPCCH}} \\ &= \frac{\alpha + 1}{\alpha} SG \cdot P_{\text{DPCCH}}, \end{aligned} \tag{10.9}$$

where $SG$ is the serving grant and the last equality holds assuming that the UE utilizes the whole serving grant.

The intention with E-DPCCH boosting is that once the E-DPCCH is detected, its symbols can be used as additional pilots. The exact mechanism is implementation specific, and, for example, both hard and soft symbol approaches can be envisioned. One observation is that the power setting for E-DPCCH and E-DPDCH relative to the DPCCH is first known once the E-DPCCH (the E-TFCI) is detected. These power ratios are important if a maximum-likelihood type of symbol combination is to be employed.

As discussed earlier, in addition to enhanced channel estimation, higher-order modulation requires increased E-DPDCH power for successful demodulation of larger transport block sizes. This can be achieved by either increasing the DPCCH operating point, or by increasing the E-DPDCH gain factors. When operating with E-DPCCH boosting, the latter approach is most desirable since the DPCCH power can be kept at a low level while still achieving reliable channel estimation. Hence, to facilitate higher-order modulation with E-DPCCH boosting configured, larger gain factors are introduced. As described in Section 9.3.3, the unquantized gain factors are essentially determined as a function of the transport block size and a number of reference values. For E-TFCIs larger than E-TFCI$_{ec,boost}$, the unquantized amplitude factor ($\beta_{ed,k}/\beta_c$) is rounded downward toward the nearest lower existing quantized value in a set of 32 values ranging between –2.7 and 14.0 dB; otherwise, the unquantized gain factor is rounded toward the nearest lower value in the legacy set, as described in Chapter 9.

A number of restrictions on the gain factors are also introduced. The main reasons for introducing these constraints are to avoid unnecessary implementation and verification efforts and to limit the cubic metric impact (see, for example, [4]). There are restrictions on the modulation orders that can be used for different gain factors, both when E-DPCCH boosting is configured and when it is not, since E-DPCCH boosting is not mandatory in order to operate higher-order modulation. Also, the highest E-DPDCH gain factors should only be used for SF2 in a 2×SF2 + 2×SF4 configuration when E-DPCCH boosting is applied.

Figure 10.11 illustrates how the E-DPCCH and E-DPDCH power settings could depend on the transport block size when E-DPCCH boosting is configured. In this example, the extrapolation scheme based on one reference point is used to set the E-DPDCH power. It can be seen that the power of both E-DPCCH and E-DPDCH

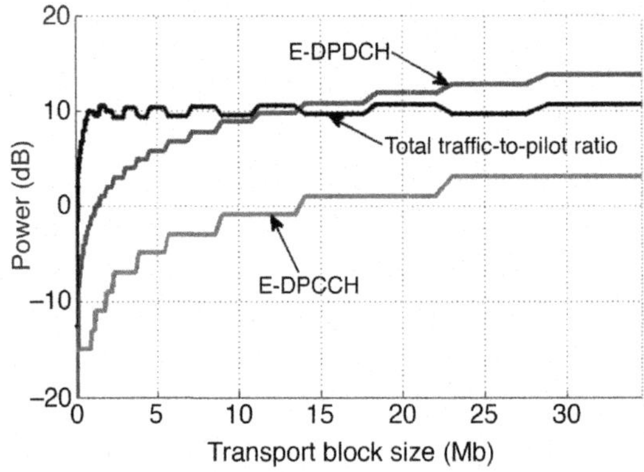


**FIGURE 10.11 Illustration of how the E-DPCCH and E-DPDCH power settings can depend on the transport-block size when E-DPCCH boosting is configured. The following parameter settings are used in this example: $\Delta_{T2TP}$ = 10 dB, a fixed DPCCH SNR target of –15 dB, and one reference value.**

increases as the transport block size increases and that the traffic-to-total-pilot ratio remains roughly constant for all transport block sizes.

### 10.3.4 CONTROL SIGNALING

Uplink out-band control signaling in release 6 consists of the *transport block size*, the *redundancy version*, and the *happy bit* (see Chapter 9). The introduction of higher-order modulation does not affect these fields and since the modulation order is implicitly given by the transport block size, there is no need to explicitly signal the modulation order. Hence, the E-DPCCH format remains the same irrespective of the configured maximum modulation order. The total number of transport block sizes that can be signaled is kept at 128, but new transport block size tables are introduced to accommodate the increased data amount.

The downlink control signaling is updated to support signaling of increased serving grants, which is required for the larger transport block sizes offered by higher-order modulation. More specifically, a new absolute grant table consisting of 32 entries with values ranging between 0 and 34.0 dB is introduced to facilitate the increased E-DPDCH gain values. Higher-layer signaling is used to determine which grant table to employ. The procedure related to a relative serving grant update is also updated taking the new E-TFCI tables and the new absolute grant table into consideration [5].

## REFERENCES

[1] 3rd Generation Partnership Project; Technical Specification Group Radio Access Network; Physical layer procedures (FDD), 3GPP TS 25.214.

[2] R1-070415, Enhanced Phase Reference for 16-QAM, QUALCOMM Europe, Ericsson, 2007.

[3] R1-070509, Bit Mapping for 16QAM for HSUPA, Ericsson, 2007.

[4] R1-071835, LS on Restricted Beta Factor Combinations for UL 16QAM.

[5] 3rd Generation Partnership Project; Technical Specification Group Radio Access Network; Medium Access Control (MAC) protocol specification, 3GPP TS 25.321.

## FURTHER READING

3rd Generation Partnership Project; Technical Specification Group Radio Access Network; Multiplexing and Channel Coding (FDD), 3GPP TS 25.212.

3rd Generation Partnership Project; Technical Specification Group Radio Access Network; Spreading and Modulation (FDD), 3GPP TS 25.213.

3rd Generation Partnership Project; Technical Specification Group Radio Access Network; Radio Resource Control (RRC), 3GPP TS 25.331.

RP-060844, Proposed WID for Higher Order Modulation in HSUPA, Ericsson, Nokia, Qualcomm Europe, 3, Cingular Wireless, KPN, SFR, TeliaSonera, T-Mobile, 2006.

CHAPTER

# Multi-antenna transmission 11

## CHAPTER OUTLINE




HSPA Evolution: The Fundamentals for Mobile Broadband. DOI: 10.1016/B978-0-08-099969-2.00011-9

Support for efficient multi-antenna transmission techniques is a fundamental requirement for any modern communication system to meet the demand of increased data rates from today's mobile broadband users. Tools for enabling improved and more efficient use of multiple transmit and receive antennas for WCDMA/HSPA have been a key area of research and standardization in 3GPP for many years, and the current standard supports a flexible toolbox to facilitate multi-antenna transmission techniques.

## 11.1 OVERVIEW

### 11.1.1 MULTI-ANTENNA TECHNIQUES

Multi-antenna techniques were discussed thoroughly in Chapter 4. Deploying multiple antennas at the receiver side is a straightforward and efficient way to increase the received *Signal-to-Interference-and-Noise-ratio* (SINR) and to combat fading, and does not require standardization. The focus of this chapter, however, is on multi-antenna transmission techniques. As discussed in Chapter 4, multi-antenna transmission techniques aim mainly at achieving one or more of the following key objectives:

1. *Transmit diversity* – Transmit diversity (also referred to as *antenna* or *spatial* diversity) combats fading by sending the transmitted signal over several independent radio channels.
2. *Directivity* – Directivity is obtained by matching the transmission to the current radio conditions to achieve coherent combining at the receiver side.
3. *Spatial multiplexing* – Spatial multiplexing is achieved by simultaneously transmitting different data streams from different (virtual) antennas.

*Transmit diversity* is typically employed when the transmitter has limited channel knowledge. The basic idea is to send the same information over several antennas. If the channels corresponding to the different antennas are sufficiently uncorrelated, this provides additional diversity that makes the transmission more robust against fading. Uncorrelated radio channels can be achieved by means of spatial antenna separation or polarization.

*Directivity* increases the SINR at the receiver side and provides an efficient means to reduce intra- and inter-cell interference, which leads to increased throughput and/or reduced transmit power. Two common forms of directivity are *classical beamforming* and *precoding* (see Chapter 4 for details).

Precoding aims at matching the multi-antenna transmission to the current channel conditions in order to obtain coherent combining at the receiver side. This is achieved by multiplying the signal with antenna-specific complex weights (phase and/or amplitude) that depend on the current channel conditions. In *closed-loop* operation, the transmitter selects appropriate weights based on feedback from the receiving side, and in *open-loop* operation the transmitter autonomously chooses the weights without dedicated feedback. One example would be TDD systems where channel reciprocity can be used to determine the appropriate weights.

Precoding can be *codebook based*, in which case a finite set of precoding vectors/matrices are defined and used for transmission. The number of vectors is a trade-off between performance and feedback overhead. The alternative to codebook-based precoding is *non-codebook*-based precoding, where any vectors/matrices can be used.

*Classic beamforming* aims at adjusting the beam direction of the signal toward a specific source, and the adjustments are typically based on direction-of-arrival estimates from the uplink. Therefore, classical beamforming does not necessarily require explicit channel knowledge. While precoding-based directivity operates at a time scale in the order of the coherence time of the channel, classic beamforming usually uses a much slower time scale.

*Spatial multiplexing* is a technique that aims at sharing the available SINR in the spatial domain. In its simplest form, several data streams are simultaneously transmitted from different antennas. The receiver can separate the streams from each other if there are a sufficient number of receive antennas and the channel conditions are suitable. Spatial multiplexing is sometimes referred to as *multiple-input multiple-output* (MIMO), although MIMO should really be seen as a broader term that includes general multi-antenna techniques.

The term *transmission rank* is used to describe the number of simultaneously transmitted data streams or layers. For example, single- and dual-stream transmissions correspond to rank-1 and rank-2 transmissions, respectively.

Spatial multiplexing transmissions rely on the availability of multiple transmit and receive antennas. In theory, the number of simultaneously transmitted streams cannot be larger than the minimum of the number of transmit and receive antennas. The *channel rank* is the number of streams that the channel matrix can support. The rank of the channel matrix needs to be larger or equal than the number of simultaneously transmitted streams in order for the receiver to be able to separate the stream. A channel is often said to be rich enough (contains enough scattering/reflections to create a well-conditioned channel matrix) if the rank constraint is satisfied. If there is insufficient scattering (for example, in line-of-sight conditions), cross-polarization of antennas can be used to create a rich enough channel.

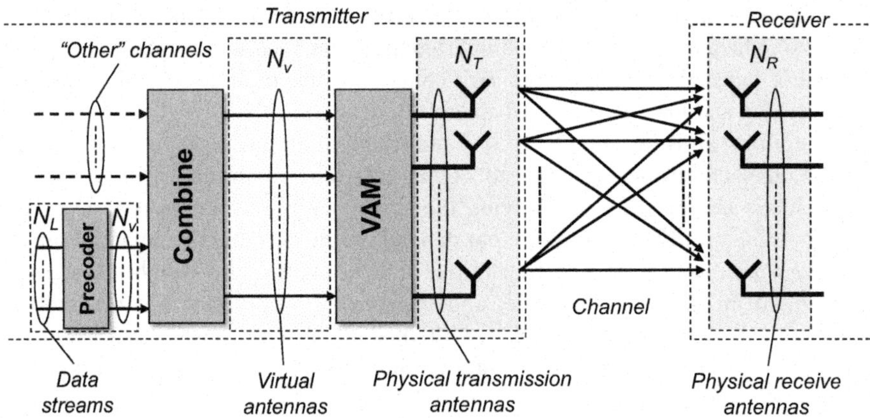


**FIGURE 11.1 Multi-antenna related notational overview.**

Spatial multiplexing can offer substantially increased peak data rates, significantly higher spectrum efficiency, and increased system capacity. However, to achieve these properties, a high SINR at the receiver is required. Spatial multiplexing is therefore mainly applicable in relatively low load scenarios close to the base station, where high SINRs are common. In situations where sufficiently high SINRs cannot be achieved to facilitate spatial multiplexing, the multiple antennas can be used for other techniques. A typical HSPA spatial multiplexing scheme employs precoding, which means that each stream is multiplied by antenna-specific weights before being transmitted from the transmit antennas. In single-stream transmission, precoding is used to achieve directivity, whereas in multi-stream transmission the aim of precoding is to optimize capacity, which generally means helping the receiver by making the *effective channel* (combination of precoding matrix and channel matrix) as orthogonal as possible. More specifically, the receiver would determine the precoding matrix such that the effective channel becomes as orthogonal as possible taking into consideration any noise/interference amplification. Also, the multiple antennas can always be used for receive diversity or for interference suppression.

In Sections 11.2–11.6, a detailed description of transmit diversity and spatial multiplexing (MIMO) for HSPA is given.

A number of multi-antenna-specific notations will be used in this chapter, such as the following (see also Figure 11.1):

- *Physical transmission* and *reception antennas* or *antenna ports* are used to describe the actual transmitting or receiving antennas. The notation $N_T \times N_R$ is used to describe a system with $N_T$ transmit antennas and $N_R$ receive antennas. For example, $2 \times 2$ and $4 \times 4$ are used to describe systems with two and four transmit and receive antennas, respectively.

- *Virtual antennas* or *virtual antenna ports* are used to denote the antennas "seen" by the receiver. Typically each virtual antenna has an associated pilot channel. The number of virtual antennas can be equal or less than the number of physical antennas. The mapping from virtual antennas to physical antennas are commonly done via a *virtual antenna mapper* (VAM) also referred to as a common precoder.
- *Streams*, *layers*, or *transmission rank* represent the number of data streams that are transmitted simultaneously. The number of streams can be equal or less than the number of virtual antennas.
- The data streams are often *precoded* before being mapped to the virtual antenna ports. The number of inputs to the precoder equals the number of streams and the number of outputs typically equals the number of virtual antennas. Other channels might or might not be subject to precoding and correspond to, for example, pilot, legacy, or common channels.
- The different virtual or physical antennas are referred to as antenna 1, 2, etc., or as the primary, secondary, etc., antenna. Similarly, the different data streams are referred to as stream 1, 2, etc., or as the primary, secondary, etc., stream.

All of these aspects will be described and discussed in more detail in this chapter.

### 11.1.2 A SHORT HISTORY OF THE DEVELOPMENT OF MULTI-ANTENNA TECHNIQUES IN 3GPP

Figure 11.2 offers an overview of multi-antenna-related WCDMA/HSPA features in 3GPP.

The first release of the UTRAN specification (release 99) already introduced support for open- and closed-loop transmit diversity for the downlink. The first release to support spatial multiplexing (MIMO) was release 7, which standardized two-branch downlink MIMO. The term two-branch is used to indicate that a maximum of two simultaneous streams are supported. This means that both the transmitting base station and the receiving UE need to have at least two antennas, but nothing prevents more than two antennas from being employed. In particular, the precoder is designed for two antenna ports. The decision to include MIMO in release 7 followed after

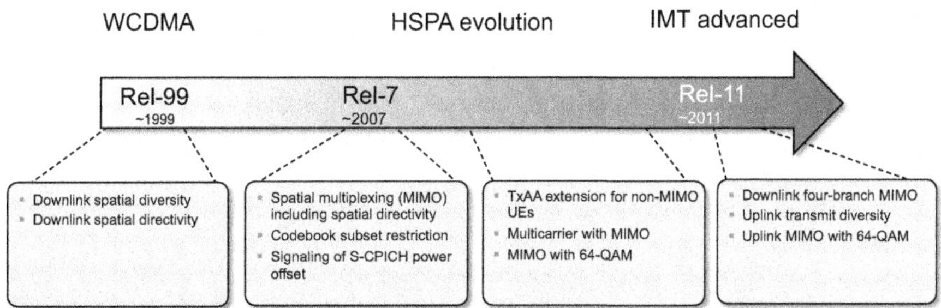


**FIGURE 11.2 Overview of multi-antenna related WCDMA/HSPA features in 3GPP.**

intensive discussions in 3GPP that lasted for several years. The main design criteria were implementation complexity, standardization impact, control information overhead, and performance. The two key drivers for inclusion of downlink MIMO were the need to keep HSPA competitive against other technologies and a more widespread availability of dual-antenna UEs. Thus, a major part of the UE functionality needed to support downlink MIMO was already present.

Between release 7 and release 11, a number of MIMO-enhancing features were added. The main motivations were to make MIMO more efficient and to enhance the performance for non-MIMO users.[1] Support of dual-stream transmission requires a UE with at least two receive antennas, whereas a single stream can be received by a UE with one or several antennas. Another key consideration was to minimize the impact on legacy users. Release 7 introduced two-branch downlink MIMO with 16-QAM as maximum modulation order and 64-QAM modulation as separate features (that is, a UE could be configured either with MIMO or with 64-QAM, but not both). In release 8 the combination of two-branch downlink MIMO and 64-QAM was standardized. In release 9, a so-called *Transmit Adaptive Arrays (TxAA) extension for non-MIMO*[1] *UEs* was introduced. The purpose was to enable scheduling of single-stream transmissions to non-MIMO UEs using the same precoding scheme as is used for MIMO UEs – this is referred to as "single-stream MIMO" or "MIMO with single-stream restriction" in the specifications. The main motivation was to facilitate directivity gains for potentially cheaper and therefore less capable UEs not necessarily equipped with two antennas. These single-stream MIMO UEs still need to receive the MIMO control channels and pilots.

Some MIMO-enhancing features were later added retrospectively into release 7, such as *codebook subset restriction* and *enhanced signaling of pilot information*. Adding functionality retrospectively is quite rare in the 3GPP specifications but was deemed necessary in order to enhance MIMO performance and minimize the impact on legacy users in MIMO-configured networks. No MIMO-capable networks were deployed and no commercial UEs were yet available. There was a broad consensus between network and UE vendors and operators to do the retrospective addition. Being able to reduce the power on the secondary pilot channel and signal the power level to the UE facilitates better coexistence of MIMO and legacy users in the same network. Similarly, codebook subset restriction is necessary to employ efficient power balancing networks (also referred to as common precoder or virtual antenna mapper). The latter is required to ensure that legacy users can make use of all available *power amplifier* (PA) power in a good manner.

Downlink MIMO also became an inherited part of many other features, such as multicarrier (see Chapter 12). During release 10, a detailed study item targeting uplink transmit diversity was initiated. A key area in this work was antenna-related aspects, such as power amplifier inefficiencies, antenna imbalances, and antenna correlation.

[1]A MIMO-capable UE supports dual-stream transmission, whereas a non-MIMO UE does not support dual-stream transmission, possibly, but not necessarily, because it has only one antenna.

**Table 11.1** Illustration of Theoretical Peak Rate for Some Multi-antenna Related Features in Different HSPA Releases

| | Feature | Release | Peak Spectral Efficiency (bits/s/Hz) | Peak Data Rate (Mbps) |
|---|---|---|---|---|
| Downlink | HSDPA | Release 5 | 2.9 | 14.4 |
| | 2×2 MIMO | Release 7 | 5.8 | 28.8 |
| | 2×2 MIMO with 64-QAM | Release 8 | 8.6 | 43.2 |
| | Up to 8 carriers with 2×2 MIMO | Release 11 | 8.6 | 345.6 |
| | Up to 4 carriers with 4×4 MIMO | Release 11 | 17.3 | 345.6 |
| Uplink | HSUPA | Release 6 | 1.2 | 5.8 |
| | 2×2 MIMO with 16-QAM | Release 11 | 4.6 | 23.0 |
| | 2×2 MIMO with 64-QAM | Release 11 | 6.9 | 34.6 |

Release 11 was the most recent release that provided a major step in the area of multi-antenna. The main motivation was a desire to make HSPA compliant with the ITU requirements for an IMT-advanced-compliant technology. These requirements included, among others, downlink and uplink peak spectral efficiencies of 15 and 6.75 bits/s/Hz, respectively. To satisfy these requirements, 4×4 downlink MIMO offering 16.9 bits/s/Hz and 2×2 uplink MIMO with 64-QAM offering 6.9 bits/s/Hz was introduced. Table 11.1 shows some different HSPA multi-antenna combinations and the corresponding peak-spectral efficiency and peak data rates.

### 11.1.3 COMPARING DESIGN CHOICES FOR DOWNLINK AND UPLINK

The same types of multi-antenna techniques can be applied in downlink and uplink, for example, diversity, directivity, and spatial multiplexing. There are, however, some fundamental differences between the downlink and uplink that affect the design of the schemes and in general make the downlink more suitable for multi-antenna transmission techniques. This has in turn impacted the release history of 3GPP, where several releases contained downlink-related multi-antenna techniques and it took until release 11 before these techniques were adopted for the uplink.

Spatial multiplexing in the uplink is generally seen as less beneficial compared to the downlink since high SINR conditions will be less common in the uplink for a number of reasons. The uplink transmit power is limited and can be several orders of magnitude lower than the downlink transmit power (typically a factor of 10–100 times lower). The uplink is also non-orthogonal, which makes it more challenging to facilitate good coverage and high rates (SINRs) simultaneously, especially in multi-user scenarios. Another key aspect is that adding uplink spatial multiplexing increases the UE complexity and thereby the cost. (Multi-antenna receive directivity at the base station, on the other hand, can be very beneficial for the uplink in order to mitigate inter-cell interference and improve coverage and does not impact UE complexity). New deployment strategies including small cells can, however, change the picture and facilitate spatial multiplexing also in the uplink, since the pathloss will be low and hence the received SINR higher, at least in low-load scenarios.

Because of the fact that the downlink contains common signals such as pilots, the MIMO design in the downlink needs to take legacy operation into consideration, which affects, for example, the pilot design. In uplink, there are no common signals and thus this is not the case for the uplink, which provides more flexibility in the physical channel design.

Performance evaluations of multi-antenna features require in general more detailed antenna and channel models, including properties such as antenna correlation, antenna power imbalances, and timing misalignment between different transmit branches. 3GPP adopted so-called *spatial-channel models* (SCM) [1] to ensure a fair system level assessment of MIMO schemes with respect to antenna and channel characteristics. Antenna-related aspects are even more significant for uplink multi-antenna transmission techniques compared to downlink because of smaller devices with closely spaced and potentially cheaper antennas, leading to imbalance and correlations between Tx antennas.

## 11.2 DOWNLINK TRANSMIT DIVERSITY

Two different downlink transmit diversity schemes have been supported since WCDMA release 99: open- and closed-loop transmit diversity. Both of these approaches rely on two transmit antennas and are briefly explained in this section.

### 11.2.1 DOWNLINK OPEN-LOOP TRANSMIT DIVERSITY

The main open-loop transmit diversity mode in UMTS is a *space time transmit diversity* (STTD) approach based on the well-known Alamouti coding scheme [2]. All downlink physical channels except the *synchronization channel* (SCH) and P/S-CPICH can be configured in the STTD mode. Support for STTD is mandatory in the UE but optional in the network. STTD can be independently configured for each radio link in the (DCH) active set.

In STTD operation, channel coding, rate matching, and interleaving are done in the same manner as in non-diversity mode. The STTD encoder operates on 4, 8, or 12 bits, respectively, depending on the modulation order, QPSK, 16-QAM, or 64-QAM. One implication of this is that full STTD is only possible if the number of bits of interest is a multiple of 4, 8, or 12, respectively. For example, for F-DPCH or F-TPICH (see Section 11.5.1.3), the TPC bits or TPI bits are not STTD encoded and the same bits are transmitted with equal power from both antennas.

The STTD encoder operates on bit outputs from the interleaver, whereas the traditional Alamouti scheme operates on symbols. This implies that the mapping of the STTD encoder differs from the traditional Alamouti mapping. This is illustrated in the following example for QPSK modulation. First, binary bits $b_k$ are mapped to real bits $r_k$, that is, 0 and 1 are mapped to 1 and –1, respectively. Two consecutive

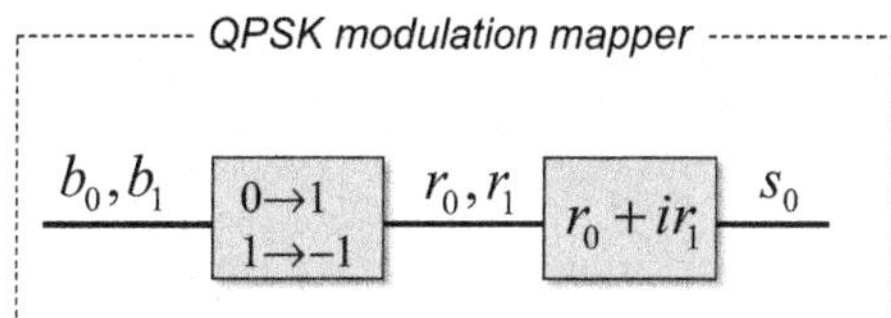


**FIGURE 11.3 Illustration of the modulation mapping for QPSK.**

real bits $r_0$ and $r_1$ are then mapped to a complex valued QPSK symbol $s_0$; even- and odd-numbered bits are mapped to I and Q, respectively (see Figure 11.3). Hence, by using the notation $\bar{b}_0 = (b_0 + 1) \bmod 2$, the relations between two consecutive binary bits and the associated QPSK symbol in Table 11.2 can be obtained and thereby also the STTD encoding relation shown in Figure 11.4. Similar arguments are used to derive the STTD encoder for 16-QAM and 64-QAM.

When the SCH is subject to open-loop transmit diversity, it employs *time-switched transmit diversity* (TSTD) in which both the primary and the secondary synchronization channels are alternatively transmitted from Antenna 1 or 2 depending on whether the slot number is even or odd. The modulation pattern for SCH depends on whether STTD is employed or not (captured by $a$ in Figure 11.5), something that can be used by the UE to know whether the *Primary Common Control Physical Channel* (P-CCPCH) employs STTD encoding. This information is essential since the P-CCPCH carries broadcast information. In particular, during initial system access, the UE needs to get this information from the SCH, but once in connected mode, higher-layer signaling, the SCH modulation, or a combination of the two can be used to deduce whether neighboring cells use STTD. SCH and P-CCPCH always use the same transmission configuration, that is, either STTD or not.

The CPICH(es) are transmitted from both antennas and use orthogonal pilot sequences in order for the receiver to be able to resolve the different channel paths.

**Table 11.2** Relation Between Bits and Symbol for QPSK

| | | | |
|---|---|---|---|
| Symbol | $s_0$ | $s_0^*$ | $-s_0^*$ |
| Real bits | $r_0, r_1$ | $r_0, -r_1$ | $-r_0, r_1$ |
| Binary bits | $b_0, b_1$ | $b_0, \bar{b}_1$ | $\bar{b}_0, b_1$ |

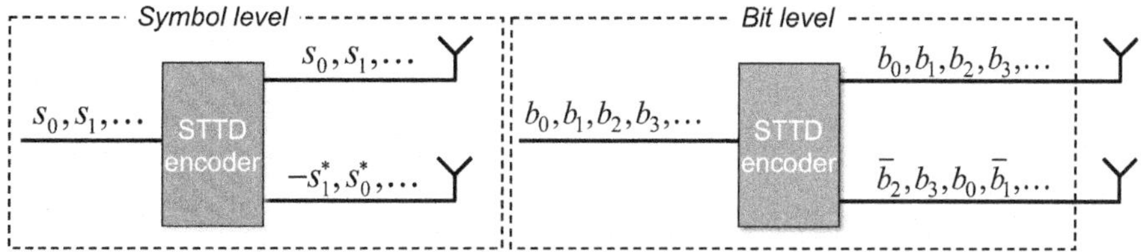


**FIGURE 11.4 Illustration of the Alamouti encoder. Traditional encoder in symbol domain (left) and encoder in bit domain (right) for QPSK modulation.**

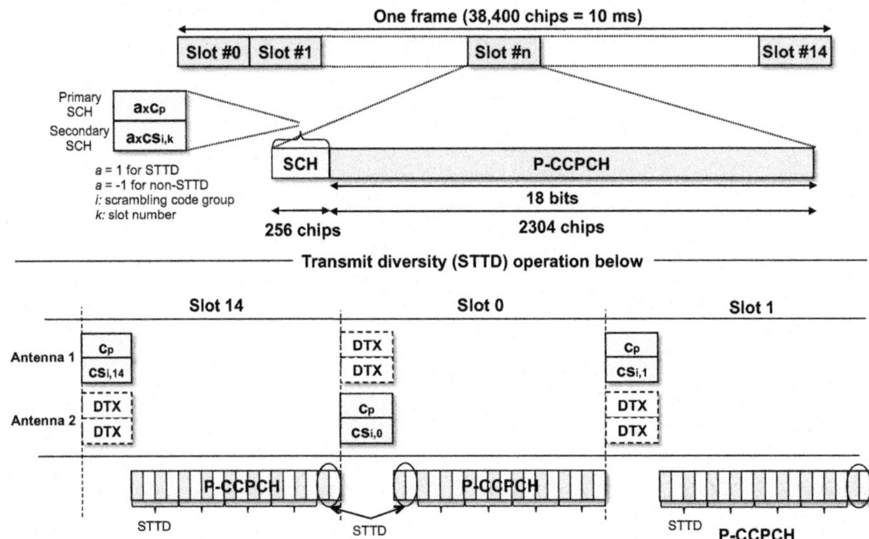


**FIGURE 11.5 Top: SCH and P-CCPCH channel layout. Bottom: Illustration of STTD operation for SCH and P-CCPCH.**

### 11.2.2 DOWNLINK CLOSED-LOOP TRANSMIT DIVERSITY

Closed-loop mode-1 transmit diversity (CL-1) can be configured for DPCH and HS-PDSCH in CELL_DCH only. This diversity mode is codebook based consisting of four codewords. The physical-layer processing up to and including spreading is identical to the non-diversity mode. The spread symbols are then multiplied by antenna-specific complex precoding weights before being sent on each antenna. A recommendation for the optimal weights is determined by the UE in order to maximize the received power and signaled to the network each slot via the *Feedback Information* (FBI) bit on the uplink DPCCH. The network is free to transmit with any weights that it chooses, but will, in general, follow the UE recommendation. The operation is illustrated in Figure 11.6.

Note that early 3GPP specifications contained two closed-loop transmit diversity modes, but mode 2 was removed in release 6. The reason was that HSDPA and F-DPCH did not support mode 2, which made it less attractive. Also, mode 2 was in general not implemented in networks and added additional complexity in UEs. One issue with supporting mode 2 for HSDPA was the power imbalance problem caused by the amplitude weights used in mode 2.

A physical channel cannot use open- and closed-loop transmit diversity simultaneously, and similar to open-loop transmit diversity, closed-loop transmit diversity for DPCH can be independently configured per radio link in a UE's active set.

The physical channels subject to CL-1 are summed together on a per UE basis and precoded before being transmitted over the air from both transmit antennas. The common pilot channel (CPICH) is not precoded and will be transmitted from

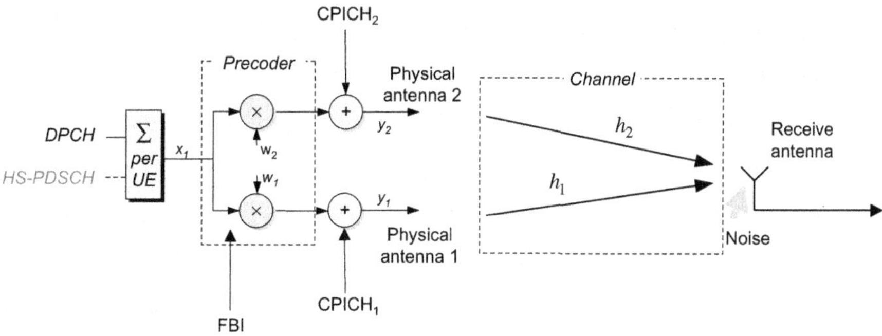


**FIGURE 11.6 Illustration of closed-loop transmit diversity.**

both antennas using orthogonal pilot patterns in order to separate the two antenna branches.

How the UE determines the preferred precoding vector $\boldsymbol{w} = [w_1\ w_2]^T$ is implementation specific, but a typical choice would be to maximize the received power

$$\boldsymbol{w} = \arg\max_{\boldsymbol{w}} \|\boldsymbol{Hw}\|^2 = \|\boldsymbol{h}_1 w_1 + \boldsymbol{h}_2 w_2\|^2, \tag{11.1}$$

where $\boldsymbol{H} = [\boldsymbol{h}_1\ \boldsymbol{h}_2]$ denotes the channel matrix, which can be estimated directly from the CPICH channel: $\boldsymbol{h}_1$ from $\text{CPICH}_1$ and $\boldsymbol{h}_2$ from $\text{CPICH}_2$, where the two CPICHs are distinguished by their respective pilot patterns. During soft handover, all links in the (DCH) active set can be included in the weight decision process. Also, if HS-PDSCH is configured, the serving cell link can be given higher priority in the weight decision process. The optimization criterion (11.1) can, for example, be extended as follows:

$$\boldsymbol{w} = \arg\max_{\boldsymbol{w}} \boldsymbol{w}^H \left( \sum_k \alpha_k \mathbf{H}_k^H \mathbf{H}_k \right) \boldsymbol{w}, \tag{11.2}$$

where $\boldsymbol{H}_k$ denotes the channel matrix associated with radio link $k$ (from base station $k$) in the active set, and $\alpha_k$ is a weight coefficient used to prioritize different links. For example, if all links are equally important $\alpha_k = 1$, but if HS-DSCH is configured and link $n$ corresponds to the HS-DSCH serving cell, a better choice would be to let $\alpha_n \gg \alpha_k$, for all $k \neq n$ in order to prioritize the link associated with the HS-DSCH serving cell. Note also that only one precoding vector that is common for all links (base stations) can be recommended since there is only one common uplink DPCCH (conveying the preferred precoding vector – FBI) for all links in the active set.

The codebook consists of four different precoding vectors,

$$\boldsymbol{w}(k) = \begin{bmatrix} w_1 \\ w_2(k) \end{bmatrix} = \frac{1}{\sqrt{2}} \begin{bmatrix} 1 \\ e^{j\frac{\pi}{4}(2k-1)} \end{bmatrix}, \quad k = 1, 2, 3, 4, \tag{11.3}$$

and the preferred precoding vector, or rather the associated phase shift on the second antenna, is signaled each slot using the FBI bit on the uplink DPCCH. To obtain four

**Table 11.3** Weight Vector, $w_2(k)$, for Different Combinations of FBI Feedback

| Received FBI Bit | | Applied Weight | |
|---|---|---|---|
| **Even Slot** | **Odd Slot** | ***k*** | **$w_2(k)$** |
| 0 | 0 | 1 | $\frac{1}{\sqrt{2}}e^{j\frac{\pi}{4}}$ |
| 0 | 1 | 4 | $\frac{1}{\sqrt{2}}e^{-j\frac{\pi}{4}}$ |
| 1 | 0 | 2 | $\frac{1}{\sqrt{2}}e^{j\frac{3\pi}{4}}$ |
| 1 | 1 | 3 | $\frac{1}{\sqrt{2}}e^{-j\frac{3\pi}{4}}$ |

different phase shifts using one feedback bit per slot, two consecutive slots of signaled values are combined. The final weights given the different FBI combinations are given in Table 11.3. The feedback signaling is Gray coded, which means that if one of two consecutive bits is in error, then the phase error is at most 90 degrees. On link setup, before any FBI bits are received, the network should use $w_2(1)$. Note that there are special considerations for compressed mode and at end of frames; see [3] for details.

Assuming that the received precoded weights are applied by the network and that the channel impulse response has not changed in the time between estimation of precoding weights and reception with the recommended precoding weights, the received precoded signal vector $y$ associated with the physical channel $x$ can be written as

$$\boldsymbol{y} = \boldsymbol{H}\boldsymbol{w}x = (\boldsymbol{h}_1 w_1 + \boldsymbol{h}_2 w_2)x = (\boldsymbol{h}_{\text{eff},1} + \boldsymbol{h}_{\text{eff},2})x = \boldsymbol{h}_{\text{eff}}x, \tag{11.4}$$

where $\boldsymbol{h}_{\text{eff}}$ represents the effective precoded channel. The channel estimate based on CPICH gives an estimate of $\boldsymbol{H}$. Hence, in order to demodulate a precoded physical channel, the correct weight vector $\boldsymbol{w}$ applied by the network needs to be known by the UE. The used weight vector is not signaled in the downlink, and instead a so-called *antenna verification* algorithm can be employed by the UE to determine the vector. Antenna verification is typically done by hypothesis testing utilizing the difference between the non-precoded CPICH pilot and the precoded DPCCH pilot. An alternative demodulation process can be employed if DCH is configured. The pilots on the DPCCH are precoded with the same weight vector as the DPDCH. Hence, the channel estimate based on DPCCH will include the precoding weights: $\text{DPCCH}_1$ yields $\boldsymbol{h}_{\text{eff},1}$ and $\text{DPCCH}_2$ gives $\boldsymbol{h}_{\text{eff},2}$. One potential problem with this approach is that the dedicated DPCCH pilots might not be transmitted with sufficient power to do accurate enough channel estimation for demodulation.

## 11.3 DOWNLINK TWO-BRANCH MIMO

### 11.3.1 OVERVIEW

The scheme used for two-branch MIMO is sometimes referred to as *dual-stream transmit adaptive arrays* (D-TxAA[2]) [5], which is a multi-codeword scheme with rank adaptation and precoding. The scheme can be seen as a generalization of the closed-loop transmit diversity mode–1 (CL-1). Two-branch MIMO is also referred to as 2×2 MIMO, which reflects that at least two transmit and two receive antennas are required and up to two streams are supported.

Each stream is subject to the same physical-layer processing in terms of coding, spreading, and modulation as the corresponding single-layer HSDPA case. After coding, spreading, and modulation, precoding is used before the result is mapped to the two transmit antennas. As discussed in Chapter 4, there are several reasons for precoding. Even if only a single stream is transmitted, it can be beneficial to exploit both transmit antennas by using (closed-loop) transmit diversity. Therefore, the precoding in the single-stream case is similar to closed-loop mode-1 transmit diversity (the difference is mainly in the details for signaling and the update rate, as will be discussed below). For dual-stream transmission, the precoding attempts to pre-distort the signal such that the two streams become close to orthogonal at the receiver side. This reduces the interference between the two streams and decreases the burden on the receiver processing. A MIMO-capable UE supports dual-stream transmission, whereas a non-MIMO UE does not support dual-stream transmission, possibly, but not necessarily, because it has only one antenna. In fact, support of TxAA for non-MIMO UEs exists and can be viewed as a special case of the general two-branch MIMO described below where the transmission rank is restricted to one. This allows a release 8 single-stream-only non-MIMO UE to enjoy beamforming gains, while keeping cost and implementation complexity lower compared to a MIMO-capable UE.

The UE determines the preferred precoding weights, preferred rank, and the channel quality for each stream and feeds this information to the network using a MIMO-specific format on the HS-DPCCH. It is then up to the network to make a final decision on what to transmit to the UE and notify the UE about the scheduling decision using the type-3 HS-SCCH. MIMO affects mainly the physical-layer processing, while the protocol layer impact is small.

### 11.3.2 HSDPA TWO-BRANCH MIMO DATA TRANSMISSION

To support dual-stream transmission, the HS-DSCH is modified to support up to two transport blocks per TTI. Each transport block is mapped to one stream. A CRC is attached to each of the transport blocks, and each transport block is individually coded. This is illustrated in Figure 11.7 (compare with Chapter 8 for the release 5

[2]Note that the term D-TxAA was used to denote something different in the MIMO TR [4].

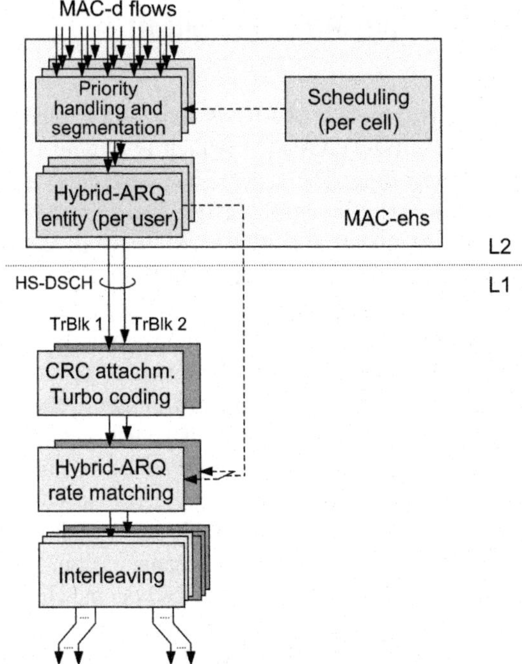


**FIGURE 11.7 HS-DSCH processing in case of MIMO transmission.**

case). Since two transport blocks are used in case of dual-stream transmission, HS-DPA-MIMO is a multi-codeword scheme (see Chapter 4 for a discussion on single vs. multi-codeword schemes) and allows for a successive-interference-cancellation receiver in the UE.

The physical-layer processing for each stream, up to and including spreading and scrambling, is identical to the release 5 case. The same set of channelization codes should be used for the two streams. At the receiver, the two streams are separated by appropriate receiver processing. A number of equalization techniques can be used, but the reuse of channelization codes between streams needs to be accounted for. One popular approach is to employ inter-stream interference cancellation (see Chapter 4).

Before each of the spread and scrambled streams are fed to the virtual antennas,[3] precoding is used, as illustrated in Figure 11.8. For each of the streams, the precoder is simply a pair of weights. Stream $i$ is multiplied with the complex weight $w_{ij}$ before being fed to virtual antenna $j$.

[3]The term *virtual antenna* here denotes the antenna as seen by the specifications, that is, each antenna from the specification perspective has a separate common pilot. The actual antenna arrangement on the base station may take any form, although the only antennas visible from the UE are those with a unique common pilot.

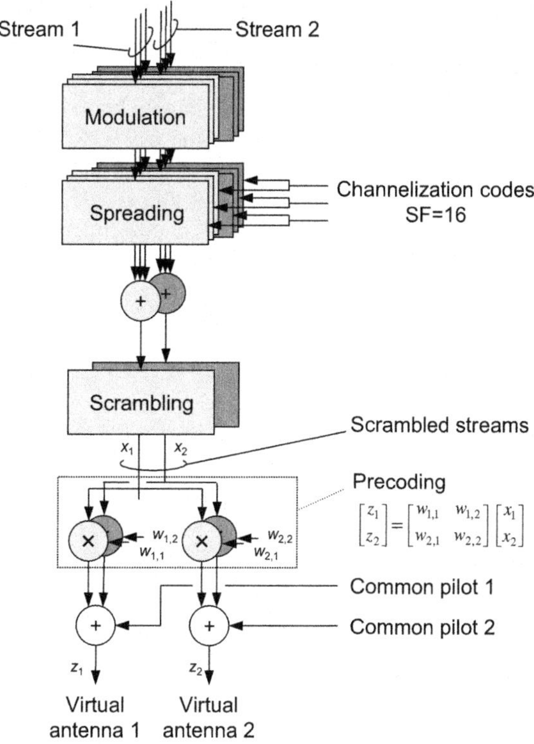


**FIGURE 11.8 Modulation, spreading, scrambling, and precoding for dual-stream transmission.**

### *11.3.2.1 Precoding*

Using precoding is beneficial for several reasons, especially in the case of single-stream transmission. In this case, precoding provides both diversity gain and maximum ratio combining gain as both transmit antennas are used and the weights are selected such that the signals from the two antennas add coherently at the receiver. This gives a higher received SINR than in absence of precoding, thus increasing the coverage for a certain data rate. Furthermore, precoding ensures that both power amplifiers are used in case of single-stream transmissions, thereby increasing the total transmit power (this is further discussed below). The precoding weights in case of single-stream transmission are chosen to be identical to those used for release 99 closed-loop transmit diversity,

$$w_{1,1} = 1/\sqrt{2},$$

$$w_{2,1} \in \left\{ \frac{1}{\sqrt{2}} e^{i\frac{\pi}{4}(2k-1)}, \quad k = 1,\ldots,4 \right\} = \left\{ \frac{1+j}{2}, \frac{1-j}{2}, \frac{-1+j}{2}, \frac{-1-j}{2} \right\}, \tag{11.5}$$

In case of dual-stream transmission, precoding can be used to aid the receiver in separating the two streams. If the precoding matrix is chosen as the eigenvectors of the channel matrix at the receiver, the two streams will not interfere with each other at the transmitter, and the inter-stream interference can be completely removed through strictly linear detection methods (see Section 4.5.2 for a thorough discussion). The drawback with this choice is that full channel knowledge is required at the transmitter, and to balance the performance and the feedback overhead, a codebook of size four is employed. The precoding matrix is chosen to be unitary and is therefore completely determined by the primary precoding vector. Once the weights for the first stream are selected, the weights $w_{1,2}$ and $w_{2,2}$ used for the second stream are given by the requirement to make the columns of the precoding matrix

$$\boldsymbol{W} = \begin{bmatrix} w_{1,1} & w_{1,2} \\ w_{2,1} & w_{2,2} \end{bmatrix} \tag{11.6}$$

orthogonal. As there are four different values possible for $w_{2,1}$, it follows that there are in total four different precoding matrices $\boldsymbol{W}$. The precoding matrix is constant over at least one sub-frame. The setting of the weights is up to the Node B implementation but follows typically the *Precoding Control Indication* (PCI) feedback from the UE.

Most commercial two-branch antennas are *dual-polarized.* Figure 11.9 illustrates a +45°/–45° dual-polarization antenna with orthogonal polarization directions. Assuming that both the transmitter and the receiver employ a +45°/–45° dual-polarization antenna and the radio channel preserves the polarization properties of the emitted radio waves, it follows that two uncorrelated signals transmitted from the two antenna ports will also be uncorrelated at the receiver because the +45° polarization port at the receiver side will be *polarization matched* to the signal transmitted from the +45° polarization port, and vice versa for the other polarization direction.

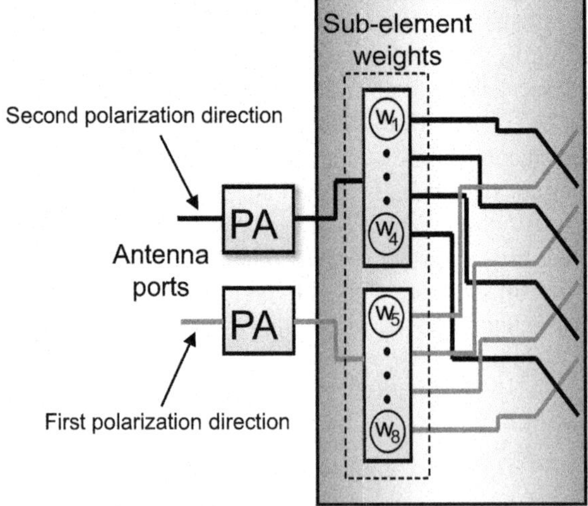


**FIGURE 11.9 Illustration of a +45°/–45° dual-polarization antenna.**

Unfortunately, the radio environment often alters the polarization characteristics of the emitted waves. Furthermore, because of user behaviors, the UE antennas are typically not fixed but rather moving, which, given a coordinate system relative the base station, affects the polarization state of the UE antennas. These aspects affect how the polarization components of the electrical field are picked up by the UE antennas. The orthogonality of the received signals is in general lost since both receive antenna ports will typically receive parts of the signals transmitted from each transmit antenna port. For rank-2, the precoding then aims at achieving *polarization matching* (as opposed to purely maximizing SINR for rank-1), that is, trying to adapt the polarization characteristics at the transmitter side such that it matches the polarization at the receiver side, thereby creating close to orthogonal received signals. Note that the overall aim of precoding is to maximize capacity; this relates to polarization matching for rank-2 and SINR optimization for rank-1, although there is no strong algorithmic difference between these two goals.

How the UE determines the precoding weights is implementation specific, and one choice would be to choose weights and transmission rank jointly in order to maximize capacity. To reduce complexity, on the other hand, weights and rank can be decided independently, and weights can be selected such that the received power on the primary stream is maximized. Needless to say, the UE needs to meet all RAN4 requirements, which put some constraints on the receiver/transmitter processing design.

It is only the HS-PDSCH that is precoded during MIMO transmissions. Common channels and HS-PDSCH to legacy users are not precoded and can be transmitted either from both transmit antennas (by employing transmit diversity) or only from the primary antenna. During development of MIMO equipment, STTD in fact was found to exhibit poor performance for HSPA UEs that are not equipped with MIMO processing receivers. The reason for the performance degradation due to STTD can be attributed to more challenging equalization. Sending the desired signal through more radio channels makes it more challenging for linear equalization to restore orthogonality. For this reason, single antenna transmission rather than Tx diversity is often preferable for serving non-MIMO-capable legacy UEs. The drawback with sending non-precoded channels from only the primary antenna is that these channels will use only half of the available power since each antenna usually has its own *power amplifier* (PA). This could cause only half of the PA power to be available for legacy UEs. A typical solution to this problem is to introduce an additional *common precoder*, also referred to as *virtual antenna mapping* (VAM), that distributes the physical channels over both antennas in a manner that is transparent to all UEs (see Figure 11.10).

For two-branch MIMO, the common precoder is represented by a 2×2 precoder matrix $C$ that satisfies the following two properties:

1. achieves power balancing and
2. not affecting the cell shape, that is, the coverage for common channels.

To achieve power balancing, the common precoder needs to ensure that a signal fed into one or both of the virtual antenna ports is evenly distributed at both physical antenna ports. The common precoder is transparent to the UE (not standardized) and

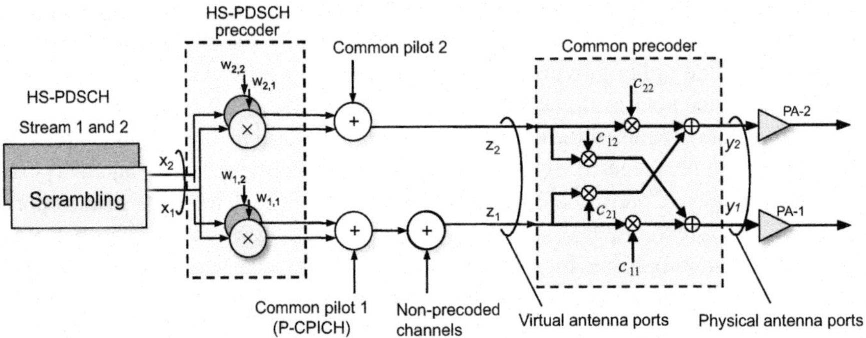


**FIGURE 11.10 Illustration of common precoder.**

many choices satisfying the above property exist. In particular, a common precoder satisfying the above property guarantees power balancing for common channels and channels for legacy users. For efficient MIMO operation, it is moreover important that the power of the HSDPA signal after feeding through both the MIMO precoder and the common precoder is still balanced between the two transmit antennas. Unfortunately, this is not always the case for all types of common precoder. The interaction between the MIMO precoder and the common precoder makes it more challenging to achieve power balancing for the MIMO precoded HS-PDSCH. As an example, consider the following common precoder:

$$\boldsymbol{C} = \frac{1}{\sqrt{2}} \begin{bmatrix} 1 & e^{i\pi/4} \\ e^{-i\pi/4} & -1 \end{bmatrix}, \tag{11.7}$$

The combined HS-PDSCH precoder (common precoder $\boldsymbol{C}$ and MIMO precoder $\boldsymbol{W}_k$) for single-stream transmission then becomes

$$\boldsymbol{CW}_k = \frac{1}{2} \begin{bmatrix} 1 & e^{i\pi/4} \\ e^{-i\pi/4} & -1 \end{bmatrix} \begin{bmatrix} 1 \\ e^{i\frac{\pi}{4}(2k-1)} \end{bmatrix} = \frac{1}{2} \begin{bmatrix} 1+e^{i\frac{\pi}{2}k} \\ e^{-i\frac{\pi}{4}}(1-e^{i\frac{\pi}{2}k}) \end{bmatrix}, \quad k=1, \ldots, 4, \tag{11.8}$$

meaning that

$$\boldsymbol{CW}_1 = \frac{1}{\sqrt{2}} \begin{bmatrix} e^{i\pi/4} \\ e^{-i\pi/2} \end{bmatrix}, \quad \boldsymbol{CW}_2 = \begin{bmatrix} 0 \\ e^{-i\pi/4} \end{bmatrix}, \quad \boldsymbol{CW}_3 = \frac{1}{\sqrt{2}} \begin{bmatrix} e^{-i\pi/4} \\ 1 \end{bmatrix}, \quad \boldsymbol{CW}_4 = \begin{bmatrix} 1 \\ 0 \end{bmatrix}, \tag{11.9}$$

showing that precoder indices $k = 2$ and 4 yield an unbalanced combined precoder and should consequently not be used.

In principle, the Node B scheduler can decide to ignore the UE recommended precoder in order to comply with the common precoder. A drawback with this approach is that the CQI report is based on the recommended precoder and it is not straightforward to predict the impact on the CQIs if the precoder would change. Hence, the standard allows *codebook subset restriction* for the case that the Node B has implemented a common precoder that interacts badly with the MIMO precoder. The restriction, if configured, only allows two of the four precoding vectors to be chosen/recommended by the UE for rank-1 transmissions,

$$w_{2,1} \in \left\{ \frac{(1 + j)}{2}, \frac{(-1 - j)}{2} \right\}. \tag{11.10}$$

The codebook restriction is configured by RRC signaling and known to the UE. Thus, the UE selects only precoding vectors from the allowed set when calculating CQIs.

The second property a common precoder should satisfy is to preserve the cell shape (coverage) for legacy users and common channels. This means that the common precoder is not allowed to affect the far-field antenna beam pattern. Employing a unitary common precoder and dual-polarized physical antennas is one way to satisfy this property. The key observation is that both physical antenna ports for a dual-polarized antenna have the same phase center, which means that no matter how the common precoder (or the HS-PDSCH MIMO precoder) phase rotates the signals, the beam width in the far-field is preserved. It can also be noted that if spatially separated antennas with the same polarization direction are used, the different antenna ports will have different phase centers and therefore the common precoder may affect the sector coverage. This also highlights one interesting aspect of rank-1 beamforming. With dual-polarized antennas, it is not possible to achieve any antenna array gain by forming a physical lobe. However, it is still possible to adjust the transmission via the MIMO precoder to match the current channel conditions in order to achieve coherent reception and thereby an increase in the received SINR. Note also that with more than two antennas, a physical lobe can be created using ports of the same polarization.

#### 11.3.2.2 Pilot design

To demodulate the transmitted data, the UE requires estimates of the channels between each of the base station scrambled streams and each of the UE's receive antennas. Hence, for dual- or single-stream transmission, a total of four or two 'channels' need to be estimated. One possibility would be to add a common pilot signal for each of the scrambled streams (before MIMO precoding). However, this would not be backwards compatible as legacy UEs assume the primary common pilot to be transmitted from the first virtual antenna. Demodulation of non-MIMO channels, for example, the control channels necessary for the operation of the system, would not be possible either as these channels are not transmitted using any precoding. Therefore, no changes to the common pilot channels are introduced. On each virtual antenna, a common pilot channel is transmitted. Either a primary common pilot channel (P-CPICH) is configured on each virtual antenna, using the same channelization and scrambling code, or a primary common pilot is configured on one virtual antenna and a secondary common pilot (S-CPICH) on the other virtual antenna. In case a primary common pilot is configured on both antennas, different mutually orthogonal pilot patterns are used on the different virtual antennas (all zeros for the first virtual antenna as in the single-antenna case and a sequence of zero and ones for the second virtual antenna). Both these schemes enable the UE to estimate the channel from each of the virtual transmit antennas to each of its receive antennas. From the UE's perspective, the common precoder is part of the channel over which the HS-PDSCH is received. Given knowledge of the precoding matrix the Node B used, the UE can form an estimate of the effective channel from each of the streams to each of the receiver antennas as $\hat{\boldsymbol{H}}\boldsymbol{W}$, where

$$\hat{\boldsymbol{H}} = \begin{bmatrix} \hat{h}_{1,1} & \hat{h}_{1,2} \\ \hat{h}_{2,1} & \hat{h}_{2,2} \end{bmatrix} \tag{11.11}$$

and $\hat{h}_{i,j}$ is the estimate of the channel between the virtual antenna $j$ at the base station and receive antenna $i$ at the UE. For this reason, the precoding matrix is signaled to the UE on the HS-SCCH. Explicit signaling of the precoding matrix greatly simplifies the UE implementation compared to estimation of the antenna weights, which is the case for closed-loop transmit diversity in release 99.

Another benefit of using common pilots is that (11.11) can directly be used in order to determine the preferred precoding weights, preferred rank, and CQI for each stream.

One problem with the introduction of a secondary pilot is a potential negative effect for legacy users. Each additional transmission path that is not decodable by a legacy UE results in interference and more demanding equalization. This is particularly challenging for a single-antenna UE that has little degrees of freedom to suppress different interference components. The losses for dual antenna UEs are relatively smaller, mainly since the spatial degree of freedom may be used to suppress the dominant interference component. The performance loss due to the secondary pilot in the presence of dispersion may be significant at higher geometries and may feel counter-intuitive, considering the modest power allocation on the second pilot (for example –13 dB). However, it is not surprising considering the noise floor modification effect. If the baseline noise floor is determined by the EVM at say –30 dB, then introducing an additional signal with an uncorrelated power spectrum at –13 dB on the average lifts that floor significantly. As a result, the effective geometry is reduced and the achievable SINR is limited to lower levels.

To reduce the problems mentioned above, the standard allows for a relatively flexible and low power setting of the secondary pilot such that demodulation performance for MIMO users and impact to the achievable SINR for legacy users can be traded off.

#### 11.3.2.3 Rate control

Rate control for each stream is similar to the release 5 HSDPA case. However, the rate-control mechanism also needs to determine the number of streams to transmit and the precoding matrix to use. Hence, for each TTI, the number of streams to transmit, the transport-block sizes for each of the streams, the number of channelization codes, the modulation scheme, and the precoding matrix are determined by the rate-control mechanism. This information is provided to the UE on the HS-SCCH, similar to the Release 5 case. As the scheduler controls the size of the two transport blocks in case of multi-stream transmission, the data rate of the two streams can be individually controlled.

Multi-stream transmission is mainly beneficial at relatively high SINRs and will consequently be used for the highest data rates. For lower data rates, single-stream transmission should be used. In this case, the precoder is used for diversity transmission and only one of the streams is present (carrying user data). Also, the implementation complexity and cost increase for a dual-stream capable UE since a lot of the processing is essentially duplicated. Because of these reasons, there is a possibility to configure downlink MIMO with *single stream restriction*, meaning that only rank-1 transmissions are allowed. This simplifies the design and reduces cost, while retaining the beamforming gain offered by rank-1 precoding.

Similar to release 6, the rate-control mechanism typically relies on UE feedback of the instantaneous channel quality. In case of dual-stream transmission, information about the supported data rate on each of the streams is required. In addition to the CQI reports, the UE also informs about the preferred rank and precoding. The UE reports either single or dual stream and associated precoding vector and CQIs depending on what choice maximizes the overall throughput. As the Node B scheduler is free to select the number of streams transmitted to a UE, the supported data rate in case of single-stream transmission is also of interest even though dual-stream is preferred. The CQI reports are therefore extended to cover both single- and dual-stream CQIs. The reason for reporting both single and dual stream is that it is non-trivial to transform the dual-stream CQI to a single-stream CQI in the Node B because the mapping is tightly coupled to the UE receiver structure and, for example, its capability to mitigate inter-stream interference.

The CQI handling is very similar to the release 5 HSDPA case. One difference is that the CQI granularity is reduced for rank-2 in order to save feedback overhead. For rank-2 there are 15 different CQI values compared to 30 for rank-1. This means that new CQI tables are introduced for rank-2 transmissions. The different CQI tables are described in [3]. Another difference is that for rank-2 CQI calculations, the UE can always assume that the overall power available for HS-PDSCH is uniformly distributed over 15 codes per stream. Still there are rank-2 scenarios where it is beneficial to use transport-block sizes corresponding to less than 15 codes. In fact, the lowest standardized TBSes in the rank-2 CQI tables would result in a code rate less than 1/3 if 15 codes were used. Hence, in this case the network will use less than 15 codes to ensure a code rate larger than 1/3 and scale the per-code power correspondingly.

Yet another difference in CQI handling between MIMO and non-MIMO is the use of the reference power adjustment $\Delta$. For release 5 HSDPA and single-stream MIMO operation, the CQI essentially says that this particular *Modulation and Coding Scheme* (MCS) can be supported with roughly 10% initial BLER if the transmission power of HS-PDSCH is

$$P_{\text{CPICH}} + \Gamma + \Delta, \tag{11.12}$$

where $P_{\text{CPICH}}$ denotes the combined power of all CPICH(es), $\Gamma$ is the measurement power offset (signaled via higher layers), and $\Delta$ represents a reference power adjustment given by the CQI table. The reference value adjustment helps the

**Table 11.4** Summary of CQI Tables C, D, and H

| CQI or CQIs Value | Table C | | | Table D | | | Table H | | |
|---|---|---|---|---|---|---|---|---|---|
| | TBS | # Codes | Δ | TBS | # Codes | Δ | TBS | # Codes | Δ |
| 0 | N/A | Out of Range | | N/A | Out of Range | | 4581 | 15 | –3.00 |
| 1 | 137 | 1 | 0 | 137 | 1 | 0 | 4581 | 15 | –1.00 |
| 2 | 173 | 1 | 0 | 173 | 1 | 0 | 5101 | 15 | 0 |
| 3 | 233 | 1 | 0 | 233 | 1 | 0 | 6673 | 15 | 0 |
| 4 | 317 | 1 | 0 | 317 | 1 | 0 | 8574 | 15 | 0 |
| 5 | 377 | 1 | 0 | 377 | 1 | 0 | 10,255 | 15 | 0 |
| 6 | 461 | 1 | 0 | 461 | 1 | 0 | 11,835 | 15 | 0 |
| 7 | 650 | 2 | 0 | 650 | 2 | 0 | 14,936 | 15 | 0 |
| 8 | 792 | 2 | 0 | 792 | 2 | 0 | 17,548 | 15 | 0 |
| 9 | 931 | 2 | 0 | 931 | 2 | 0 | 20,617 | 15 | 0 |
| 10 | 1262 | 3 | 0 | 1262 | 3 | 0 | 23,370 | 15 | 0 |
| 11 | 1483 | 3 | 0 | 1483 | 3 | 0 | 23,370 | 15 | 1.50 |
| 12 | 1742 | 3 | 0 | 1742 | 3 | 0 | 23,370 | 15 | 2.50 |
| 13 | 2279 | 4 | 0 | 2279 | 4 | 0 | 23,370 | 15 | 4.00 |
| 14 | 2583 | 4 | 0 | 2583 | 4 | 0 | 23,370 | 15 | 5.00 |
| 15 | 3319 | 5 | 0 | 3319 | 5 | 0 | | | |
| 16 | 3565 | 5 | 0 | 3565 | 5 | 0 | | | |
| 17 | 4189 | 5 | 0 | 4189 | 5 | 0 | | | |
| 18 | 4664 | 5 | 0 | 4664 | 5 | 0 | | | |
| 19 | 5287 | 5 | 0 | 5287 | 5 | 0 | | | |
| 20 | 5887 | 5 | 0 | 5887 | 5 | 0 | | | |
| 21 | 6554 | 5 | 0 | 6554 | 5 | 0 | | | |
| 22 | 7168 | 5 | 0 | 7168 | 5 | 0 | | | |
| 23 | 9719 | 7 | 0 | 9719 | 7 | 0 | | | |
| 24 | 11,418 | 8 | 0 | 11,418 | 8 | 0 | | | |
| 25 | 14,411 | 10 | 0 | 14,411 | 10 | 0 | | | |
| 26 | 17,237 | 12 | 0 | 17,237 | 12 | 0 | | | |
| 27 | 17,237 | 12 | –1 | 21,754 | 15 | 0 | | | |
| 28 | 17,237 | 12 | –2 | 23,370 | 15 | 0 | | | |
| 29 | 17,237 | 12 | –3 | 24,222 | 15 | 0 | | | |
| 30 | 17,237 | 12 | –4 | 25,558 | 15 | 0 | | | |

UE to be able to cover the same dynamic range between different UE categories (CQI tables) and indicates when the quality is superior or inferior to the maximum or minimum tabulated MCS. To illustrate this, consider CQI Tables C and D (summarized in Table 11.4). The maximum supported TBS in Table C is 17,237, which can be signaled by multiple CIQ values (26 to 30). The difference between these CQI values is that different Δ values are used. For example, if the UE reports CQI 30 (corresponding to Δ = –4), this means that the received power is roughly

4 dB higher than what is required to decode TBS = 17,237 with an initial BLER of 10%. In CQI Table D, on the other hand, each CQI value corresponds to different TBSes and $\Delta = 0$ dB is always used. In fact, since the symbol SNR step between each CQI value is roughly 1 dB, it follows that the UE can use the same CQI algorithm for calculating CQIs for Table C and D. For example, the SNR required for CQI 30 in Table D and the SNR required for CQI 30 in Table C with $\Delta = -4$ are the same.

For rank-2, $\Delta$ is removed from (11.12) and the interpretation of $\Delta$ is instead changed to indicate an equivalent AWGN SINR (as opposed to power in the single-stream) difference between the actual SINR and the SINR required to meet the 10% initial BLER target. The reason is that for rank-2, the SINR does not, in general, scale linearly with the power (SNR) even for an AWGN channel due to the inter-stream interference term. More specifically, depending on channel conditions and receiver type, each stream can be subject to very different SINRs, for example if an IC capable receiver is employed.

Similarly as discussed above, consider CQI Table H (see Table 11.4). The maximum supported TBS (per stream) in Table H is 23,370, which can be signaled by multiple CIQ values (10–14). These CQI values have different associated $\Delta$ values. For example, CQI 14 has $\Delta = 5$ dB, which means that the AWGN symbol SINR (for this particular stream) is 5 dB larger than what is required to meet the 10% BLER target for this TBS. Note that a positive rank-2 $\Delta$ value indicates that the quality is better than required, while for the rank-1 (or release 5) case, a negative $\Delta$ value indicates that the quality is better than required.

A second observation is that CQI values 0 and 1 are associated with negative $\Delta$ values. The reason is that for CQI reporting, it is always assumed that 15 codes are used (as discussed above). As an example, the per-code SINR for CQI 1 is 1 dB too low, assuming that the total transmission power is distributed evenly between 15 codes. Hence, instead of scheduling 15 codes, it is more reasonable that the Node B schedules this TBS with 15/[1 dB in linear domain] $\approx$12 codes. Also, scheduling the TBS corresponding to CQI 1 using 12 codes gives a code rate above 1/3, while 15 codes gives a code rate below 1/3.

Rank-2 is mainly beneficial for large transport-block sizes. Hence, for a maximum modulation order of 16-QAM, the SINR step size between different TBSes is kept at roughly 1 dB even though the number of CQIs are decreased from 30 to 15 values. This is illustrated in Figure 11.11, which shows the SINR step size for CQI Table I in an AWGN scenario. However, similar as for 64-QAM, for MIMO with 64-QAM the SINR step size for rank-2 is increased to roughly 2 dB. This is illustrated in Figure 11.12, which shows the SINR step size for CQI Table K in an AWGN scenario.

#### *11.3.2.4 Hybrid-ARQ with soft combining*

For each stream, the physical-layer hybrid-ARQ processing and the use of multiple hybrid-ARQ processes are identical to the single-stream case. Hence, there is one hybrid-ARQ entity per HS-DSCH with a number of configurable parallel hybrid-ARQ processes. This will in essence double the number of hybrid-ARQ processes required for MIMO compared to non-MIMO. However, as the multiple streams are

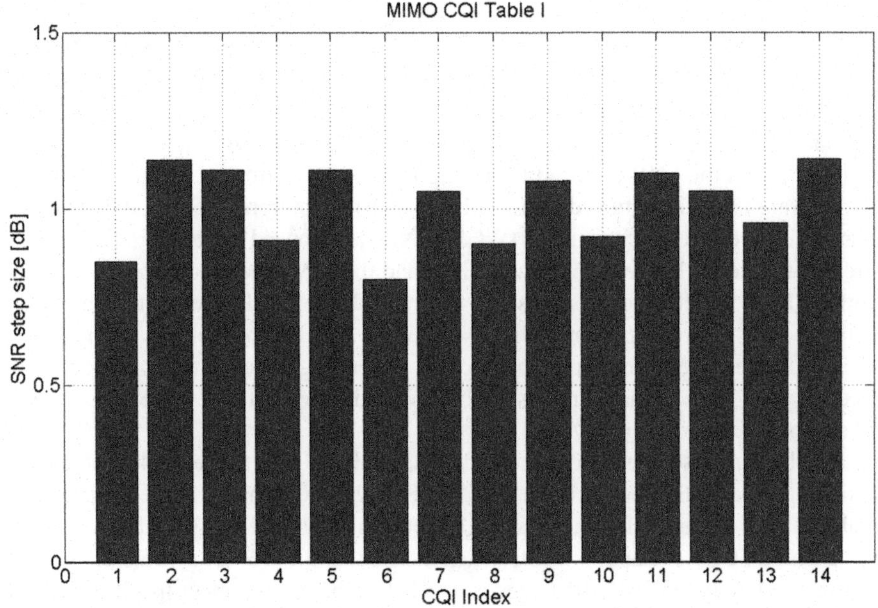


**FIGURE 11.11 SINR step size as a function of CQI index for CQI Table I in an AWGN scenario.**

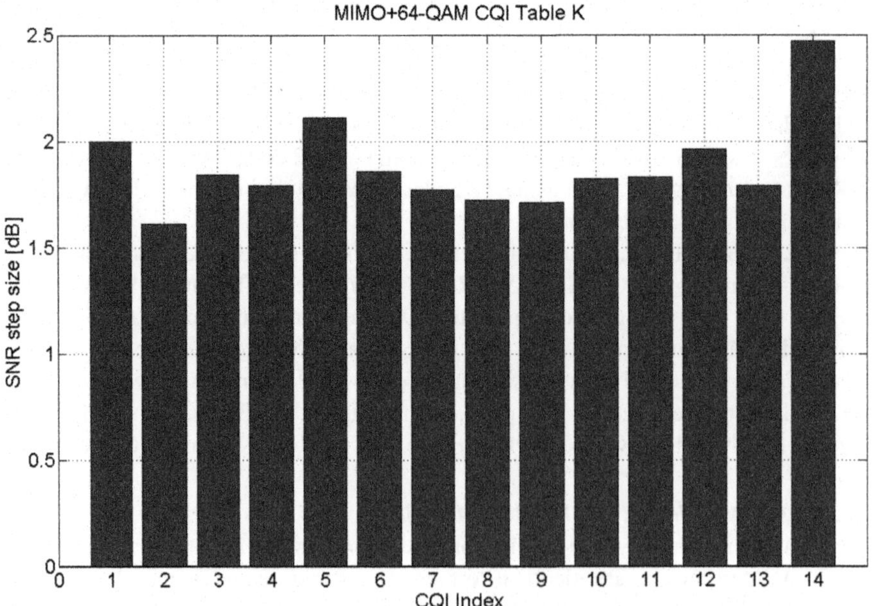


**FIGURE 11.12 SINR step size as a function of CQI index for CQI Table K in an AWGN scenario.**

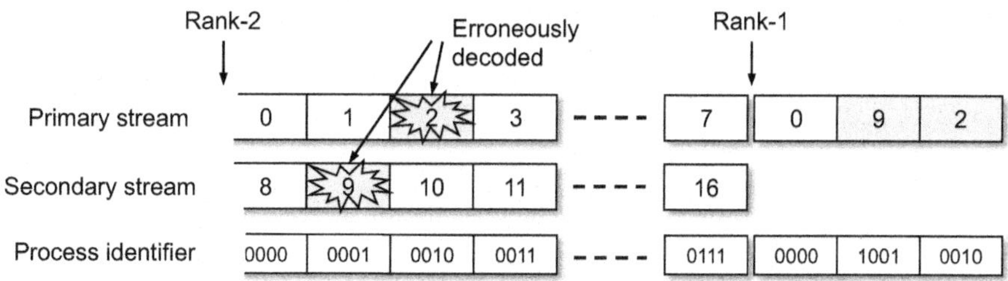


**FIGURE 11.13 Illustration of hybrid-ARQ processing, with 16 configured processes. Processes {0, 8}, {1, 9},..., {7, 16} are coupled, and only the primary stream process number is signaled using the process identifier field on the HS-SCCH. In this example, retransmissions have highest priority.**

transmitted over different antennas, one stream may be correctly received while another stream requires retransmission of the transport block.[4] Therefore, one hybrid-ARQ acknowledgment per stream is sent from the UE to the Node B. To reduce the amount of hybrid-ARQ signaling for rank-2 transmissions, two hybrid-ARQ process numbers are coupled and only the process number associated with the primary stream needs to be signaled (the process number on the secondary stream can then be inferred; see Figure 11.13). Note, in particular, that a transport block originally sent on one stream may be retransmitted on the other stream. This is illustrated in Figure 11.13, where the transport block associated with hybrid-ARQ process number 9 sent from the secondary stream in the original transmission is retransmitted on the primary stream. Note also how the HS-SCCH process identifier always indicates the hybrid-ARQ process number associated with the primary stream.

### 11.3.3 CONTROL CHANNEL STRUCTURE

To support MIMO, the out-band control signaling needs to be modified accordingly. Also, to efficiently support the higher data rates provided by MIMO, MAC-ehs is employed in order to facilitate flexible segmentation and reordering of larger data amounts.

The downlink out-band control signaling is carried on the HS-SCCH. In case MIMO support is enabled, a new format of the HS-SCCH (type 3) is used to accommodate the additional information required (see Figure 11.14). The division of the HS-SCCH into two parts (see Chapter 8) is maintained. Part one is extended to include information about the number of streams transmitted to the UE, one or two, and their respective modulation scheme as well as which of the four precoding matrices the Node B used for the transmission. Modulation and rank information are combined into one field. To save bits when introducing 64-QAM, one of the bits in the CCS field ($x_{ccs,7}$) is used to indicate modulation and rank information in certain

[4]Typically, if the first stream is erroneously received, decoding of the second stream is likely to fail if successive interference cancellation is used.

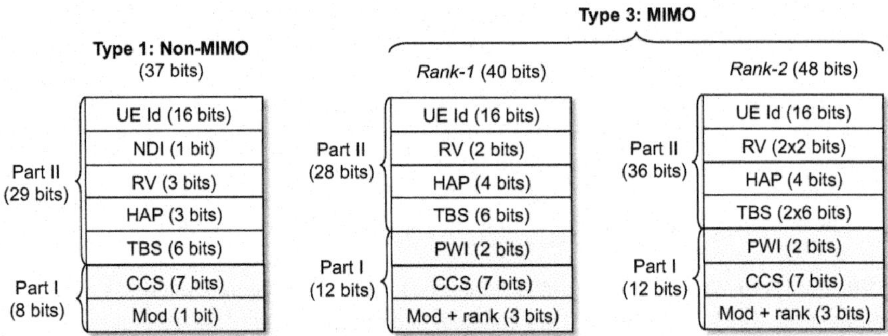


**FIGURE 11.14 Illustration of HS-SCCH information for type 1 (non-MIMO) and type 3 (MIMO). CCS, channelization code set; Mod, modulation; TBS, transport-block size; HAP, hybrid-ARQ process number; RV, redundancy version; NDI, new data indicator; and PWI, precoding weight indicator.**

cases (field value 101 in Table 11.5). This is made possible by letting the HS-SCCH number (even or odd) affect the channelization code-set start point in these cases; see Chapter 10 and [6] for further details.

For the second part of the HS-SCCH, the format depends on whether one or two streams are scheduled to the UE. In the latter case, additional bits are transmitted on part two to convey the hybrid-ARQ and transport-block size information also for the second stream. Despite the slight increase in the number of bits on the HS-SCCH in case MIMO is enabled in the UE, the spreading factor is kept at 128. The additional bits are fitted onto the physical channel by adjusting the rate matching of the two parts appropriately. Furthermore, the new-data indicator, the redundancy version, and the constellation-rearrangement fields are combined into a joint field. In this process, some less-frequently used combinations from previous releases of HSPA are removed, and the mapping is made dependent on the code rate. Also, the new data

**Table 11.5** Modulation and Rank Mapping for MIMO

| Mapping | Configuration | | |
|---|---|---|---|
| | **Modulation Primary Stream** | **Modulation Secondary Stream** | **Rank** |
| 111 | 16-QAM | 16-QAM | 2 |
| 110 | 16-QAM | QPSK | 2 |
| 101 | 64-QAM | Indicated by $x_{ccs,7}$ | Indicated by $x_{ccs,7}$ |
| 100 | 16-QAM | n/a | 1 |
| 011 | QPSK | QPSK | 2 |
| 010 | 64-QAM | 64-QAM | 2 |
| 001 | 64-QAM | 16-QAM | 2 |
| 000 | QPSK | n/a | 1 |

indicator has essentially been removed and is replaced by letting $(s,r,b)$ equal (1,0,0) for new data, where $s$ and $r$ are the redundancy version parameters and $b$ denotes the constellation version parameter. As discussed earlier, only the hybrid-ARQ process number for the primary stream, *HAP*, is signaled during rank-2, and the process number for the second stream is given by

$$(HAP + N/2)\,\mathrm{mod}(N), \tag{11.13}$$

where $N$ is the total number of hybrid-ARQ processes configured by higher layers.

Uplink out-band control signaling consists of hybrid-ARQ ACK/NACK, PCI, and CQI and is transmitted on the HS-DPCCH. In case of single-stream transmission, only a single hybrid-ARQ acknowledgment bit is transmitted and the format is identical to release 5. In case of dual-stream transmission, the two hybrid-ARQ acknowledgment bits are jointly coded to 10 bits and transmitted in one slot on the HS-DPCCH. In single-cell operation, the release 7/8 MIMO hybrid-ARQ ACK/NACK codebook consists of in total eight codewords including POST and PRE (see Table 11.6).

The PCI consists of two bits, indicating which one of the four precoding matrices best matches the channel conditions at the UE. The CQI indicates the data rate the UE recommends in case transmission is done using the recommended PCI and nominal HS-PDSCH transmit power. Both single-stream and dual-stream CQI reports can be useful as the scheduler may decide to transmit a single stream, even if the channel

**Table 11.6** Hybrid-ARQ ACK/NACK Codebook

| **Hybrid-ARQ ACK Message to Be Transmitted** | | $w_0$ | $w_1$ | $w_2$ | $w_3$ | $w_4$ | $w_5$ | $w_6$ | $w_7$ | $w_8$ | $w_9$ |
|---|---|---|---|---|---|---|---|---|---|---|---|
| **Hybrid-ARQ ACK in response to a single transport block** | | | | | | | | | | | |
| ACK | | 1 | 1 | 1 | 1 | 1 | 1 | 1 | 1 | 1 | 1 |
| NACK | | 0 | 0 | 0 | 0 | 0 | 0 | 0 | 0 | 0 | 0 |
| **Hybrid-ARQ ACK in response to two transport blocks** | | | | | | | | | | | |
| **Response to primary stream** | **Response to secondary stream** | | | | | | | | | | |
| ACK | ACK | 1 | 0 | 1 | 0 | 1 | 1 | 1 | 1 | 0 | 1 |
| ACK | NACK | 1 | 1 | 0 | 1 | 0 | 1 | 0 | 1 | 1 | 1 |
| NACK | ACK | 0 | 1 | 1 | 1 | 1 | 0 | 1 | 0 | 1 | 1 |
| NACK | NACK | 1 | 0 | 0 | 1 | 0 | 0 | 1 | 0 | 0 | 0 |
| **PRE/POST indication** | | | | | | | | | | | |
| PRE | | 0 | 0 | 1 | 0 | 0 | 1 | 0 | 0 | 1 | 0 |
| POST | | 0 | 1 | 0 | 0 | 1 | 0 | 0 | 1 | 0 | 0 |

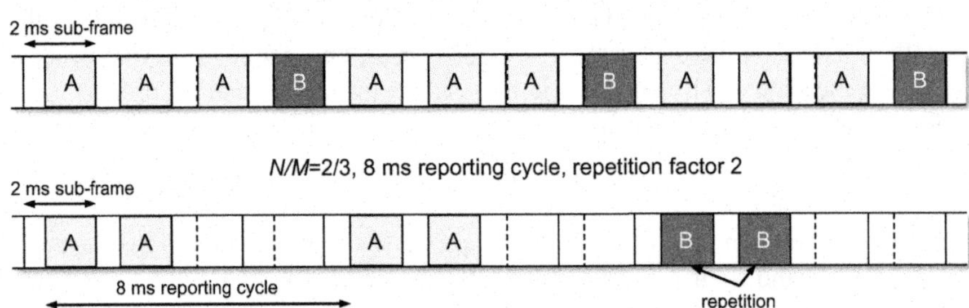


**FIGURE 11.15 Two examples of type-A and type-B CQI reporting configurations.**

conditions permit two streams, for example, if the amount of data to transmit is small or if the Node B is power limited. In this case, a single-stream CQI report is required, while in the case of dual-stream transmission obviously one CQI report per stream is useful. As it is not straightforward to deduce the single-stream quality from a dual-stream report, two types of CQI reports are defined:

1. *Type A reports*, containing the PCI and the recommended number of streams, one or two, as well as the CQI for each of these streams.
2. *Type B reports*, containing the PCI and the CQI in case of single-stream transmission.

For the type-B reports, the same 5-bit CQI report as in release 5 is used, while for type A the CQI consists of eight bits. In both cases, the PCI and CQI reports are concatenated and coded into 20 bits using a block code and transmitted in two slots on the HS-DPCCH. To allow for a flexible adaptation to different propagation environments, the first $N$ out of $M$ PCI/CQI reports are type-A reports and the remaining $M$–$N$ reports are type-B reports. The ratio $N/M$ is configured by RRC signaling. Two examples are shown in Figure 11.15. Single- and dual-stream transmissions use different CQI granularities: 30 and 15 CQI values, respectively, for single- and dual-stream transmissions. Type-A CQI reports are constructed using a compound CQI value that is computed according to

$$\text{CQI} = \begin{cases} 15\times\text{CQI}_1 + \text{CQI}_2 + 31 & \text{rank-2} \\ \text{CQI}_S & \text{rank-1} \end{cases}, \tag{11.14}$$

where $\text{CQI}_1$ and $\text{CQI}_2$ indicate the supported transport format for the primary and secondary streams, respectively, during rank-2 and $\text{CQI}_S$ is the CQI during rank-1. Hence, the preferred rank is implicitly signaled in the CQI report. The mapping between CQI and MCS depends on the UE category and transmission rank. For type-A reports, the two PCI bits and eight CQI bits are jointly coded using a (20, 10) block code, whereas for type-B reports, the two PCI bits and five CQI bits are jointly coded using a (20, 7) block code; see [6] for details. Figure 11.16 provides an overview of HS-DPCCH for two-branch MIMO.

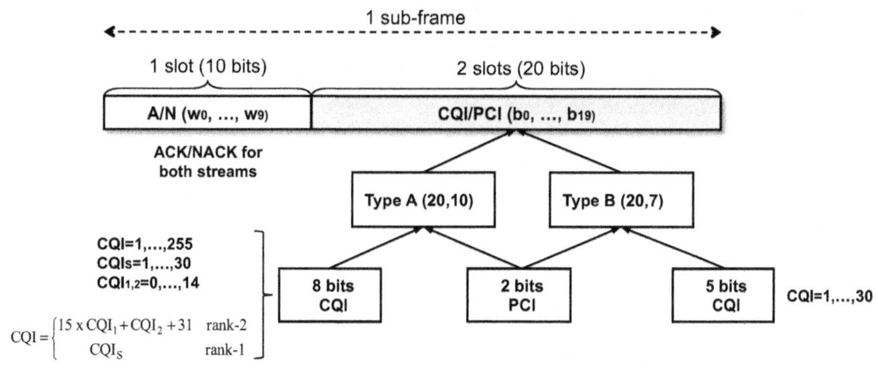


**FIGURE 11.16 Overview of HS-DPCCH operation for release 7 MIMO.**

The power setting of the HS-DPCCH is updated to account for the increased information in each sub-frame. The power offsets $\Delta_{CQI}$, $\Delta_{ACK}$, and $\Delta_{NACK}$ are signaled by the network and they can take on values between zero and eight (see Section 8.3.9). Based on the signaled values, the UE determines the appropriate quantized amplitude ratio ($\beta_{hs}/\beta_c$) based on the number of transport blocks that is detected by the UE on the downlink (for hybrid-ARQ ACK transmissions) and the CQI type that is to be transmitted by the UE; see Table 11.7. From the table it is also evident that the quantized amplitude ratios employed by the UE should be $\Delta_{CQI} + 1$ and/or $\Delta_{ACK} + 1$ for certain cases (for example, if type-A CQI/PCI should be transmitted). Hence, the list of quantized amplitude ratios that the UE needs to support has been extended with one additional value $A_{hs} = 38/15$ (signaled value 9).

## 11.3.4 UE CAPABILITIES, RRC SIGNALING AND CONFIGURATIONS

To allow for a wide range of different UE implementations, MIMO support is not mandatory for all UEs. Furthermore, as multi-stream transmission is mainly a tool

**Table 11.7** Power Offsets for HS-DPCCH With Release 7 MIMO Configured

| Condition | $A_{hs}$ Equals the Quantized Amplitude Ratio Translated From | | | | |
|---|---|---|---|---|---|
| | CQI/PCI Message | | Hybrid-ARQ ACK Message | | |
| | Type A | Type B | Only ACK(s) | Only NACK(s) | Contains Both ACK and NACK or a PRE or POST |
| Single stream | $\Delta_{CQI} + 1$ | $\Delta_{CQI}$ | $\Delta_{ACK}$ | $\Delta_{NACK}$ | MAX ($\Delta_{ACK}$, $\Delta_{NACK}$) |
| Dual stream | $\Delta_{CQI} + 1$ | $\Delta_{CQI}$ | $\Delta_{ACK} + 1$ | $\Delta_{NACK} + 1$ | MAX ($\Delta_{ACK} + 1$, $\Delta_{NACK} + 1$) |

Table 11.8 Illustration of Release 7/8 HS-DSCH Categories Supporting MIMO

| HS-DSCH Category | Modulation Scheme | Max Data Rate (Mbit/s) | |
|---|---|---|---|
| | | Single Stream | Dual Stream |
| 15 | QPSK, 16-QAM | 11.7 | 23.4 |
| 16 | QPSK, 16-QAM | 14 | 28 |
| 17[a] | QPSK, 16-QAM (64-QAM) | 11.7 (17.6) | 23.4 |
| 18[a] | QPSK, 16-QAM (64-QAM) | 14 (21) | 28 |
| 19 | QPSK, 16-QAM, 64-QAM | 17.6 | 35.3 |
| 20 | QPSK, 16-QAM, 64-QAM | 21.1 | 42.2 |

[a]*Can support either 64-QAM or MIMO with 16-QAM.*

for increasing the supported peak data rates, MIMO is seen as a high-end UE category. Therefore, a UE supporting rank-2 MIMO is, for example, always capable of receiving 15 channelization codes. The supported data rates with and without MIMO for these UEs are listed in Table 11.8. For single carrier operation, a MIMO-capable UE supporting up to 16-QAM is category 15-18, while a MIMO-capable UE simultaneously supporting up to 64-QAM is either category 19 or 20. Furthermore, some capabilities differ depending on whether the UE is configured in MIMO mode or not. Categories 17 and 18 support either 16-QAM with MIMO or 64-QAM without MIMO. A UE supporting MIMO restricted to single-stream operation can belong to any HS-DSCH physical layer category not supporting MIMO or category 17 or 18.

A number of MIMO-related parameters are set by higher-layer signaling. These include [7]

- configuration of type-A and type-B CQIs; the ratio of type-A CQIs and total number of CQIs can range between 1/2 and 1;
- precoding weight set restriction;
- pilot configuration for secondary antenna, either transmit diversity enabled P-CPCIH or S-CPICH based; and
- power offset for S-CPICH relative P-CPICH (–6,…,0 dB).

## 11.4 DOWNLINK FOUR-BRANCH MIMO

### 11.4.1 OVERVIEW

Downlink four-branch MIMO, referred to as HSDPA-MIMO mode with four transmit antennas in the specifications, enables transmission of up to four simultaneous streams. The design builds to a large extent on the existing release 7 MIMO mode and corresponding LTE MIMO principles.

To minimize the required signaling/feedback overhead, a design based on maximum two codewords with symbol interleaving is adopted. Feedback and signaling

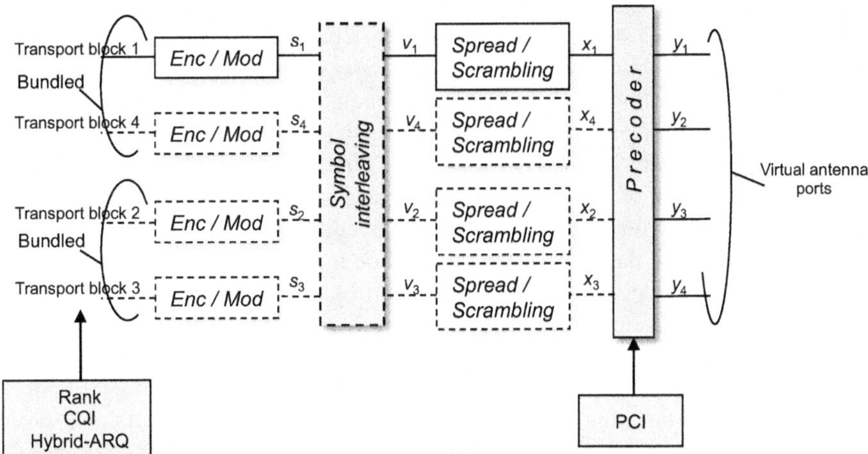


**FIGURE 11.17 Layer-1 HS-PDSCH processing for HSDPA-MIMO mode with four transmit antennas.**

are per codeword and each codeword can be mapped to up to two transport blocks. Whenever more than two streams are transmitted, pairs of streams are bundled. For example, the transport-block size and hybrid-ARQ information are the same for bundled streams. Each stream is subject to the same physical-layer processing in terms of coding, modulation, and spreading as the corresponding single-layer HSDPA case. After coding, spreading, and modulation, precoding is applied to map the streams to the four virtual antennas. Between the modulation and spreading functionality, a symbol interleaver is introduced to ensure that coupled streams experience the same average channel quality. See Figure 11.17 for an illustration of the layer-1 HS-PDSCH processing.

A common pilot solution is adopted for channel estimation purposes. Four common pilots, one from each virtual antenna branch, are configured. The pilots sent on virtual antennas three and four use very low power to limit the impact on legacy users. Two additional demodulation pilots can be enabled for virtual antennas three and four when HS-PDSCH is sent in order to ensure sufficient demodulation quality.

Similarly to the two-branch MIMO mode, the UE determines the preferred precoding weights, preferred rank, and the channel-quality for each of the streams (or each of the bundled streams in case of more than two streams) and feeds this information back to the network using a special type of the HS-DPCCH based on convolutional coding and spreading factor 128. It is then up to the network to make a final decision on what to transmit to the UE and notify the UE about the scheduling decision using the type-4 HS-SCCH. HSDPA-MIMO mode with four transmit antennas is designed from the beginning to support multicarrier operation with up to four carriers. This is different compared to two-branch MIMO, where multicarrier was added in later releases.

## 11.4.2 HSDPA FOUR-BRANCH MIMO TRANSMISSION

A key design question during the standardization phase was whether to support a maximum of two or four codewords, where a codeword is defined as the basic unit associated with signaling and feedback. For example, hybrid-ARQ and CQI feedback and MCS signaling are per codeword. A scheme based on four codewords offers more flexibility in terms of MCS per stream and optimal SIC receiver processing but comes at the expense of increased complexity and signaling/feedback overhead. It turned out that the performance advantage of four codewords compared to two codewords was small, and therefore a design based on max two codewords, where each codeword can be mapped to up to two transport blocks, was adopted. When more than two streams are transmitted, pairs of streams become bundled; see Figure 11.18. For rank-1 and rank-2 transmissions, the streams are assigned independent TBSes based on stream-independent CQI reports and receive independent hybrid-ARQ acknowledgments. For rank-3 transmissions, stream 1 is associated with codeword one and streams 2 and 3 are bundled and associated with codeword two. For rank-4 transmissions, streams 1 and 4 are associated with codeword one and streams 2 and 3 are associated with codeword two. There is one transport block per stream, but

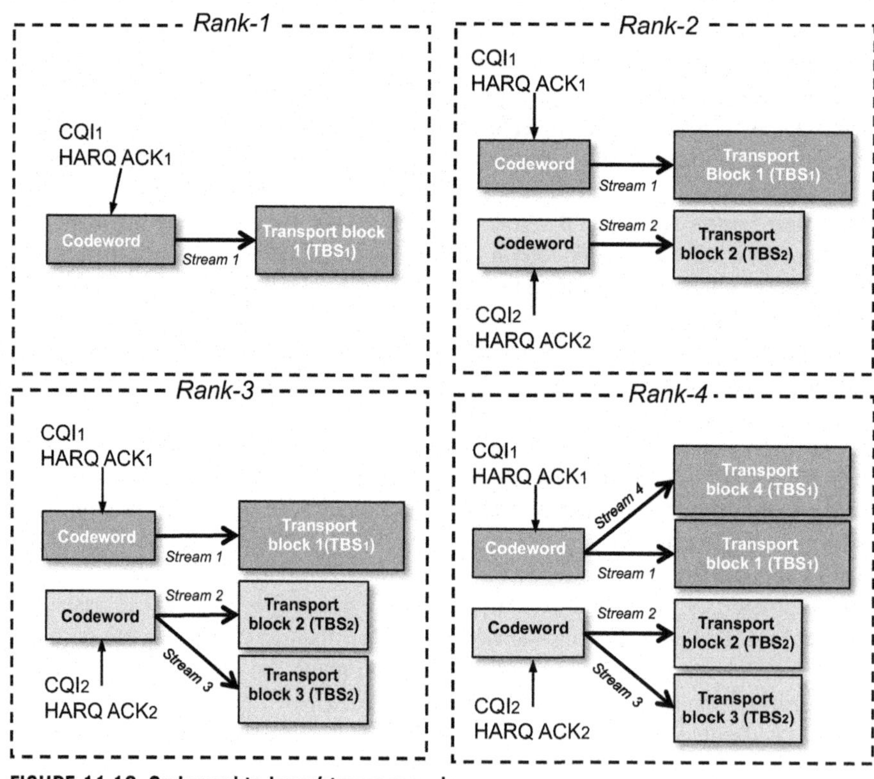


**FIGURE 11.18 Codeword to layer/stream mapping.**

bundled streams use the same transport-block size and receive a common hybrid-ARQ and CQI feedback indication. No new transport-block sizes are introduced for four-branch MIMO.

A CRC is attached to each transport block before being encoded and modulated. A symbol interleaver is introduced after modulation and before spreading, scrambling, and precoding (see Figure 11.17). All these processing blocks except the symbol interleaver and the precoder are identical to the release 5 case. The purpose of the symbol interleaver is to ensure that bundled streams experience roughly the same average channel quality and therefore have a similar probability of being successfully encoded. This is important since only one hybrid-ARQ feedback message is sent per bundled stream. Hence, as long as one stream fails to be decoded, then both of the bundled steams are NACKed irrespectively of whether the other stream was successfully decoded. Also, since only one CQI per bundled stream is sent, it is advantageous if both streams have similar channel quality. The symbol interleaver can be expressed as in Table 11.9. The interleaver operates on bundled streams and is consequently transparent for rank-1 and rank-2 transmissions.

### 11.4.2.1 Precoding

The purpose of precoding for four-branch MIMO is the same as for two-branch MIMO. For high-rank transmissions, precoding helps the receiver by trying to orthogonalize the channel, whereas for low-rank transmissions, the intention is to provide an array gain from coherent reception at the receiver.

Two main precoder design alternatives were evaluated during the standardization phase, *Householder based* or *product based.* The product based precoder consists of the product of an inner and an outer precoder, where the inner precoder captures long-term properties of the channel characteristics and can therefore be updated less

**Table 11.9** Operation of Symbol-Level Interleaving

| **Number of Transport Blocks** | **1** | **2** |
|---|---|---|
| Symbol-level interleaving output | $v_1(k)=s_1(k)$ | $v_1(k)=s_1(k)$<br>$v_2(k)=s_2(k)$ |
| **Number of Transport Blocks** | **3** | **4** |
| Symbol-level interleaving output | $v_1(k)=s_1(k)$<br>$v_2(k)=\begin{cases} s_2(k), k \text{ is odd} \\ s_3(k), k \text{ is even} \end{cases}$<br>$v_3(k)=\begin{cases} s_3(k), k \text{ is odd} \\ s_2(k), k \text{ is even} \end{cases}$ | $v_1(k)=\begin{cases} s_1(k), k \text{ is odd} \\ s_4(k), k \text{ is even} \end{cases}$<br>$v_4(k)=\begin{cases} s_4(k), k \text{ is odd} \\ s_1(k), k \text{ is even} \end{cases}$<br>$v_2(k)=\begin{cases} s_2(k), k \text{ is odd} \\ s_3(k), k \text{ is even} \end{cases}$<br>$v_3(k)=\begin{cases} s_3(k), k \text{ is odd} \\ s_2(k), k \text{ is even} \end{cases}$ |

frequently, while the outer precoder adapts to short-term properties of the channel and needs more frequent updates. Evaluations suggested that the Householder precoder performed slightly better and was thereby adopted by the standard. For each rank, there are 16 different weight matrices to choose from. The precoding matrices are based on the release 8 LTE Householder codebook [8],

$$W_n = I - 2u_n u_n^H / (u_n^H u_n), \quad n = 0, \ldots, 15, \tag{11.15}$$

where $u_n$ is given in Table 11.10. A Householder matrix is Hermitian and unitary. Depending on the transmission rank (ν), different scaling and columns of $W_n$ should be used to form the final precoding matrix $\tilde{W}_{\nu,n}$. The quantity $W_n^{\{s\}}$ denotes the matrix defined by the columns given by the set $\{s\}$ from $W_n$. This information is also given in Table 11.10. For example, for $n = 14$ and rank-4, the final precoding matrix $\tilde{W}_{4,14}$ consists of columns 3, 2, 1, and 4 of $W_{14}$ (in that order) and is scaled by ½. The preferred rank and precoding matrix is determined by the UE and the associated index is signaled to the network using the HS-DPCCH. It is then up to the network to make a final decision about the precoder and this information is signaled to the UE in the HS-SCCH. The precoded HS-DPSCH signal is given by

$$y = \tilde{W}_{\nu,k} x_\nu, \tag{11.16}$$

where the 4x$\nu$ matrix $\tilde{W}_{\nu,k}$ and the $\nu$-by-1 vector $x_\nu$ are the precoding matrix and scrambled signal, respectively, for rank $\nu$ and precoding index $k$ (see also Figure 11.20).

Similar to that for two-branch MIMO, a common precoder can be used also for four-branch MIMO in order to use the power amplifiers more efficiently for legacy channels. Compared to two-branch MIMO, the number of weight matrices is significantly larger (16 compared to 4) for four-branch MIMO, which makes the precoder more efficient. On the other hand, this also makes it more challenging to find a suitable common precoder that yields a balanced power for all codewords and relevant ranks. To make this process easier, there is an option to reduce the number of elements by using codebook subset restriction. The procedure is very flexible and any codewords can be excluded. A UE is restricted to report rank and PCI within a precoder codebook subset specified by a bitmap parameter configured by higher layer signaling. The bitmap forms the bit sequence $a_{63}, \ldots, a_0$, where $a_0$ is the LSB and $a_{63}$ is the MSB and where a bit value of zero indicates that the PMI and RI reporting associated with the bit is not allowed. The association of bits to precoders follows the LTE design principles [9] and are given as follows: bit $a_{16(\upsilon-1)+i_c}$ is associated with rank $\upsilon$ and codebook index $i_c$ in Table 11.10.

Another aspect of finding a good common precoder is that the antenna architecture for four-branch MIMO is different from that for two-branch MIMO. A two-column array with dual-polarized elements, as illustrated in Figure 11.19, is a common antenna arrangement for four-branch MIMO deployments. Here it can be seen that antenna ports one and three, and two and four, respectively, are dual-polarized and hence have the same phase center, whereas antenna ports one and two, and three

**Table 11.10** Precoding Weight Information

| $i_c$ | $u_n$ | Number of Transport Blocks (Rank) | | | |
|---|---|---|---|---|---|
| | | 1 | 2 | 3 | 4 |
| 0 | $u_0 = [\ 1\ \ -1\ \ -1\ \ -1\ ]^T$ | $W_0^{\{1\}}$ | $W_0^{\{14\}}/\sqrt{2}$ | $W_0^{\{124\}}/\sqrt{3}$ | $W_0^{\{1234\}}/2$ |
| 1 | $u_1 = [\ 1\ \ -j\ \ 1\ \ j\ ]^T$ | $W_1^{\{1\}}$ | $W_1^{\{12\}}/\sqrt{2}$ | $W_1^{\{123\}}/\sqrt{3}$ | $W_1^{\{1234\}}/2$ |
| 2 | $u_2 = [\ 1\ \ 1\ \ -1\ \ 1\ ]^T$ | $W_2^{\{1\}}$ | $W_2^{\{12\}}/\sqrt{2}$ | $W_2^{\{123\}}/\sqrt{3}$ | $W_2^{\{3214\}}/2$ |
| 3 | $u_3 = [\ 1\ \ j\ \ 1\ \ -j\ ]^T$ | $W_3^{\{1\}}$ | $W_3^{\{12\}}/\sqrt{2}$ | $W_3^{\{123\}}/\sqrt{3}$ | $W_3^{\{3214\}}/2$ |
| 4 | $u_4 = [\ 1\ \ (-1-j)/\sqrt{2}\ \ -j\ \ (1-j)/\sqrt{2}\ ]^T$ | $W_4^{\{1\}}$ | $W_4^{\{14\}}/\sqrt{2}$ | $W_4^{\{124\}}/\sqrt{3}$ | $W_4^{\{1234\}}/2$ |
| 5 | $u_5 = [\ 1\ \ (1-j)/\sqrt{2}\ \ j\ \ (-1-j)/\sqrt{2}\ ]^T$ | $W_5^{\{1\}}$ | $W_5^{\{14\}}/\sqrt{2}$ | $W_5^{\{124\}}/\sqrt{3}$ | $W_5^{\{1234\}}/2$ |
| 6 | $u_6 = [\ 1\ \ (1+j)/\sqrt{2}\ \ -j\ \ (-1+j)/\sqrt{2}\ ]^T$ | $W_6^{\{1\}}$ | $W_6^{\{13\}}/\sqrt{2}$ | $W_6^{\{134\}}/\sqrt{3}$ | $W_6^{\{1324\}}/2$ |
| 7 | $u_7 = [\ 1\ \ (-1+j)/\sqrt{2}\ \ j\ \ (1+j)/\sqrt{2}\ ]^T$ | $W_7^{\{1\}}$ | $W_7^{\{13\}}/\sqrt{2}$ | $W_7^{\{134\}}/\sqrt{3}$ | $W_7^{\{1324\}}/2$ |
| 8 | $u_8 = [\ 1\ \ -1\ \ 1\ \ 1\ ]^T$ | $W_8^{\{1\}}$ | $W_8^{\{12\}}/\sqrt{2}$ | $W_8^{\{124\}}/\sqrt{3}$ | $W_8^{\{1234\}}/2$ |
| 9 | $u_9 = [\ 1\ \ -j\ \ -1\ \ -j\ ]^T$ | $W_9^{\{1\}}$ | $W_9^{\{14\}}/\sqrt{2}$ | $W_9^{\{134\}}/\sqrt{3}$ | $W_9^{\{1234\}}/2$ |
| 10 | $u_{10} = [\ 1\ \ 1\ \ 1\ \ -1\ ]^T$ | $W_{10}^{\{1\}}$ | $W_{10}^{\{13\}}/\sqrt{2}$ | $W_{10}^{\{123\}}/\sqrt{3}$ | $W_{10}^{\{1324\}}/2$ |
| 11 | $u_{11} = [\ 1\ \ j\ \ -1\ \ j\ ]^T$ | $W_{11}^{\{1\}}$ | $W_{11}^{\{13\}}/\sqrt{2}$ | $W_{11}^{\{134\}}/\sqrt{3}$ | $W_{11}^{\{1324\}}/2$ |
| 12 | $u_{12} = [\ 1\ \ -1\ \ -1\ \ 1\ ]^T$ | $W_{12}^{\{1\}}$ | $W_{12}^{\{12\}}/\sqrt{2}$ | $W_{12}^{\{123\}}/\sqrt{3}$ | $W_{12}^{\{1234\}}/2$ |
| 13 | $u_{13} = [\ 1\ \ -1\ \ 1\ \ -1\ ]^T$ | $W_{13}^{\{1\}}$ | $W_{13}^{\{13\}}/\sqrt{2}$ | $W_{13}^{\{123\}}/\sqrt{3}$ | $W_{13}^{\{1324\}}/2$ |
| 14 | $u_{14} = [\ 1\ \ 1\ \ -1\ \ -1\ ]^T$ | $W_{14}^{\{1\}}$ | $W_{14}^{\{13\}}/\sqrt{2}$ | $W_{14}^{\{123\}}/\sqrt{3}$ | $W_{14}^{\{3214\}}/2$ |
| 15 | $u_{15} = [\ 1\ \ 1\ \ 1\ \ 1\ ]^T$ | $W_{15}^{\{1\}}$ | $W_{15}^{\{12\}}/\sqrt{2}$ | $W_{15}^{\{123\}}/\sqrt{3}$ | $W_{15}^{\{1234\}}/2$ |

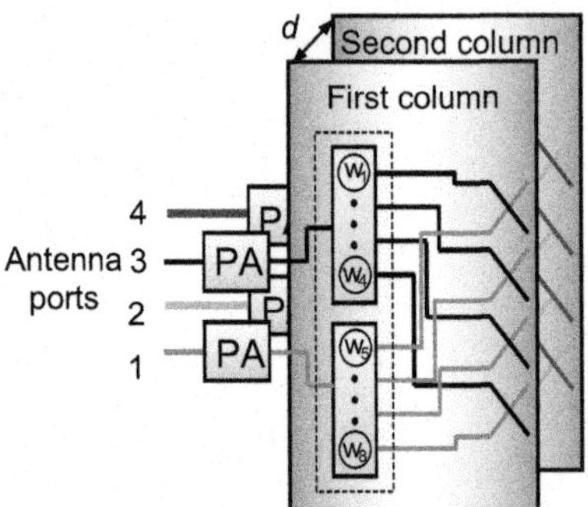


**FIGURE 11.19 Illustration of a two-column array with dual-polarized elements.**

and four, respectively, have the same polarization direction and are physically separated typically yielding different phase centers. A common precoder design for the four-branch MIMO scenario should preferably fulfill the following properties:

1. achieve power balancing;
2. no or minimal impact on legacy UEs;
3. avoiding spatial impact on the pilots for first and second port; and
4. no or low correlation between created ports.

There is a possibility to configure four-branch MIMO with dual-stream restriction. However, the current standard does not contain a single stream restriction.

### *11.4.2.2 Pilot design*

An accurate channel estimate is needed for many purposes, for example, demodulation of data, determination of preferred rank and precoding matrix, and channel quality indication reports. Needless to say, the required quality of the channel estimate varies depending on the use-case, for example, demodulation of HS-PDSCH with high code-rate, and higher-order modulation requires much better channel estimates than determining the preferred precoding matrix.

A key question was whether to adopt dedicated pilots or use only common pilots. A number of observations formed the basis for this decision:

- Existing common pilots, that is, P-CPICH and S-CPICH, are required in order to support legacy operation.
- Common pilots are needed in order to sound the channel, for example, determine appropriate transmission rank and precoding vectors.
- Each new pilot results in increased power overhead and changed interference characteristics. This affects legacy user performance, and the degree of impact depends on receiver type.
- A dedicated pilot design provides a number of benefits, for example, pilot overhead scales with the transmission rank. Hence, for rank-1, one new pilot is sufficient whereas for rank-4, four new pilots are needed. Also, advanced non-codebook-based beamforming is facilitated.

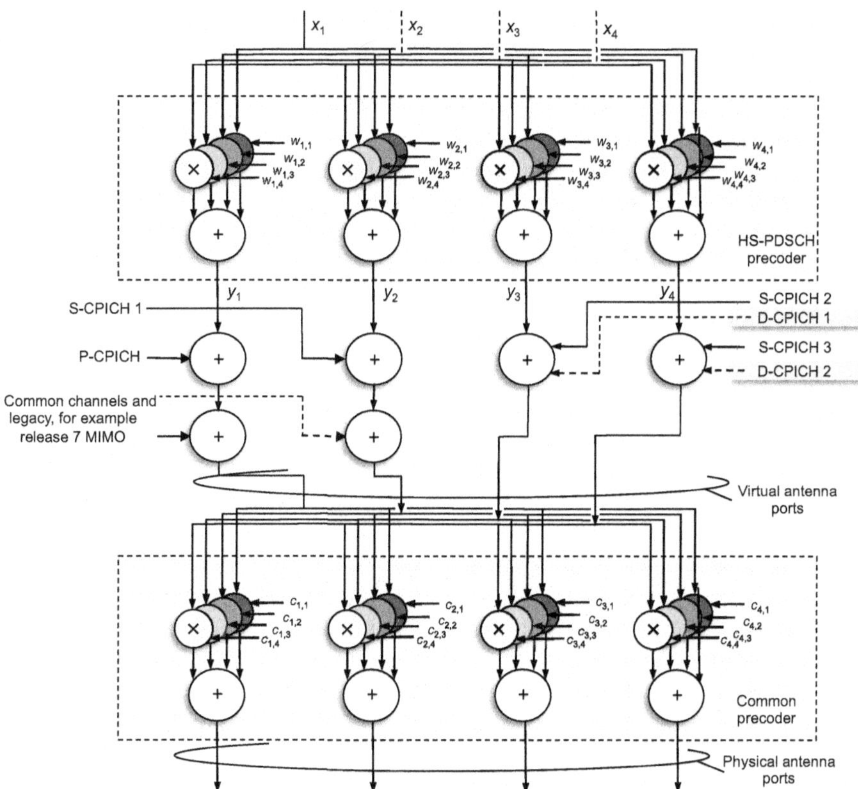


**FIGURE 11.20 Precoding and pilot layout for downlink four-branch MIMO.**

Four-branch MIMO adopts a common pilot design where existing P-CPICH and S-CPICH remain untouched. The P-CPICH power needs to be set such that sufficient coverage can be ensured and the S-CPICH can be configured with a lower power than the P-CPICH. Two additional very low power S-CPICHes are introduced to facilitate rank, precoding, and CQI estimation. These new common pilots are transmitted from virtual antennas three and four, respectively, and they use a common power offset relative P-CPICH signaled via higher-layers. To aid the demodulation process, especially for higher-order modulation, two additional *common demodulation pilots* (D-CPICHes) transmitted from antennas three and four can be dynamically scheduled. Enabling and disabling of these two pilots are done via HS-SCCH orders (the order needs to be acknowledged before being applied). When enabled, the pilots are always scheduled together with HS-PDSCH. The two D-CPICHes are transmitted with equal power and the power offset for D-CPICH relative P-CPICH is set via higher-layer signaling. See Figure 11.20 for an overview of the pilot design.

This pilot design provides a flexible framework for channel estimation, but the exact use of the toolbox is implementation specific. Some aspects to consider are how to minimize impact on legacy users and how to maximize the potential of the design. For example, the power setting of the additional common pilots is a trade off between impact on legacy users and channel sounding quality for four-branch MIMO.

#### 11.4.2.3 Rate control

The scheduling and rate control mechanism is similar to two-branch MIMO. For each TTI, the number of streams to transmit, the transport-block sizes for each stream, the number of channelization codes, the modulation scheme, and the precoding matrix are determined by the rate-control mechanism. One difference is the codeword to the layer-mapping process for transmission ranks larger than two, where streams become bundled and use the same transport-block size, modulation, and hybrid-ARQ information. Another difference is that the scheduler needs to assess the need for enabling additional demodulation pilots (D-CPICH).

Similar to release 6, the rate-control mechanism typically relies on UE feedback of the instantaneous channel quality. In case of dual-stream transmission, information about the supported data rate on each of the codewords (bundled streams) is required. In addition to the CQI reports, the UE also informs about the preferred rank and precoding. As the Node B scheduler is free to select the number of streams transmitted to a UE, the supported data rate in case of transmission of a rank different than the preferred one is of interest. It is non-trivial to transform a CQI report for a given rank into CQIs for another rank in the Node B because the mapping is tightly coupled to the UE receiver structure and, for example, its capability to mitigate inter-stream interference. In theory, there is a need to know the CQI for all possible ranks, but in order to limit the feedback, two different CQI reports can be configured: CQI reports for the preferred rank and rank-1, respectively. The ratio between the percentage of preferred-rank CQI reports and rank-1 CQI reports can be configured by higher-layer signaling.

#### 11.4.2.4 Hybrid-ARQ retransmission procedure

The hybrid-ARQ retransmission procedure becomes a bit more complex because of the codeword to layer mapping and the many cases that may arise depending on hybrid-ARQ and rank status. Compared to release 5 HSDPA, the maximum number of transport blocks and associated buffer memory is quadrupled (×4). Hybrid-ARQ signaling/feedback information is, however, associated with a codeword rather than a stream. Hence, a maximum of two hybrid-ARQ information elements can be signaled or fed back over HS-SCCH or HS-DPCCH. Similarly, there are up to two *hybrid-ARQ process identifiers*, one per codeword. Bundled streams rely on a single hybrid-ARQ acknowledgment. Hence, a NACK needs to be signaled as long as one of the bundled streams is unsuccessfully decoded irrespective of whether the other stream was successfully decoded or not. Note, though, that the receiver can make use of the fact that streams have independent CRCs. If one of the bundled streams was successfully decoded but not the other, the UE can internally remember this and does not need to decode it again when the retransmission occurs. In particular, having a CRC for each stream facilitates effective interference-cancellation receivers.

The hybrid-ARQ retransmission design is based on a number of principles:

- The number of layers/streams (and transport blocks) per codeword must be maintained between transmission attempts. This applies for NACKed codewords.
- The order of transport blocks within codewords must be maintained between transmission attempts, and a negatively acknowledged transport block cannot be mapped to another codeword. This is different compared to two-branch MIMO, where a codeword or transport block can change stream between transmission attempts.
- Bundled transport blocks belonging to the same hybrid-ARQ process identifier are retransmitted on the same layers.
- A special DTX codeword can be used to indicate that no transport blocks corresponding to the codeword are transmitted. This is useful, for example, when all transport blocks associated with the primary codeword are positively acknowledged, but at least one transport block associated with the second codeword is negatively acknowledged. When retransmitting this transport block, the primary stream needs to be "sent" as well, but it can be sent without any data payload and thereby reduces the interference and increases the likelihood that the retransmitted transport block becomes successfully received. An empty stream can be sent by indicating DTX for that codeword on the HS-SCCH. In this case, the UE shall consider the DTXed transport blocks as successfully decoded and send a positive acknowledgment for the associated codeword on the HS-DPCCH.

Given these principles, a number of retransmission cases occur depending on hybrid-ARQ status and preferred rank. Table 11.11 summarizes a number of the possible cases. Cases 1 to 7 correspond to scenarios when the original transmission was rank-2. For cases 1 to 3, stream 1 was positively acknowledged, while stream 2 was negatively acknowledged. It is interesting to see that for this case the retransmission needs to be rank-2 because of the rules described above. In particular, the number of streams associated with the second codeword cannot change and must be one. Also, the rank cannot be reduced since the negatively acknowledged transport block on stream 2 must be associated with codeword 2. However, it is up to the network to decide what to schedule on stream 1, that is, sending no information (DTX) or scheduling new data. For cases 4 to 7, stream 1 was negatively acknowledged, while stream 2 was positively acknowledged. In this case, the second codeword can be adjusted since it corresponds to a positively acknowledged transport block; for example, a third stream can be scheduled or the codeword can be completely DTXed. Still, a fourth stream cannot be transmitted since the status of the primary codeword cannot change. Similar arguments apply for the other cases 8 to 21.

The hybrid-ARQ process identifiers for the two codewords are coupled in order to reduce control information, which implies that only one of the hybrid-ARQ process identifiers needs to be signaled on the HS-SCCH. This is similar to that for two-branch MIMO.

**Table 11.11** A Summary of Hybrid-ARQ Retransmission Cases

| Case | | 1 | 2 | 3 |
|---|---|---|---|---|
| | **Original Tx** | **Retransmission** | | |
| **Stream** | **Rank-2** | **Rank-2** | **Rank-2** | **Ranks 1, 3, 4** |
| 1 | ACK | DTX | New data | |
| 2 | NACK | Re-Tx | Re-Tx | |

| Case | | 4 | 5 | 6 | 7 |
|---|---|---|---|---|---|
| | **Original Tx** | **Retransmission** | | | |
| **Stream** | **Rank-2** | **Rank-1** | **Rank-2** | **Rank-3** | **Rank-4** |
| 1 | NACK | Re-Tx | Re-Tx | Re-Tx | |
| 2 | ACK | – | New data | New data | |
| 3 | – | – | – | New data | |

| Case | | 8 | 9 | 10 | 11 |
|---|---|---|---|---|---|
| | **Original Tx** | **Retransmission** | | | |
| **Stream** | **Rank-3** | **Rank-1** | **Rank-2** | **Rank-3** | **Rank-4** |
| 1 | NACK | Re-Tx | Re-Tx | Re-Tx | |
| 2 | ACK | – | New data/DTX | New data/DTX | |
| 3 | ACK | – | – | New data/DTX | |

| Case | | 12 | 13 | 14 |
|---|---|---|---|---|
| | **Original Tx** | **Retransmission** | | |
| **Stream** | **Rank-3** | **Ranks 1, 2** | **Rank-3** | **Rank-4** |
| 1 | ACK | | New data/DTX | New data/DTX |
| 2 | NACK | | Re-Tx | Re-Tx |
| 3 | NACK | | Re-Tx | Re-Tx |
| 4 | – | | – | New data/DTX |

| Case | | 15 | 16 | 17 |
|---|---|---|---|---|
| | **Original Tx** | **Retransmission** | | |
| **Stream** | **Rank-4** | **Ranks 1, 2, 3** | **Rank-4** | **Rank-4** |
| 1 | NACK | | Re-Tx | Re-Tx |
| 2 | ACK | | DTX | New data |
| 3 | ACK | | DTX | New data |
| 4 | NACK | | Re-Tx | Re-Tx |

| Case | | 18 | 19 | 20 | 21 |
|---|---|---|---|---|---|
| | **Original Tx** | **Retransmission** | | | |
| **Stream** | **Rank-4** | **Ranks 1, 2** | **Rank-3** | **Rank-4** | **Rank-4** |
| 1 | ACK | | DTX/new data | DTX | New data |
| 2 | NACK | | Re-Tx | Re-Tx | Re-Tx |
| 3 | NACK | | Re-Tx | Re-Tx | Re-Tx |
| 4 | ACK | | – | DTX | New data |

*Note: Bundled streams use the same shading. DTX denotes the special DTX codeword, indicating that no payload is associated with this codeword.*

Four-branch MIMO requires MAC-ehs with TSN field extension (14 contiguous bits) to handle the possibly large amount of data.

### 11.4.3 CONTROL CHANNEL STRUCTURE

The downlink out-band control signaling is carried on the HS-SCCH. In case of four-branch MIMO, a new format of the HS-SCCH (type 4) is used to accommodate the additional information required (see Figure 11.21). The division of the HS-SCCH into two parts (see Chapter 8) is maintained. Part one is extended to include information about the transmission rank (one to four), modulation scheme for each codeword, and which of the 16 precoding matrices the Node B used for the transmission. Modulation and rank information are combined into one field. Four additional bits in Part I are added compared to the two-branch MIMO. To accommodate the increased number of bits, rate 1/2 convolutional coding instead of rate 1/3 is used for Part I. Contrary to two-branch MIMO, there is no special consideration in the channelization code-set mapping in order to facilitate 64-QAM. Two additional bits are introduced to ensure that all combinations of modulation and rank can be signaled. However, bundled transport blocks need to use the same modulation order, which reduces the total number of combinations.

The second part of the HS-SCCH is identical to the type-3 two-branch MIMO case. The format depends on whether one or two codewords were scheduled to the UE. In the latter case, additional bits are transmitted on part two to convey the hybrid-ARQ and transport-block size information also for the second codeword. The additional bits are fitted onto the physical channel by adjusting the rate matching appropriately. Furthermore, the new-data indicator, the redundancy version, and the constellation-rearrangement fields are combined into a joint field ($X_{rv}$), where the

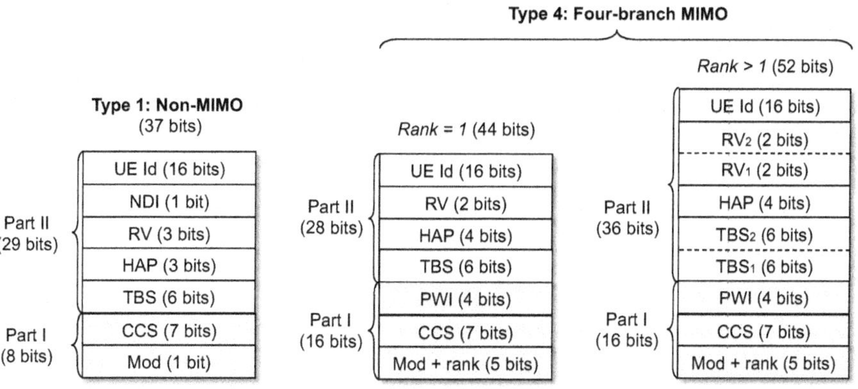


**FIGURE 11.21 Illustration of HS-SCCH information for type 1 (non-MIMO) and type 4 (four-branch MIMO). CCS, channelization code set; Mod, modulation; TBS, transport-block size; HAP, hybrid-ARQ process information; RV, redundancy version; NDI, new data indicator; and PWI, precoding weight indicator.**

mapping depends on the code-rate. Also, the new data indicator has essentially been removed and is replaced by letting $(s, r, b)$ equal (1, 0, 0), where $s$ and $r$ are the redundancy version parameters and $b$ denotes the constellation version parameter. As discussed earlier, only the hybrid-ARQ process identifier for the primary codeword is signaled during rank-2, and the process identifier for the second codeword $HAP_2$ is given by

$$HAP_2 = (HAP_1 + N/2)\,\mathrm{mod}(N), \tag{11.17}$$

where $HAP_1$ is the process identifier for the primary codeword, and $N$ is the total number of hybrid-ARQ process identifiers.

For efficient hybrid-ARQ retransmission processing, a DTX indication has been added to inform the UE that no transport blocks associated with the codeword are transmitted. This is indicated by letting the transport-block size bit sequence equal "111111" and bit zero in the field $X_{rv}$ equal 0. See Section 11.3.2.4 for further details when a DTX indication may be needed.

The channel coding structure for type-4 HS-SCCH is identical to the type-1 non-MIMO case (see Section 8.3.7) with the exception that 1/2 convolutional coding is used for Part I and that the puncturing is updated to match the 40 and 80 bits for Parts I and II. The spreading factor for HS-SCCH is still 128.

Uplink out-band control signaling consists of hybrid-ARQ acknowledgments, preferred rank, PCI, and CQI, respectively, and is transmitted on the HS-DPCCH. The processing of uplink out-band control signaling differs from two-branch MIMO in a number of aspects. The main difference is that more information needs to be conveyed and that the basic design includes multicarrier operation with up to four carriers. For example, the preferred rank is explicitly signaled and not implicitly via the CQI as for two-branch MIMO. The structure of the HS-DPCCH with one slot of hybrid-ARQ-related information and two slots of CQI/PCI/rank-related information per sub-frame is kept. The spreading factor of the HS-DPCCH is always 128, resulting in 20 bits per slot.

The two-branch MIMO hybrid-ARQ codebook is reused, and it consists of in total eight codewords, including POST and PRE. Repetition is employed to map the 10 message bits into 20 channel bits. Hybrid-ARQ feedback is per codeword, meaning that only one hybrid-ARQ message per bundled transport blocks is sent; see Table 11.12.

The rank field consists of two bits, indicating the preferred rank (1–4). The PCI comprises four bits, specifying which one of the 16 precoding matrices for the preferred rank best matches the channel conditions at the UE. The CQI indicates the data rate the UE recommends in case transmission is done using the recommended rank and PCI, and the nominal HS-PDSCH power. It would be beneficial for the network to know the supported data rate for all ranks, since the scheduler may decide to transmit with another rank than the preferred one. The standard allows for two different CQI reports. Type-A reports indicate the CQI and PCI for the preferred rank, while type-B reports contain the CQI and PCI for single-stream transmission. For type-B

**Table 11.12** Illustration of the Hybrid-ARQ Codebook

| Hybrid-ARQ ACK Message to Be Transmitted | | $w_0$ | $w_1$ | $w_2$ | $w_3$ | $w_4$ | $w_5$ | $w_6$ | $w_7$ | $w_8$ | $w_9$ |
|---|---|---|---|---|---|---|---|---|---|---|---|
| **Hybrid-ARQ ACK in response to a single scheduled transport block** | | | | | | | | | | | |
| ACK | | 1 | 1 | 1 | 1 | 1 | 1 | 1 | 1 | 1 | 1 |
| NACK | | 0 | 0 | 0 | 0 | 0 | 0 | 0 | 0 | 0 | 0 |
| **Hybrid-ARQ ACK in response to more than one scheduled transport block** | | | | | | | | | | | |
| **Response to transport blocks (TBs) 1 and 4** | **Response to transport blocks (TBs) 2 and 3** | | | | | | | | | | |
| All TBs are ACK | All TBs are ACK | 1 | 0 | 1 | 0 | 1 | 1 | 1 | 1 | 0 | 1 |
| All TBs are ACK | Any TB is NACK | 1 | 1 | 0 | 1 | 0 | 1 | 0 | 1 | 1 | 1 |
| Any TB is NACK | All TBs are ACK | 0 | 1 | 1 | 1 | 1 | 0 | 1 | 0 | 1 | 1 |
| Any TB is NACK | Any TB is NACK | 1 | 0 | 0 | 1 | 0 | 0 | 1 | 0 | 0 | 0 |
| **PRE/POST indication** | | | | | | | | | | | |
| PRE | | 0 | 0 | 1 | 0 | 0 | 1 | 0 | 0 | 1 | 0 |
| POST | | 0 | 1 | 0 | 0 | 1 | 0 | 0 | 1 | 0 | 0 |

reports, the release 5 CQI report is used, while for type-A reports, the CQI consists of eight bits and is computed according to

$$\mathrm{CQI} = \begin{cases} \mathrm{CQI_S} & \text{rank-1, where } \mathrm{CQI_S} = 1,\ldots,30 \\ 15 \times \mathrm{CQI_1} + \mathrm{CQI_2} & \text{otherwise, where } \mathrm{CQI_1}, \mathrm{CQI_2} = 0,\ldots,14 \end{cases} \tag{11.18}$$

where $\mathrm{CQI_S}$ denotes the supported transport format for rank-1 transmissions, and $\mathrm{CQI_1}$ and $\mathrm{CQI_2}$ indicate the supported transport format for the primary and secondary codewords, respectively, for rank-2, rank-3, or rank-4 transmission. Note that (11.18) is slightly different compared to the corresponding formula for two-branch MIMO (11.14). The reason is that rank is explicitly signaled for four-branch MIMO, while for two-branch MIMO rank is implicitly signaled in the CQI report. Hence, the interpretation of the CQI report (11.18) will depend on the rank. It can be noted that single- and multi-stream transmissions use different CQI granularities; 30 and 15 CQI values, respectively, for single- and multi-stream transmissions. CQI, PCI, and rank information are concatenated and coded into 40 bits using convolutional coding and transmitted in two slots on the HS-DPCCH; see Figure 11.22. To allow for a flexible adaptation to different propagation environments, the first $N$ out of $M$ PCI/CQI reports are Type-A reports and the remaining $(M–N)$ reports are type-B reports. The ratio $N/M$ is configured by RRC signaling.

The mapping between CQI and *Modulation and Coding Scheme* (MCS) depends on the UE category and transmission rank.

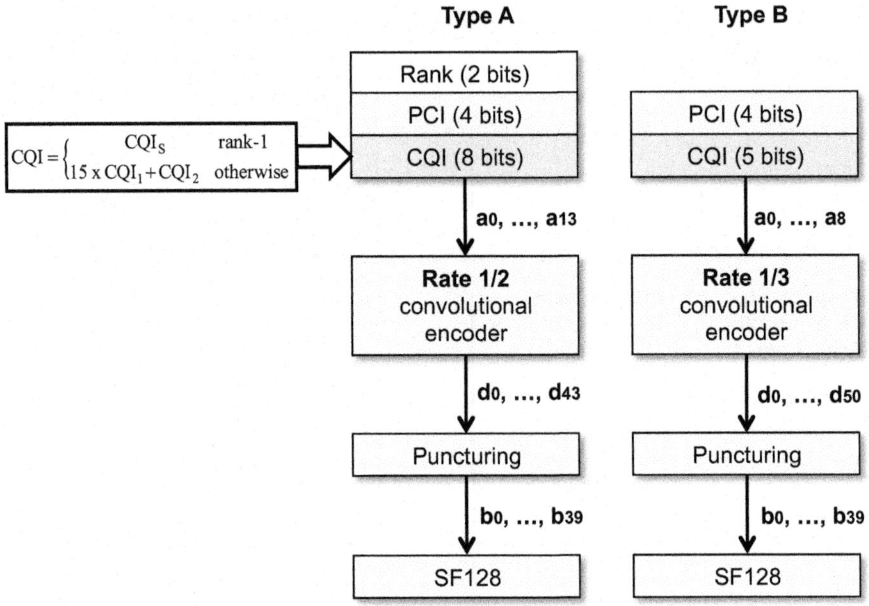


**FIGURE 11.22 Coding of rank, PCI, and CQI information for type A and B.**

Up to four downlink carriers are supported and the HS-DPCCH coding is independent per carrier. The HS-DPCCH processing rules are very similar to the 4C-HSDPA and 8C-HSDPA scenario with some exceptions (see Chapter 12 for a description of multicarrier design details). The main principles include the following (see Figures 11.23 and 11.24 for illustrations):

- The serving cell (carrier 1) and the first secondary serving cell (carrier 2) are associated with HS-DPCCH 1, and the second and third secondary serving cells (carriers 3 and 4) are associated with HS-DPCCH 2.
- Remapping of carriers between HS-DPCCHs is not allowed.
- HS-DPCCH slot format 1 (SF128) is always employed if at least one carrier associated with the HS-DPCCH is configured with four-branch MIMO.
- If no carrier associated with an HS-DPCCH is configured with four-branch MIMO, then HS-DPCCH slot format 0 (SF256) is used and the ACK/NACK and CQI handling reverts to the release 8 or release 9 dual-cell format, depending on whether $2 \times 2$ MIMO is configured on any of the carriers. This is different compared to 8C-HSDPA.
- If only one of the carriers associated with an HS-DPCCH is configured with four-branch MIMO, the other carrier employs the release 5 or release 7 CQI/PCI encoding dependent on whether two-branch MIMO is configured, and the information is repeated in two consecutive slots. The ACK/NACK coding can, however, use the release 11 codebook for both carriers. It can be noticed that the release 11 codebook coincides with the

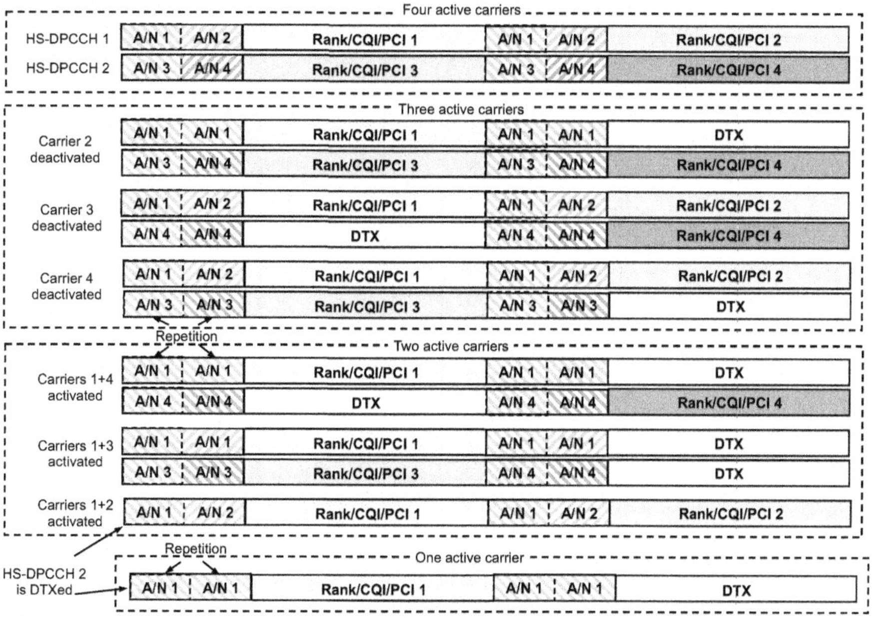

**FIGURE 11.23 Illustration of ACK/NACK and rank/CQI/PCI information mapping for different configurations of number of active carriers. All carriers are configured with four-branch MIMO.**

release 7 codebook, and the release 5 codebook is a subset of the release 11 codebook.

- The HS-DPCCH slot format (spreading factor) and spreading code (number) depend on the configuration.

Figure 11.24 shows two examples of different multicarrier configurations. In the upper example, carriers 1 and 2 are not configured with two-branch or four-branch MIMO, carrier 3 is configured with four-branch MIMO, and carrier 4 is either non-MIMO or configured with two-branch MIMO. In the lower example, carrier 1 is configured with four-branch MIMO, carrier 2 is non-MIMO or configured with two-branch MIMO, and carriers 3 and 4 are not configured with four-branch MIMO but at least one of them is configured with two-branch MIMO.

### 11.4.4 UE CAPABILITIES, RRC SIGNALING AND CONFIGURATIONS

The four-branch MIMO capability is defined per frequency band. Two new UE categories are introduced for four-branch MIMO. Categories 37 and 38 support four-branch MIMO with up to two and four carriers, respectively; see Table 11.13. Four-branch MIMO is only supported in CELL_DCH and requires MAC_ehs with TSN field extension (14 contiguous bits). Up to four HS-DSCH cells can be configured and the use of DPCH as associated channel is supported. When using F-DPCH as associated channel, only single antenna transmission or STTD can be used for the F-DPCH.

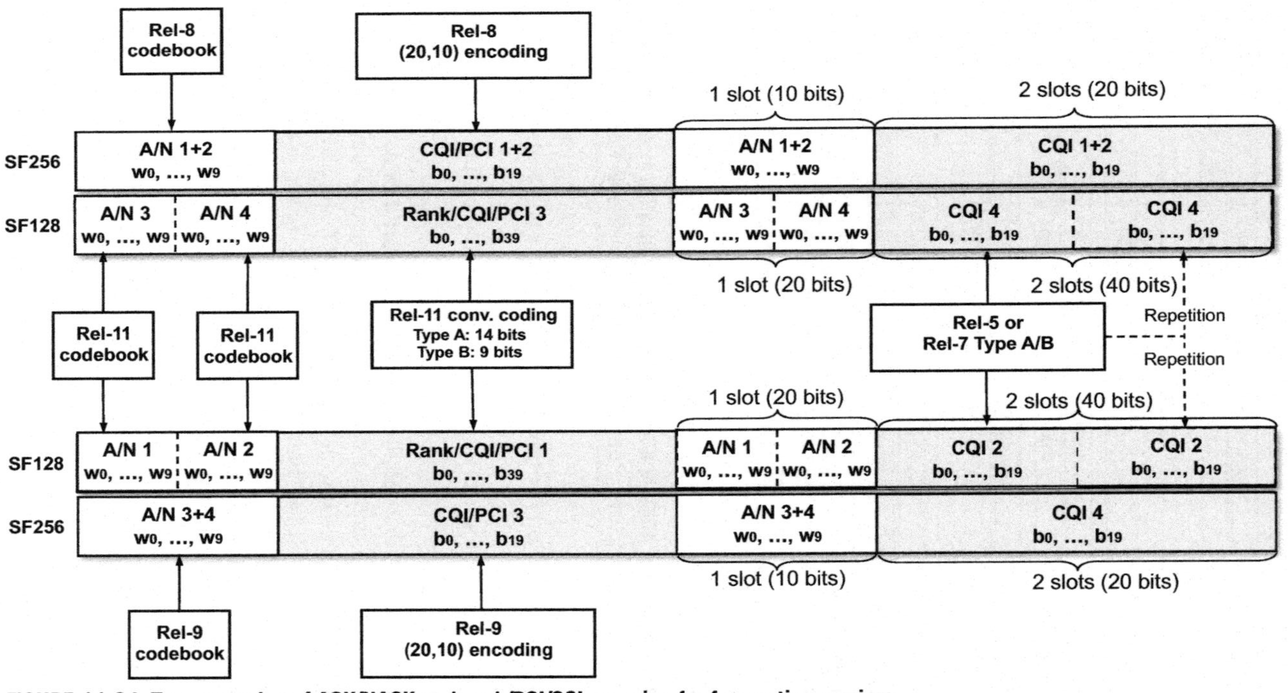


**FIGURE 11.24** Two examples of ACK/NACK and rank/PCI/CQI mapping for four active carriers.

**Table 11.13** HS-DSCH Categories Supporting Four-Branch MIMO

| HS-DSCH Category | Modulation Scheme | Number of Supported Carriers | Max Data Rate Per Carrier (Mbit/s) | | | |
|---|---|---|---|---|---|---|
| | | | One Stream | Two Streams | Three Streams | Four Streams |
| 37 | QPSK, 16-QAM, 64-QAM | 2 | 21.1 | 42.2 | 63.3 | 84.4 |
| 38 | QPSK, 16-QAM, 64-QAM | 4 | 21.1 | 42.2 | 63.3 | 84.4 |

A number of MIMO related parameters are set by higher-layer signaling, including [7,10]

- Configuration of type-A and type-B CQIs. The ratio of type-A CQIs and total number of CQIs can take the values {1/2, 2/3, 3/4, 4/5, 5/6, 6/7, 7/8, 8/9, 9/10, 1/1}.
- Precoding weight set restriction (64 bits).
- Pilot configuration
  - Power offset relative P-CPICH in dB for S-CPICH on antenna 2 (–6,..., 0).
  - Common power offset relative P-CPICH in dB for S-CPICH on antennas 3 and 4 (–12,..., 0) and spreading codes for S-CPICH on antennas 3 and 4.
  - Common power offset relative P-CPICH in dB for D-CPICH on antennas 3 and 4 (–12,..., 0), spreading codes for D-CPICH on antennas 3 and 4, and initial status of D-CPICH.

## 11.5 UPLINK TRANSMIT DIVERSITY

The uplink supports two types of transmit diversity: open- and closed-loop transmit diversity. Both these schemes will be described in this section.

### 11.5.1 CLOSED-LOOP TRANSMIT DIVERSITY

Closed-loop transmit diversity consists of three different modes. Mode 1 offers codebook-based precoding where the network indicates the preferred precoding vector. All uplink physical channels are subject to precoding, and an additional pilot channel is introduced to facilitate channel sounding. Modes 2 and 3 are based on so-called fixed *switched antenna transmit diversity* (SATD) where all physical channels are sent from one of the two antennas.[5] It is mainly layer-1 processing that is affected by closed-loop transmit diversity.

[5]Similarly as for the downlink, the uplink can employ a common precoder. Hence, even though the discussion here focuses on physical antennas, a better term would in fact be virtual antennas.

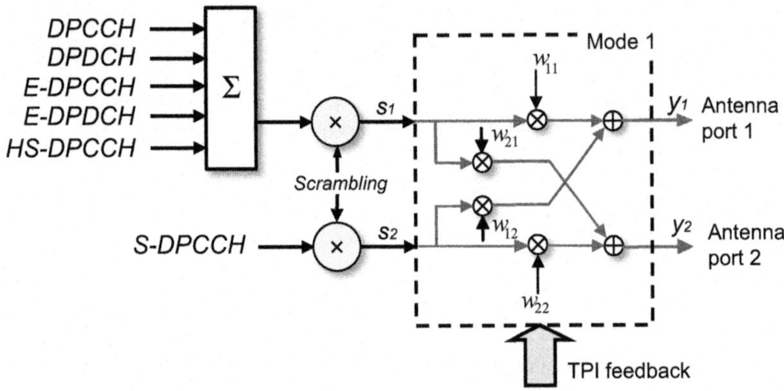


**FIGURE 11.25 Closed-loop transmit diversity mode 1.**

The physical-layer processing in terms of coding, modulation, spreading, and scrambling is untouched. The scrambled signal is then precoded before the result is mapped to the two transmit antennas. Mathematically, this can be expressed as

$$\begin{bmatrix} x_1 \\ x_2 \end{bmatrix} = \begin{bmatrix} w_{11} \\ w_{21} \end{bmatrix} s_1, \tag{11.19}$$

where $s_1$ is the scrambled signal, $\{w_{k1}\}$ denotes the precoding weights, and $x_k$ is the resulting signal sent on antenna $k$. The preferred precoding weights are signaled by the network using the downlink physical channel referred to as *Fractional Transmitted Precoding Indicator Channel* (F-TPICH). All existing (pre-release 11) uplink physical channels are subject to precoding. To allow the Node B to decide the preferred precoding weights for Mode 1, a second pilot channel referred to as the *Secondary Dedicated Physical Control Channel* (S-DPCCH) is introduced that is precoded with weights $w_{12}$ and $w_{22}$ (see Figure 11.25). It should be noted that Modes 2 and 3 can also be described in the framework of (11.19) with special weight vectors $w_{11} = 1$ and $w_{21} = 0$ for Mode 2 and $w_{11} = 0$ and $w_{21} = 1$ for Mode 3 (see Figure 11.26).

### 11.5.1.1 Pilot design for Mode 1

A new uplink physical pilot channel is introduced for Mode 1 in order to enable an efficient precoder choice by the network. It should be noted that this channel is not configured for Modes 2 and 3. The pilot channel is called the *Secondary Dedicated Physical Control Channel* (S-DPCCH) and it is very similar to the existing DPCCH; see Figure 11.27. The spreading factor is 256, and it employs a slot format with eight pilots and two fixed bits. The pilot bit pattern follows that of the DPCCH with eight pilots. The S-DPCCH always uses the spreading code $c_{sc} = C_{ch,256,31}$ and is mapped to the Q-branch. The S-DPCCH is precoded with a weight vector that is orthogonal to the weight vector used for all other physical channels. By extending (11.19) the following model is obtained

$$\begin{bmatrix} x_1 \\ x_2 \end{bmatrix} = \begin{bmatrix} w_{11} & w_{12} \\ w_{21} & w_{22} \end{bmatrix} \begin{bmatrix} s_1 \\ s_2 \end{bmatrix} = \begin{bmatrix} w_{11} \\ w_{21} \end{bmatrix} s_1 + \begin{bmatrix} w_{12} \\ w_{22} \end{bmatrix} s_2 = \mathbf{w}_1 s_1 + \mathbf{w}_2 s_2, \tag{11.20}$$

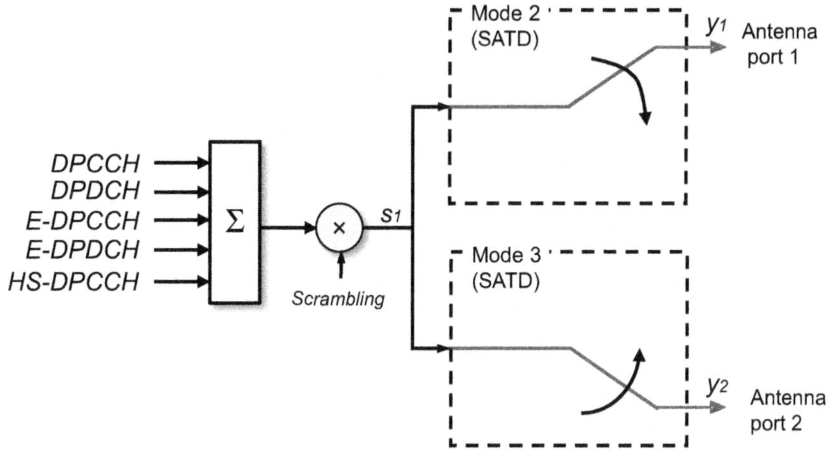


**FIGURE 11.26 Closed-loop transmit diversity modes 2 and 3.**

where $s_2$ is the scrambled S-DPCCH signal, $s_1$ contains the sum of all other scrambled uplink physical channels, and $\boldsymbol{w}_1$ and $\boldsymbol{w}_2$ are orthogonal ($\boldsymbol{w}_1^H \boldsymbol{w}_2 = 0$).

The power setting of the S-DPCCH depends on whether E-DPCCH boosting is employed (see Section 10.3.3 for a description of E-DPCCH boosting). Strictly speaking, for Tx diversity, boosting of S-DPCCH is not necessary, while for MIMO operation where two streams are transmitted, boosting of S-DPCCH is essential (see Section 11.6). When no E-DCH transmission is ongoing or when E-DPCCH boosting is configured and E-TFCI ≤ $E\text{-}TFCI_{ec,\text{boost}}$, then the power of the S-DPCCH is given by a fixed offset $\beta_{sc}$ relative DPCCH, where the offset is derived based on higher-layer information. When E-TFCI > $E\text{-}TFCI_{ec,\text{boost}}$, the power offset is derived based on the *traffic to secondary pilot power offset* ($\Delta_{\text{T2SP}}$), configured by higher layers, according to

$$\beta_{sc,i,uq} = \beta_c \cdot \sqrt{\max\left( A_{sc}^2, \frac{\sum_{k=1}^{K_{\max,i}} \left( \frac{\beta_{ed,i,k}}{\beta_c} \right)^2}{10^{\frac{\Delta_{T2SP}}{10}}} \right)} \tag{11.21}$$

which is then rounded toward the closest quantized value as specified in [7,10].

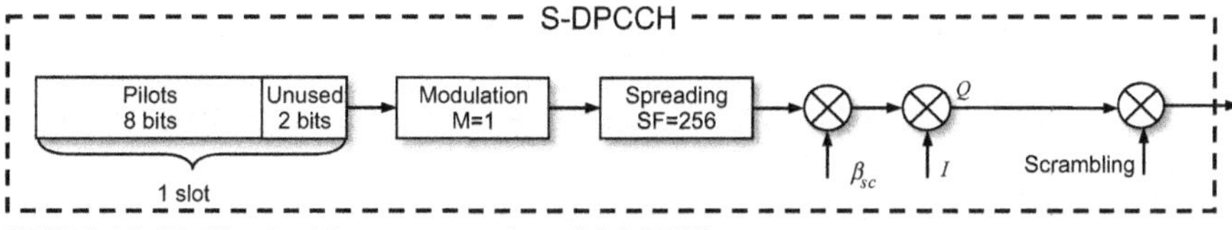


**FIGURE 11.27 Physical layer processing of S-DPCCH.**

The intention is to maintain a fixed pilot to data power ratio for all E-TFCIs satisfying the boosting criteria. Hence, as the E-DPDCH power offset (or similarly the E-TFCI) increases the power offset for S-DPCCH-relative DPCCH will increase. Note the minor difference (a term –1) between (11.21) and the corresponding E-DPCCH boosting formula (10.1), which reflects that for the primary stream the total pilot power is given by DPCCH and E-DPCCH together while for the secondary stream the total power is given by S-DPCCH. (The notion of streams is used more in the context of MIMO which will be described in detail in Section 11.6.)

One of the main design choices during the standardization was whether pilots should be precoded or not. For the downlink, non-precoded pilots are used, but for the uplink a precoded structure was adopted. Some of the reasons for this choice were as follows:

- For the uplink, there is no issue with backwards compatibility as for the downlink.
- Minimizing any misalignment or PA inaccuracies between transmission branches. This is the main reason why the common precoder is needed for the downlink. Still, however, a common precoder can also be employed for the uplink.
- The power ratio between a physical channel and the DPCCH remains the same irrespective of the precoder vector. Existing parameters can be reused and the impact on, for example, the scheduler is minimized.
- SHO operation with nodes not aware of the currently used precoder is possible since the DPCCH based channel estimate gives the effective precoded channel.
- It is not necessary to signal the precoding weights in the uplink.

The precoded channel structure has an impact on the receiver processing. The exact receiver processing is implementation specific, but some general aspects will be covered here. To illustrate the receiver processing, a simplified received signal model is used

$$y = \underbrace{\begin{bmatrix} h_1 & h_2 \end{bmatrix}}_{H} \underbrace{\begin{bmatrix} w_1 & w_2 \end{bmatrix}}_{W} \begin{bmatrix} s_1 \\ s_2 \end{bmatrix} = \underbrace{\begin{bmatrix} h_{\text{eff},1} & h_{\text{eff},2} \end{bmatrix}}_{H_{\text{eff}}} \begin{bmatrix} s_1 \\ s_2 \end{bmatrix} = h_{\text{eff},1} s_1 + h_{\text{eff},2} s_2, \tag{11.22}$$

where $y$ is the received signal vector, $H$ denotes the channel matrix, $W$ represents the precoding matrix, and $s_k$ is the transmitted signal precoded by weight vector $k$. Also, the relation between the channel matrix $H$ and the *effective* (precoded) channel matrix $H_{\text{eff}}$ is given by

$$H_{\text{eff}} = HW. \tag{11.23}$$

Since the pilots are precoded, the channel estimation procedure will give an estimate of the effective channel matrix. In fact, the pilots in DPCCH precoded with the primary precoding vector $w_1$ will give an estimate of $h_{\text{eff},1}$, whereas pilots in S-DPCCH precoded with $w_2$ yield an estimate of $h_{\text{eff},2}$. Consequently, a receiver not aware of the precoding will automatically estimate the effective channel $h_{\text{eff},1}$. Hence, demodulation of a user employing CLTD can essentially use legacy processing procedures, which is one of the reasons why SHO works without explicit knowledge of the used precoding weights. A potential issue is that the estimated effective channel changes whenever the precoding weights are changed. Hence, if the channel estimator uses aggressive time-averaging, the quality might in fact degrade if the precoding weights change. However, typically the precoder needs to be updated at the same rate as the fast-fading changes. Hence, if Doppler-based averaging is used, the problem should be small. Another solution would be to remove the effect of the precoder before doing the time-averaging. This, however, requires accurate knowledge of the precoder matrix. A related consideration is that the S-DPCCH power offset relative DPCCH will vary depending on E-TFCI if E-DPCCH boosting is enabled. This can have a detrimental effect on the estimate of $\boldsymbol{h}_{\text{eff},2}$ if no counteractions are taken. However, this affects only the magnitude of the estimate and not the phase. Hence, this is of less concern unless higher-order modulated symbols need to be detected.

A second aspect of the pilot design is the channel sounding that is required in order to determine the appropriate precoding weights. A common approach to deciding the preferred precoding weights would be to maximize the SNR (or the post-equalizing SINR) of the primary stream

$$\boldsymbol{w}_1 = \arg\max_{w} \|\boldsymbol{H}\boldsymbol{w}\|^2 = \arg\max_{w} \|\boldsymbol{H}_{\text{eff}}\boldsymbol{W}_{\text{prev}}^{-1}\boldsymbol{w}\|^2, \tag{11.24}$$

where the last equality follows since the channel estimator yields an estimate of the effective channel that needs to be converted into the non-precoded channel by removing the impact of the currently used (or previously signaled) precoding matrix.

This approach can be made more robust by taking into consideration that the network does not explicitly know the precoder the user employed. The receiver can therefore try to verify the actually used precoding weights or rather take into account a priori information about the reliability of the downlink F-TPICH link. The reason is that the used precoder is not signaled in the uplink and that signaling errors might occur.

#### 11.5.1.2 Precoding for Mode 1

An enhanced symmetric precoding approach is adopted, where the precoding weight matrix $\boldsymbol{W}$ is given by

$$\boldsymbol{W} = \begin{bmatrix} w_{11} & w_{12} \\ w_{21} & w_{22} \end{bmatrix} = \frac{1}{\sqrt{2}} \begin{bmatrix} e^{-j\theta/2} & e^{-j\theta/2} \\ e^{j\theta/2} & -e^{j\theta/2} \end{bmatrix}, \tag{11.25}$$

where the phase shift $\theta$ has a resolution of $\pi/2$. The symmetric part reflects that half of the total phase shift is applied to each antenna branch, and the enhanced part was introduced in order to make channel estimation more robust against signaling errors. The key observation is that even though two phases that are $2\pi$ radians apart are equivalent in terms of total precoding phase shift, the end result may differ because of the symmetric implementation where half of the phase shift is applied to each antenna branch. In essence, the enhanced algorithm ensures that the phase difference between two consecutive transmissions does not exceed $\pi$ radians. The idea of enhanced symmetric precoding is best illustrated by a few examples (see, for example, [11] for further details).

Consider a single receive antenna scenario where the channel components from transmit antenna one and two are denoted $h_1$ and $h_2$, respectively, and the effective channel $h_{\text{eff}}$ is given by the channel $\boldsymbol{h}$ multiplied by the precoder ($\boldsymbol{w}$)

$$h_{\text{eff}} = \boldsymbol{hw} = \begin{bmatrix} h_1 & h_2 \end{bmatrix} \begin{bmatrix} w_1 \\ w_2 \end{bmatrix} = (h_1 w_1 + h_2 w_2). \tag{11.26}$$

The first example aims at illustrating why a symmetric precoder, denoted $\boldsymbol{w}$, has benefits compared to an asymmetric precoder, denoted $\boldsymbol{v}$. The precoders are given by

$$\boldsymbol{w} = \begin{bmatrix} w_1 \\ w_2 \end{bmatrix} = \frac{1}{\sqrt{2}} \begin{bmatrix} e^{-j\theta/2} \\ e^{j\theta/2} \end{bmatrix}, \qquad \boldsymbol{v} = \begin{bmatrix} v_1 \\ v_2 \end{bmatrix} = \frac{1}{\sqrt{2}} \begin{bmatrix} 1 \\ e^{j\theta} \end{bmatrix}. \tag{11.27}$$

Assume that the channel components are given by

$$h_1 = \frac{1}{\sqrt{2}} e^{j\alpha}, \quad h_2 = \frac{1}{\sqrt{2}} e^{j\beta}, \quad \alpha > \beta. \tag{11.28}$$

Then it follows that the phase of the effective channel using the symmetric precoder is given by

$$\begin{cases} \dfrac{\alpha+\beta}{2} & \text{if } |\alpha-\beta-\theta-2\pi k| < \pi \quad \text{for some } k \\ \pi - \dfrac{\alpha+\beta}{2} & \text{otherwise} \end{cases} \tag{11.29}$$

Hence, for channel components of roughly equal magnitude and small to moderate phase shifts, the symmetric precoder does not affect the phase of the resulting effective channel by much. This property does not hold in general for the asymmetric precoder. Figure 11.28 illustrates this aspect for the symmetric precoder. All the sub-figures show the channel components $h_1$ and $h_2$ in (11.28) with $\alpha = 5\pi/8$ and $\beta = \pi/8$. In Figure 11.28a, the optimal precoder $w$ with $\theta = \pi/2$ is applied. It is seen that the channel components are added constructively (rotated toward each

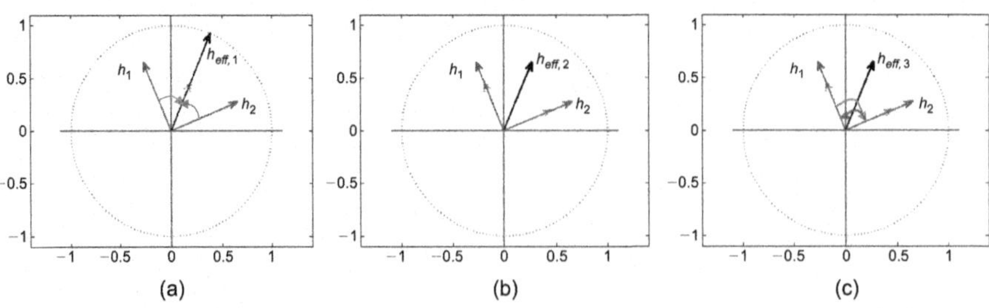


**FIGURE 11.28 Illustration of the symmetric precoding procedure.**

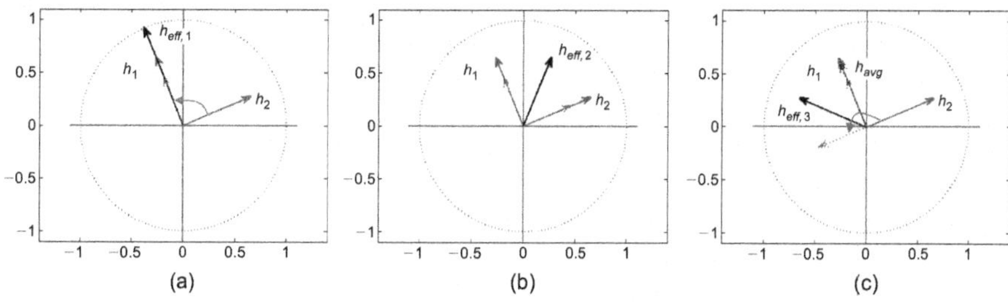


**FIGURE 11.29 Illustration of the asymmetric precoding procedure.**

other and summed) giving a stronger effective channel. In Figure 11.28b and c sub-optimal phase shifts $\theta = 0$ and $\theta = \pi$ are applied. Even though the magnitude of the effective channel becomes less compared to the optimal precoder in these cases, it is seen that the phase remains equal for all cases (a)–(c). Figure 11.29 shows the same scenarios using the asymmetric precoder $v$. In this case, it is observed that the effective channels corresponding to different precoders can have significantly different phases. This fact can have a negative impact on demodulation performance if time-averaged channel estimation is used. In general, the demodulation performance of E-DPDCH is less sensitive to changes in the amplitude compared to the phase, at least for BPSK modulation. In particular, it is important that estimated symbols end up in the correct half-plane of the constellation diagram. For example, if the UE used the $h_{eff,2}$ precoder, but the Node B receiver uses a channel estimate based on the average of $h_{eff,1}$, $h_{eff,2}$, and $h_{eff,3}$ (denoted $h_{avg}$ in Figure 11.29c) in the demodulation process, it is likely that detected I-branch symbols will have the wrong sign.

The examples above show that the symmetric precoder can make channel estimation more robust against erroneous precoder choices. This observation should, however, be put into perspective:

- The problem occurs when pilots are precoded. If pilots are not precoded, as for downlink operation, then channel estimation is not impacted by the choice of the precoder. Hence, for downlink transmit diversity and downlink MIMO, an asymmetric precoder works well.
- If the receiver has full knowledge about the precoder choice made by the UE, then the problem can be alleviated. The receiver can then compensate for the precoder and perform channel estimation on the non-precoded channel (similar to the previous bullet).
- The previous bullets show that a symmetric precoder is particularly useful when operating in open-loop or when the receiver has limited knowledge of the employed precoder choice. Here it can be noticed in what order features were discussed in 3GPP. Uplink open-loop transmit diversity was first discussed extensively, followed by closed-loop transmit diversity, and finally UL MIMO. The two latter features both rely on a precoded channel layout. Hence, the choice of a symmetric precoder makes sense.

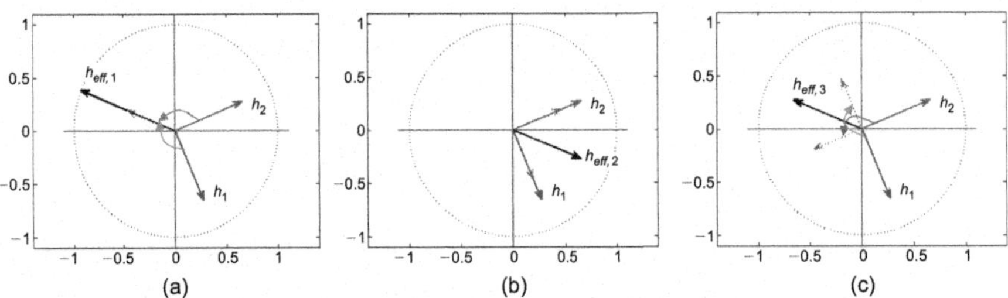


**FIGURE 11.30 Illustration of the enhanced symmetric precoding procedure.**

The second question is why an "enhanced" symmetric precoder is needed. This is also best understood by considering an example. Figure 11.30 shows an example where the channel components $h_1$ and $h_2$ are separated by $3\pi/2$. Figure 11.30a shows the effective channel after applying the optimal symmetric precoder $w$ with $\theta = 3\pi/2$. It is seen that the channel components are added constructively (rotated toward each other and summed), giving a stronger effective channel. In Figure 11.30b, a suboptimal phase shift $\theta = 0$ is applied due to, for example, a downlink signaling error. This gives an effective channel that is $\pi$ radians off from the optimal one. This is clearly not good if time-averaged channel estimation is applied, and it seems to contradict the previous discussion that a symmetric precoder does not affect the phase of the effective channel. However, the reason is that the erroneous phase ($\theta = 0$ compared to $\theta = 3\pi/2$) is larger than $\pi$ radians and this is in fact closely related to (11.29).

Depending on the sign of the previous phase shift, the enhanced symmetric beamforming scheme essentially ensures that the phase between two consecutive phase shifts does not differ by more than $\pi$ radians. If the previous phase was negative, then the default new phase is the signaled phase minus $2\pi$ radians.

If the signaled phase (or signaled phase minus $2\pi$ radians) differs by more than $\pi$ radians compared to the previous phase, then the algorithm will change the applied phase by $2\pi$ radians. Evidently, a change of $2\pi$ radians does not affect the total precoder phase, but because of the symmetric implementation where half of the phase is applied on both antennas, the precoder as such is changed. This can be seen as there are four different precoding phases for the network to choose between, but the UE can actually apply one out of eight precoders. Hence, the effective precoder codebook is twice the size of the signaled precoder codebook. This is illustrated in Table 11.14. Figure 11.30c illustrates the operation of the enhanced symmetric precoder when the previously applied phase was $\theta = 3\pi/2$ and the signaled phase is 0 radians. It is seen that the enhanced symmetric precoder will apply the phase $\theta = 0–2\pi$ radians instead, that is, shift the signaled phase by $2\pi$ radians.

The enhanced symmetric precoding procedure can be described as follows. Let $\tilde{\theta}^{(n+1)} \in \{0,\ \pi/2,\ \pi,\ 3\pi/2\}$ be the currently received *Transmitted Precoding Indicator* (TPI) phase calculated by the Node B and signaled to the UE via the F-TPICH bit pattern and $\theta^{(n)}$ the phase applied in the previous TPI update period. Then the final

**Table 11.14** The Applied Precoding Phase for the Enhanced Symmetric Precoder as a Function of the Previously Applied Phase and the Currently Signaled Phase Shift

| Signaled \ Previous | $\theta = 0$ | $\theta = \pi/2$ | $\theta = \pi$ | $\theta = 3\pi/2$ | $\theta = -\pi/2$ | $\theta = -\pi$ | $\theta = -3\pi/2$ | $\theta = -2\pi$ |
|---|---|---|---|---|---|---|---|---|
| $\theta = 0$ | 0 | 0 | 0 | $-2\pi$ | 0 | $-2\pi$ | $-2\pi$ | $-2\pi$ |
| $\theta = \pi/2$ | $\pi/2$ | $\pi/2$ | $\pi/2$ | $\pi/2$ | $-3\pi/2$ | $-3\pi/2$ | $-3\pi/2$ | $-3\pi/2$ |
| $\theta = \pi$ | $\pi$ | $\pi$ | $\pi$ | $\pi$ | $-\pi$ | $-\pi$ | $-\pi$ | $-\pi$ |
| $\theta = 3\pi/2$ | $-\pi/2$ | $3\pi/2$ | $3\pi/2$ | $3\pi/2$ | $-\pi/2$ | $-\pi/2$ | $-\pi/2$ | $3\pi/2$ |

*A shadowed cell corresponds to a case where the "enhanced" scheme has changed the signaled phase.*

beamforming phase $\theta^{(n+1)} \in \{0, \pm\pi/2, \pm\pi, \pm 3\pi/2, -2\pi\}$ applied for the current TPI update period is calculated as

$$\theta^{(n+1)} = \begin{cases} \phi_1 & \text{if } \left|\phi_1 - \theta^{(n)}\right| \le \pi \\ \phi_2 & \text{otherwise} \end{cases} \tag{11.30}$$

where

$$\phi_1 = \begin{cases} \tilde{\theta}^{(n+1)} - 2\pi, & \text{if } \theta^{(n)} < 0 \\ \tilde{\theta}^{(n+1)}, & \text{otherwise} \end{cases} \tag{11.31}$$

and

$$\phi_2 = \begin{cases} \tilde{\theta}^{(n+1)} - 2\pi, & \text{if } \theta^{(n)} \ge 0 \\ \tilde{\theta}^{(n+1)}, & \text{otherwise} \end{cases} \tag{11.32}$$

This algorithm can alternatively be written as:

| If $\theta^{(n)} \ge 0$ | If $\theta^{(n)} < 0$ |
|---|---|
| $\theta^{(n+1)} = \begin{cases} \tilde{\theta}^{(n+1)} & \text{if } \left|\tilde{\theta}^{(n+1)} - \theta^{(n)}\right| \le \pi \\ \tilde{\theta}^{(n+1)} - 2\pi & \text{otherwise} \end{cases}$ | $\theta^{(n+1)} = \begin{cases} \tilde{\theta}^{(n+1)} - 2\pi & \text{if } \left|\tilde{\theta}^{(n+1)} - 2\pi - \theta^{(n)}\right| \le \pi \\ \tilde{\theta}^{(n+1)} & \text{otherwise} \end{cases}$ |

#### *11.5.1.3 Control channel structure for Mode 1*

The preferred precoding weight is updated every sub-frame (2 ms) and is determined by the network and signaled to the UE using the *Fractional Transmitted Precoding Indicator Channel* (F-TPICH). The F-TPICH design is based on the F-DPCH.

The preferred phase shift is converted to a *Transmitted Precoding Indicator* (TPI) message, which is conveyed on the F-TPICH over the first two slots in a sub-frame using one of ten available slot formats ($i$ = 0,…, 9). The corresponding bits in the third slot are reserved for future use. Two TPI information bits (one QPSK symbol) are transmitted in 1/10th of a slot, using a spreading factor 256, and the rest of the

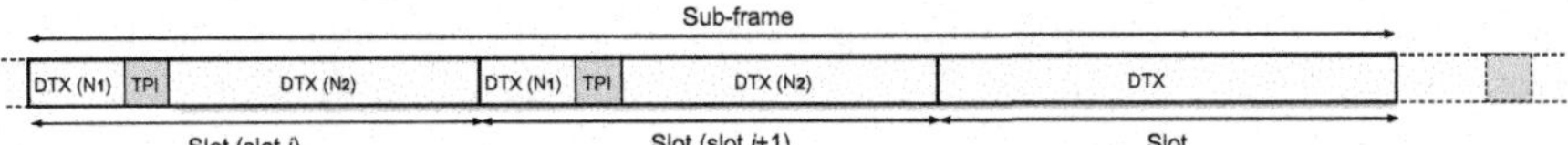


**FIGURE 11.31 The F-TPICH channel layout. The number of DTX bits is given as $N_1 = (2i + 2)\text{mod}(20)$ and $N_2 = 18 - N_1$, where $i = 0,\ldots, 9$ denotes the slot format.**

slot remains unused (see Figure 11.31). By using different slot formats, up to 10 UEs can share a single channelization code. This can be seen as time-multiplexing TPI commands to several users on one shared channelization code. The mapping between TPI and phase is given by Table 11.15 and it can be observed that TPI bits between slots are Gray coded in order to minimize the impact of signaling errors.

The standard allows for some flexibility when it comes to which node(s) signals the preferred precoding weights. One reason is that CLTD can be configured without a HS-DPCCH (for example, only DCH traffic), in which case it is less obvious which link is the most important to enhance when in soft handover.

If the UE is configured with an HS-DPCCH, higher-layer signaling can be used to set whether only one cell (the HS-DSCH serving cell if available) transmits the F-TPICH or all cells from the serving radio link set transmit the F-TPICH. The UE can assume that all TPI bits in a TPI combining period are identical.

The user should only respond to reliably received TPI commands. Hence, the UE measures the reliability of the TPI bits. If the quality is better than what is required then the precoding weights are updated, otherwise the last received reliable phase shift is used. The quality threshold is implicitly defined by RAN4 test cases.

The precoding weights used by the user are not explicitly signaled to the network in the uplink. This is done implicitly via the precoded channel structure.

#### 11.5.1.4 RRC signaling and configurations

CLTD is configured via higher-layer signaling (UL_CLTD_Enabled is true or false) and the current mode is kept in the parameter UL_CLTD_Active (0, 1, 2, or 3). When UL_CLTD_Enabled initially is set to TRUE, UL_CLTD_Active shall be set to 1 or 2 as indicated by higher-layer signaling. If activation state 1 is configured, or a transition from a different activation state to activation state 1 occurs, the TPI is initially set to the fixed precoder weight $\tilde{\theta} = 3\pi/2$. This weight is used until F-TPICH

**Table 11.15** Mapping of TPI Information to Phase

| $\tilde{\theta}^{(n+1)}$ | **TPI Bit Pattern for Slot** *i* | *i* + 1 |
|---|---|---|
| 0 | 00 | 00 |
| $\pi/2$ | 00 | 11 |
| $\pi$ | 11 | 11 |
| $3\pi/2$ | 11 | 00 |

satisfies the quality target. HS-SCCH orders are used to change between CLTD modes. Mode 1 offers codebook based precoding where the network indicates the preferred precoding vector, and Modes 2 and 3 are based on so-called fixed antenna switching where all physical channels are sent from one of the two antennas. CLTD operation can only be configured in CELL_DCH and is not supported in combination with uplink multicarrier.

A number of higher-layer information elements are used to configure CLTD, including the following:

- S-DPCCH/DPCCH power offset.
- Initial CLTD activation state; mode 1 or mode 2.
- If only DCH is configured, the cell that determines the precoding weights is notified.
- A number of F-TPICH information, that is, slot format, code number, and frame offset. Furthermore, information about which cells should transmit F-TPICH is configured.

## 11.5.2 OPEN-LOOP TRANSMIT DIVERSITY

The operation of *open-loop transmit diversity* (OLTD) is transparent to the network and does not affect the layer-1 specifications. It is the network that configures OLTD since OLTD might not give gains in all circumstances and the network needs to retain control of the UE behavior. OLTD operation is only allowed in CELL_DCH and with single-carrier operation.

The physical-layer processing in terms of coding, modulation, spreading, scrambling, and precoding follows the closed-loop transmit diversity counterpart; see Figure 11.32. The main difference is that the UE autonomously needs to decide the preferred precoding weight without any dedicated feedback. No new physical channels or signaling are supported for OLTD. The exact precoding operation is implementation specific, but the weights need to be chosen based on readily available

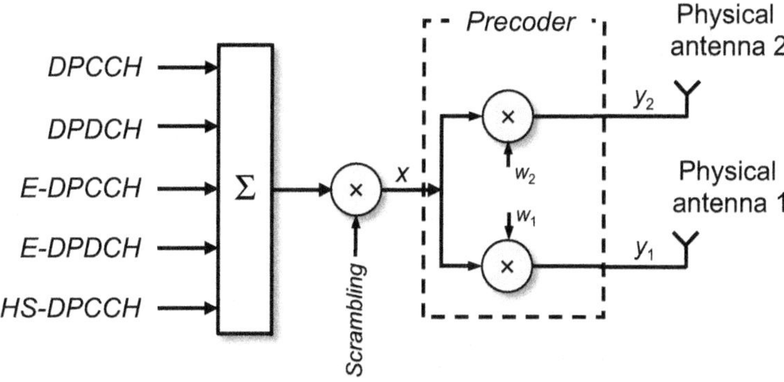


**FIGURE 11.32 Physical channel layout for open-loop transmit diversity.**

uplink-related information. One example would be to employ a numerical search approach based on uplink transmit power control commands. Different precoding weights are essentially evaluated by considering the TPC feedback response. A surplus of TPC DOWN in response to a changed weight might be a good indication that the chosen weight was better than the previous choice, and vice versa. RAN4 performance requirements are introduced in order to ensure that the UE does something sensible.

## 11.6 UPLINK TWO-BRANCH MIMO

### 11.6.1 OVERVIEW

Uplink two-branch MIMO offers the possibility to transmit up to two streams simultaneously. Similarly to downlink operation, a dual-codeword approach with stream-independent coding and hybrid-ARQ processing is adopted where the transmission rank can be dynamically adjusted each TTI. CLTD is required in order to operate with MIMO, meaning that the CLTD precoded physical channel layout is employed, and in particular the CLTD precoding framework and the S-DPCCH are used also for MIMO. *Secondary Dedicated Physical Data Channel for E-DCH* (S-E-DPDCH) and *Secondary Dedicated Physical Control Channel for E-DCH* (S-E-DPCCH) are introduced to support the transmission of the secondary stream; see Figure 11.33 for an overview of the physical channel layout. The primary stream can be viewed as a legacy stream, whereas the secondary stream is more a best-effort stream. This means, for example, that power control functionalities operate on the primary stream. The E-TFC selection for the secondary stream is based on rate adaptation by means of network signaled quality measures. A number of hybrid-ARQ retransmission related rules are introduced.

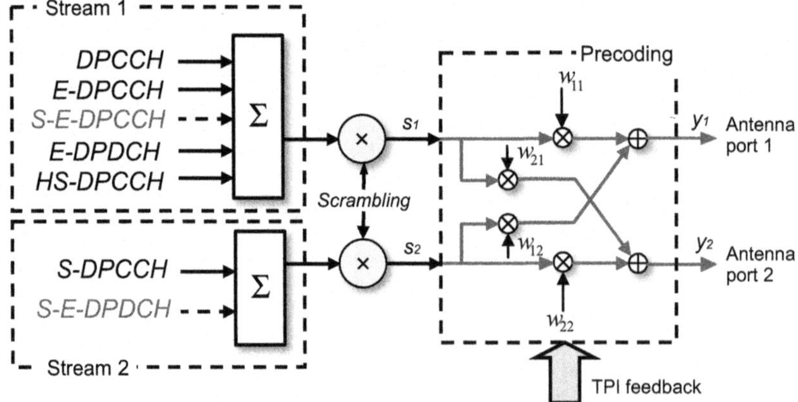


**FIGURE 11.33 Physical channel layout for uplink MIMO. Shadowed channels are only present during rank-2.**

As for non-MIMO operation, the network determines the appropriate SIR target for inner-loop power control and the serving grant used for scheduling and signals related information by means of TPC and grant commands. Furthermore, the network determines the preferred precoding weights and signals them to the UE via the F-TPICH. To support MIMO, the network also needs to determine the preferred rank and a suitable transport-block size for the secondary stream. This information is sent to the user by means of the physical channel referred to as *E-DCH Rank and Offset Channel* (E-ROCH).

### 11.6.2 E-DCH MIMO TRANSMISSION

Up to two streams can be transmitted simultaneously and the transmission rank and the transport-block size of each stream can be dynamically adjusted on a TTI basis. The design is based on a dual codeword approach, where each stream performs independent coding, modulation, spreading, and scrambling according to the non-MIMO case. After scrambling, a precoder is added that maps the streams to the antenna branches. The precoder design introduced for CLTD is reused for MIMO, and CLTD is therefore a pre-requisite in order to operate MIMO. Similarly, uplink MIMO builds on the physical channel layout used for CLTD. All physical channels are precoded, and a second pilot channel, the S-DPCCH, is used to facilitate channel sounding and data demodulation.

During the 3GPP standardization process, there were initial discussions whether or not to interleave the two streams after spreading. This would imply that both streams experience the same radio channel which would simplify the design of some functionality. However, simulations showed significant performance losses using this approach, so interleaving was not adopted by the standard. It was also discussed to use identity precoding, that is, map each stream to different physical antennas, during rank-2 in order to reduce the PAPR impact. However, in the end it turned out that the loss by having identity precoding during rank-2 was larger than the gain in reduced PAPR, so the Mode-1 CLTD precoder is employed both for rank-1 and rank-2 transmissions. It should, however, be noticed that rank-1 transmissions are possible with any CLTD mode, whereas rank-2 transmissions are only possible in Mode-1.

An additional uplink physical channel type, the *Secondary Dedicated Physical Data Channel for E-DCH* (S-E-DPDCH), is introduced in order to carry transport blocks associated with the secondary stream. An associated uplink physical control channel, the *Secondary Dedicated Physical Control Channel for E-DCH* (S-E-DPCCH), is also defined. See Figure 11.34 for an overview of the physical layer processing for S-E-DPCCH and S-E-DPDCH.

The S-E-DPDCH is only transmitted during rank-2 transmissions. The S-E-DPDCH can use slot formats corresponding to BPSK, 4-PAM, or 8-PAM with SF2 or SF4. In order to use a linear receiver structure, the secondary stream needs to use a code configuration that equals (or is a subset of) the primary stream code configuration. Otherwise the different OVSF codes are not orthogonal and there will be leakage between different symbols. For simplicity, it was agreed to always use

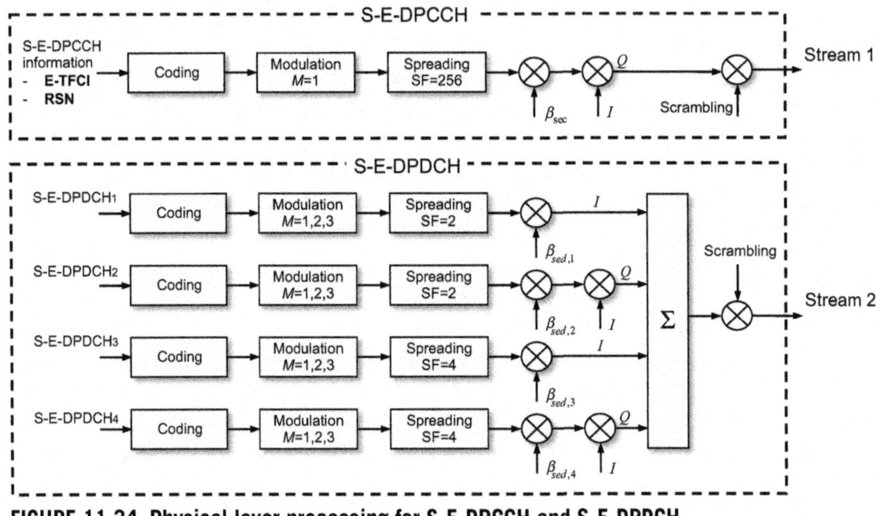


**FIGURE 11.34 Physical layer processing for S-E-DPCCH and S-E-DPDCH.**

the 2 × SF2 + 2 × SF4 OVSF configuration for both E-DPDCH and S-E-DPDCH during rank-2 transmissions. Strictly speaking, it would not be MIMO (spatial-multiplexing) if the streams use different OVSF codes. S-E-DPDCH$_1$ and S-E-DPDCH$_3$ are both mapped to the I-branch, and S-E-DPDCH$_2$ and S-E-DPDCH$_4$ are both mapped to the Q-branch.

Rank-2 transmissions are only allowed if the transport-block size of each stream is larger than a configured value, the *minimum E-TFCI for Rank-2 transmission*. The rate matching functionality for determining SF, modulation, and number of physical channels is affected since rank-2 only is allowed for the 2×SF2+2×SF4 OVSF configuration, and the minimum E-TFCI for rank-2 transmissions can be set arbitrarily small; see Figure 11.35 for an illustration.

The S-E-DPCCH has the same structure as E-DPCCH and uses rate 1/3 Reed–Muller coding. The S-E-DPCCH is used to indicate the format used on S-E-DPDCH, that is, E-TFCI and RSN. The happy bit is not used and is set to 1. The S-E-DPCCH is only sent during rank-2 transmissions, and its presence/detection at the Node B is used to indicate rank-2. The S-E-DPCCH is mapped to the primary stream since it is seen as the most robust stream as it is power controlled and precoding is typically determined in order to maximize the SNR/SINR of the primary stream. The S-E-DPCCH uses SF256 and is mapped to the Q-branch (E-DPCCH and S-E-DPCCH are I/Q multiplexed).

#### *11.6.2.1 Power settings*

In theory, there are benefits of setting the power of each stream independently, but because of simplicity reasons it was agreed to allocate the same data power to both streams during rank-2 transmissions. As alternatives, water-filling schemes were

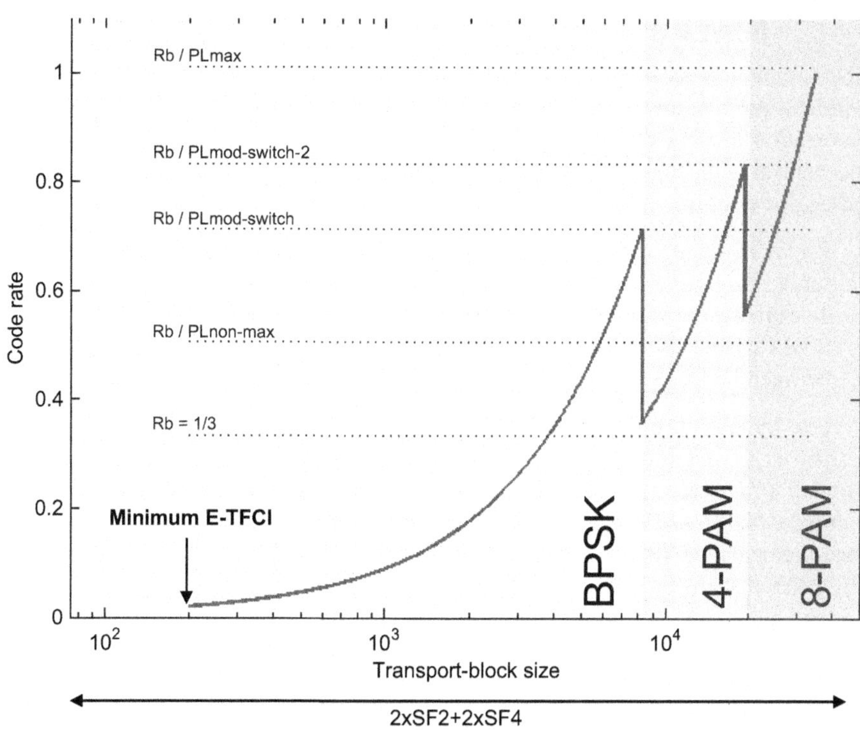


**FIGURE 11.35 Illustrations of the process of determining spreading factor (SF), modulation, and code configuration during rank-2. $PL_{max}$ = 1/3.**

discussed, but in general they imply increased signaling overhead, and the potential gains are hard to realize in practical scenarios. The gain factors of the S-E-DPDCHs equal the gain factors of the E-DPDCHs, that is, the gain factors for E-DPDCH and S-E-DPDCH using the same spreading factor (slot format) are equal ($\beta_{sed} = \beta_{ed}$).

The S-E-DPCCH power is set relative DPCCH via $\beta_{sec}$, which is signaled over RRC by the RNC. The S-E-DPCCH is never subject to power boosting as compared to E-DPCCH.

For rank-1 transmissions, E-DPCCH boosting follows release 8 rules and S-DPCCH boosting follows CLTD rules. For rank-2, on the other hand, E-DPCCH boosting is always applied irrespectively of the E-TFCI, and the S-DPCCH gain factor equals the E-DPCCH gain factor. The primary stream (E-DPDCH) E-TFCI is used in boosting formulas.

Whenever the total transmit power exceeds the maximum allowed value, all E-DPDCH and S-E-DPDCH gain factors $\beta_{ed,k}$ and $\beta_{sed,k}$ are scaled down equally until reaching the values $\beta_{ed,k,\text{reduced,min}}$ or the total transmit power equals the maximum allowed power. If further scaling is still required after reaching $\beta_{ed,k,\text{reduced,min}}$ then all channels are scaled equally. See Section 9.3.3.3 and [11] for further details.

### 11.6.3 POWER CONTROL

The variation in received power from the first stream and from the second stream will be somewhat uncorrelated due to the varying channel statistics. The 3GPP standardization process discussed whether to have a single inner-loop power control mechanism or one loop per stream. Several arguments against having two loops were raised, for example

- For CLTD (rank-1 transmission), one TPC loop operating on the DPCCH makes sense.
- To estimate the wireless channel, the relative power between the DPCCH and S-DPCCH needs to be known. Having a fixed or very deterministic offset between DPCCH and S-DPCCH simplifies this procedure significantly.

One inner-loop power control (ILPC) based on the (primary stream) DPCCH quality target and one outer-loop power control (OLPC) targeting the quality of the primary stream are used for MIMO operation. Hence, there are no modifications compared to non-MIMO operation. This also implies that current definitions of measurements are kept unchanged. For example, *uplink power headroom* (UPH) is still defined as the ratio between maximum UE transmission power and the DPCCH power.

The second stream can be viewed as "best effort," meaning that the quality of the received secondary stream will vary. This infers a need to adapt rank and, in particular, the transport-block size (TBS) of the second stream in a dynamic fashion. See Figure 11.36 for an illustration of the power control procedure.

### 11.6.4 DOWNLINK CONTROL CHANNEL STRUCTURE

The preferred rank and secondary stream transport-block size related information are sent in the downlink using a new common/shared physical channel, the so-called *E-DCH Rank and Offset Channel* (E-ROCH). The E-ROCH has the same sub-frame structure and coding as the E-AGCH. The E-ROCH is sent from the serving E-DCH cell and a specific user is addressed by means of a user-specific E-ROCH E-RNTI CRC.

$$r = HWPs = Hw_1\sqrt{p_1}s_1 + Hw_2\sqrt{p_2}s_2 = h_{eff,1}s_1 + h_{eff,2}s_2$$

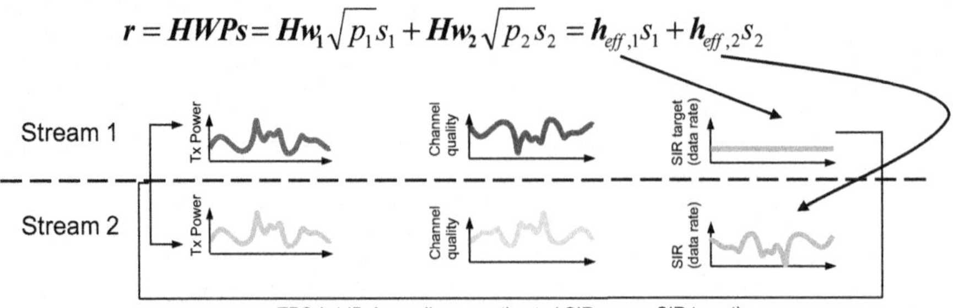


**FIGURE 11.36 Illustration of the power control procedure.**

Table 11.16 Mapping Between the Signaled Offset Index and the S-ETFC Offset Value

| S-ETFC Offset Value | Index | S-ETFC Offset Value | Index |
|---|---|---|---|
| 60/30 | 31 | 16/30 | 15 |
| 55/30 | 30 | 15/30 | 14 |
| 50/30 | 29 | 14/30 | 13 |
| 45/30 | 28 | 13/30 | 12 |
| 40/30 | 27 | 12/30 | 11 |
| 35/30 | 26 | 11/30 | 10 |
| 30/30 | 25 | 10/30 | 9 |
| 28/30 | 24 | 9/30 | 8 |
| 26/30 | 23 | 8/30 | 7 |
| 24/30 | 22 | 7/30 | 6 |
| 22/30 | 21 | 6/30 | 5 |
| 21/30 | 20 | 5/30 | 4 |
| 20/30 | 19 | 4/30 | 3 |
| 19/30 | 18 | 3/30 | 2 |
| 18/30 | 17 | 2/30 | 1 |
| 17/30 | 16 | 1/30 | 0 |

The E-ROCH carries in total six information bits, where five bits are used to signal the S-ETFC offset value and one bit is used to signal the rank (rank-2 allowed or rank-2 not allowed). The mapping between the signaled offset index and the S-ETFC offset value is given in Table 11.16. The S-ETFC offset is used to calculate the virtual serving grant that the UE is allowed to use in the next transmission of the secondary stream. The S-ETFC offset and rank information elements correspond to the absolute grant value and the absolute grant scope information elements for the E-AGCH.

The E-AGCH and E-ROCH can be independently configured with a channelization code and E-RNTI. The channels can be code multiplexed or time multiplexed using one code with different E-RNTIs. Similarly as for the E-AGCH, it is up to the serving E-DCH cell to decide how often the E-ROCH needs to be transmitted, and received E-ROCH information is kept until a new E-ROCH message is received.

Hybrid-ARQ-related information for both streams needs to be conveyed in the downlink for rank-2 transmissions. Two independent E-HICH signatures are configured in order to convey hybrid-ARQ ACK feedback for each stream independently.

The F-TPICH is used to carry TPI as for CLTD without modifications, and E-AGCH and E-RGCH are used to signal serving grant updates.

### 11.6.5 GRANT AND E-TFC SELECTION

The grant, defined as the power ratio between the E-DPDCH and DPCCH, gives the user an "upper bound" on how much data power it may use and is essentially a power measure which allows the network to control the interference (received power or RoT) a certain UE is allowed to create. The grant is also tightly coupled with the E-TFC selection procedure. For uplink MIMO, it was discussed whether to have one common grant or one grant per stream, and a common grant was adopted because of a number of reasons:

- From the network's perspective it is the total interference level that is of interest and it is of less concern whether it comes from the primary or secondary stream.
- It was agreed to have one ILPC operating on the primary stream. Hence, since the S-DPCCH is not power controlled and its receive power will vary, it makes little sense trying to define a grant for that stream.
- It was agreed to have equal gain values for both streams, that is, the potential gain from a water-filling procedure was seen as rather limited.
- It simplifies scheduler implementation, as well as rank and precoder adaptation.
- It yields less downlink feedback overhead (that is, E-AGCH/E-RGCH).

Also, it was discussed during the standardization phase whether to keep the existing definition of the grant or to redefine it as the power ratio between the total data power (E-DPDCH and S-E-DPDCH) and DPCCH.

- Both alternatives will result in a fluctuating received power for rank-2 transmissions since the S-DPCCH is not power controlled. Hence, the network needs to take this into account to ensure a stable system and possibly reduce the grant or schedule rank-1 transmissions if RoT becomes too large.
- Existing definition essentially implies that grant and rank decisions are coupled to each other. This means that whenever the rank changes, it might be necessary to issue a new grant since otherwise the RoT target will be over- or under-shot. Redefining the grant would mean that it is not necessary to give a new grant when switching from rank-1 to rank-2 since the total grant is shared between the two streams. Note, though, that it still might be necessary to update the grant when changing from rank-2 to rank-1 since the primary stream in general uses the stronger eigenmode of the channel.
- Given the existing definition of grant, the full RoT will not be utilized whenever the UE needs to fallback to rank-1, that is, the data power is reduced by 3 dB.
- From a more fundamental point of view, it might be beneficial to decouple grant and rank from each other since grant is a scheduling resource affecting system stability such as RoT and hence possibly changed on a slow basis. Rank, on the other hand, is a more instantaneous measure that depends on channel characteristics, and hence it might need more frequent updating.

In the end, it was decided to keep the existing definition of the grant, that is, the power ratio between the E-DPDCH and DPCCH. The interpretation of the serving grant remains the same irrespectively of the rank. The serving grant can be updated via existing E-AGCH and E-RGCH mechanisms.

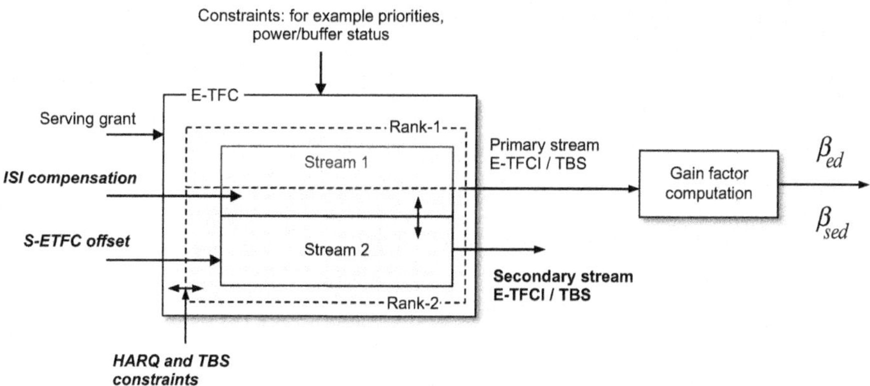


**FIGURE 11.37 Illustration of parameters affecting the E-TFC selection procedure. Shadowed parameters are explicitly related to MIMO operation.**

### 11.6.5.1 E-TFC selection procedure

The E-TFC selection procedure for uplink MIMO is highly influenced by some fundamental design choices. Existing power control mechanisms are kept unchanged, that is, the ILPC and OLPC control the quality of the primary stream. The grant definition remains the same as for non-MIMO, that is, the grant is given by the power ratio between the E-DPDCH and DPCCH. Both E-DPDCH and S-E-DPDCH have the same gain values during rank-2 transmissions. Also, existing procedures remain untouched to a large extent. For example, the serving grant, buffer/power limitations, priorities, and reference value settings affect the procedure in a similar manner as for the non-MIMO case. See Figure 11.37 for a schematic overview of the E-TFC selection procedure for uplink MIMO.

The E-TFC selection for the primary stream during rank-1 transmission is identical to CLTD. For rank-2 transmissions, there are some additional constraints that need to be taken into consideration:

- Rank-2 transmissions are always using 2×SF2+2×SF4 on both streams, and any scheduled transport block needs to satisfy the *minimum E-TFCI for Rank-2 transmission*; otherwise rank-1 is employed.
- The *normalized remaining power marginal* (NRPM) depends on rank. The UE estimates the NRPM for E-TFC candidate $j$ according to

$$\text{NRPM}_j = \frac{1}{2}\frac{P_{\max,j} - P_{\text{DPCCH}} - P_{\text{S-DPCCH},j} - P_{\text{HS-DPCCH}} - P_{\text{E-DPCCH},j} - P_{\text{S-E-DPCCH}}}{P_{\text{DPCCH}}}$$

for rank-2, and

$$\text{NRPM}_j = \frac{P_{\max,j} - P_{\text{DPCCH}} - P_{\text{S-DPCCH},j} - P_{\text{DPDCH}} - P_{\text{HS-DPCCH}} - P_{\text{E-DPCCH},j}}{P_{\text{DPCCH}}}$$

otherwise, where $P_{\max,j}$ is the maximum UE transmitter power for E-TFC-$j$ and $P_x$ denotes the power for physical channel $x$.

- Since there is code-reuse between the streams during rank-2 transmissions, there will inevitably be inter-stream interference. The degree of inter-stream interference will be scenario dependent as well as receiver dependent. The legacy E-TFC selection mechanism operates on the grant (power) and hence it does not explicitly take inter-stream interference into account. This will obviously affect the dual-stream performance. In many cases, too poor received SINRs are obtained in order to support the TBS provided via the grant and E-TFC selection procedure. This will trigger the OLPC, which will increase the SIR target and eventually increase the SINR and/or reduce the transport-block size to a level that the receiver can support. This is a slow process and it wastes DPCCH power. A better solution is to take the inter-stream interference explicitly into account in the E-TFC selection procedure. Because of the inter-stream interference, a specific grant should generally correspond to a lower transport-block size compared to if no interference was present. The transmit power should, however, not be affected. Optimally, the compensation of inter-stream interference (or any interference) in the E-TFC selection process needs to be done dynamically. However, having a dynamic compensation of the inter-stream interference would require some major redesign of existing functionalities, for example, control channels. Hence, a semi-static compensation mechanism is introduced. The standard supports rank dependent reference amplitude values $A_{ed}$ for the primary stream. The $A_{ed}$ for the primary stream is modified during rank-2 by an additional $A_{ISI}$ parameter

$$A_{ed} = A_{ed,non\text{-}ISI} \cdot A_{ISI}, \tag{11.33}$$

  where $A_{ISI}$ is configured via higher layer signaling. Rank-dependent reference values for the secondary stream are not needed since the secondary stream anyway employs dynamic rate adaptation.

The E-TFC selection for the secondary stream is based on rate-control principles since the secondary stream is not power controlled and the quality will vary dynamically depending on channel conditions, etc. This is achieved by using a *virtual serving grant* as input to the E-TFC selection mechanism, where the virtual serving grant is based on the effective data power used for the primary stream (effective gain values) and the S-ETFC offset,

$$\text{Virtual\_Serving\_Grant} = \sum_{k=1}^{4} \left( \frac{\beta_{ed,k}}{\beta_c} \right)^2 x\,\text{offset}. \tag{11.34}$$

The offset is signaled using the E-ROCH and is adjusted by the Node B depending on the quality of the secondary stream.

A key parameter when determining the E-TFCI for the secondary stream is the S-ETFC offset. The offset is determined by the network in order to achieve a design target. The exact solution is implementation specific, but a typical choice would be to maximize the capacity while maintaining a quality target, where the quality target can, for example, be a certain BLER target. Hence, if the BLER on the secondary stream is below the target the offset is increased, and if the BLER is above the target the offset is decreased. The BLER target can be controlled by the serving cell alone, or by all MIMO-capable cells in the active set (similarly, the OLPC affects the SIR target used for ILPC). Alternatively, or additionally, the SINR of the secondary stream can be used to determine the appropriate offset.

Some other considerations that influence the E-TFC selection procedure for uplink MIMO that were discussed during the standardization phase include the following:

- Non-scheduled data is mapped to the primary stream.
- *Power-limited scenario* – as discussed above, the normalized remaining power marginal will depend on the rank since during rank-2 the remaining power needs to be divided equally between the two streams.
- *Buffer-limited scenario* – fill the primary stream before filling the secondary stream.
- TBS restriction – if any of the resulting transport-block sizes during rank-2 are less than the minimum allowed TBS, then the UE falls back to rank-1 and has to redo the E-TFC selection procedure.
- There is a close interaction between the E-TFC selection and hybrid-ARQ functionality.

### 11.6.6 HYBRID-ARQ RETRANSMISSIONS

There is one hybrid-ARQ entity per E-DCH with a number of parallel hybrid-ARQ processes. The number of hybrid-ARQ processes is doubled for uplink MIMO (16 processes). Processes 0 to 7 are always mapped to the primary stream and processes 8 to 15 are for an initial transmission mapped to the secondary stream but may be remapped to the primary stream because of a retransmission. Each stream is subject to independent coding and hybrid-ARQ processing, but a number of retransmission constraints apply. Because of the synchronous and non-adaptive hybrid-ARQ operation in the uplink, the Node B knows when a retransmission is supposed to occur and with what RSN. Hence, there is no explicit hybrid-ARQ process identifier, and two hybrid-ARQ processes are in principle tightly connected. During rank-2 transmission, process number $k$ and $(k + 8)\text{mod}(16)$ are sent simultaneously. The UE and the network need to keep track of which hybrid-ARQ process or buffer should be targeted. This becomes more intricate when the preferred rank changes between transmission attempts. In particular, there are a number of cases where the preferred rank is overridden by the UE, and there are occasions where a process originally sent over the secondary stream is retransmitted over the primary stream – a so-called *stream switch*. Table 11.17 summarizes the different retransmission rules. It can be seen that the preferred rank is used whenever possible. Two exceptions are when one or two transport blocks need to be retransmitted, and the previous transmission was rank-1 or rank-2, respectively. Also, if the E-TFC selection does not allow for rank-2, then rank-1 needs to be used irrespective of the preferred rank. It is also seen that a stream switch occurs in a number of cases.

**Table 11.17** Retransmission Rules for Uplink MIMO

| Case | Last Transmission | | | Behavior Upon Retransmission | | |
|---|---|---|---|---|---|---|
| | | ACK/NACK | | | | |
| | Rank at Transmission | First Stream Transport Block | Second Stream Transport Block | Maximum Allowed Rank at the Time of Retransmission | Used Rank | Information |
| 1 | 1 | NACK | – | 2 | 1 | Retransmit data from the primary stream |
| 2 | 2 | NACK | NACK | 1 or 2 | 2 | Retransmit failed data from both streams |
| 3 | 2 | ACK | NACK | 1 | 1 | Retransmit failed data over the primary stream |
| | | NACK | ACK | | | |
| 4 | 2 | ACK | NACK | 2 | 1 or 2 | If rank-2 is allowed according to the E-TFC selection, retransmit failed data on the same stream and new data on the other stream. Otherwise retransmit failed data in rank-1 over the primary stream |
| | | NACK | ACK | | | |

The Node B needs to keep track of retransmission events and potential error cases using available information. In particular, the Node B needs to detect that a stream switch has occurred in order to soft combine with the correct hybrid-ARQ buffer. Some additional intelligence taking RSN and E-TFCI into account might be necessary. Figure 11.38 provides an example of hybrid-ARQ retransmission procedures.

The power control loop targets the primary stream. Hence, the SIR target set by the OLPC should only be affected by the quality of the primary stream. The number of *total transmission attempts* (TTA) associated with the secondary stream should consequently not be sent to the RNC. This can be achieved by, for example, not sending the *number of hybrid-ARQ retransmissions* (NHR) information element for the secondary stream, or using value 13, indicating that this value should not be used by the OLPC.

### 11.6.7 RANK AND SHO MANAGEMENT

It is the serving E-DCH cell that determines the preferred rank and S-ETFC offset value and signals this information to the UE via the E-ROCH. Having a single entity in the network controlling rank, S-ETFC offset, and precoding simplifies the design and increases the predictability of the system. The exact approach to determine rank, S-ETFC offset, and precoding is implementation specific. The optimal approach would be to jointly choose these parameters in order to maximize the capacity. Since there is no explicit rank indication in the uplink, the Node B needs to assess the rank by means of the number of detected E-DPCCHs. As discussed in the previous section, there are some scenarios where the UE needs to override the network-signaled preferred rank, for example, because of hybrid-ARQ retransmission or buffer limitation.

One advantage of the precoded physical channel structure is that a non-serving cell can always demodulate the primary stream without explicit knowledge of the used precoding or rank. To demodulate the secondary stream, on the other hand, the Node B needs to be aware of MIMO related information, for example, S-DPCCH, S-E-DPCCH, rank, etc. This will obviously consume significantly more resources, and it is questionable how much to gain. In general, it is less likely to schedule rank-2 in SHO since this often implies that the user is close to cell-edge with worse SNR conditions. Needless to say, this is something that can change when considering, for example, heterogeneous network deployments. The MIMO design allows both rank-1 and rank-2 during soft/softer handover. The serving cell and non-serving cells send independent hybrid-ARQ ACK acknowledgments for each transport block during rank-2.

### 11.6.8 UE CAPABILITIES, RRC SIGNALING AND CONFIGURATIONS

Two new UE categories are introduced for uplink MIMO. Categories 11 and 12 support uplink MIMO with maximum modulation order equal to 16-QAM and 64-QAM, respectively (see Table 11.18).

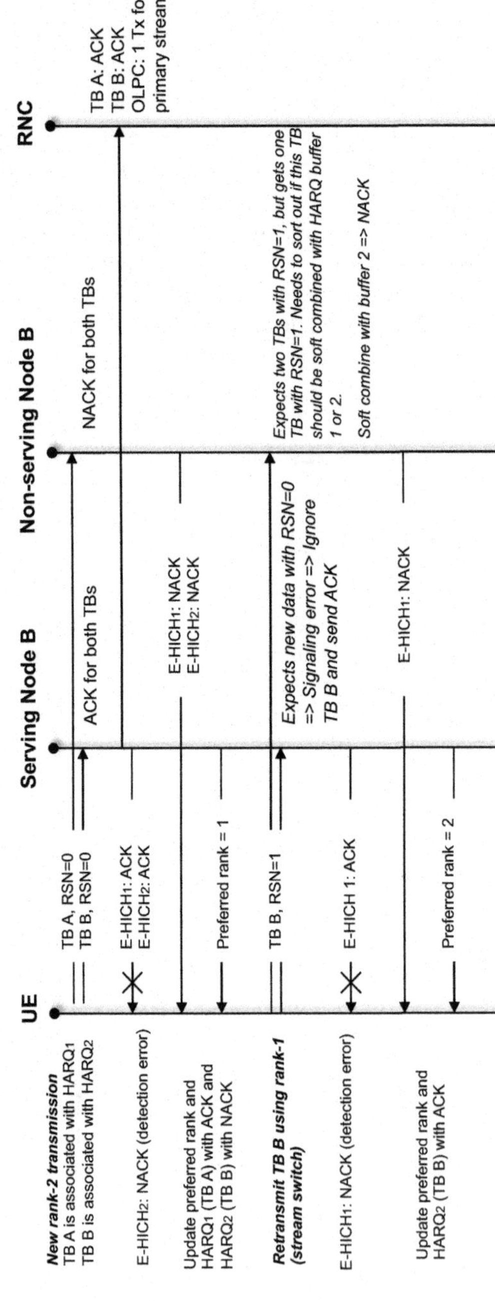


**FIGURE 11.38** An example of hybrid-ARQ retransmission procedures.

**Table 11.18** E-DCH Categories Supporting Uplink MIMO

| E-DCH Category | Modulation Scheme | Max Data Rate (Mbit/s) | |
|---|---|---|---|
| | | One Stream | Two Streams |
| 11 | QPSK, 16-QAM | 11.5 | 23.0 |
| 12 | QPSK, 16-QAM, 64-QAM | 17.3 | 34.5 |

Uplink MIMO is supported per frequency band, and relies on a number of pre-requisites:

- the UE is in CELL_DCH state;
- the UE is configured with CLTD;
- the UE is configured with E-DCH and HS-DSCH serving cells;
- the UE is configured with MAC-i/is with TSN field extension;
- the UE is configured with E-ROCH and E-DPCCH boosting;
- the UE is configured with 2ms TTI; and
- dual-cell operation is not configured.

A number of uplink MIMO-related parameters are set by higher-layer signaling, including [7,10]

- the secondary transport-block signature sequence – second E-HICH signature used to acknowledge the secondary stream;
- S-E-DPCCH power offset (0,…, 17) – the $\Delta_{S\text{-}E\text{-}DPCCH}$ index that translates into the S-E-DPCCH/DPCCH amplitude ratio;
- minimum E-TFCI for rank-2 transmission (0,…, 127) – specifying the minimum allowed transport-block size for rank-2 transmission;
- inter-stream interference compensation index (0,…, 15) – the $\Delta_{ISI}$ index that translates into $A_{ed,\ ISI}$; and
- E-RNTI and channelization code for the E-ROCH.

## REFERENCES

[1] 3rd Generation Partnership Project; Technical Specification Group Radio Access Network; Spatial channel model for Multiple Input Multiple Output (MIMO) simulations, 3GPP, 3GP TS 25.306.

[2] S.M. Alamouti, A simple transmit diversity technique for wireless communications, IEEE J. Select. Areas Commun. 16 (1998) 1451–1458.

[3] 3rd Generation Partnership Project; Technical Specification Group Radio Access Network; Physical layer procedures (FDD), 3GPP TS 25.214.

[4] 3rd Generation Partnership Project; Technical Specification Group Radio Access Network; Multiple Input Multiple Output in UTRA, 3GPP TS 25.876.

[5] J. Pautler, M. Ahmed, K. Rohani, On application of multiple-input multiple-output antennas to CDMA cellular systems, IEEE Vehicular Technology Conference, Atlantic City, NJ, USA, October 7–11, 2001.

[6] 3rd Generation Partnership Project; Technical Specification Group Radio Access Network; Multiplexing and Channel Coding (FDD), 3GPP TS 25.212.

[7] 3rd Generation Partnership Project; Technical Specification Group Radio Access Network; Radio Resource Control (RRC); Protocol Specification, 3GPP TS 25.331.

[8] 3rd Generation Partnership Project; Technical Specification Group Radio Access Network; Evolved Universal Terrestrial Radio Access (E-UTRA); Physical channels and modulation, 3GPP TS 36.211.

[9] 3rd Generation Partnership Project; Technical Specification Group Radio Access Network; Evolved Universal Terrestrial Radio Access (E-UTRA); Physical layer procedures, 3GPP TS 36.213.

[10] 3rd Generation Partnership Project; Technical Specification Group Radio Access Network; UE Radio Access capabilities, 3GPP, 3GP TS 25.306.

[11] R1-110664, Link analysis of mechanisms to improve impact of phase discontinuity due to CLTD on Node B receiver, Qualcomm Incorporated, 3GPP TSG RAN WG1 Meeting #64, Taipei, Taiwan, February 21–25, 2011.

## FURTHER READING

3rd Generation Partnership Project; Technical Specification Group Radio Access Network; Spreading and Modulation (FDD), 3GPP TS 25.213.

V. Tarokh, H. Jafarkhani, A.R. Calderbank, Space–time block codes from orthogonal designs, IEEE Trans. Info. Theory 45 (1999) 1456–1467.

CHAPTER

# Multi-carrier techniques 12

## CHAPTER OUTLINE



Modern mobile broadband traffic is typically bursty in nature. When a user has a burst to transmit or receive, the user's experience is defined by the time taken to transfer the burst, which is dependent on the maximum achievable data rate. User experience can be improved by increasing the bandwidth that is available to the user.

Increasing the user bandwidth could in principle be achieved by redefining the underlying WCDMA structure to operate with a higher chip rate and bandwidth. Such an approach would require contiguous spectrum, which is not always available, and also it would not be compatible with legacy UEs.

An attractive approach is to keep the concept of 5 MHz WCDMA carriers that can be accessed by all UEs but to enable new UEs to transmit and receive on multiple carriers simultaneously. This is the basis of dual-carrier and multi-carrier HSPA evolutions. Multi-carrier support was introduced into HSPA over several releases. This chapter outlines how multi-carrier support has evolved by means of explaining which features are supported in each release. It should be noted that in the 3GPP specifications, the term "multi-cell" is used to refer to multi-carrier transmissions. For the specific case of two carriers, the term "dual-cell" is used.[1]

[1]In this chapter, *dual-cell* refers to dual-carrier transmission. The term *dual-cell* is also used to refer to multi-flow transmissions from two sites or sectors.

HSPA Evolution: The Fundamentals for Mobile Broadband. DOI: 10.1016/B978-0-08-099969-2.00013-2

## 12.1 EVOLUTION OF MULTI-CARRIER HSPA

The first step in evolving multi-carrier HSDPA was taken in release 8, when *Dual-Cell HSDPA* (DC-HSDPA) was introduced. As its name implies, supporting DC-HSDPA enables a UE to receive data from two cells simultaneously. Release 8 restricted the two cells to belong to the same Node B, be mapped to adjacent 5 MHz carriers in the same frequency band, and be tightly time synchronized. Furthermore, for complexity reasons, DC-HSDPA in release 8 was a separate capability from MIMO and the two could not be configured simultaneously. Only dual-carrier in the downlink was standardized; thus, a single uplink carrier carries physical layer signaling for both downlink carriers.

In release 9, the capability of DC-HSDPA was extended. First, the combination of DC-HSDPA and MIMO (with MIMO operating on either of the carriers or both carriers) was standardized; this involved revision of the uplink control signaling. Second, an ability to configure the carriers in different frequency bands was introduced. (Chapter 16 discusses the concept of frequency bands and the bands defined for HSPA.) DC-HSDPA operating with carriers in different bands is referred to as *Dual-Band, Dual-Cell HSDPA* (DB-DC-HSDPA). Dual band capability is attractive for operators that operate with fragmented spectrum. Furthermore, in release 9 it was recognized that the uplink could also benefit from multi-carrier operation, and *Dual-Cell HSUPA* (DC-HSUPA) was introduced into the specifications.

After release 9, it was recognized that some operators would have access to even more than two carriers and it would be desirable to utilize these carriers for multi-carrier transmission. *Four-Carrier HSDPA* (4C-HSDPA) was therefore introduced. In release 10, the four carriers can be configured in one band or two bands. In a single band, the carriers need to be configured to be contiguous. After configuration, it is, however, possible to deactivate a carrier such that not all carriers are contiguous. 4C-HSDPA functionality was extended in release 11 to enable *non-contiguous* (NC) carriers to be allocated.

In release 11, emphasis was placed on meeting the requirements for IMT-Advanced (see Chapter 18). In order to provide sufficient spectrum, *Eight-Carrier HSDPA* (8C-HSDPA) was defined, in which up to eight simultaneous carriers can be configured for a UE from the same Node B.

Figure 12.1 depicts the evolution of multi-carrier HSDPA. An overview of the bit rates achieved with multi-carrier HSDPA is depicted in Figure 7.2 in Chapter 7.

### 12.1.1 BENEFITS OF MULTI-CARRIER HSDPA

Deploying multiple carriers within a single band or different bands at each base station site allows for users to be distributed across carriers, which can either improve user throughput for a given total load or improve capacity while still maintaining the desired minimum user throughput. Multiple carrier base stations can be operated with legacy users, provided that the users support the band to which they are allocated. Each legacy user can transmit and receive on only one 5 MHz carrier though.

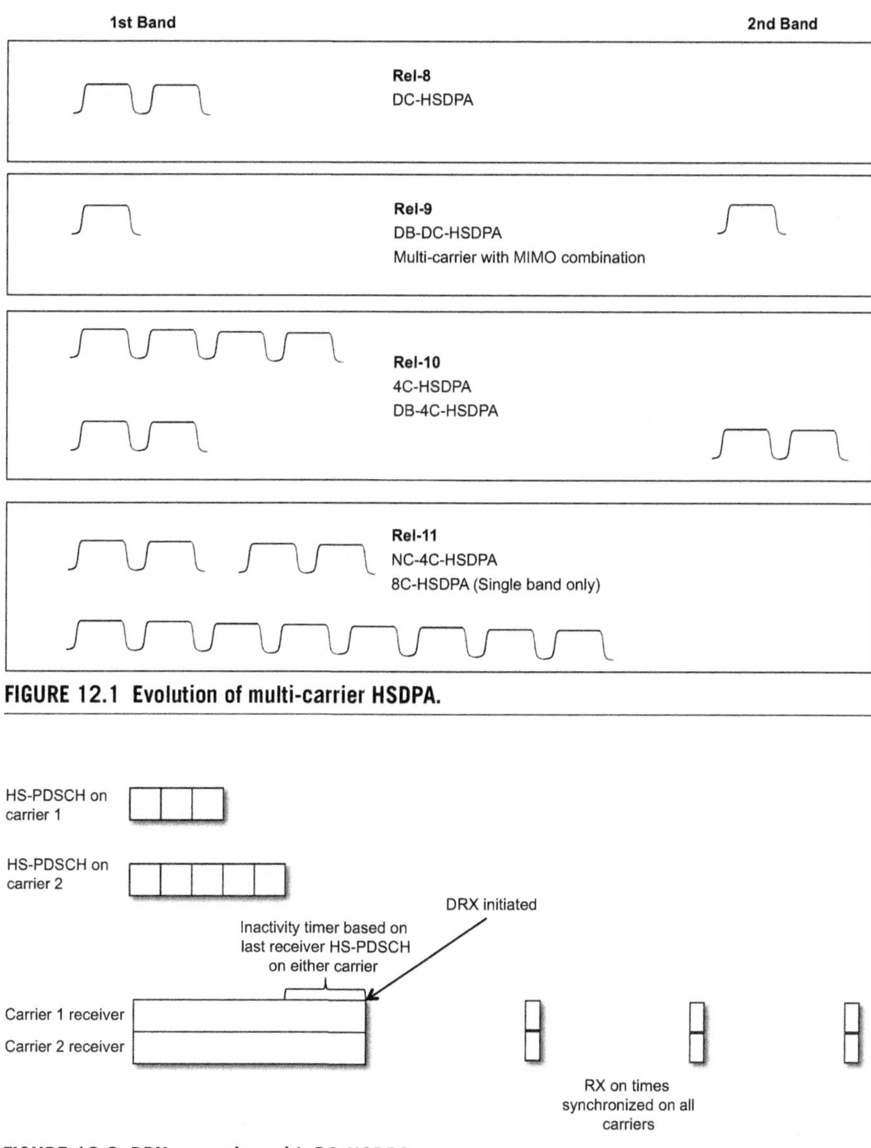


**FIGURE 12.1 Evolution of multi-carrier HSDPA.**

**FIGURE 12.2 DRX operation with DC-HSDPA.**

However, if a UE can transmit or receive on more than one carrier, then in bursty traffic conditions, when there is a burst in the buffer to transmit to a user, the Node B can allocate all carriers that are free during a scheduling TTI to the user. The average amount of carriers that are allocated will usually be larger than a single carrier, and thus users will in effect experience larger bandwidth and increased burst speeds. From a network perspective, the combination of freedom to schedule users on carriers that are free and the opportunity to schedule users on multiple carriers leads to significant trunking gains.

In addition, it is possible for a Node B to rapidly reallocate traffic between carriers. RNC-based inter-carrier handover is replaced by Node B scheduling. In effect, the basic step in release 5 of moving the management of radio resources from RNC to Node B is extended to moving the management of carriers to the Node B.

When the cell load is high and multiple UEs can be scheduled on multiple carriers, then it is possible for a Node B scheduler to perform carrier specific scheduling and schedule UEs on carriers that have an instantaneously better radio link. This extra dimension of freedom in scheduling can in particular bring advantages in flat fading channels.

In this manner, multi-carrier HSPA brings benefits both in terms of user experience and network capacity.

## 12.2 DOWNLINK MULTI-CARRIER HSDPA IN DETAIL

### 12.2.1 DOWNLINK PHYSICAL LAYER AND SCHEDULING OPERATION OF MULTI-CARRIER HSDPA

Similar basic principles apply to the operation of multi-carrier HSDPA, regardless of the number of carriers. Multi-carrier HSPA can be configured for UEs that are in the CELL_DCH state only. From a downlink physical layer perspective, the carriers are treated as completely independent. Thus, HS-SCCH is transmitted separately on each carrier. An HS-SCCH transmitted on a particular carrier carries scheduling information for that carrier only; cross-carrier scheduling is not possible. In release 5 HSDPA, a UE can be configured to monitor a maximum of four HS-SCCH codes. This is extended to a maximum of six HS-SCCH codes for UEs capable of DC-HSDPA across the two carriers, with a maximum of four HS-SCCH codes on any one carrier. For more than two cells, the maximum number of HS-SCCHs that a UE should monitor across all carriers is $3n$, where $n$ is the number of carriers that the UE is capable of receiving.

One of the carriers is named as the HSDPA serving cell (as defined in Chapter 8) and the other carriers secondary serving cells. The serving cell is always enabled and activated. The number of potentially enabled secondary serving cells is controlled via RRC signaling. HS-SCCH orders can be used to activate or deactivate secondary serving cells. Deactivating a secondary cell allows the UE to reduce its receiver bandwidth or deactivate a receive chain, which can reduce current consumption and extend battery life.

HS-SCCH less operation, as discussed in Chapter 14, can only be configured in the serving cell; it is not configurable in secondary serving cells. The Node B can send an HS-SCCH order on up to two carriers simultaneously.

All cells are transmitted from the same Node B and must be tightly time synchronized to within five chips if there is no MIMO or a half chip with MIMO configured. Configuration of a carrier as the serving cell is UE specific; thus, it is possible that a carrier that is a serving cell for one UE is a secondary cell for other UEs. It is possible to configure all UEs with the same serving cell, in which case it is strictly only necessary to transmit P-CPICH, HS-SCCH, and HS-PDSCH in the secondary serving cells; other control overhead such as P-CCPCH can be spared. Since the carriers are synchronized, synchronization information is not required either. However, it is usually desirable to configure different serving cells between different UEs, since this distributes the uplink signaling overhead between carriers.

The uplink signaling for all downlink carriers is carried on the uplink carrier that is duplexed with the serving downlink carrier.

Each cell generates independent CQI and ACK/NACK signaling. However, all uplink signaling is carried on a single uplink carrier (even in the case that DC-HSUPA is configured), and to accommodate independent uplink signaling relating to multiple carriers, HS-DPCCH design has been evolved, as described in Section 12.2.3.

The UE reads broadcast information from the serving cell and camps on the serving cell if it returns to CELL_FACH, CELL_PCH, or idle mode.

When there is no data to transfer to a UE, the UE may enter DRX if DRX is activated. DRX activation and DRX status are common across all active cells. This implies that if DRX is activated, carriers will enter DRX once the inactivity timer expires following reception of the last HS-PDSCH on any cell. During DRX, the time intervals at which the UE is expected to receive HS-SCCH are common across all of the cells. DRX is illustrated in Figure 12.2.

*Space–Time Transmit Diversity* (STTD), two-branch MIMO, and four-branch MIMO can be configured independently per cell. Release 8 UEs are only capable of STTD (or no STTD) operation when configured with dual-cells, and release 9 UEs are able to be configured with dual-stream MIMO (where the UE is capable) or STTD (or no STTD). In releases 10 and 11, capable UEs can be configured with four-branch MIMO on up to four cells. Note that configuration of more than four carriers with four-branch MIMO is not possible (although configuration of two-branch MIMO is).

## 12.2.2 HIGHER-LAYER OPERATION OF MULTI-CARRIER HSDPA

Figure 12.3 shows the MAC structure for multi-carrier HSDPA in the Node B and in the UE. In the Node B, separate hybrid-ARQ entities exist for each of the cells. It is possible for the scheduling to be separate for each of the cells; however, it is possible

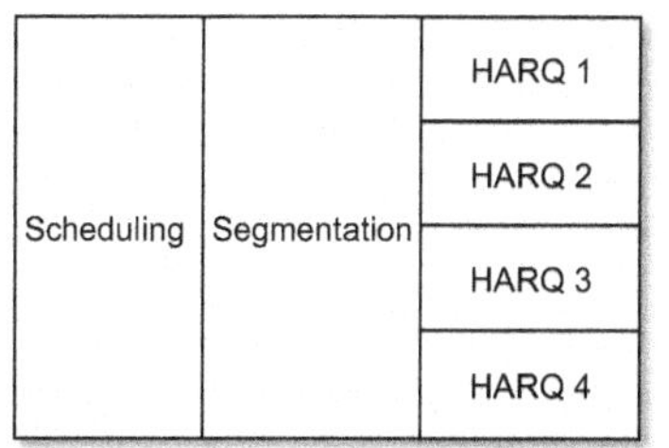


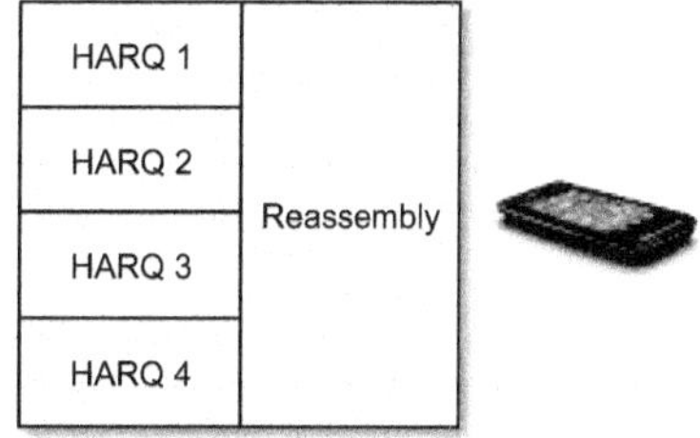


**FIGURE 12.3 Multi-carrier HSDPA MAC structure.**

and more beneficial to operate a combined scheduler for all cells, as indicated in the figure. Hybrid-ARQ processes are separated between cells and it is not possible to schedule a retransmission of a MAC PDU from one cell in a different cell. MAC segmentation (and reassembly in the UE) for a UE configured with multi-carrier is common across all the cells. Thus, the Node B is responsible for splitting the MAC data flow between carriers and scheduling transmission of the data in each of the cells. The multiple flows across multiple cells are reassembled by the UE. In order to accommodate the larger number of MAC PDUs across a larger number of hybrid-ARQ processes with multi-carrier, the length of the *Transmission Sequence Number* (TSN) used for identifying where MAC PDUs belong in the sequence of transmitted blocks is extended to 14 bits for multi-carrier, except for the case of DC-HSDPA operating without MIMO.

Mobility procedures such as measurement events are based on the primary carrier only. Basing mobility on the primary carrier enables operation of a secondary carrier without control channels, simplifies RRM operations for the UE, and resolves situations in which there could be mobility conflicts between different carriers by prioritizing the primary carrier. Radio link failure can be declared based on the primary carrier only.

### 12.2.3 UPLINK L1/L2 CONTROL SIGNALING FOR MULTI-CARRIER HSDPA

A key design area for each multi-carrier extension has been the HS-DPCCH feedback design. HSDPA-related feedback information is always sent on the primary uplink carrier irrespective of whether DC-HSUPA is configured. The more downlink carriers that are enabled, the more feedback information needs to be conveyed in the uplink. Consequently, the HS-DPCCH format needs to adapt to the configured multi-carrier scenario, resulting in a number of different HS-DPCCH formats. A description of the release 5 HS-DPCCH can be found in Section 8.3.9 while the extensions needed for multi-carrier HSPA are discussed below.

The main design objectives for the extensions to the HS-DPCCH feedback channel include the following:

- minimize the *peak-to-average-power-ratio* (PAPR) for power efficiency;
- minimize the standards impact by keeping original key design choices relating to, for example, the physical channel layout and power settings; and
- minimize potential ACK/NACK error cases and design ACK/NACK codebooks with a focus on the most important error cases that lead to RLC retransmissions.

The number of enabled and activated secondary serving cells determines how ACK/NACK and CQI/PCI feedback information is handled.

In the following, the enhancements introduced are described on a per-release basis.

#### *12.2.3.1 Release 8 DC-HSDPA*

Release 8 DC-HSUPA did not introduce any new slot formats to the HS-DPCCH; that is, SF256 with hybrid-ARQ acknowledgement transmitted in the first slot (10 channel bits) of the sub-frame and the CQI transmitted in the second and third slots (20 channel bits) of the sub-frame were retained. To accommodate the hybrid-ARQ feedback information for the two downlink carriers, a new ACK/NACK (A/N) codebook was designed. The A/N information for the two carriers is jointly encoded according to the codebook shown in Table 12.1. Six additional ACK/NACK codewords were introduced compared to release 5, and the release 5 codebook is a subset of the new codebook. The procedure for detecting and decoding the ACK/NACK information in the Node B is implementation specific. The Node B can, however, exploit information about the number of carriers that were scheduled in the detection process. For example, if the Node B has scheduled transport blocks on both downlink carriers, the Node B needs to consider all eight codewords in the decoding process, whereas if only one carrier was scheduled only two codewords are in principle of relevance. By limiting the considered code-space of the ACK/NACK codebook in the decoding process, the decoding and detection performance can be improved as compared to a case where the Node B receiver disregards the number of transport blocks that were scheduled, that is, always assumes the worst case.

**Table 12.1** ACK/NACK Codewords for DC-HSDPA (Release 8)

| **Hybrid-ARQ ACK Message to be Transmitted** | | $w_0$ | $w_1$ | $w_2$ | $w_3$ | $w_4$ | $w_5$ | $w_6$ | $w_7$ | $w_8$ | $w_9$ |
|---|---|---|---|---|---|---|---|---|---|---|---|
| **Hybrid-ARQ ACK in response to a single transport block on the serving HS-DSCH cell** | | | | | | | | | | | |
| ACK | | 1 | 1 | 1 | 1 | 1 | 1 | 1 | 1 | 1 | 1 |
| NACK | | 0 | 0 | 0 | 0 | 0 | 0 | 0 | 0 | 0 | 0 |
| **Hybrid-ARQ ACK in response to a single transport block on the secondary serving HS-DSCH cell** | | | | | | | | | | | |
| ACK | | 1 | 1 | 1 | 1 | 1 | 0 | 0 | 0 | 0 | 0 |
| NACK | | 0 | 0 | 0 | 0 | 0 | 1 | 1 | 1 | 1 | 1 |
| **Hybrid-ARQ ACK in response to two transport blocks, one for the serving HS-DSCH cell and one for the secondary serving HS-DSCH cell** | | | | | | | | | | | |
| **Response in Serving HS-DSCH Cell** | **Response in Secondary Serving HS-DSCH Cell** | | | | | | | | | | |
| ACK | ACK | 1 | 0 | 1 | 0 | 1 | 0 | 1 | 0 | 1 | 0 |
| ACK | NACK | 1 | 1 | 0 | 0 | 1 | 1 | 0 | 0 | 1 | 1 |
| NACK | ACK | 0 | 0 | 1 | 1 | 0 | 0 | 1 | 1 | 0 | 0 |
| NACK | NACK | 0 | 1 | 0 | 1 | 0 | 1 | 0 | 1 | 0 | 1 |
| **PRE/POST Indication** | | | | | | | | | | | |
| PRE | | 0 | 0 | 1 | 0 | 0 | 1 | 0 | 0 | 1 | 0 |
| POST | | 0 | 1 | 0 | 0 | 1 | 0 | 0 | 1 | 0 | 0 |

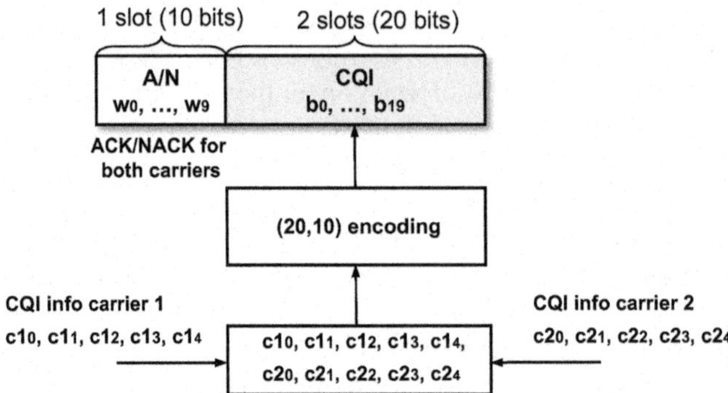


**FIGURE 12.4 HS-DPCCH overview for release 8.**

The CQI values for each carrier are jointly encoded into a single CQI report. The CQI information for each carrier consists of one out of 30 different CQI values mapped to five bits. The five bits per carrier are then combined into ten bits which are coded into 20 bits, according to Figure 12.4. The release 7 Type A CQI (20,10) encoding block code is used also for DC-HSDPA (see Section 11.3.3).

If the secondary serving cell is deactivated, then the ACK/NACK and CQI handling reverts to release 5 operation. Hence, there are only two ACK/NACK codewords and the CQI coding employs a (20,5) linear block code. The change in HS-DPCCH format takes place exactly 12 slots after the end of the HS-SCCH subframe delivering the order. The fact that the release 5 ACK/NACK codebook is a subset of the DC-HSDPA codebook provides some robustness against possible mismatches between the UE and the Node B. For example, if the UE has one carrier enabled, while the network has two, the ACK/NACK message for the serving cell can still be decoded.

The power setting of the HS-DPCCH is updated to account for the increased information in each sub-frame. The signaled Δ for dual-cell operation is essentially increased by one step in the quantization table compared to release 5 (see Table 12.2). No changes to the RRC signaled parameters $\Delta_{ACK}$, $\Delta_{NACK}$, or $\Delta_{CQI}$ were introduced for DC-HSDPA. Similarly, no changes to repetition for ACK/NACK or CQI were made. Since ACK/NACK and CQI information is encoded jointly for both the downlink carriers, the ACK/NACK and CQI repetition factors are common parameters that apply for both configured downlink carriers.

### *12.2.3.2 Release 9 DC-HSDPA with 2×2 MIMO*

Similarly to release 8, for DC-HSDPA with 2×2 MIMO no changes to the HS-DPCCH slot format were introduced. The ACK/NACK codebook was updated to support simultaneous acknowledgment of up to four transport blocks. In total, 48 codewords (excluding POST/PRE) are needed to cover all possible ACK/NACK messages. This is a significant increase compared to release 8 DC-HSDPA, and maintaining sufficiently good code properties (for example, minimum Hamming distance) without changing the slot format required some special considerations. One observation

**Table 12.2** Power Offsets for HS-DPCCH in DC-HSDPA (Release 8)

| Condition | $A_{hs}$ Equals the Quantized Amplitude Ratio Translated From | | | |
|---|---|---|---|---|
| | CQI Message | Hybrid-ARQ ACK Message Sent in One Time Slot | | |
| | | Contains at Least One ACK but no NACK | Contains at Least One NACK but no ACK | Contains Both ACK and NACK or is a PRE or a POST |
| Secondary serving HS-DSCH cell deactivated | $\Delta_{CQI}$ | $\Delta_{ACK}$ | $\Delta_{NACK}$ | $\max(\Delta_{ACK}, \Delta_{NACK})$ |
| Secondary serving HS-DSCH cell activated | $\Delta_{CQI} + 1$ | $\Delta_{ACK} + 1$ | $\Delta_{NACK} + 1$ | $\max(\Delta_{ACK} + 1, \Delta_{NACK} + 1)$ |

affecting the codebook design was that since the HS-SCCH is CRC protected, it is very likely that the correct information is obtained if the UE detects the HS-SCCH (CRC checks). Hence, the UE will either miss the HS-SCCH altogether or detect the correct number of transport blocks.

Four different codeword subsets are formed based on the number of scheduled/detected transport blocks on each carrier, and each subset is optimized in order to have good code properties. This is illustrated in Figure 12.5, where A, N, and D stand for ACK, NACK, and DTX, respectively. The notation AN/A means that two transport blocks (two streams) were received on carrier 1 (serving cell), where the first transport block was successfully decoded (ACK) and the secondary transport block unsuccessfully decoded (NACK), and one successfully decoded transport block was received on carrier 2 (secondary serving cell). A codeword containing DTX on one of the carriers needs to be included in two subsets since the UE cannot know if one or two streams were scheduled on a carrier for which the HS-SCCH was not detected. It can be seen that some codewords are reused and have different meanings in different groups. For example, the codewords representing A/N and A/NA are identical. The codebook contains the release 5 codebook as a subset but is in general incompatible with the release 8 DC-HSDPA codebook. For example, different codewords are used in releases 8 and 9 to describe a positive acknowledgement of one transport block per carrier (A/A).

The Node B ACK/NACK detector should use information about the number of scheduled transport blocks in the detection and decoding phase. More specifically, the Node B will choose different subsets of the codebook in the detection/decoding process depending on the number of transport blocks that were scheduled for each carrier (as illustrated in Figure 12.5).

To report CQI/PCI information for up to four streams (two per carrier) using in total 20 channel bits, the minimum feedback cycle was extended to two sub-frames

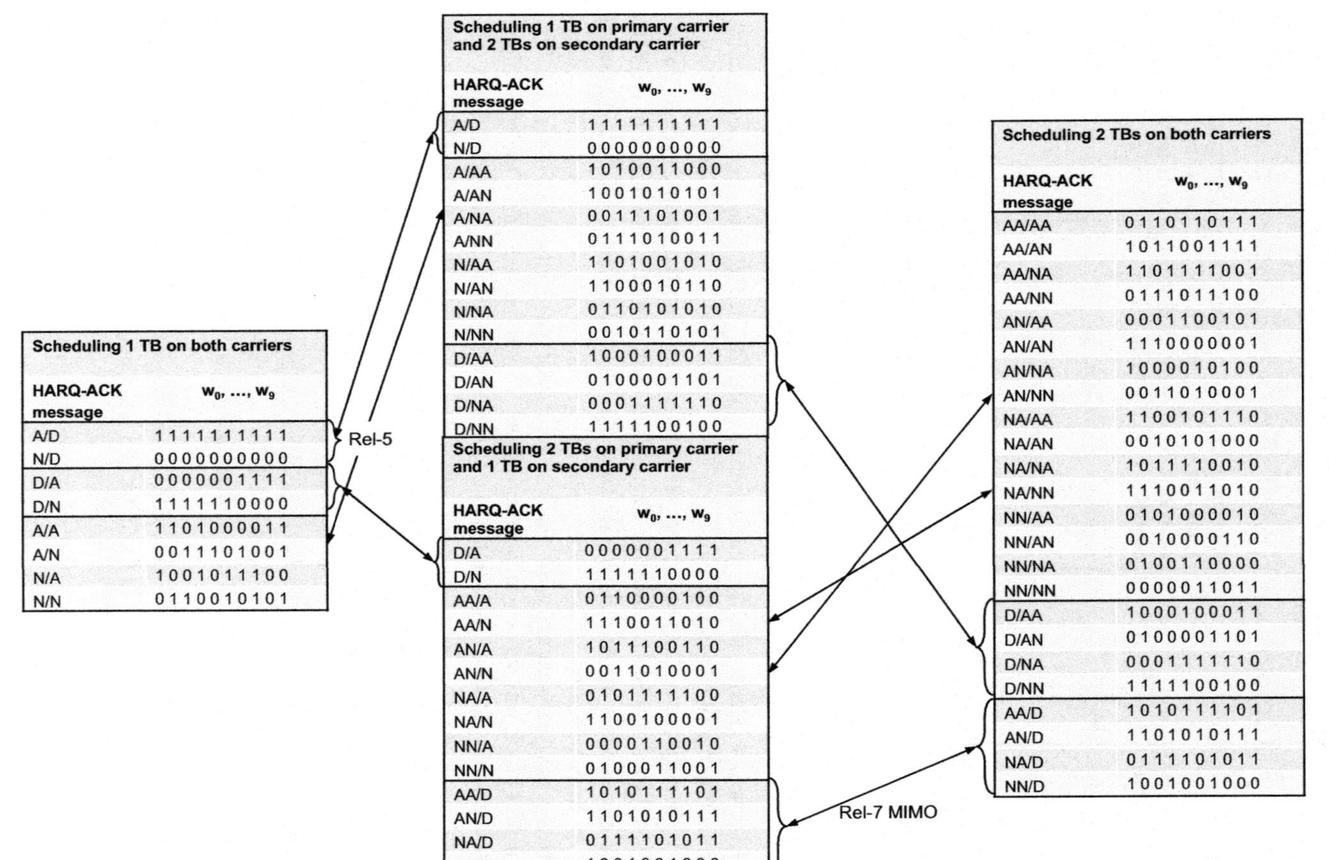


**FIGURE 12.5** Illustration of the ACK/NACK codebook and decoding constraints for DC-HSDPA with MIMO (release 9).

(4 ms). The CQI report corresponding to the serving cell is sent in the first sub-frame, and the CQI report associated with the secondary serving cell is sent in the second sub-frame (see Figure 12.6 for an illustration). Increasing the minimum CQI feedback cycle to 4 ms instead of 2 ms has, in general, only a very small performance impact. Unlike for DC-HSDPA, the CQI information for each carrier is independently encoded and carriers configured in SISO mode rely on the release 5 CQI encoding whereas carriers configured in MIMO mode utilize the release 7 Type A/Type B encoding of the CQI/PCI information.

Deactivation of the secondary serving cell reverts ACK/NACK handling to release 5 or release 7 depending on whether MIMO is configured in the serving cell. It can be seen in Figure 12.5 that the release 5 and release 7 codewords are added with some intelligence as a subset of the DC-HSDPA with MIMO codebook. The release 5 codewords correspond to A/D and N/D and the additional release 7 codewords correspond to AA/D, AN/D, NA/D, and NN/D, which makes the ACK/NACK decoding more robust against a potential mismatch between network and UE with respect to number of activated carriers. As for release 8, if the ACK/NACK repetition factor is larger than one, the UE repeats the ACK/NACK information field in *N_acknack*-1 consecutive sub-frames (during which the UE does not need to decode HS-SCCHs). Unlike release 8, however, if the CQI repetition factor is larger than one, the UE will first transmit the CQI information for the serving HS-DSCH cell (a total of *N_cqi_transmit* times) and then immediately transmit the CQI information for the secondary serving HS-DSCH cell. The impact of the ACK/NACK repetition factor and the CQI repetition factor is illustrated in Figure 12.7, where both repetition factors are set to two, and the CQI feedback cycle is 8 ms.

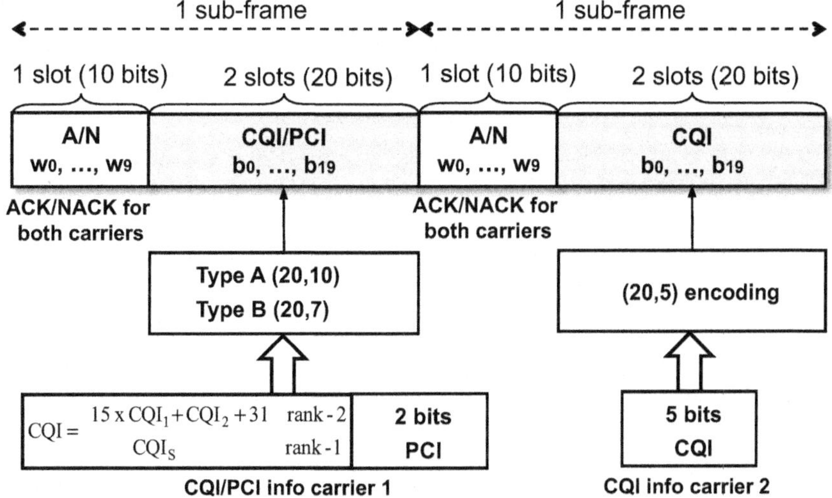


**FIGURE 12.6 Illustration of HS-DPCCH for DC-HSDPA with MIMO. In this example, the serving HS-DSCH cell is configured with 2×2 MIMO, while the secondary serving HS-DSCH cell is SIMO (non-MIMO) configured.**

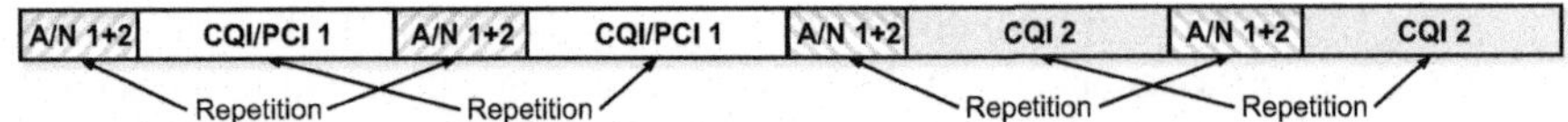


**FIGURE 12.7 Illustration of HS-DPCCH with ACK/NACK and CQI repetition factors equal to two. The serving cell (carrier 1) is configured with MIMO, whereas the secondary serving cell (carrier 2) is not configured with MIMO.**

The power offset is updated according to Table 12.3. One observation is that the power offset does not depend on the number of scheduled transport blocks. This is different from release 7 but consistent with release 8.

### *12.2.3.3 Release 10 four-carrier HSDPA*

Four-carrier HSDPA (4C-HSDPA) in combination with 2×2 MIMO allows a Node B to transmit up to eight transport blocks to a UE in a single sub-frame. Hence, the hybrid-ARQ ACK/NACK and CQI feedback information that needs to be transmitted on HS-DPCCH is doubled as compared to release 9. The increase in feedback information motivated a new HS-DPCCH slot format. Two main design alternatives were considered, 2xSF256 and SF128, and the latter was adopted because of superior PAPR characteristics. A slot format using SF128 is employed for most carrier combinations, but for the special case of three active carriers without MIMO, SF256 is used. Thus, two different HS-DPCCH slot formats can be used for 4C-HSDPA (see Table 12.4). Slot format 0 is used if the UE is configured according to release 5 to release 9 or if the UE is configured with three carriers without MIMO. If the UE is configured with three carriers whereof at least one is configured in MIMO mode or if the UE is configured with four downlink carriers, slot format 1 is used.

**Scenario I: Employing slot format 1**

The first scenario, in which slot format 1 with SF128 is used, consists of two cases:

- Three carriers are enabled, where at least one has MIMO configured.
- Four carriers are enabled.

Employing slot format 0 with SF128 results in twice as many channel bits per slot compared to previous releases (20 bits per slot). The ACK/NACK slot is divided into two subslots, each consisting of 10 bits. To simplify the ACK/NACK design, carriers are grouped two and two, and each group employs the release 9 ACK/NACK codebook. The serving cell and the first secondary serving cells constitute the first group and the ACK/NACK message for these carriers is encoded according to the release 9 codebook, resulting in 10 bits transmitted in the first subslot. Similarly, the second and third secondary serving cells constitute the second group, and the jointly encoded bits are transmitted in the second subslot. This procedure is illustrated in Figure 12.8.

Two key aspects for the ACK/NACK design were to ensure a sufficient detection/decoding performance irrespective of the number of scheduled carriers and to avoid having the UE change the transmit power (or even DTX) within a slot. For

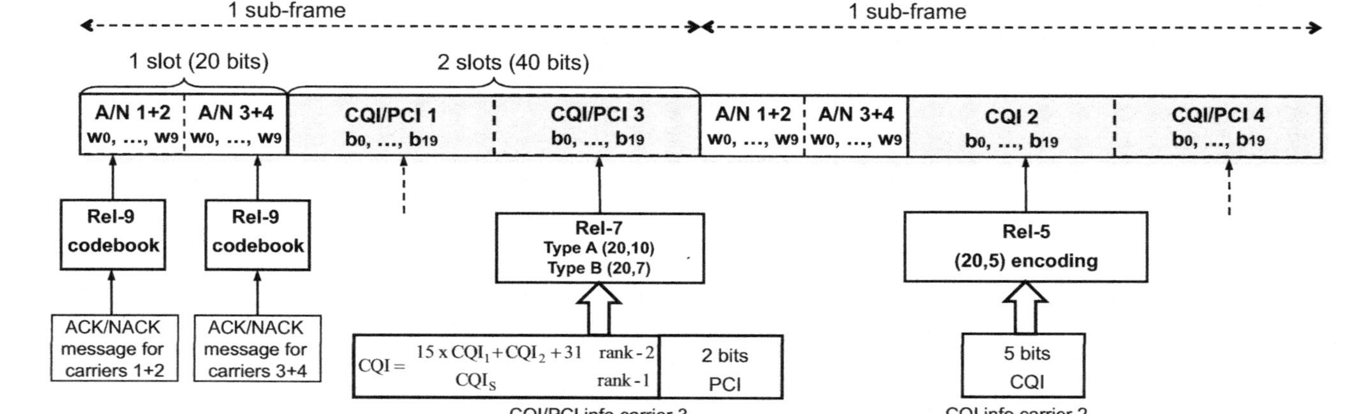


**FIGURE 12.8 Illustration of the HS-DPCCH for 4C-HSDPA (release 10). In this example, the serving cell (carrier 1) and the second and third secondary serving cells (carriers 3 and 4) have MIMO configured whereas the first secondary serving cell (carrier 2) does not have MIMO configured.**

**Table 12.3** Power Offsets for HS-DPCCH in DC-HSDPA with MIMO (Release 9)

| Condition | $A_{hs}$ Equals the Quantized Amplitude Ratio Translated From | | | | | |
|---|---|---|---|---|---|---|
| | CQI Message | | | HARQ-ACK message sent in one time slot | | |
| | UE not Configured in MIMO Mode in a Cell | UE Configured in MIMO Mode in a Cell | | Contains at Least One ACK but no NACK | Contains at Least One NACK but no ACK | Contains Both ACK and NACK or a PRE or POST |
| | | Type A | Type B | | | |
| Secondary serving HS-DSCH cell deactivated | $\Delta_{CQI}$ | $\Delta_{CQI} + 1$ | $\Delta_{CQI}$ | $\Delta_{ACK}$ | $\Delta_{NACK}$ | $\max(\Delta_{ACK}, \Delta_{NACK})$ |
| Secondary serving HS-DSCH cell activated | $\Delta_{CQI}$ | $\Delta_{CQI} + 1$ | $\Delta_{CQI}$ | $\Delta_{ACK} + 1$ | $\Delta_{NACK} + 1$ | $\max(\Delta_{ACK} + 1, \Delta_{NACK} + 1)$ |

**Table 12.4** Channelization Code Used for Transmitting the HS-DPCCH

| | Channelization Code $c_{hs}$ | |
|---|---|---|
| $N_{max\text{-}dpdch}$ | HS-DPCCH Slot Format #0 | HS-DPCCH Slot Format #1 |
| 0 | $C_{ch,256,33}$ | $C_{ch,128,16}$ |
| 1 | $C_{ch,256,64}$ | $C_{ch,128,32}$ |
| 2, 4, 6 | $C_{ch,256,1}$ | N/A |
| 3, 5 | $C_{ch,256,32}$ | N/A |

*Note:* $N_{max\text{-}dpdch}$ *denotes the actual number of configured DPDCHs and* $C_{ch,SF,n}$ *denotes the nth channelization code with spreading factor SF.*

**Table 12.5** DTX Codeword Introduced in Release 10

| Hybrid-ARQ ACK Message to be Transmitted | $w_0$ | $w_1$ | $w_2$ | $w_3$ | $w_4$ | $w_5$ | $w_6$ | $w_7$ | $w_8$ | $w_9$ |
|---|---|---|---|---|---|---|---|---|---|---|
| D/D | 0 | 0 | 1 | 1 | 0 | 1 | 1 | 0 | 1 | 0 |

example, assume that all four carriers are active, but the Node B schedules only on the serving cell and the first secondary serving cell. Hence, the second ACK/NACK subslot would be empty, resulting in an abrupt power change within a slot, which complicates transmitter and receiver design. Furthermore, an empty subslot would essentially mean 3 dB less detection energy for the receiver processing, which would significantly reduce the detection performance. To avoid this situation, a DTX codeword was introduced (see Table 12.5) that is transmitted whenever at least one carrier within an ACK/NACK group is active but has no ACK/NACK information to send (both carriers are DTXed). The DTX codeword should, however, not be confused with a "true" DTX (nothing is sent), which is used if no HS-SCCH associated with the active carriers is detected. The receiver will typically use the whole ACK/NACK slot in the detection process in order to accumulate as much energy as possible, but the decoding is done independently per subslot, that is, independently for the two ACK/NACK groups.

CQI/PCI information is coded independently for each active carrier and uses the release 5 or release 7 Type A/Type B format depending on whether MIMO is configured on the carrier. As for release 9, the minimum CQI feedback cycle is 4 ms and all CQI-related parameters (for example, ACK/NACK and CQI repetition factors) are common for all enabled carriers and independent of the carriers' activation status. The order in which the CQI fields corresponding to the different downlink carriers are transmitted is not the same as for the hybrid-ARQ ACK transmissions. This is illustrated in Figure 12.8, where it can be seen that CQI/PCI information for the serving HS-DSCH cell (carrier 1) is followed by the CQI/PCI information for the second secondary serving HS-DSCH cell (carrier 3). This order was adopted to minimize the remapping needed if the second and third secondary serving cells (carriers 3 and 4) are deactivated.

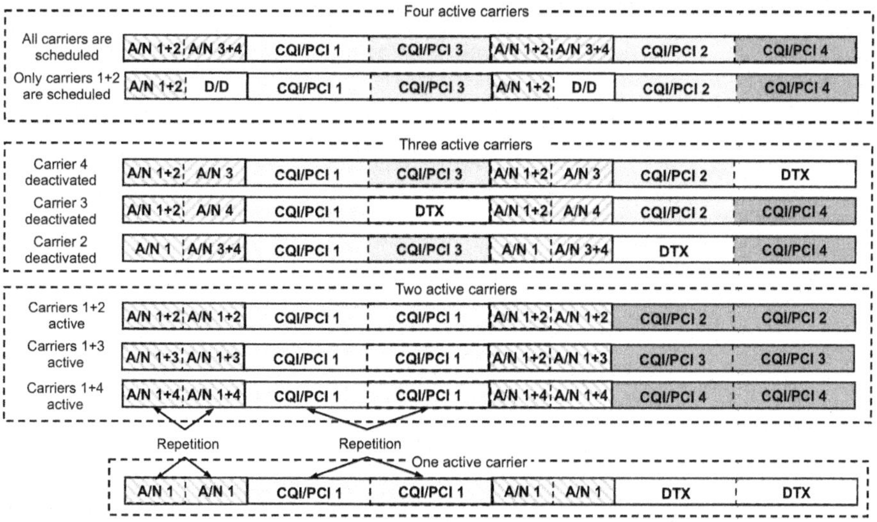

**FIGURE 12.9 Illustration of ACK/NACK and CQI/PCI information mapping for different configurations of number of active carriers.**

Needless to say, the number of possible configurations, which depends on the permutations of secondary serving cells' activation statuses, increases compared to previous releases. This causes some additional complexity. The ACK/NACK coding always uses the release 9 codebook, irrespective of the number of active carriers. The following rules apply (see Figure 12.9 for an illustration):

- Three active carriers: The CQI field corresponding to the deactivated carrier is DTXed.
- Two active carriers: Repetition of the information fields is employed.
- One active carrier: Repetition of the information fields is used, and the second CQI sub-frame is DTXed.

The HS-DPCCH power offset parameters are shown in Table 12.6. Compared to previous releases, the power has been increased for some cases in order to compensate for the reduction in spreading factor to SF128. A new HS-DPCCH amplitude ratio $A_{hs}$ has been added with the value 48/15 since in some cases the RLC signaled $\Delta$ value is increased by 2. In total, 10 different amplitude ratios are defined for release 10, but the allowed range for the signaled $\Delta$ is still 0 to 8.

**Scenario II: Employing slot format 0**

The second scenario, in which slot format 0 with SF256 is used, consists of the case of two enabled secondary serving cells (three carriers) without MIMO configured. This scenario was seen as an important special case where an optimized design was justified. In total 26 ACK/NACK codewords (excluding POST/PRE) are required, which can be compared with the maximum 24 for release 9 with four streams scheduled. The release 9 HS-DPCCH design is based on SF256, which has

**Table 12.6** Power Offsets for HS-DPCCH in Scenario I

| Number of Active Secondary Serving Cells | $A_{hs}$ Equals the Quantized Amplitude Ratio Translated From | | | | | |
|---|---|---|---|---|---|---|
| | CQI Message | | | Composite Hybrid-ARQ ACK Message Sent in One Time Slot | | |
| | UE not Configured in MIMO Mode in a Cell | UE Configured in MIMO Mode in a Cell | | Contains at Least One ACK but no NACK | Contains at Least One NACK but no ACK | Contains Both ACK and NACK or a PRE or POST |
| | | Type A | Type B | | | |
| 0 | N/A | $\Delta_{CQI} + 1$ | $\Delta_{CQI}$ | $\Delta_{ACK} + 1$ | $\Delta_{NACK} + 1$ | $\max(\Delta_{ACK} + 1, \Delta_{NACK} + 1)$ |
| 1 | $\Delta_{CQI}$ | $\Delta_{CQI} + 1$ | $\Delta_{CQI}$ | $\Delta_{ACK} + 1$ | $\Delta_{NACK} + 1$ | $\max(\Delta_{ACK} + 1, \Delta_{NACK} + 1)$ |
| 2 | $\Delta_{CQI} + 1_{I}$ | $\Delta_{CQI} + 2$ | $\Delta_{CQI} + 1$ | $\Delta_{ACK} + 2$ | $\Delta_{NACK} + 2$ | $\max(\Delta_{ACK} + 2, \Delta_{NACK} + 2)$ |
| 3 | $\Delta_{CQI} + 1$ | $\Delta_{CQI} + 2$ | $\Delta_{CQI} + 1$ | $\Delta_{ACK} + 2$ | $\Delta_{NACK} + 2$ | $\max(\Delta_{ACK} + 2, \Delta_{NACK} + 2)$ |

better PAPR properties compared to SF128, and therefore SF256 was also adopted for Scenario II.

A new ACK/NACK codebook was introduced for Scenario II (see Table 12.7). The notations A, N, and D are used to represent ACK, NACK, and DTX, respectively, and the notation A/N/D means that the hybrid-ARQ messages for carrier 1, carrier 2, and carrier 3 are ACK, NACK, and DTX, respectively. The codebook is used irrespective of the number of activated carriers. The Node B receiver can, however, use a subset of all codewords when carriers are deactivated in order to enhance the detection/decoding performance.

CQI coding is based on a combination of releases 5 and 8 principles. More specifically, the serving cell is encoded based on the release 5 format, whereas the first and second secondary serving cells are jointly encoded using the release 8 format. ACK/NACK encoding is always based on the new Scenario II codebook irrespective of the number of active carriers. When deactivating one of the secondary serving cells, the CQI encoding for the remaining secondary serving cell employs the release 5 format. If both secondary serving cells are deactivated, the second CQI sub-frame is DTXed. See Figure 12.10 for an illustration.

The power setting for scenario II is shown in Table 12.8. It can be seen that release 8 CQI encoding uses 2 dB (+1) more power than the release 5 CQI encoding.

#### *12.2.3.4 Release 11 eight-carrier HSDPA*

The eight-carrier HSDPA (8C-HSDPA) design needs to support ACK/NACK and CQI feedback information for up to 16 streams (eight carriers with 2×2 MIMO). In order to achieve this, two HS-DPCCHs employing slot format 1 (SF128) are used, where HS-DPCCH$_1$ is mapped to the Q-branch and HS-DPCCH$_2$ is mapped to the

**Table 12.7** ACK/NACK Codebook for Three Carriers Without MIMO Configured (Release 10)

| Hybrid-ARQ ACK Message to be Transmitted | $w_0$ | $w_1$ | $w_2$ | $w_3$ | $w_4$ | $w_5$ | $w_6$ | $w_7$ | $w_8$ | $w_9$ |
|---|---|---|---|---|---|---|---|---|---|---|
| A/D/D | 1 | 1 | 1 | 1 | 1 | 1 | 1 | 1 | 1 | 1 |
| N/D/D | 0 | 0 | 0 | 0 | 0 | 0 | 0 | 0 | 0 | 0 |
| D/A/D | 1 | 1 | 1 | 1 | 1 | 0 | 0 | 0 | 0 | 0 |
| D/N/D | 0 | 0 | 0 | 0 | 0 | 1 | 1 | 1 | 1 | 1 |
| D/D/A | 1 | 1 | 0 | 0 | 0 | 1 | 1 | 0 | 0 | 0 |
| D/D/N | 0 | 0 | 1 | 1 | 1 | 0 | 0 | 1 | 1 | 1 |
| A/A/D | 1 | 0 | 1 | 0 | 1 | 0 | 1 | 0 | 1 | 0 |
| A/N/D | 1 | 1 | 0 | 0 | 1 | 1 | 0 | 0 | 1 | 1 |
| N/A/D | 0 | 0 | 1 | 1 | 0 | 0 | 1 | 1 | 0 | 0 |
| N/N/D | 0 | 1 | 0 | 1 | 0 | 1 | 0 | 1 | 0 | 1 |
| A/D/A | 1 | 0 | 1 | 1 | 0 | 1 | 1 | 0 | 0 | 1 |
| A/D/N | 0 | 1 | 0 | 1 | 1 | 0 | 1 | 0 | 0 | 1 |
| N/D/A | 0 | 0 | 0 | 1 | 1 | 1 | 1 | 0 | 1 | 0 |
| N/D/N | 1 | 0 | 0 | 1 | 1 | 1 | 0 | 1 | 0 | 0 |
| D/A/A | 0 | 1 | 1 | 1 | 0 | 1 | 0 | 0 | 1 | 0 |
| D/A/N | 1 | 0 | 1 | 0 | 0 | 1 | 0 | 1 | 1 | 0 |
| D/N/A | 0 | 1 | 1 | 0 | 0 | 0 | 1 | 0 | 1 | 1 |
| D/N/N | 0 | 0 | 0 | 0 | 1 | 0 | 1 | 0 | 1 | 1 |
| A/A/A | 1 | 1 | 0 | 1 | 0 | 0 | 1 | 1 | 1 | 0 |
| A/A/N | 0 | 1 | 1 | 0 | 1 | 1 | 1 | 1 | 0 | 0 |
| A/N/A | 1 | 0 | 0 | 1 | 0 | 0 | 0 | 0 | 1 | 1 |
| A/N/N | 0 | 0 | 1 | 0 | 1 | 1 | 0 | 0 | 0 | 1 |
| N/A/A | 1 | 1 | 1 | 0 | 0 | 0 | 0 | 1 | 0 | 1 |
| N/A/N | 0 | 1 | 0 | 0 | 1 | 0 | 0 | 1 | 1 | 0 |
| N/N/A | 1 | 0 | 0 | 0 | 1 | 0 | 1 | 1 | 0 | 1 |
| N/N/N | 1 | 1 | 1 | 1 | 0 | 1 | 0 | 1 | 0 | 0 |

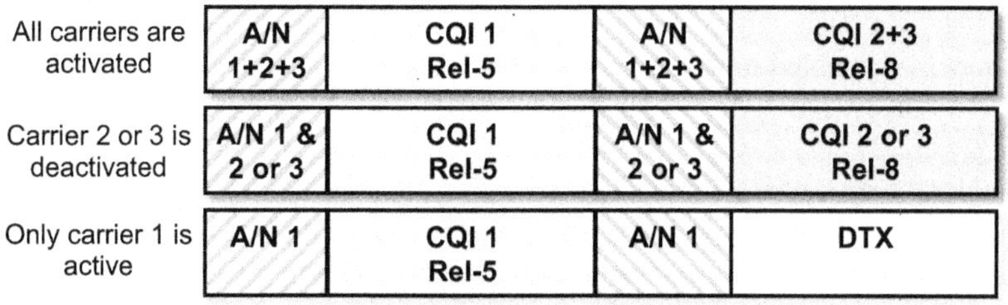

**FIGURE 12.10** Illustration of ACK/NACK and CQI/PCI information mapping for different configurations of number of active carriers.

**Table 12.8** Power Offsets for HS-DPCCH in Scenario I

| Number of Active Secondary Serving Cells | $A_{hs}$ Equals the Quantized Amplitude Ratio Translated From | | | | | |
|---|---|---|---|---|---|---|
| | CQI Message | | | Composite Hybrid-ARQ ACK Message Sent in One Time Slot | | |
| | UE not Configured in MIMO Mode in a Cell | UE Configured in MIMO Mode in a Cell | | Contains at Least One ACK but no NACK | Contains at Least One NACK but no ACK | Contains Both ACK and NACK or a PRE or POST |
| | | Type A | Type B | | | |
| 0 | N/A | $\Delta_{CQI}+1$ | $\Delta_{CQI}$ | $\Delta_{ACK}+1$ | $\Delta_{NACK}+1$ | $\max(\Delta_{ACK}+1, \Delta_{NACK}+1)$ |
| 1 | $\Delta_{CQI}$ | $\Delta_{CQI}+1$ | $\Delta_{CQI}$ | $\Delta_{ACK}+1$ | $\Delta_{NACK}+1$ | $\max(\Delta_{ACK}+1, \Delta_{NACK}+1)$ |
| 2 | $\Delta_{CQI}+1$ | $\Delta_{CQI}+2$ | $\Delta_{CQI}+1$ | $\Delta_{ACK}+2$ | $\Delta_{NACK}+2$ | $\max(\Delta_{ACK}+2, \Delta_{NACK}+2)$ |
| 3 | $\Delta_{CQI}+1$ | $\Delta_{CQI}+2$ | $\Delta_{CQI}+1$ | $\Delta_{ACK}+2$ | $\Delta_{NACK}+2$ | $\max(\Delta_{ACK}+2, \Delta_{NACK}+2)$ |

I-branch. The hybrid-ARQ ACK/NACK and CQI information associated with the serving HS-DSCH cell, and the first, second, and third secondary serving HS-DSCH cells are always transmitted on HS-DPCCH$_1$. The hybrid-ARQ ACK/NACK and CQI information associated with the fourth, fifth, sixth, and seventh secondary serving HS-DSCH cells are transmitted on HS-DPCCH$_2$. For each HS-DPCCH, ACK/NACK and CQI information is handled and remapped according to release 10 principles. No remapping of ACK/NACK or CQI information is performed between the two HS-DPCCHs. Repetition factors (ACK/NACK and CQI) and the CQI feedback cycle are common for all configured downlink carriers.

The release 9/10 ACK/NACK codebook is used to encode the ACK/NACK information according to release 10 principles. Note, though, that the special case of three carriers without MIMO is not applicable for 8C-HSDPA (that configuration is only applicable when secondary_cell_enabled equals two and no carrier is configured with MIMO). The release 10 principles are also used for CQI/PCI encoding. In particular, the minimum CQI feedback cycle is 4 ms and CQI/PCI encoding is independent for each carrier. An illustration of the HS-DPCCH configuration when the number of enabled and active carriers coincides is shown in Figure 12.11. As stated previously, handling of deactivated carriers obeys release 10 principles per HS-DPCCH.

One set of $\Delta_{ACK}$, $\Delta_{NACK}$, and $\Delta_{CQI}$ values is signaled to the UE and is used for both HS-DPCCHs. The power setting for the two HS-DPCCHs follows release 10 principles and is done independently. Thus, when selecting the power offsets to use in a certain sub-frame, the UE shall consider the number of activated carriers on HS-DPCCH$_1$ and HS-DPCCH$_2$ independently. Existing release 10 RRC signaling for conveying the $\Delta_{ACK}$, $\Delta_{NACK}$, and $\Delta_{CQI}$ can be reused without modifications. The power

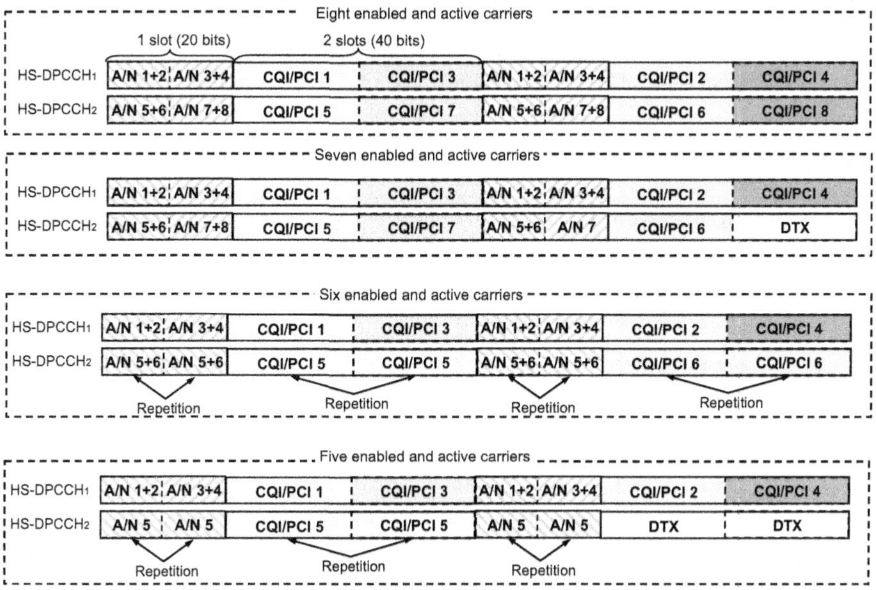

FIGURE 12.11 Illustration of ACK/NACK and CQI/PCI information mapping for different configurations of number of active carriers.

offsets used on the two HS-DPCCHs may be different as a result of different amounts of repetition of the ACK/NACK or CQI fields.

Some special considerations for 8C-HSDPA include the following:

- If the fourth, fifth, sixth, and seventh secondary serving HS-DSCH cells are deactivated, then $HS\text{-}DPCCH_2$ is not transmitted.
- The PRE/POST procedure is updated so that PRE/POST is sent on both HS-DPCCHs.
- If Secondary_Cell_Active is larger than three and the UE does not detect HS-SCCH for any downlink carrier associated with one of the HS-DPCCHs, and at the same time at least one HS-SCCH is detected for a carrier associated with the other HS-DPCCH, then the UE shall repeat the DTX codeword in the hybrid-ARQ acknowledgment field of the HS-DPCCH for which it did not detect any HS-SCCH transmissions.

## 12.2.4 FREQUENCY BANDS FOR MULTI-CARRIER HSDPA

The core specifications for multi-carrier HSDPA are agnostic to the frequency bands and carriers for which multiple cells are configured, with the exception that in release 10, non-contiguous carriers for 4C-HSDPA cannot be configured, and in release 11, non-contiguous carriers for 8C-HSDPA cannot be configured. The fact that the core specifications enable any carrier or band combination does not imply that an arbitrary combination is supported by the 3GPP specifications, since it is necessary for the RF

**Table 12.9** Band Combinations for Multi-carrier HSDPA

| | Aggregation Type | Band Combination | Release 8 | Release 9 | Release 10 | Release 11 |
|---|---|---|---|---|---|---|
| DC-HSDPA | Single band | Bands 1–14 | X | X | X | X |
| | Single band | Bands 15–21 | | X | X | X |
| | Single band | Bands 22–25 | | | X | X |
| | Single band | Band 26 | | | | X |
| | Dual band | Bands 1 + 8, 2 + 4, 1 + 5 | | X | X | X |
| | Dual band | Bands 2 + 11, 2 + 5 | | | X | X |
| 4C-HSDPA | Dual band | Band 1 (3 carrier) + band 8 (1 carrier) | | | X | X |
| | Dual band | Bands 2 + 4, 1 + 5[1] | | | X | X |
| | Single band, contiguous | Band 2 (3 or 4 contiguous carriers) | | | | X |
| | Dual band | Band 2 (1 carrier) + band 5 (2 contiguous) | | | | X |
| | Single band, non-contiguous | Band 1[2] Band 5[3] | | | | X |
| 8C-HSDPA | Single band, contiguous | Band 1 | | | | X |

*[1]Any configuration of one or two contiguous carriers in each band.*
*[2]Two sub-blocks can be configured in Band 1; one sub-block with up to three contiguous carriers and the other with max one carrier.*
*[3]Two sub-blocks can be configured in Band 4; each can have one or two contiguous carriers.*

conformance requirements to be updated with values specifically relevant to each combination. Thus, in practice the configurable carriers for multi-carrier operation are restricted, as illustrated in Table 12.9.

Other combinations of carriers and bands are not covered by the 3GPP specifications. Further carrier combinations are likely to be standardized as the need arises based on specific operator scenarios.

## 12.3 UPLINK DC-HSUPA IN DETAIL

Dual-cell HSUPA was introduced in release 9. Since then, there has been no further evolution of DC-HSUPA to include non-contiguous carriers, more than two carriers, etc. DC-HSUPA is only supported together with 2 ms TTI operation on each carrier.

### 12.3.1 PHYSICAL LAYER AND SCHEDULING OPERATION OF DC-HSUPA

Similarly to DC-HSDPA, one of the carriers is termed the serving E-DCH cell while the other carrier is the secondary cell. The serving uplink carrier must correspond to the FDD duplex of the serving downlink carrier. The secondary uplink cell can be deactivated by means of an HS-SCCH order. Deactivation not only reduces UE battery consumption but also reduces the overhead on the network, since no uplink DPCCH is then transmitted in the secondary cell. It is not necessary to deactivate the downlink cell that is associated with the secondary uplink cell when the secondary uplink cell is deactivated. However, if the FDD duplexed downlink cell is deactivated, then the secondary uplink cell must be deactivated; the secondary uplink cell cannot operate as a standalone. When the secondary uplink cell is deactivated, its serving grant is lost; thus, when a secondary cell is activated or reactivated it always starts with the initial serving grant as configured by higher layers. When the UE enters the CELL_DCH state with two uplink cells configured, the secondary cell is initially not activated and must be activated by the serving Node B.

HS-DPCCH is configured on the serving uplink cell only; it is not transmitted on the secondary cell. Furthermore, non-scheduled transmissions are only possible on the serving uplink cell. However, a minimum E-TFC set can be configured independently by higher layers for each cell.

The uplink cells are treated as independent, apart from the fact that they must share the maximum transmit power of the UE. All downlink signaling relating to an uplink cell (for example, F-DPCH, E-AGCH, E-HICH, E-RGCH) is carried on the FDD duplexed downlink cell.

Soft handover can be configured and operates independently on each carrier. It is possible to configure the active set sizes to be different for each carrier. This is useful, for example, in a scenario in which a UE is being served by a hotspot that operates two carriers, while the surrounding macro network operates on a single carrier only. Scheduling information is calculated independently for each cell.

DTX operation can be configured but must be set as on or off jointly for both of the cells. There is an underlying assumption that the UE uses a single wider bandwidth power amplifier to transmit both cells, and enabling DTX for one cell but not the other would not be a sensible configuration as it would not allow for the power amplifier to be switched off during DTX periods. However, the DTX status can differ between cells. Thus, if one cell is not scheduled and the inactivity timer expires since its last transmission, it may enter DTX even if transmission is continuing on the other cell in order to reduce DPCCH overhead. DTX operation is illustrated in Figure 12.12.

If compressed mode is configured, the compressed mode pattern is the same between the primary and secondary cells, since it is necessary during the compressed mode period to be able to switch off the transmitter completely.

As discussed in Chapter 9, scheduling of system resources between UEs is performed by the Node B, while scheduling of which MAC PDUs to transmit is performed by the UE E-TFC selection. The Node B scheduling is implementation

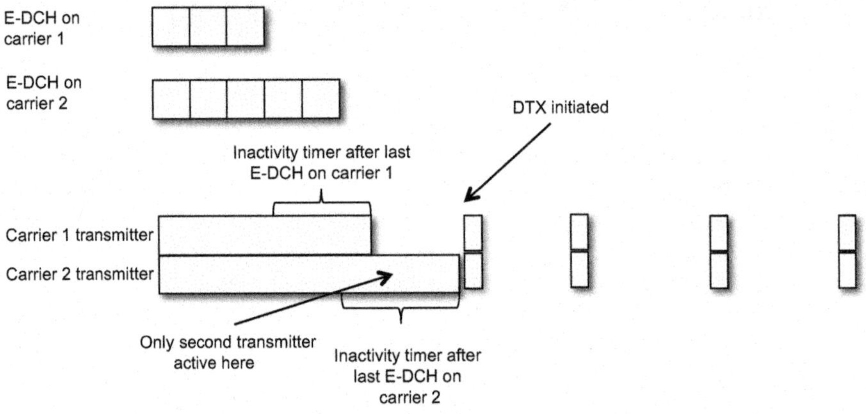


**FIGURE 12.12 Multi-carrier HSUPA DTX operation.**

specific, and it is entirely possible for Node Bs to schedule on uplink carriers independently, without considering that UEs are DC-HSUPA capable. Alternatively, however, joint scheduling can be performed across carriers. Joint scheduling is likely to be particularly important when there is a mixture of DC-HSUPA UEs and non-DC-HSUPA-capable UEs in order to ensure fairness of resource allocation. Scheduling allocations are signaled to the UEs by means of E-AGCH and E-RGCH on each associated downlink carrier.

E-TFC selection in the UE must take into account the fact that the two cells must share a common maximum UE transmit power. It is possible that the scheduler allocations on either or both of the uplink cells could cause the UE to require more than its maximum transmit power, in which case power must be allocated between the cells. In order to allocate power between the cells, the specification requires the UE to use the following algorithm. First, the UE estimates the power required for transmitting any non-scheduled MAC-d flows and removes this from the total power available for E-DCH transmission. Then, the remaining power is allocated in relation to the DPCCH levels and serving grant levels, as illustrated in Figure 12.13.

Thus, a cell with a higher DPCCH level is prioritized in allocating power. This is necessary because the E-DPDCH power is an offset to DPCCH; thus, a cell with a higher DPCCH level will require a greater amount of power to reach its serving grant.

After the relative levels of available power have been calculated, E-TFC selection continues individually on each cell in the same manner, as described in Chapter 9. Allocation is sequential; if the UE is buffer limited then first an E-TFC will be allocated for the secondary cell and then any remainder to the primary cell. E-TFCs that cannot be supported with the given power allocation to a cell are considered blocked for that cell. The UE prioritizes hybrid-ARQ transmissions and MAC-d flows as in the case of single-carrier transmission.

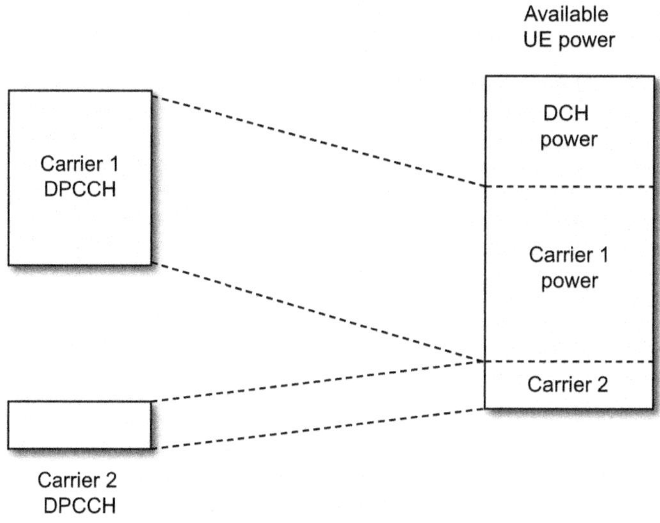


**FIGURE 12.13 DC-HSUPA power allocation.**

The happy bit is calculated individually on each cell considering the amount of power that is available for E-TFC selection on that cell. Thus, if there is excess available power on the cell when the maximum allowable E-TFC is transmitted and if the UE could continue to transmit for several more milliseconds considering the amount of data in its buffer and the combined data rate on both of the cells, the happy bit is set to "not happy" on that cell; otherwise it is set to happy.

It is possible that because of fast fading in the channel and the inner loop power control, following E-TFC selection and during transmission of the E-TFCs on the two cells the UE may reach its maximum allowed transmit power, in which case it will be unable to increase its transmit power when receiving TPC UP commands. Similarly to the single-carrier case, rules are defined on how the UE is to behave in such a situation, in order to provide a consistent UE behavior such that the outer loop power control algorithm implemented in the network does not function incorrectly because of an unpredictable UE behavior. When the maximum UE transmit power is reached with two cell transmission, the UE applies the following steps:

- The DPCCH levels on both carriers are adjusted independently according to the received TPC commands.
- If there is insufficient power, the E-DPDCH on the cell with the highest DPCCH level is scaled down. The E-DPDCH may be scaled down until either the required power no longer exceeds the maximum transmit power or the DPCCH power reaches a minimum level $\beta_{ed,k,reduced,min}$. The minimum level is configurable for each carrier by higher-layers.
- If the E-DPDCH power on the higher power cell has been scaled down to its minimum value and the required power still exceeds the UE's maximum

transmit power, then the E-DPDCH power is scaled down on the other cell until either the TX power is not greater than the UE's maximum or the configured $\beta_{ed,k,reduced,min}$ for the other cell is reached.

- If the E-DPDCH power on both cells has been scaled down to the minimum level and the required TX power is still greater than the UE's maximum power, then the power on all channels on both carriers is scaled down equally.

The rationale behind these steps is as follows: first, it is better to reduce E-DPDCH power and preserve power on DPCCH and the signaling channels E-DPCCH and HS-DPCCH since E-DPDCH is subject to hybrid-ARQ and thus there is some chance that even though E-DPDCH was transmitted with a reduced power, it can still be received successfully after combining of multiple hybrid-ARQ transmissions. It is preferable to start by scaling down E-DPDCH on the cell with the highest DPCCH level because it is likely to be this cell that is causing the excess power requirement, and scaling down E-DPDCH/DPCCH where DPCCH is at a high level will have a larger impact on the total TX power than scaling the other cell, while the impact on receiver Rx SINR of scaling down on either of the two cells will be the same.

### 12.3.2 HIGHER-LAYER OPERATION OF DC-HSUPA

Figure 12.14 shows the MAC structure for DC-HSUPA. When operating with HSUPA, only the flexible L2 MAC i/is can be utilized. In the Node B, MAC is untouched as a logical entity, although it is possible within a Node B implementation to perform joint scheduling across multiple carriers. In the UE and RNC, MAC PDU segmentation, reassembly, reordering, and duplicate removal is performed jointly across all of the cells. In order to provide a sufficient reordering space considering the potential for a larger number of received MAC PDUs across two carriers, the MAC *Transmission Sequence Number* (TSN) is increased to 14 bits in a similar manner to multi-carrier HSDPA.

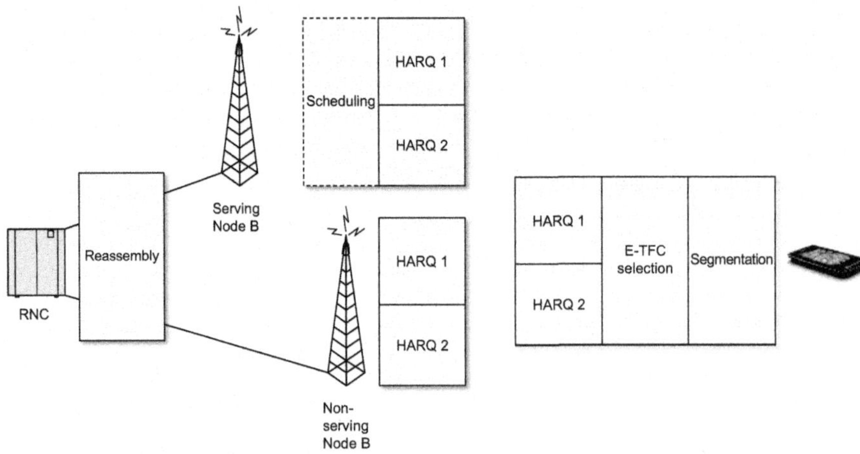


**FIGURE 12.14 DC-HSUPA MAC structure.**

As discussed in Section 12.2.1, it is possible for a Node B to activate and deactivate the secondary cell, for a UE using HS-SCCH orders. Activation and deactivation of a secondary uplink cell however, causes a problem for handling a UE in soft handover, since non-serving Node Bs are unaware of whether a cell is active. It is undesirable for the non-serving Node Bs to continuously monitor the UE scrambling code to discover whether the cell is active from a baseband resource perspective. To mitigate the problem, on sending an HS-SCCH order, the serving Node B also sends NBAP signaling to the RNC, which in turn uses NBAP to inform other Node Bs in the active set of the (de)activation of the cell. The activation procedure for non-serving Node Bs is similar to an active set update. The activation time for the HS-SCCH order is much shorter than the general round-trip time to the RNC.

If synchronization is lost with the secondary uplink carrier, it is expected that this will be handled by means of deactivation (and possible reactivation).

## FURTHER READING

3rd Generation Partnership Project; Technical Specification Group Radio Access Network; Multiplexing and Channel Coding (FDD), 3GPP TS 25.212.

D.M. de Andrade, A. Klein, H. Holma, I. Viering, G. Liebl, Performance evaluation on dual-cell HSDPA operation. IEEE VTC Fall, 2009, pp. 1–5.

K. Johansson, J. Bergman, D. Gerstenberger, M. Blomgren, A. Wallen, Multi-carrier HSPA evolution. IEEE VTC Spring, 2009, pp. 1–5.

M. Kazmi, H. Nyberg, O. Drugge, F. Ghasemzadeh, Handling of uplink transmitted carrier power difference in DC-HSUPA. IEEE VTC Spring, 2011, pp. 1–5.

I. Repo, K. Aho, S. Hakola, T. Chapman, F. Laakso, Enhancing HSUPA system level performance with dual-carrier capability. ISWPC, 2010, pp. 579–583.

D. Zhang, P.K. Vitthaladevuni, B. Mohanty, J. Hou, Performance analysis of dual-carrier HSDPA. IEEE VTC Spring, 2010, pp. 1–5.

CHAPTER

# Multi-flow transmission

# 13

## CHAPTER OUTLINE



*Multi-flow transmission*, also referred to as *HSDPA multi-flow data transmission*, was introduced in release 11. It enables improved user experience for cell-edge UEs located in the soft or softer handover coverage region of two cells on the same carrier frequency by allowing HS-DSCH to be scheduled from both cells. Such operation allows more scheduling opportunities for the network and aids system load balancing.

## 13.1 OVERVIEW

The popularity of HSPA-based mobile broadband offerings and increasing data usage led to the deployment of HSPA with more than one transmit antenna and/or more than one carrier. In release 7, the single-cell downlink MIMO feature described in

HSPA Evolution: The Fundamentals for Mobile Broadband. DOI: 10.1016/B978-0-08-099969-2.00013-2

Chapter 11 was standardized, allowing the Node B to transmit two transport blocks to a UE from the same cell using a pair of transmit antennas, thereby improving data rates at high geometries and providing a beamforming gain at low geometries. In releases 8, 9, 10, and 11, multi-carrier features were introduced, allowing the Node B to simultaneously serve one or more users by operating HSDPA on up to eight different carrier frequencies in geographically overlapping cells having the same timing, thereby improving the end user experience in the cell coverage region. The development of multicarrier HSPA is described in Chapter 12.

Multi-flow operation relies on simultaneous reception of up to four HS-DSCH transport channels in the CELL_DCH state on up to two frequencies, of which a maximum of two HS-DSCH transport channels may reside at the same frequency and belong either to the same or different Node Bs. MIMO transmission can be configured for each of the cells.

Many of the multi-flow design aspects are similar to the multi-carrier feature. The key difference to multi-carrier transmission is that two cells belonging to the same or even different Node Bs can be simultaneously configured as serving HS-DSCH cells. Multi-flow operation may lead to an improved end user experience for UEs in the soft or softer handover region when the serving HS-DSCH cell is capacity or coverage limited, while other cells in the active set have available resources. However, it should be noted that in order to achieve gains with multi-flow operation, the involved UEs should have advanced interference mitigating receivers (Type 3i). Such receivers (discussed further in Chapter 3) are also used for MIMO and multi-carrier operation.

### 13.1.1 SUPPORTED CONFIGURATIONS

Simultaneous HSDPA transmission from a pair of geographically non-overlapping cells operating on the same carrier frequency in any given TTI to a particular user is referred to as *Single-Frequency Dual-Cell* (SF-DC) aggregation. In addition to the serving HS-DSCH cell, a cell on the same frequency, where the UE is configured to simultaneously monitor an HS-SCCH set and receive HS-DSCH if it is scheduled in that cell, is referred to as assisting serving HS-DSCH cell. This is the simplest form of multi-flow operation, and the participating cells may either belong to the same or different Node Bs. Such configurations are referred to as intra- and inter-Node B multi-flow operation, respectively. An example of inter-Node B SF-DC multi-flow transmission is illustrated in Figure 13.1.

An extension of the above, enabling operation in a dual carrier configuration with two cell pairs, each on their respective carrier frequencies, is referred to as *Dual-Frequency Four-Cell* (DF-4C) aggregation. In addition to the assisting serving HS-DSCH cell, a cell on the secondary downlink frequency, for which the UE is configured to simultaneously monitor an HS-SCCH set and receive HS-DSCH if it is scheduled in that cell, is referred to as assisting secondary serving HS-DSCH cell. Both intra- and inter-Node B configurations are allowed and the HS-DSCH channels can also be configured on three cells: two cells on the primary carrier and one cell on the secondary carrier. This operation mode is referred to as DF-3C aggregation.

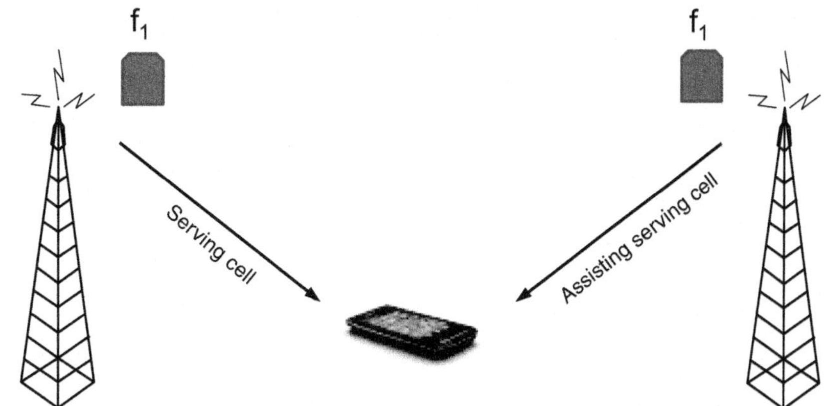


**FIGURE 13.1 Inter-Node B SF-DC operation. A user is configured with two different cells (on the same frequency) from two different sites.**

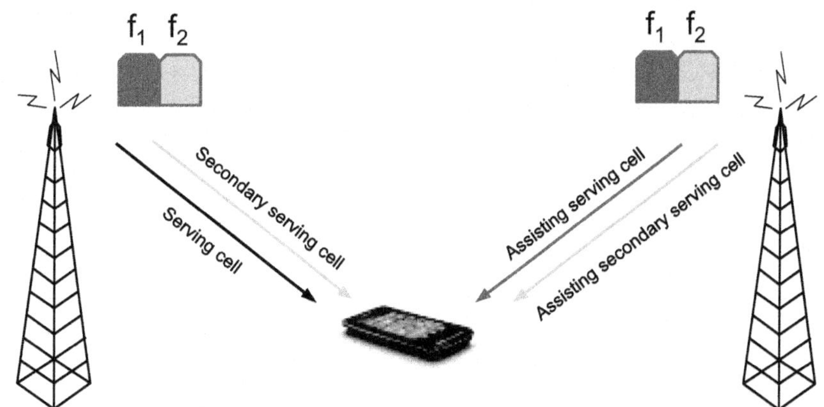


**FIGURE 13.2 Inter-Node B DF-4C operation. A user is configured with two different cells from different Node Bs on each frequency.**

In Figure 13.2, an inter-Node B DF-4C scenario is illustrated. Note that no more than one cell pair can be operated on each frequency.

Multi-flow operation is supported both for configurations with and without 2×2 downlink MIMO (single- or dual-stream), and the participating cells may potentially belong to different frequency bands. A summary of possible multi-flow configurations is provided in Table 13.1.

### 13.1.2 ARCHITECTURE

To enable scheduling different HS-DSCH PDUs from different cells, a split of the data may be required at the RLC level, depending on whether the two cells in the cell pair reside in the same or different Node Bs, that is, whether multi-flow is operated in

**Table 13.1** Supported Multi-flow Configurations

| Name | Configuration | Carrier Frequency A | Carrier Frequency B |
|---|---|---|---|
| SF-DC | One frequency, two cells | The serving HS-DSCH cell, the assisting serving HS-DSCH cell | N/A |
| DF-3C | Two frequencies, three cells | The serving HS-DSCH cell, the assisting serving HS-DSCH cell | The secondary serving HS-DSCH cell |
| DF-3C | Two frequencies, three cells | The serving HS-DSCH cell, the assisting serving HS-DSCH cell | The assisting secondary serving HS-DSCH cell |
| DF-4C | Two frequencies, four cells | The serving HS-DSCH cell, the assisting serving HS-DSCH cell | The secondary serving HS-DSCH cell, the assisting secondary serving HS-DSCH cell |

intra- or inter-site mode. In case of inter-site transmission, this necessitates changes in the MAC-layers, whereas for intra-site operation, the data split is performed at MAC-ehs level, similar to the multi-carrier operation.

#### *13.1.2.1 Intra-Node B multi-flow operation*

For intra-Node B multi-flow operation, one common MAC-ehs entity is used at the UE. This entity handles up to four HS-DSCH transport channels, and functions such as reordering, duplicate detection, segmentation, and reassembly are common for these HS-DSCH transport channels and hybrid-ARQ entities. Each of the HS-DSCH transport channels has its own associated uplink and downlink signaling and hybrid-ARQ entity. The architecture of intra-Node B SF-DC transmission is shown in Figure 13.3. For this configuration, multi-flow can be compared, from a MAC perspective, to multi-carrier operation.

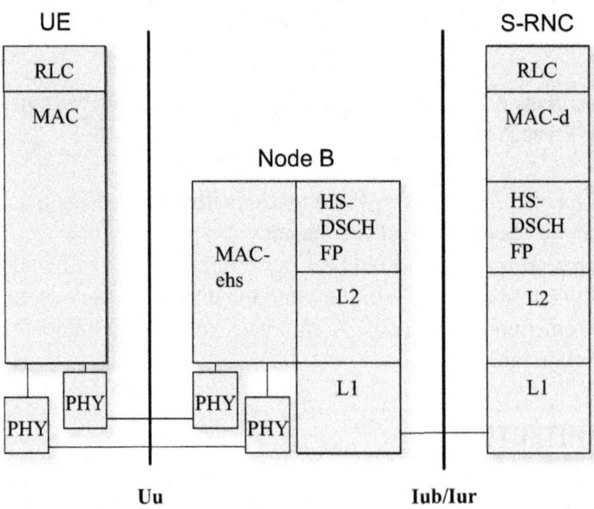


**FIGURE 13.3** Protocol architecture of intra-Node B SF-DC configuration (without MAC-c/sh).

### 13.1.2.2 Inter-Node B multi-flow operation

For inter-Node B multi-flow operation, there are two MAC-ehs entities in the UE, one each for every Node B. Up to two HS-DSCH transport channels are supported per transmitting and receiving MAC-ehs entity and each of them independently performs reordering, duplicate detection, segmentation, and reassembly for the corresponding HS-DSCH transport channels and hybrid-ARQ entities. As for intra-Node B operation, each of these HS-DSCH transport channels has its own associated uplink and downlink signaling and hybrid-ARQ entity. The architecture of inter-Node B DF-4C transmission is shown in Figure 13.4.

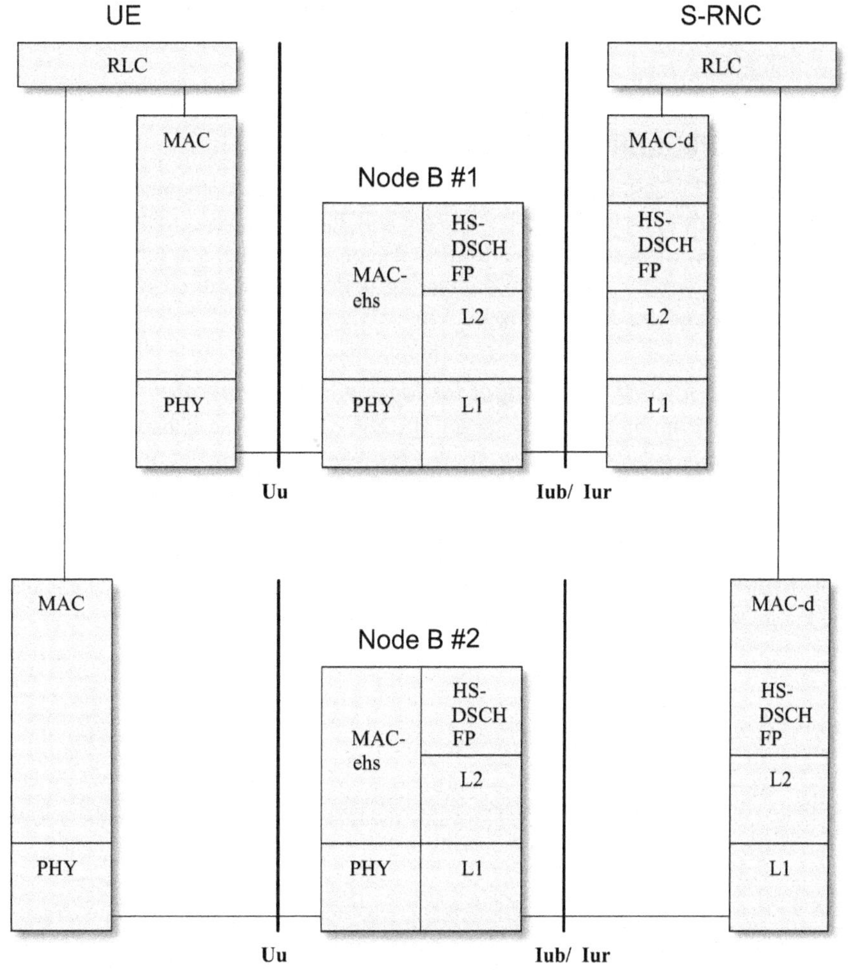


**FIGURE 13.4** Protocol architecture of inter-Node B DF-4C configuration (without MAC-c/sh).

## 13.2 DETAILS OF MULTI-FLOW TRANSMISSION

There are several similarities in the design of multi-flow and multi-carrier HSDPA but also some fundamental differences. Unlike multi-carrier operation, where all cells on which downlink transmissions occur to a particular UE have identical timing, the slot boundaries in multi-flow HSDPA for any two cells belonging to different sites or sectors do not necessarily coincide. The timing delay $T_{\text{cell}}$, which specifies the relative time difference between the *Node B Frame Number* (BFN) and *Cell System Frame Number* (SFN), and thus when the P/S-SCH are transmitted, has a granularity of 256 chips, corresponding to 1/10th of a slot (the SCH burst duration). To avoid overlapping P/S-SCH transmissions, $T_{cell}$ may be different for cells belonging to different sectors. Also, clocks at different cells can have different sources and thereby drift relative to each other. Apart from that, different sites are likely to be asynchronous. Much of the design of multi-flow HSDPA therefore focuses around these timing differences.

### 13.2.1 UPLINK CONTROL CHANNEL

When multiple HS-DSCH transmissions take place to a UE, corresponding feedback information needs to be conveyed back to the node/nodes active in multi-flow operation. This is done using a common HS-DPCCH control channel and jointly encoding the feedback information for all participating nodes. The required changes to the HS-DPCCH design are described next.

#### *13.2.1.1 Hybrid-ARQ time budget for legacy HSDPA operation*

The hybrid-ARQ timing budget for legacy HSDPA operation (see Chapter 8 for more details) for a reference scenario of six hybrid-ARQ processes requires the UE to transmit the ACK/NACK information 7.5 slots after receiving the HS-PDSCH subframe, giving the Node B 4.5 slots to decode the feedback information. This is illustrated in Figure 13.5. Note that the propagation delay (in slots), $T_p$, is not included in this figure; if one also were to account for the propagation delay, the hybrid-ARQ time budget of the Node B would be $4.5 - 2T_p$ slots.

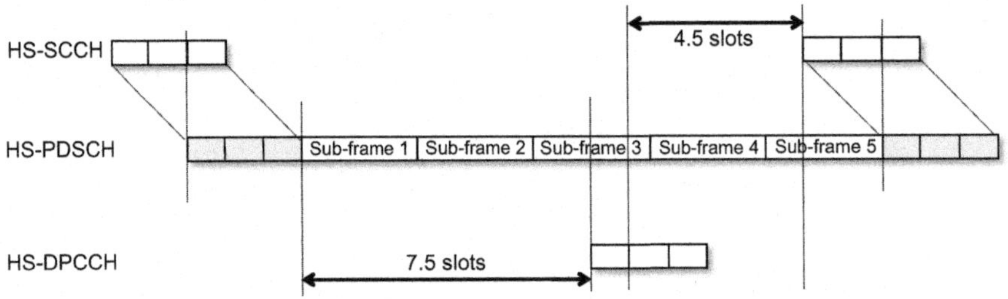


**FIGURE 13.5 Hybrid-ARQ time budget for the UE and Node B for a reference scenario of six hybrid-ARQ processes.**

#### 13.2.1.2 Hybrid-ARQ time budget for multi-flow operation

Similar to multi-carrier operation, a single common HS-DPCCH is used for cells participating in multi-flow HSDPA transmission (intra- and inter-Node B) and the *hybrid-ARQ* (HARQ) and *Channel Quality Information* (CQI) is jointly encoded. This design is motivated by a lower cubic metric compared to a dual HS-DPCCH solution and also thereby reduced multi-user interference. Note that using this solution, only one midslot power change takes place in the uplink, whereas two power changes would be required if a dual HS-DPCCH structure would be employed. This is because the cells participating in multi-flow operation do not necessarily have the same timing and therefore the timing for the corresponding HS-DPCCH channels would be different.

By design, the serving and secondary serving HS-DSCH cells have the same downlink timing, and the same applies to the assisting serving and assisting secondary serving HS-DSCH cells. Thus, only two different downlink timings may be present for the cells involved in multi-flow operation, allowing cells in a certain sector to simultaneously support multi-carrier and multi-flow UEs.

The UE always pairs sub-frames corresponding to the HS-PDSCHs having the largest overlap, resulting in a maximum time difference between the sub-frames of 1.5 slots. This pairing rule is used regardless of the number of hybrid-ARQ processes configured. See Figure 13.6 for an illustration. In addition, a guard region corresponding to 20 chips is used to avoid unnecessary reconfigurations. The resulting time difference $\tau_{diff}$ between the P-CCPCHs for the paired sub-frames is thus in the range $-20 \leq \tau_{diff} \leq 3840 + 20$ chips. If the value of $\tau_{diff}$ drifts beyond 3840 + 20 chips or beyond −20 chips, the UE hybrid-ARQ ACK reporting may be compromised. The UE will however remain in CELL_DCH state. The information regarding the time difference between the links may be reported by the UE by means of the *Information Element* (IE) "Cell synchronization information" in an RRC Measurement Report (if so configured by the network), when triggered.

Since the feedback information for all cells active in multi-flow transmission is jointly encoded and the timing of the paired sub-frames can differ up to 1.5 slot, a reduction in the hybrid-ARQ time budget is needed either at the Node B or UE side. At the Node B, the existing hybrid-ARQ time budget of 4.5 slots is used for HARQ-ACK decoding, scheduling of the upcoming TTI, and allocation of hardware resources. Since it is desirable for operators to reuse the existing hardware

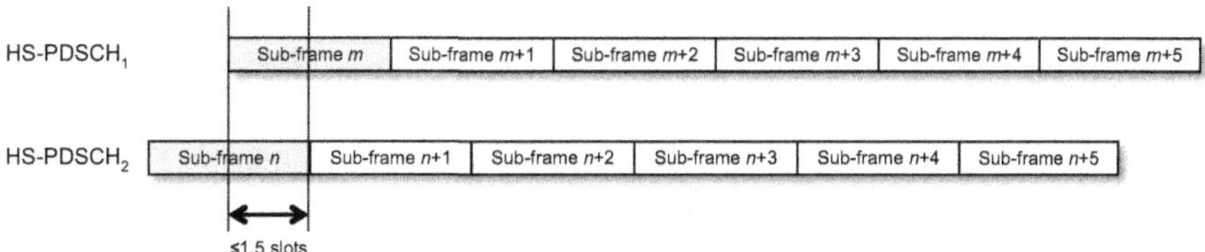


**FIGURE 13.6 Sub-frame pairing when HS-PDSCH$_1$ and HS-PDSCH$_2$ have the largest overlap corresponding to 1.5 slots.**

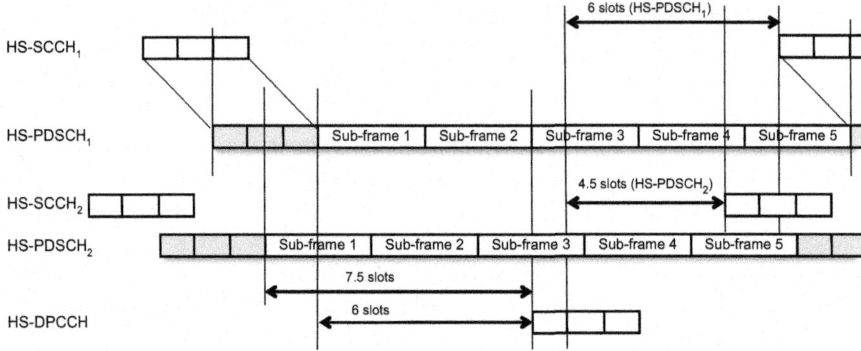


**FIGURE 13.7 Hybrid-ARQ time budget for the jointly encoded HS-DPCCH when the network has configured a UE with six hybrid-ARQ processes.**

infrastructure and do software upgrades only, while UEs anyway get replaced, a reduced hybrid-ARQ time budget at the Node B is not wanted. Therefore, the timing compression is handled entirely by reducing the UE processing time by 1.5 slot, at least in the non-MIMO cases. For MIMO, UE capability bits are introduced to indicate that the UE does/does not require an additional hybrid-ARQ process and supports a larger soft buffer in order to support single- or dual-stream MIMO with multi-flow. No change is made to the soft buffer size per hybrid-ARQ process. The hybrid-ARQ time budget for a jointly encoded HS-DPCCH for a network configured with six hybrid-ARQ processes is shown in Figure 13.7.

If the UE requires an additional hybrid-ARQ process to support MIMO, the network may configure seven hybrid-ARQ processes. The RNC informs all the cells active in multi-flow operation about the MIMO configurations in each of the cells.

#### 13.2.1.3 HARQ-ACK and CQI encoding

For SF-DC multi-flow operation, the *Dual-Carrier HSDPA* (DC-HSDPA) CQI formats are used, whereas for more than two cells, CQI formats based on 4C-HSDPA are used. The cells characterized by the same CPICH timing constitute a cell group and the group cells' CQI reports are combined together in one HS-DPCCH sub-frame.

The cells are reported in the order defined for 4C-HSDPA. The information of which CQI belongs to which cell group is provided by the RNC and the UE group CQIs of one group adjacent to each other.

The HARQ-ACK and CQI encoding is similar to that of multi-carrier HSDPA discussed in Chapter 12. For dual-cell non-MIMO multi-flow configurations, the release 8 DC-HSDPA HARQ-ACK encoding is used, whereas a new codebook was standardized for the case when MIMO is enabled on at least one of the cells. This stems from the need to decode the HS-DPCCH at each Node B without requiring prior knowledge about the number of streams employed by the other Node B. The codebook standardized for DC-HSDPA with MIMO would not allow such decentralized decoding. For three-cell configurations without MIMO, the release 10 3C-HSDPA non-MIMO encoding is used, and 3C-HSDPA MIMO encoding is used when MIMO is enabled on at least one carrier. For DF-4C multi-flow configurations, with or without MIMO, the release 10 4C-HSDPA encoding is used.

For each of these configurations, the spreading factor and location of ACK/NACK and CQI is not affected through L1 signaling for inter-Node B multi-flow configurations with MIMO enabled on at least one cell. The ACK/NACKs are grouped in the same way as CQI and the same codebook is used for intra- and inter-Node B scenarios. Note that this HS-DPCCH design requires the presence of both serving and assisting cells on the primary frequency since the HS-DPCCH is fed back on the primary carrier.

#### 13.2.1.4 HARQ-ACK and CQI repetition

HARQ-ACK and CQI can be repeated to allow feedback information to be received even when the UE is headroom limited. Also for multi-flow, repetition can be configured to improve the control channel performance. For SF-DC non-MIMO configuration, the HARQ-ACK is not repeated and a single CQI repetition factor is configured for both cells. In the case of SF-DC with MIMO and DF-3C without any MIMO configuration, HARQ-ACK is not repeated and a CQI repetition factor is configured for each cell group. For DF-3C with MIMO or a DF-4C configuration, a HARQ-ACK repetition factor and a CQI repetition factor are configured for each cell group. The rules for HARQ-ACK and CQI repetition are the same for both intra- and inter-Node B configurations. The different encoding formats are illustrated in Figure 13.8.

When DF-3C or DF-4C is configured, the CQI reports are repeated whenever a blank slot is available in the CQI field on the HS-DPCCH channel. CQIs are repeated

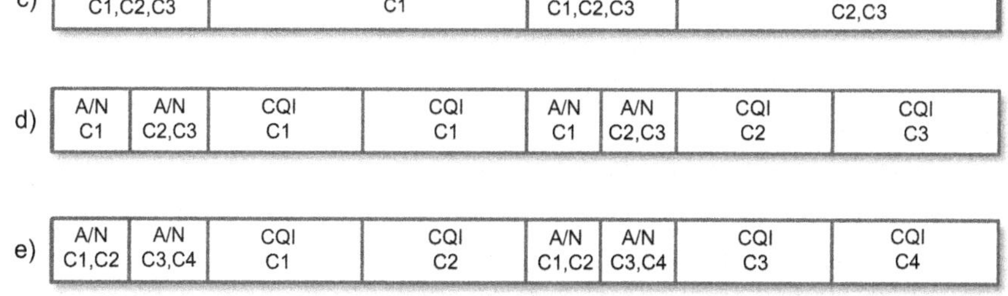

a)

| ACK/NACK C1,C2 | CQI C1,C2 | ACK/NACK C1,C2 | CQI C1,C2 |
|---|---|---|---|

b)

| ACK/NACK C1,C2 | CQI C1 | ACK/NACK C1,C2 | CQI C2 |
|---|---|---|---|

c)

| ACK/NACK C1,C2,C3 | CQI C1 | ACK/NACK C1,C2,C3 | CQI C2,C3 |
|---|---|---|---|

d)

| A/N C1 | A/N C2,C3 | CQI C1 | CQI C1 | A/N C1 | A/N C2,C3 | CQI C2 | CQI C3 |
|---|---|---|---|---|---|---|---|

e)

| A/N C1,C2 | A/N C3,C4 | CQI C1 | CQI C2 | A/N C1,C2 | A/N C3,C4 | CQI C3 | CQI C4 |
|---|---|---|---|---|---|---|---|

**FIGURE 13.8 HS-DPCCH slot formats for the different multi-flow configurations. Cx denotes cell x. (a) HS-DPCCH format for SF-DC non-MIMO multi-flow; (b) HS-DPCCH format for SF-DC MIMO multi-flow; (c) HS-DPCCH format for 3C non-MIMO multi-flow; (d) HS-DPCCH format for 3C MIMO multi-flow; (e) HS-DPCCH format for DF-4C multi-flow.**

even when blank CQI slots arise as a result of deactivation. The CQI power offset is set as if repetition did not occur when CQI reports are repeated.

#### 13.2.1.5 HS-DPCCH timing

The HS-DPCCH timing is determined by one of the downlink links (configured by the RRC) and follows the release 5 rules relative to the downlink acting as time reference. The devices capable of and configured for simultaneous multi-flow and MIMO operation and not requiring additional processing time for hybrid-ARQ ACK generation follow the same timing rules as in the multi-flow without MIMO case. However, devices capable of and configured for simultaneous multi-flow and MIMO operation at least in the non-time reference cell and requiring additional processing time for ACK generation follow the same timing rules as multi-flow without MIMO case, with the exception that the ACK of the non-time reference cell is transmitted one HS-DPCCH sub-frame later.

Similar to the update of cell pairing, the reconfiguration of the downlink time reference is handled by the RNC, and the UE can report the time difference between the links. The uplink DPCCH timing is, however, not affected by changes to the time-reference cell. Furthermore, the time-reference cell is not changed because of out-of-sync conditions and neither is the reporting of downlink synchronization primitives CPHY-Sync-IND and CPHY-Out-of-Sync-IND due to multi-flow operation.

#### 13.2.1.6 HS-DPCCH power setting

Further, for multi-flow, uplink follows power-control commands from the instantaneously best uplink cell (because of the "or of downs" rule for power control in soft handover), and the HS-DPCCH is transmitted with a fixed power offset with respect to the corresponding DPCCH. To ensure that a sufficient HS-DPCCH quality/SINR is maintained at all Node Bs in the active set, especially when there is an imbalance in the uplink, the existing range of $\Delta_{ACK}$, $\Delta_{NACK}$, and $\Delta_{CQI}$ values that the RNC can signal to the UE was extended in release 11. This allows handling larger differences in the uplink SINR, in particular when the Event 1a (cell entering the reporting range) and Event 1b (cell leaving the reporting range) thresholds are set rather aggressively to obtain system-level gains with multi-flow operation. Note that the imbalance in the uplink is more of a challenge when operating in multi-flow mode than during legacy operation, since the uplink quality is not only affected by fast-fading variations but also by the difference in path gain resulting from a larger imbalance region associated with multi-flow operation. Also, in legacy systems, non-serving cells do not need to receive the HS-DPCCH.

#### 13.2.1.7 DTX/DRX operation

In general, an HS-SCCH order sent from one Node B cannot affect the operation of the other Node B in case of inter-Node B configuration. Therefore, DTX/DRX activation/deactivation through HS-SCCH orders is not allowed for inter-Node B multi-flow operation. For the intra-Node B case, because of the timing offset existing between the serving and the assisting serving cell, there might be issues related to when the received order should take effect. Therefore, DTX/DRX activation/deactivation through HS-SCCH orders is not allowed for the intra-Node B multi-flow operation either. The UE maintains a common DTX/DRX status for all the HS-DSCH cells, and HS-SCCH-less operation is restricted to the serving HS-DSCH cell only.

### 13.2.2 ACTIVATION AND DEACTIVATION OF MULTI-FLOW OPERATION

The activation and deactivation of multi-flow operation is controlled by the network through RRC signaling as well as HS-SCCH orders (as described in Chapter 14). Only the secondary serving HS-DSCH cells (assisted or assisting) can be activated/deactivated using HS-SCCH orders and orders targeting a particular secondary serving HS-DSCH cell can be sent only from cells in the same cell group (or from the Node B containing the target cell). In case of DF-3C, only the single secondary serving HS-DSCH cell can be deactivated, while in case of DF-4C, both secondary serving HS-DSCH cells can be deactivated.

### 13.2.3 RLC AND MAC ASPECTS

Explicit discard indications from RNC to Node B are introduced for multi-flow. The RNC knows via RLC status reports when an acknowledged mode RLC PDU is received by the UE. The Node B, in turn, reads the discard indication from the RNC and, if such data is stored, discards it. The HS-DSCH data frame Type 2 is updated accordingly. This mechanism allows the RNC to send data to both Node Bs without causing the unnecessary duplicate data to be sent over the Uu interface.

An explicit drop indication is also introduced from Node B to RNC. This aids RNC in detecting imbalances in the Iub/Iur and, if the Node B detects a drop, it can inform the RNC and request a quicker RLC retransmission without having to wait for an RLC STATUS report.

It should be noted that multi-flow is specified for acknowledged mode transmission, in particular for SRB2, SRB3, and SRB4. Bicasting of SRBs from the RNC to the serving and assisting serving cell is allowed in order to improve the SRB robustness.

#### *13.2.3.1 Skew handling*

During inter-Node B multi-flow operation, a skew (PDUs received out of sequence) in data split may arise for the acknowledged mode RLC PDUs. There may be different causes of the skew, for example, packets being sent in batches, Uu fluctuations, or different queue times in the Node B. This is illustrated in Figure 13.9.

It should be further noted that the different RCL PDUs may be received out of order by the UE due to hybrid-ARQ retransmissions or physical-layer losses and fluctuation due to channel or loading conditions.

When detecting a gap corresponding to a potentially missing PDU, the UE may trigger STATUS PDU requesting a retransmission, even though the missing PDU might be queued in a Node B but not yet transmitted. To limit the number of retransmissions, when a gap is detected, the UE starts the timer *Timer_Reordering* (when

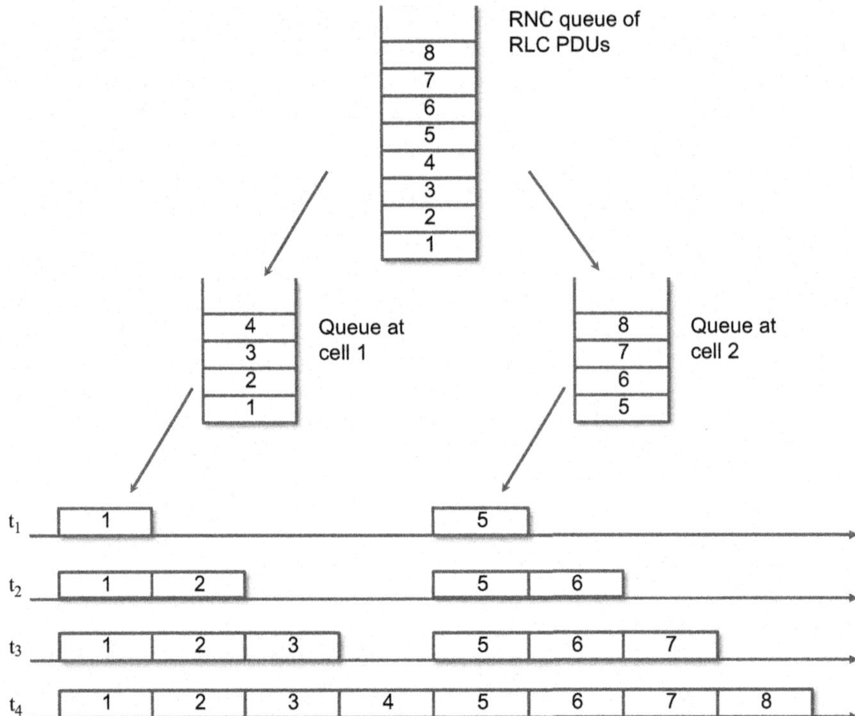


**FIGURE 13.9 RLC packets arriving at UE at different time instants, without packet loss over the air.**

configured by higher-layer). The STATUS report is sent by the UE only if this timer has expired. If the gap is filled, *Timer_Reordering* is reset.

A network centric solution can also be deployed, where the *serving RNC* (S-RNC) records to which cell different RLC PDUs are sent. When a STATUS PDU is received, the S-RNC checks whether an RLC packet transmitted to the same Node B with a higher sequence number has been identified. If so, then the packet loss is genuine. If missing RLC packets have been transmitted to the other Node B, the S-RNC starts a timer. If the packets still are missing when the timer expires, the S-RNC retransmits the RLC packet.

The handling of STATUS PDUs with "erroneous Sequence Number" is similar to the behavior in E-UTRA, where the UE ignores such RLC STATUS PDU. This is different compared to legacy UMTS operation, where a "RLC unrecoverable error" would be triggered.

### 13.2.4 MOBILITY

All cells involved in multi-flow operation belong to the UE's active set. The multi-flow mobility procedures (such as serving cell change and active set update) are based on legacy mechanisms, such as Event1d for the change of best cell and Event1a, 1b, and 1c for cells entering and/or leaving the reporting range. These events are based on measurements on the primary frequency, as for multicarrier HSDPA. The mobility for DF-3C and DF-4C is hence the same as for SF-DC. It should be noted that legacy mobility mechanisms allow determining the best cell through Event1d but not the second best cell. No new mobility events are introduced for multi-flow operation to determine the second best cell in the active set. As the standard does not specify how to determine the second best cell, it is up to the network to make this choice. The network may use any available information about radio link quality, load information, etc.

For *enhanced serving cell change* (eSCC), it is possible to reuse the legacy procedures (based on target cell pre-configuration and Event 1d) without further enhancements.

### 13.2.5 FEATURE DEPENDENCIES

Multi-flow can be operated simultaneously with DC-HSUPA, where two carriers are used for uplink transmission. The uplink feedback information for the cells involved in multi-flow operation is then carried on the primary carrier.

Multi-flow can also be operated simultaneously with CLTD or uplink MIMO. The HS-DPCCH precoding weights can be signaled either by the serving or assisting serving cell.

### 13.2.6 UE CATEGORIES AND CAPABILITIES

Multi-flow reuses the multi-carrier HS-DSCH physical-layer category extensions with the addition of a multi-flow capability indication. An SF-DC multi-flow UE shall support DC-HSDPA and a DF-3C and DF-4C UE shall support 3C- and 4C-HSDPA, respectively. The multi-flow support is per band.

For multi-flow operation with MIMO, single-stream transmission is supported on a per UE basis (that is, if supported it must be supported in all the multi-flow bands) while dual-stream transmission is supported on a per band basis (that is, the UE may support dual-stream in one band but not in another). A UE supporting multi-flow and MIMO shall indicate MIMO support (single- or dual-stream) in all the supported multi-flow bands.

### 13.2.7 RECEIVER IMPACT

From the *radio frequency* (RF) point of view, multi-flow can be considered similar to single or multicarrier operation. UEs whose RF only supports a single carrier can decode a normal single carrier signal as well as a multi-flow single carrier signal (SF-DC) with no additional losses, filters, or noise sources. Similarly, dual-receiver UE architecture can be used in general for decoding a dual-carrier (or band) signal as well as a multi-flow dual-carrier signal with no additional losses, filters, or noise sources. Hence, single-carrier RF-capable UEs should be able to handle SF-DC operation, and DC-HSDPA RF-capable UEs should be able to handle DF-4C operation. The UE under multi-flow operation does not receive any new type of signal, and the maximum interference is lower than for full-buffer transmissions. Therefore, there is no need to introduce new RF requirements for UEs supporting multi-flow operation, and multi-flow is only tested via demodulation requirements.

From the baseband perspective, the UE should be capable of demodulating two, three, or four streams (maximum two streams per frequency) depending on the multi-flow configuration. In order to limit the effect of inter-cell interference, an interference mitigating Type 3i receiver is required for all the branches.

## FURTHER READING

3GPP TS 25.212 Multiplexing and Channel Coding (FDD).

3GPP TR 25.872, High Speed Packet Access (HSDPA) Multipoint Transmission (Release 11), September 2011.

V. Hytonen, O. Puchko, T. Hohne, T. Chapman, Introduction of multiflow for HSDPA, NTMS (2012) 1–5.

D. Petrov, I. Repo, M. Lampinen, Overview of single frequency multipoint transmission concepts for HSDPA and performance evaluation of intra-site multiflow aggregation scheme, IEEE VTC Spring (2012) 1–5.

A. Yaver, T. Hohne, J. Moilanen, V. Hytonen, Flow control for multiflow in HSPA+, IEEE VTC Spring (2013) 1–5.

CHAPTER

# Connectivity enhancements

# 14

## CHAPTER OUTLINE



Packet-data traffic is often highly bursty, with occasional periods of transmission activity. From a user performance perspective, it is advantageous to keep the HS-DSCH and E-DCH configured to rapidly be able to transmit any user data. At the same time, maintaining an active connection with HS-DSCH and E-DCH configured comes at a cost. From a network perspective, there is a cost in uplink interference from the necessary DPCCH transmission even in absence of data transmission. For UEs, power consumption is the main concern; even when no data is received, the UE needs to transmit the DPCCH and monitor the HS-SCCH.

The rapid increase in smartphone traffic and associated smartphone apps in recent years has led to some migration in the profile of UE interactions with the network.


HSPA Evolution: The Fundamentals for Mobile Broadband. DOI: 10.1016/B978-0-08-099969-2.00014-4

Although smartphones may still download or upload large quantities of data (such as file transfers) or require continuous data flow with a well-defined QoS (such as voice or video streaming), as envisaged and handled efficiently by the basic HSDPA and HSUPA, new types of application have emerged in which the size of application-level packets is small and where the regularity with which packets occur can vary from occasionally to frequently. Examples of such applications can include keep alive signals (which are very small and occasional), "pull" e-mail polling messages, weather and news updates, updates on social media applications, such as Facebook and Twitter, location-related information, and so on. A typical smartphone may run a variety of applications that may interact with the network at differing times. Furthermore, many applications operate without user interaction, and thus continuously impact the network, even during times at which the user is not actively engaged with the device. All of this requires that modern networks be able to cope with an ever-increasing number of connected devices that can interact with small data sizes on an unpredictable basis. Growth of smartphone and tablet apps and, in the future, *Machine to Machine* (M2M) interactions are likely to further increase the complexity of managing traffic effectively.

To reduce UE energy consumption and use of network resources, WCDMA, like most other cellular systems, has several states: URA_PCH, CELL_PCH, CELL_FACH, CELL_DCH, and idle mode. The different states are illustrated in Figure 14.1.

The lowest energy and resource consumption is achieved when the UE is in one of the two paging states specified for WCDMA, namely CELL_PCH and URA_PCH. In these states, the UE sleeps and only occasionally wakes up to check for paging messages. The difference is that in CELL_PCH, the UE location is known at the cell level, whereas in URA_PCH, the location is only known within a paging

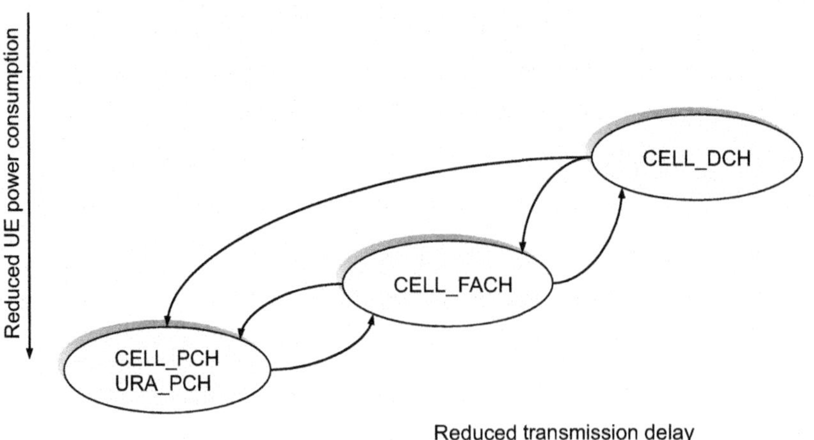


**FIGURE 14.1 WCDMA state model.**

area, which consists of several cells. Thus, in URA_PCH, the UE needs to update its location less often. The paging mechanism is mainly intended for longer periods of inactivity. For exchange of data, the UE needs to be moved to the CELL_FACH or CELL_DCH states.

In CELL_FACH, the UE can transmit small amounts of data as part of the random-access procedure. The UE also monitors common downlink channels for small amounts of user data and RRC signaling from the network.

The high transmission activity state is known as CELL_DCH. In this state, the UE can use HS-DSCH and E-DCH for exchanging data with the network, as described in Chapters 8 and 9, respectively. This state allows for rapid transmission of large amounts of user data but also has the highest UE energy consumption.

RRC signaling is used to move the UE between the different states. Thus, from a delay perspective, it is preferable to keep the UE in CELL_DCH while there are still likely to be substantial bursts arriving for the UE, whereas from the network resource and energy-consumption perspective, one of the paging states is preferred. In order to reduce the impact of CELL_DCH on energy consumption and network resources, release 7 introduced the so-called *Continuous Packet Connectivity* (CPC) feature, which is described in Section 14.1.

CPC reduces the impact of CELL_DCH, but for some types of smartphone interaction, even when CPC is applied, CELL_DCH is still a relatively expensive state in terms of energy consumption and network resources. In release 8, therefore, enhancements to the CELL_FACH state were introduced for handling of less intensive smartphone traffic. The CELL_FACH enhancements are described in Section 14.2.

## 14.1 CONTINUOUS PACKET CONNECTIVITY (CPC)

The release 99 framework for UMTS, although focused on data transfer, was connection oriented in that it required long-term allocation of dedicated resources in the downlink and uplink by the RNC for maintaining data transfer. Many types of data transmissions, however, are characterized by the fact that the activity is bursty and resources are needed on a sporadic basis. As described in Chapters 8 and 9, release 5 and release 6 introduced HSDPA and HSUPA, respectively, which provided enhanced handling of packet data by, among other things, moving the short-term management of resources from the RNC to the Node B scheduler. Maintaining an ongoing session with HSDPA and HSUPA still requires a connection to be sustained and for the UE to operate in the CELL_DCH state. Remaining in CELL_DCH prior to release 7 required a continuous allocation of physical layer control channels and continuous monitoring of downlink control channels by the UE, even during periods in which data transfer was instantaneously not occurring. To reduce the associated impact on network resources and UE battery life, CPC was introduced into the release 7 specifications. The techniques introduced with CPC enable efficient handling of many types of data transfer and interaction with the network. CPC consists of three main building blocks:

1. *Discontinuous transmission* (DTX), to reduce the uplink interference and thereby increase the uplink capacity, as well as to lengthen battery life.
2. *Discontinuous reception* (DRX), to allow the UE to periodically switch off the receiver circuitry and save battery power.
3. *HS-SCCH-less operation*, to reduce the control signaling overhead for transmission of small amounts of data per TTI, as will be the case for services, such as VoIP.

The intention with these features is to provide an "always-on" experience for the end user by keeping the UE in CELL_DCH for a longer time and avoiding frequent state changes to the low-activity states, as well as improving the capacity for services, such as VoIP. Because they mainly relate to packet-data support, they are only supported in combination with HSPA; thus, if a DCH is configured, the CPC features cannot be used. In the following, the three building blocks and the interaction between them are described.

### 14.1.1 DISCONTINUOUS TRANSMISSION (DTX) – REDUCING UPLINK OVERHEAD

The shared resource in the uplink is, as discussed in Chapter 9, the interference headroom in the cell. During periods when no data transmission is taking place in the uplink, the interference generated by a UE is due to the uplink DPCCH (and potentially HS-DPCCH), which is continuously transmitted as long as E-DCH is configured. Any reduction in unnecessary DPCCH activity would therefore directly reduce the uplink interference, thereby lowering the cost in terms of system capacity of keeping the UE connected. Clearly, from an interference reduction perspective, the best approach would be to completely switch off the DPCCH when no data transmission is taking place. However, this would have a serious impact on the possibility to maintain uplink synchronization, as well as negatively impact the power control operation. Therefore, occasional slots of DPCCH activity, even if there is no data to transmit, are beneficial to maintain uplink synchronization and to maintain a reasonably accurate power control. This is the basic idea behind uplink *Discontinuous Transmission* (uplink DTX). Obviously, the burstier the data traffic, the larger the benefits with discontinuous transmission.

Basically, if there is no E-DCH transmission in the uplink, the UE automatically stops continuous DPCCH transmission and regularly transmits a DPCCH burst according to a UE DTX cycle. The UE DTX cycle, configured in the UE and in the Node B by the RNC, defines when to transmit the DPCCH even if there is no E-DCH activity. This is illustrated in Figure 14.2. The length of the DPCCH burst can be configured. Note that the DPCCH is transmitted whenever there is activity on the uplink regardless of the UE DTX cycle. There is also a possibility to set UE-specific offsets to spread the DPCCH transmission occasions from different UEs in time.

To adapt the UE DTX cycle to the traffic properties, two different cycles are defined, *UE DTX cycle 1 & UE DTX cycle 2*, where the latter is an integer multiple

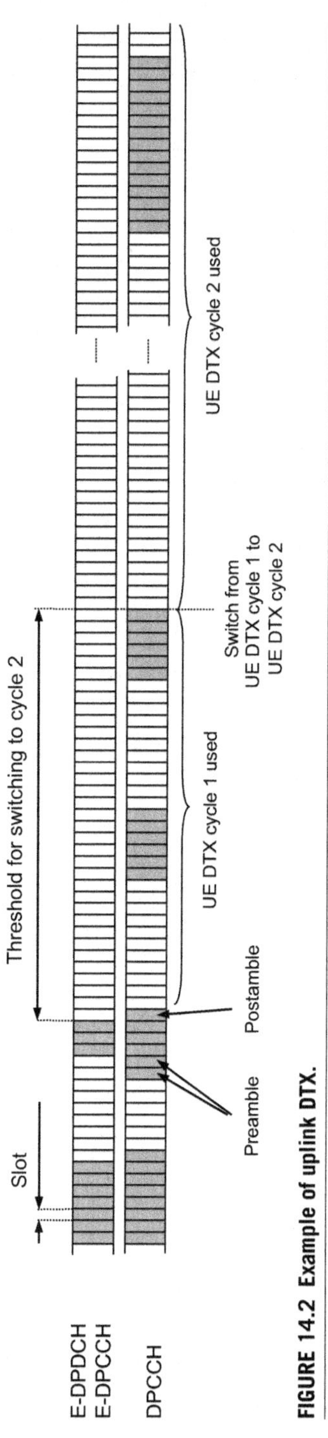


**FIGURE 14.2 Example of uplink DTX.**

of the former. After a certain configurable period of inactivity on the E-DCH, the UE switches from UE DTX cycle 1 to UE DTX cycle 2, which has less frequent DPCCH transmission instants.

Discontinuous reception in the Node B is possible thanks to the use of uplink DTX and can be useful to save processing resources in the Node B as it does not have to continuously process the received signal from all users. To enable this possibility, the network can configure the UE to allow E-DCH transmissions to start only in certain (sub)frames. A certain time after the last E-DCH transmission, the restriction takes effect and the UE can only transmit in the uplink according to the MAC DTX cycle.

During slots where the DPCCH is not transmitted, the Node B cannot estimate the uplink signal-to-interference ratio for power-control purposes, and there is no reason for transmitting a power control bit in the downlink. Consequently, the UE shall not receive any power control commands on the F-DPCH in downlink slots corresponding to inactive uplink DPCCH slots. For improved channel-estimation performance and more accurate power control, *preambles* and *postambles* are used. For UE DTX cycle 1, the UE starts DPCCH transmission two slots prior to the start of E-DPDCH, as well as ends the DPCCH transmission one slot after the E-DPDCH transmission. This can be seen in Figure 14.2. For UE DTX cycle 2, the preamble can be extended to 15 slots. The preamble and postamble is used also for the DPCCH bursts due to data transmission as well as any HS-DPCCH transmission activity, as discussed below.

Until now, the discussion has concerned user-data transmission on the E-DCH and not the control signaling on the HS-DPCCH, which also represents a certain overhead. With CPC enabled, the hybrid-ARQ operation remains unchanged and the UE transmits a hybrid-ARQ acknowledgment after each HS-DSCH reception, regardless of the UE DTX cycle. Clearly, this is sensible as hybrid-ARQ acknowledgment signaling is important for the HS-DSCH performance. It also does not conflict with the possibilities for Node B discontinuous reception as the Node B knows when to expect any acknowledgments.

For the CQI reports, the transmission depends on whether there has been a recent HS-DSCH transmission or not. If any HS-DSCH transmission has been directed to the UE within at most *CQI DTX Timer* sub-frames, where *CQI DTX Timer* is configured via RRC signaling, the CQI reports are transmitted according to the configured CQI feedback cycle in the same way as in releases 5 and 6. However, if there has not been any recent HS-DSCH transmission, CQI reports are only transmitted if they coincide with the DPCCH bursts. Expressed differently, the uplink DTX pattern overrides the CQI reporting pattern in this case (Figure 14.3).

### 14.1.2 DISCONTINUOUS RECEPTION (DRX) – REDUCING UE POWER CONSUMPTION

In "normal" HSDPA operation, the UE is required to monitor up to four HS-SCCHs in each sub-frame. Although this allows for full scheduling flexibility, it also requires the UE to continuously have its receiver circuitry switched on, leading to a non-negligible power consumption. Therefore, to reduce the power consumption, CPC

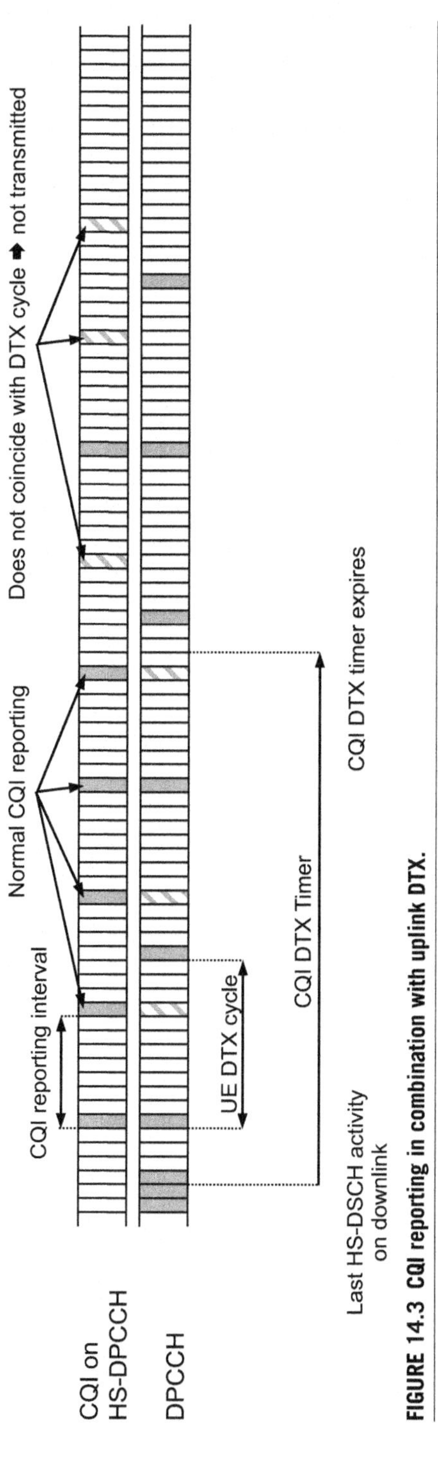


**FIGURE 14.3 CQI reporting in combination with uplink DTX.**

introduces the possibility for downlink *Discontinuous Reception* (downlink DRX). With discontinuous reception, which always is used in combination with discontinuous transmission, the network can restrict in which sub-frames the UE shall monitor the downlink HS-SCCH, E-AGCH, and E-RGCH by configuring a UE DRX cycle to be used after a certain period of HS-DSCH inactivity. Note that in this case, the UE can only be scheduled in a subset of all the sub-frames, which limits the scheduling flexibility somewhat, but for many services such as VoIP with regular packet arrival approximately once per 20 ms, this is not a major problem.

The E-HICH is not subject to DRX as this would not make sense, since it is only transmitted in response to uplink data for which the acknowledgment (ACK) or negative acknowledgment (NACK) need to be known. Hence, whenever the UE has transmitted data in the uplink, it shall monitor the E-HICH in the corresponding downlink sub-frame to receive the ACK (or NACK).

For proper power-control operation, the UE needs to receive the power control bits on the F-DPCH in all downlink slots corresponding to uplink slots where the UE does transmit. This holds, regardless of any UE DRX cycle in the downlink. Therefore, to fully benefit from downlink DRX operation, the network should use uplink DTX in combination with downlink DRX and configure the UE DTX and UE DRX cycles to match each other. An example of simultaneous use of DTX and DRX is shown in Figure 14.4.

### 14.1.3 HS-SCCH-LESS OPERATION: DOWNLINK OVERHEAD REDUCTION

In the downlink, each user represents a certain overhead for the network in terms of code usage and transmission power. The *fractional DPCH*, (F-DPCH), introduced already in release 6, addresses this issue by significantly reducing the channelization code space overhead. Another source of overhead is the HS-SCCH, used for downlink scheduling. In case of medium-to-large payloads on the HS-DSCH, the HS-SCCH overhead is small relative to the payload; however, for services such as VoIP with frequent transmissions of small payloads, the overhead compared to the actual payload may not be insignificant. Therefore, to address this issue and increase the capacity for VoIP, the possibility for *HS-SCCH-less operation* is introduced in release 7. The basic idea with HS-SCCH-less operation is to perform HS-DSCH transmissions *without* any accompanying HS-SCCH. As the UE in this case is not informed about the transmission format, it has to revert to blind decoding of the transport format used on the HS-DSCH.

When HS-SCCH-less operation is enabled, the network configures a set of predefined formats that can be used on the HS-DSCH. To limit the complexity of the blind detections in the UE, the number of formats is limited to four and all formats are limited to QPSK and at most two channelization codes. This is well matched to the small transport-block sizes, in the order of a few hundred bits, for which HS-SCCH-less operation is intended. Furthermore, the UE knows which channelization code(s) may be used for HS-SCCH-less transmission.

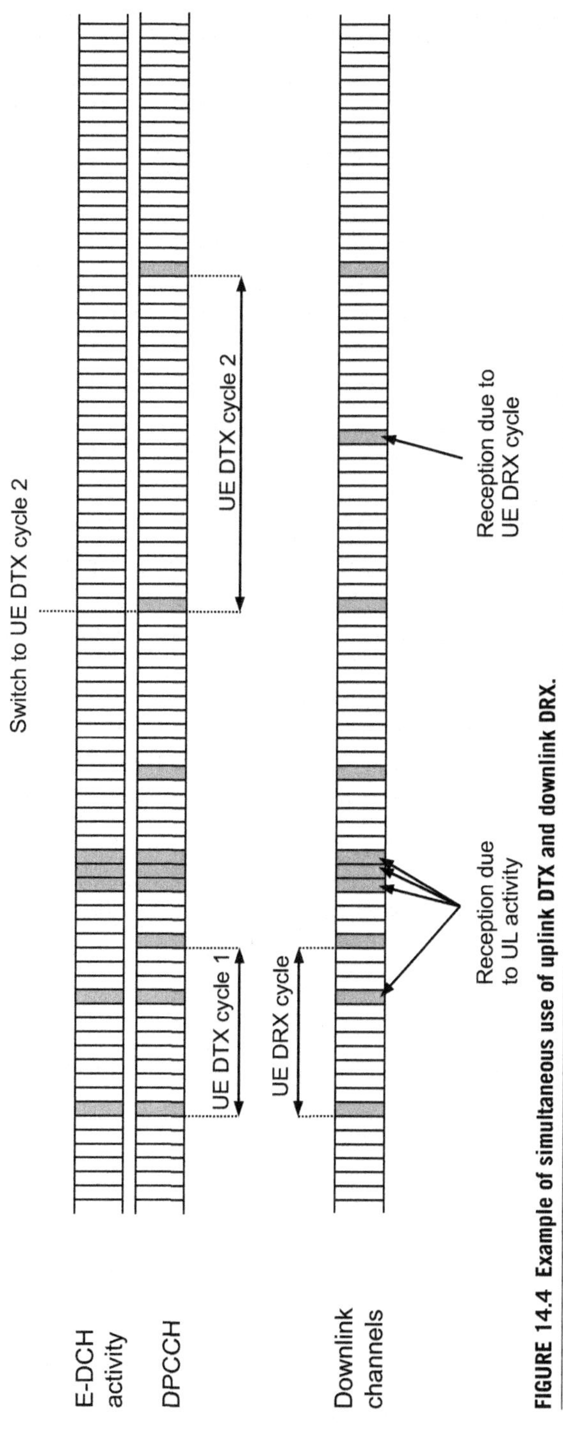


**FIGURE 14.4 Example of simultaneous use of uplink DTX and downlink DRX.**

In each sub-frame where the UE has not received any HS-SCCH control signaling, the UE tries to decode the signal received according to each of the preconfigured formats. If decoding of one of the formats is successful, the UE transmits a positive hybrid-ARQ acknowledgment on the HS-DPCCH and delivers the transport block to higher-layers. If the decoding was not successful, the UE stores the received soft bits in a soft buffer for potential later retransmissions. Note that no explicit NACK is transmitted in this case. Clearly, this would not be possible as the UE does not know whether the unsuccessful decoding was the result of the UE being addressed, but the transmission received in error, or the UE not being addressed at all. In "normal" operation, these two cases can be differentiated as there is an HS-SCCH transmission detected in the former but not in the latter case, but in HS-SCCH-less operation this is not possible.

Normally, the HS-SCCH carries the identity of the UE being scheduled. However, in case of HS-SCCH-less operation, this is obviously not possible and the identity of the scheduled UE must be conveyed elsewhere. This is solved by masking the 24-bit CRC on the HS-DSCH with the UE ID using the same general procedure as for the HS-SCCH. Because the UE knows its identity, it can take this into account when checking the CRC and will thus discard transmissions intended for other UEs.

It is possible to jointly operate HS-SCCH-less operation with "normal" transmissions. If the UE receives the HS-SCCH in a sub-frame for an initial transmission, it obeys the HS-SCCH and does not try to perform blind decoding. Only if no HS-SCCH directed to this UE is detected will the UE attempt to blindly decode the data. For backward compatibility reasons, the same procedure as in previous releases is used for CRC attachment; only for HS-SCCH-less operation is the HS-DSCH CRC masked with the UE ID.

Unlike the initial transmissions discussed so far, hybrid-ARQ retransmissions are accompanied with an HS-SCCH. The HS-SCCH is transmitted using the same coding and modulation as for normal HS-DSCH transmissions; however, a specific pattern is used for the normal HS-DSCH transport block size field, which indicates that the HS-SCCH carries information about a retransmission of an HS-SCCH-less transmission. Then, the remaining bits are reinterpreted to provide the UE with

- An indication of whether it is the first or second retransmission.
- The transport-block size from a reduced set of maximum four transport-block sizes.
- A pointer to the previous transmission attempt that the retransmission should be soft combined with. The pointer has three bits and points to TTIs offset by six or more TTIs from the current TTI.

The mapping of bits to the HS-SCCH for retransmissions is shown in Figure 14.5

The reason for this information is to guide the UE in how to perform soft combining; if this information would not have been provided to the UE, the UE would have been forced to blindly try different soft combining strategies and take a hit in complexity. Furthermore, to reduce complexity, at most two retransmissions are supported and the redundancy version to use for each of them is preconfigured.

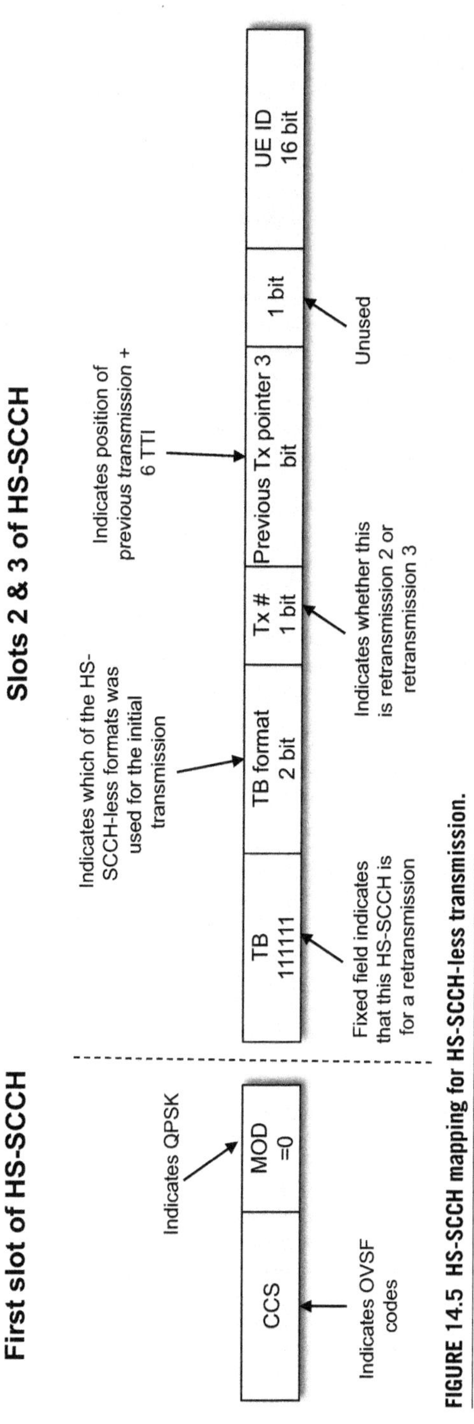


**FIGURE 14.5** **HS-SCCH mapping for HS-SCCH-less transmission.**

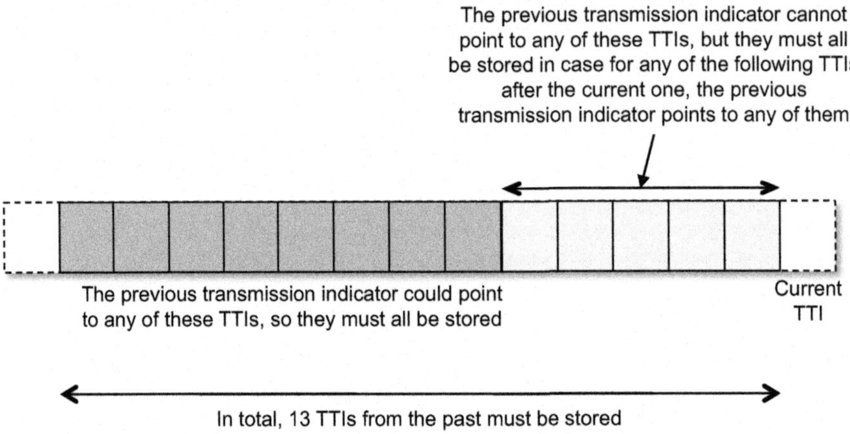


**FIGURE 14.6 Previous transmission buffering for HS-SCCH-less transmission.**

To be able to perform soft combining, the UE needs to store the soft bits from the previous decoding attempts. With a maximum of three transmissions, one initial and two retransmissions, a total of 13 sub-frames of soft buffering memory is required assuming 8 hybrid-ARQ processes (see Figure 14.6). Keeping the amount of soft buffering to a reasonable size is one of the reasons for limiting the number of retransmission attempts to a maximum of two and limiting the payload sizes for HS-SCCH-less operation.

HS-SCCH-less operation in combination with retransmissions is illustrated in Figure 14.7.

### 14.1.4 CONTROL SIGNALING

Higher-layer signaling is the primary way of setting up and controlling the CPC features. UE DTX and UE DRX cycles are configured and activated by RRC signaling. However, they are not activated immediately after call setup but only after a configurable time (known as the *Enabling Delay*) to allow synchronization and power control loops to stabilize. HS-SCCH-less operation, on the other hand, can be activated immediately at call setup.

In addition to RRC signaling, there is also a possibility for the serving Node B to switch on or off uplink DTX and downlink DRX by using reserved HS-SCCH bit patterns known as HS-SCCH orders, not used for normal scheduling operation.

HS-SCCH orders were initially introduced as part of CPC as a means of activating and deactivating CPC features. In subsequent releases, the use of HS-SCCH orders has evolved such that they can be used to provide a form of L1 control signaling. HS-SCCH orders can be seen as simplified and specific RRC signaling operations that can be controlled from the Node B. They do not replace RNC-based RRC signaling but rather complement it by enabling Node B autonomous control over RRM aspects that influence the physical layer performance.

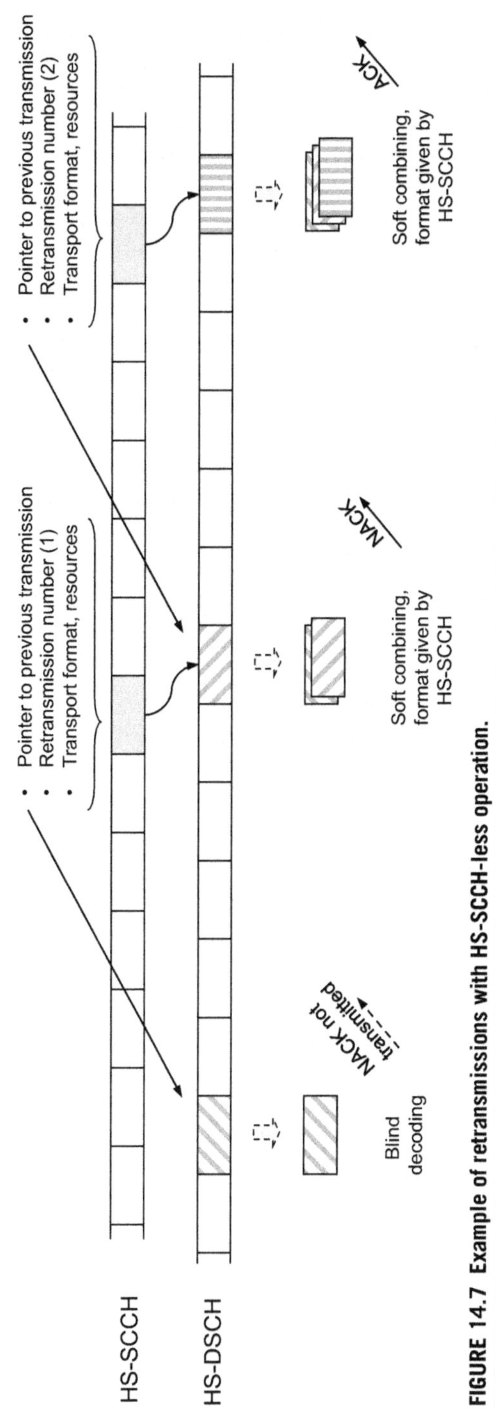


**FIGURE 14.7 Example of retransmissions with HS-SCCH-less operation.**

#### 14.1.4.1 HS-SCCH orders-basic concept

HS-SCCH orders, as the name suggests, are based on the use of the HS-SCCH. Conventionally, the HS-SCCH is used for scheduling of HS-PDSCH transmissions. However, when used for an HS-SCCH order, the HS-SCCH is not followed by an HS-PDSCH transmission, but rather the UE implements the configuration change implied by the HS-SCCH order.

Figure 14.8 illustrates the HS-SCCH order procedure. HS-SCCH orders can be transmitted using HS-SCCH type 1 (for non-MIMO) or HS-SCCH types 3 or 4. HS-SCCH may be received simultaneously in different cells, but only one HS-SCCH order may be received per cell during a TTI.

The UE decodes the order and generates an ACK to indicate that it has received the order correctly. The ACK is transmitted with the same HS-DPCCH timing as would be used for ACK/NACK for an HS-PDSCH transmission subsequent to a conventional HS-SCCH transmission with the same timing as the order, that is, 9.5 slots after the end of the HS-SCCH containing the order. The configuration change implied by the order is applied at a defined time after receiving the order; the application time depends on the type of order. It is for network implementation to decide what to do if an ACK is not received; this may be because the order has not been received by the UE or because the Node B has failed to decode the ACK correctly, although the latter case is less likely.

For releases 7-10, an HS-SCCH order is differentiated from HS-SCCH containing scheduling information by means of setting the *channelization coding set* (CCS) bit combination to 1110000, the *modulation scheme* (MS) bit to 0, and the *transport block size* (TBS) bits to 111101. On detecting this combination of CCS, MS, and TBS bits, the UE becomes aware that the HS-SCCH contains an order and reinterprets the following bits in the HS-SCCH as two groups of three bits. The first three-bit combination refers to an HS-SCCH *order type* and the second three-bit combination to a so-called HS-SCCH *order mapping*. The type and order mapping together encode the instruction to the UE.

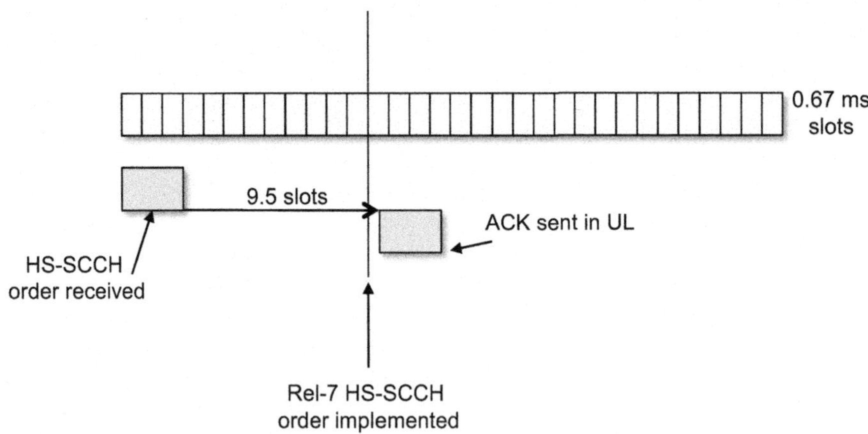


**FIGURE 14.8 HS-SCCH order procedure.**

In release 11, two additional bits, known as *extended order type,* are used together with the "order type." The two bits are mapped to the final two bits of the "transport block size" field of the HS-SCCH in place of the combination 01 used for HS-SCCH orders in the previous releases.

Figure 14.9 illustrates the bit mapping for HS-SCCH orders in releases 7-10 and release 11.

It should be noted that in addition to the extended order type, order type, and order mapping, the configuration of the UE and identity of the cell sending the order may be used for identifying the order type. Thus, the same bit configuration for an order may have different meanings depending on configuration and the status of the cell sending the order. For example, the combination of order type 000 and mapping 000 is used both for CPC-related orders in the serving cell and an enhanced serving cell change order from a non-serving cell; however, the status of the cell sending the order disambiguates the order type.

HS-SCCH orders were originally developed as part of CPC, but an overview of HS-SCCH order usage for all features up to release 11 is provided below.

#### 14.1.4.2 Usage of HS-SCCH orders in releases 7-11

This section describes the applications of HS-SCCH orders in releases 7-11. For a more complete description of the different features, the reader is referred to the appropriate chapters.

*HS-SCCH orders for CPC functionality*

CPC orders are identified by the order type 000. HS-SCCH orders can be used to activate and deactivate DRX, DTX, and HS-SCCH-less transmission for CPC. The first-order mapping bit is used for DTX, the second for DRX, and the third for HS-SCCH-less transmission (see Figure 14.9).

Activation/Deactivation according to the order is applied 12 slots after the start of the HS-SCCH order.

*HS-SCCH order for enhanced serving cell change*

An HS-SCCH order is sent by a non-serving cell as part of the enhanced serving cell change procedure to indicate that the UE should start to listen to the non-serving cell as a new cell. The combination of order type 000 and mapping 000 is used. This combination is the same as that used for deactivating DTX, DRX, and HS-SCCH-less transmission in a serving cell, but because this type of order is only sent from a non-serving cell that is a target for enhanced serving cell change, the order can be differentiated from CPC-related orders.

*HS-SCCH orders for multicarrier functionality*

The primary serving cell in uplink and downlink cannot be deactivated, but secondary serving cells may be activated or deactivated using HS-SCCH orders. It is not possible to activate an uplink carrier if the corresponding FDD duplexed downlink carrier is not active.

When a downlink carrier is activated or deactivated within a band, the order is applied 12 slots after the start of the HS-SCCH order. However, if a downlink carrier is activated or deactivated in a different band, the change in activation takes place after 18

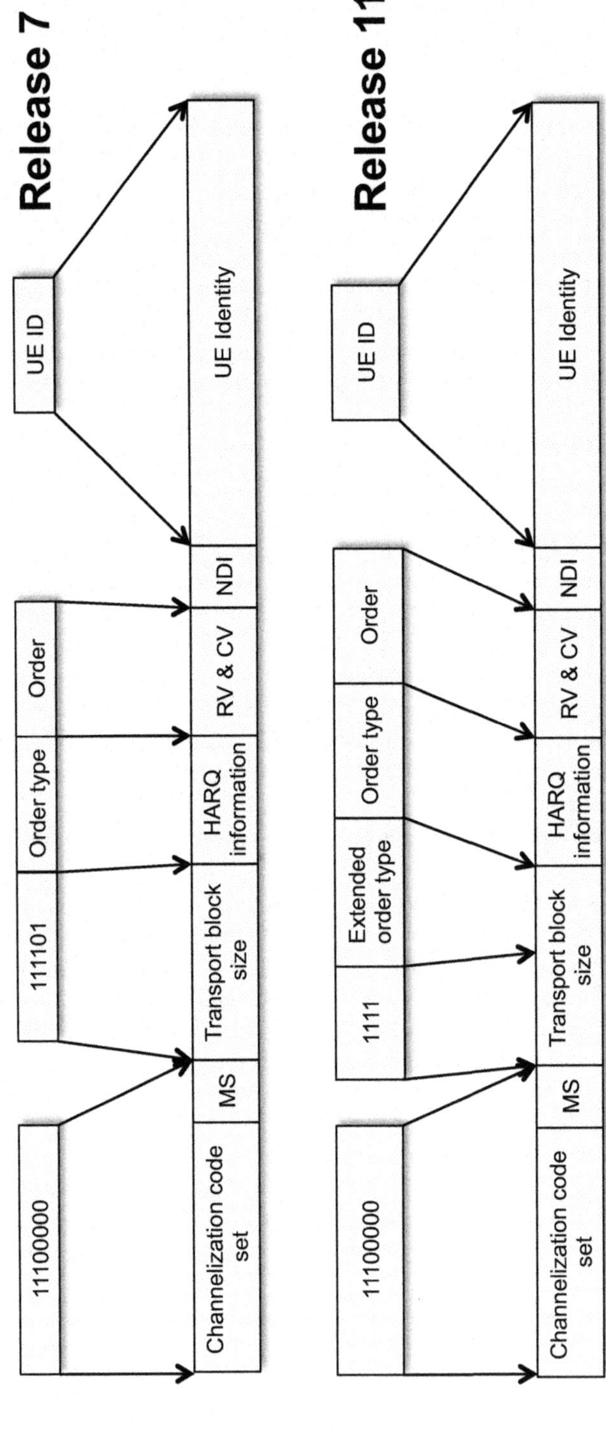


MS = Modulation Scheme; RV = Redundancy Version; CV = Constellation Version; NDI = New Data Indicator

**FIGURE 14.9 HS-SCCH order mapping.**

slots in order to allow time for activation of an additional receiver chain. Furthermore, one slot of interruption time (during which the UE may not receive in the downlink) is allowed.

When an uplink carrier is activated or deactivated, the activation or deactivation is applied at the first E-DCH TTI boundary after the HS-DPCCH that carries the ACK relating to the HS-SCCH order.

The set of orders relating to multicarrier functionality has been built up over several releases. In release 8, a single order allowed for (de)activation of the secondary serving carrier in the downlink. In release 9, this was supplemented by (de)activation of the secondary uplink carrier. In release 10, 4-carrier HSDPA was introduced and an interpretation of two potential settings for order type bits and the order mapping bits enables the UE to interpret which downlink carriers should be active and whether the uplink carrier should be active.

In release 11, 8-carrier HSDPA was introduced, and several combinations of extended order type, order type, and order mapping bits enable the UE to interpret which downlink carriers should be active and whether the uplink carrier should be active. The order tables for release 10 and release 11 enable downlink carriers to be specified as active or inactive individually.

*HS-SCCH orders for multiflow functionality*

In multiflow, the UE can receive HS-PDSCH from more than one geographical area. Carriers belonging to the same area as the serving cell are called the serving cell and assisting serving cell; carriers belonging to the second geographical area are called secondary serving cells and assisting secondary serving cells. Carriers in both the assisted and assisting region can be activated and deactivated using the order mappings for multicarrier described above. When received from the assisted cells (serving cell or the secondary serving cell), the orders are interpreted as applying to the carriers in the same geographical area as the serving cell. When received from any of the assisting cells, the orders are interpreted as applying to carriers from the assisting area.

*HS-SCCH orders for uplink CLTD and MIMO functionality*

An uplink CLTD-capable UE possesses two transmit antennas. The UE may be configured, by means of HS-SCCH orders, either to transmit from antenna 1 or from antenna 2, or to apply closed-loop Tx diversity and transmit from both antennas. Switching between antenna 1 and antenna 2 may be useful for implementing some types of slow Tx antenna switching.

The order type 011 and extended order type 01 are used for indicating that the order mapping relates to uplink CLTD. The order is applied by the UE at the first E-DCH boundary following the HS-DPCCH that contains the ACK relating to the order.

*HS-SCCH orders for 4-branch MIMO functionality*

An HS-SCCH order may be used to activate or deactivate use of the demodulation common pilots for 4-branch MIMO. The order type 010 and extended order type 11 are used for indicating that the order relates to 4-branch MIMO.

*HS-SCCH orders in CELL_FACH and CELL_PCH*

In general, HS-SCCH orders are not applicable when the UE is operating in the CELL_FACH state. However, the release 11 CELL_FACH enhancements enable a Node B triggered activation of HS-DPCCH transmission. The order is identified by the order type 000 and the extended order type 01 (this is the same as the order type used for CPC-related orders when the UE is in CELL_DCH state). As soon as a UE in CELL_FACH or CELL_PCH receives this order, it must make a RACH access in the next available RACH access slot and then start transmitting HS-DPCCH. Note that when in CELL_PCH, a paging indicator must be used to indicate to the UE that it should listen to HS-SCCH, and then the HS-SCCH order should be transmitted during one of the next five TTIs.

## 14.2 ENHANCED CELL_FACH

Some types of data traffic pose new challenges for both devices and networks, challenges that are not properly addressed by CPC:

- Even with CPC features such as DTX and DRX, maintaining devices in CELL_DCH for long periods of time because of potential and unpredictable interactions is not feasible within reasonable constraints of network resource usage and UE modem power impact.
- The most efficient states during times at which the UE has no data to transfer are idle mode and CELL_PCH. However, moving devices from idle or CELL_PCH to CELL_DCH is associated with latency, which reduces user experience, and a disproportionate signaling overhead, taking into account the sizes of individual packets that are transferred.
- Maintaining a large number of devices in connected mode states, and/or managing frequent transitions between states when aggregated across a large number of cells, leads to a potentially very high processing load within RNCs.
- The CELL_FACH state, which was traditionally envisaged for handling small interactions with the network, does not provide sufficient capacity for smartphone-like packets. Furthermore, maintaining users in the release 99 CELL_FACH state is still not network resource or UE power consumption efficient enough.

HSDPA, HSUPA, and CPC are enhancements of the CELL_DCH state. In order to provide sufficient flexibility within the specifications for handling all types of traffic, 3GPP embarked during release 7 on a process of enhancing the functionality available to the UE when in the CELL_FACH and CELL_PCH states. Downlink operation was addressed as first priority in release 7 and was followed by uplink operation in release 8. Both downlink and uplink operations in CELL_FACH were further improved during release 11.

### 14.2.1 DOWNLINK ENHANCEMENTS: HSDPA IN CELL_FACH

In the CELL_FACH state, control-plane information may consist of *broadcast information* (BCCH), *common control information* (CCCH), or *dedicated control information* (DCCH). In addition to control information, small amounts of user-plane information can be transferred using the *Dedicated Traffic Channel* (DTCH). In the release 99 specifications and until release 7, BCCH is mapped to the BCH transport channel, which is mapped to the P-CCPCH physical channel. CCCH, DCCH, and

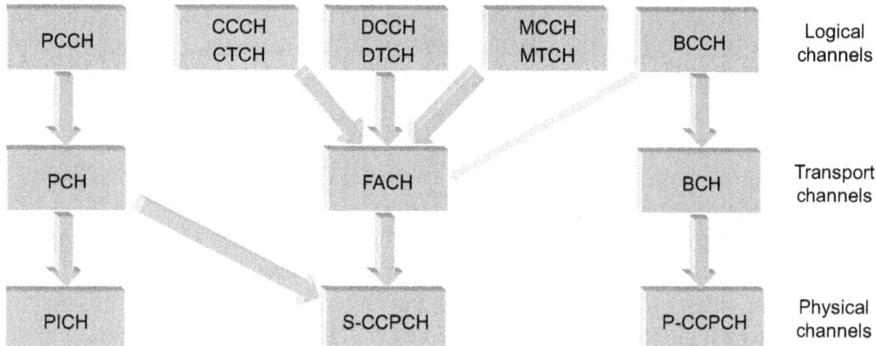


**FIGURE 14.10 Transport and physical channel mappings for rel-99 CELL_FACH.**

DTCH may be mapped to the FACH transport channel, which in turn is mapped to one or more S-CCPCH physical channels, as indicated in Figure 14.10.

Traditionally, the CELL_FACH state was used as a transitional state and control information or very low volumes of user-plane information were transferred. For handling of smartphone packet data, however, using the release 99 architecture carries several disadvantages:

- Both S-CCPCH and HS-PDSCH must use the same downlink OVSF code tree. Allocation of S-CCPCH codes and the total code space allocated to the Node B scheduler for HS-PDSCH is managed by the RNC on a semi-static basis. The RNC must estimate the relative loads in CELL_DCH and CELL_FACH and allocate code space appropriately. Semi-static management of the code space inevitably leads to resource under-utilization, as illustrated in Figure 14.11.

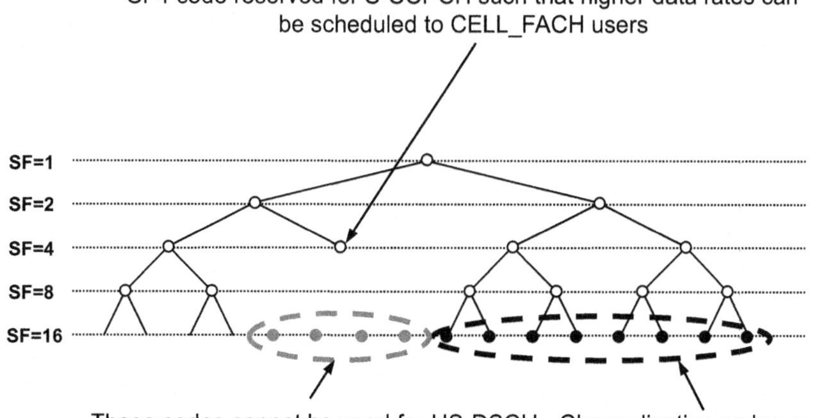


**FIGURE 14.11 Code tree blocking due to S-CCPCH.**

In the figure, an SF4 code is reserved for S-CCPCH to allow for higher data rates on CELL_FACH. This means that four SF16 codes are not available for HS-PDSCH transmission during S-CCPCH transmissions, even at time instances during which only low rates are being transmitted on S-CCPCH. Note that to reduce the need for UE despreading in CELL_FACH, the UE is only required to receive one or at most two S-CCPCHs in CELL_FACH; thus, configuring multiple small S-CCPCHs is also not an option.

- The FACH transport channel uses a fixed TTI size and cannot take advantage of Node B–based advanced resource allocation techniques and hybrid-ARQ.
- In the pre–release 7 CELL_FACH state, the UE is required to always monitor FACH, which implies a continuously activated UE receiver and an associated battery drain.

In developing methods for enhancing functionality of the CELL_FACH state, 3GPP took into account that efficient methods for Node B–based resource management and UE scheduling have been developed in release 5 in HSDPA for the CELL_DCH state. This functionality was reused to a large extent in developing enhancements to the CELL_FACH state, both in order to make good reuse of existing standards and implementations and in order to maximize the potential for the Node B to optimally use common resources for both CELL_DCH and CELL_FACH users.

The basic premise of the release 7 enhanced CELL_FACH state is that it enables the UE to be scheduled with HS-PDSCH while it is in the CELL_FACH or CELL_PCH states.

#### 14.2.1.1 Physical layer operation

From a UE physical layer perspective, in the CELL_FACH state, the basic scheduling mechanism is the same as in CELL_DCH; that is, the HS-SCCH is used for indicating scheduling information to the UE and then data is transferred using HS-PDSCH. Some restrictions exist when scheduling a UE in CELL_FACH. In particular, unlike CELL_DCH, in which up to four HS-SCCH codes can be configured that the UE must monitor, the UE monitors only one HS-SCCH code. Only one transport block may be scheduled per TTI; single- and dual-stream MIMO cannot be used in CELL_FACH.

From a network perspective, the same set of HS-PDSCH codes can be reused between CELL_DCH and CELL_FACH UEs, and it is a task for the Node B scheduler implementation to make instantaneous decisions on which codes should be used for scheduling which users, as shown in Figure 14.12.

In release 7, it is not possible for the UE to transmit HS-DPCCH in the uplink when scheduled with CELL_FACH in the downlink. Thus no CQI and ACK/NACK signaling is available. However, the RNC may include an indication of UE measured CPICH information in the Iub frame protocol when sending MAC-SDUs to the Node B MAC-hs so that the Node B can make an estimate of the supported data rate or required transmit power (see Figure 14.13). The UE may send measurement reports on RACH to the RNC. If the UE is configured with a new H-RNTI (see

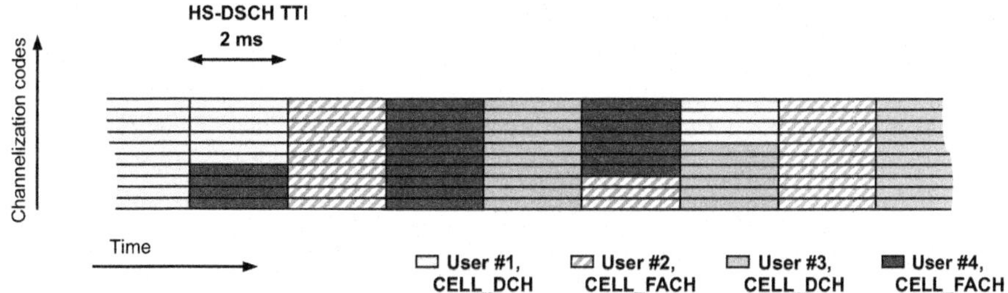


**FIGURE 14.12 Sharing of HS-PDSCH codes between CELL_DCH users and CELL_FACH users.**

Section 14.2.1.2), then it should immediately send measurement results on RACH for use by the network, because being configured with H-RNTI is a prelude to user-plane data transfer. Because no ACK/NACK signaling is available, the Node B must blindly decide whether to schedule hybrid-ARQ retransmissions.

In release 8, CELL_FACH enhancements for uplink were introduced to the specifications, as described in Section 14.2.2. If a release 8 UE has data to send in the up-

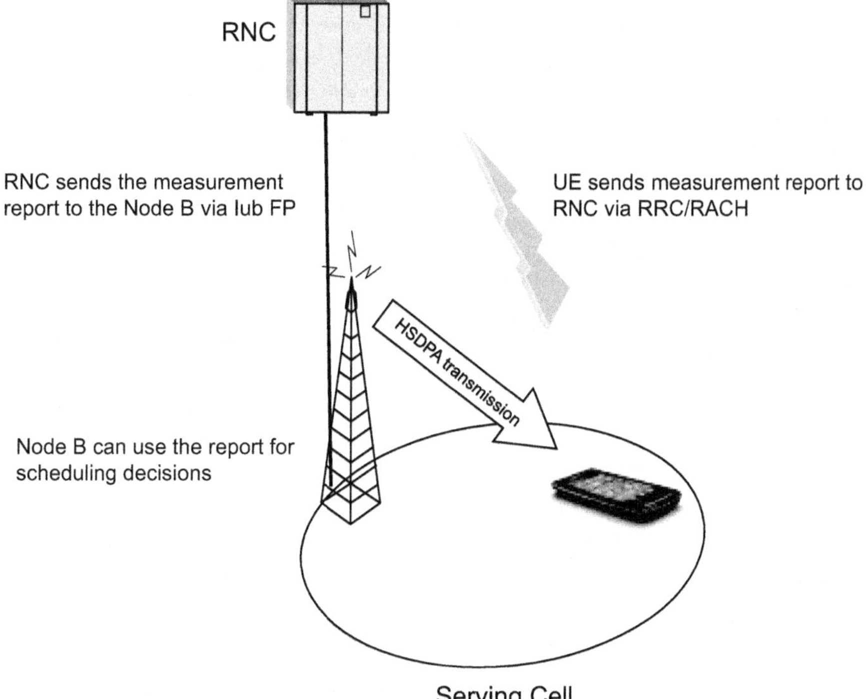


**FIGURE 14.13 Use of RRC measurements for Node B scheduling with release 7 Enhanced CELL_FACH.**

link, then it will initiate the Enhanced RACH procedure and transmit uplink E-DCH. If the UE has started transmitting on uplink E-DCH, then it can also be configured to send HS-DPCCH relating to the downlink. It should be noted, however, that HS-DPCCH transmission is only possible when the UE has initiated E-DCH because of having uplink data and only until the point in time at which the common E-DCH resources are released. For TCP/IP-type applications, it is possible that the application will generate IP ACK packets in response to receiving packets in the downlink, in which case uplink data transfer will be initiated. Also, as indicated above, once allocated with an H-RNTI, the UE will need to initiate a RACH procedure for sending measurement results.

Because it is not in CELL_DCH state, the UE does not need to transmit uplink DPCCH or receive downlink DPCCH or F-DPCH. In release 7, however, the UE is required to continuously monitor HS-SCCH. For making neighbor-cell measurements for mobility, the release 99 FACH measurement occasions functionality is reused. During FACH measurement occasions, the UE does not need to monitor HS-SCCH. Continuous monitoring of HS-SCCH requires that the UE receiver is continuously active when the UE is in CELL_FACH state. For some types of applications in which the duration between interactions with the network is relatively short, it may be desirable to hold the UE in the CELL_FACH state for extended periods of time, in which case continuous activation of the receiver requires continuous power usage by the receiver part of the UE modem. However, in general, it is not necessary for the UE to be able to receive downlink data in the FACH state within a few TTIs of the data arriving at the Node B, as the data is not delay sensitive. Thus, in release 8, the concept of DRX cycles was introduced to the Enhanced CELL_FACH operation.

With DRX operation, after the last time a UE is scheduled with HS-PDSCH, an inactivity timer is started. Once the timer expires, the UE is no longer expected to continuously monitor HS-SCCH. The UE is only required to monitor HS-SCCH periodically, and in the intervening time, the UE is free to deactivate its receiver. The inactivity timer, the periodicity of the UE requirement on monitoring HS-SCCH, and the duration of time during each period for which the UE must monitor HS-SCCH are configurable by the network. A network operator will optimize these parameters to trade-off responsiveness of the UE, flexibility for the Node B scheduler (which can only schedule the UE while it is monitoring HS-SCCH), and the potential for UE battery saving. If a release 8 UE is configured with DRX, then it is expected to make mobility-related inter-frequency and inter-RAT measurements during the time periods in which it is not required to monitor HS-SCCH, as illustrated in Figure 14.14. UE modem vendors are able to plan when to make measurements and when to deactivate the receiver in order to meet the performance requirements for mobility measurements and optimize the model power reduction potential. It should be noted that DRX can only be active when the UE is configured with a dedicated H-RNTI (see Section 14.2.1.2).

In release 7, the functionality relating to operation in CELL_PCH state was also extended. Similarly to CELL_FACH state, HS-PDSCH can be scheduled to a release 7 UE in CELL_PCH. However, the details of the scheduling operation and link

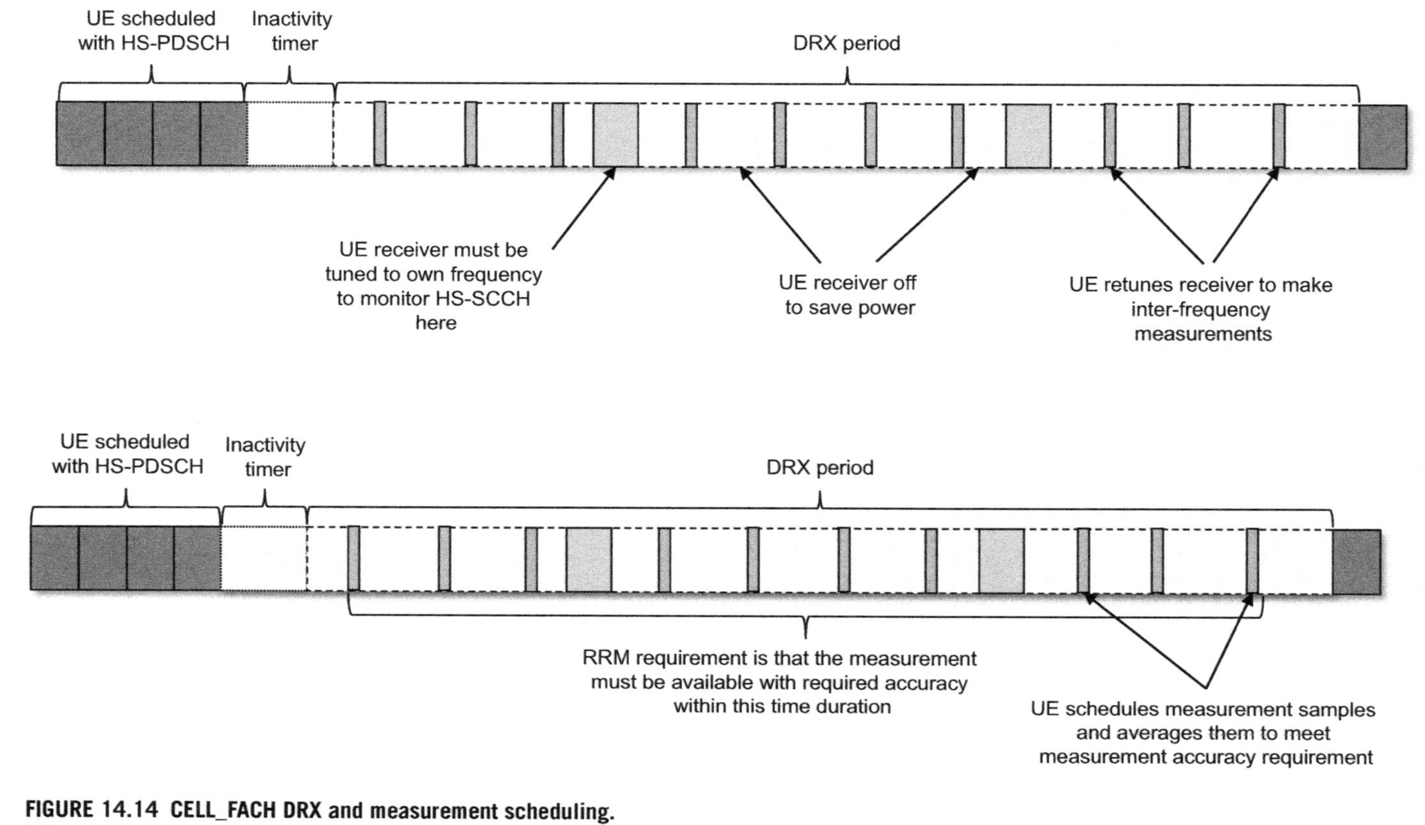


**FIGURE 14.14 CELL_FACH DRX and measurement scheduling.**

management differ for CELL_PCH. Because CELL_PCH is intended for holding the UE in a reduced power consumption mode, the release 99 principle is maintained that a UE in CELL_PCH periodically monitors a paging indicator. If the paging indicator is set for the group to which the UE corresponds, then the action that the UE takes depends on whether the UE has been configured with a dedicated H-RNTI or is operating with a common H-RNTI (see Section 14.2.1.2).

If the UE is operating with a common H-RNTI then, as shown in Figure 14.15a, the fact that a paging indicator has been set indicates that an HS-PDSCH transmission will follow with a defined time between the paging indicator and the HS-PDSCH. Furthermore, after the first TTI of the HS-PDSCH, the HS-PDSCH transmission can be repeated in consecutive sub-frames for *M* sub-frames, where *M* is configurable by the network, and the UE can combine the received HS-PDSCH in such sub-frames.

If the UE is configured with a dedicated H-RNTI then, as shown in Figure 14.15b, the UE must monitor the HS-SCCH for *N* consecutive sub-frames (*N* configurable by

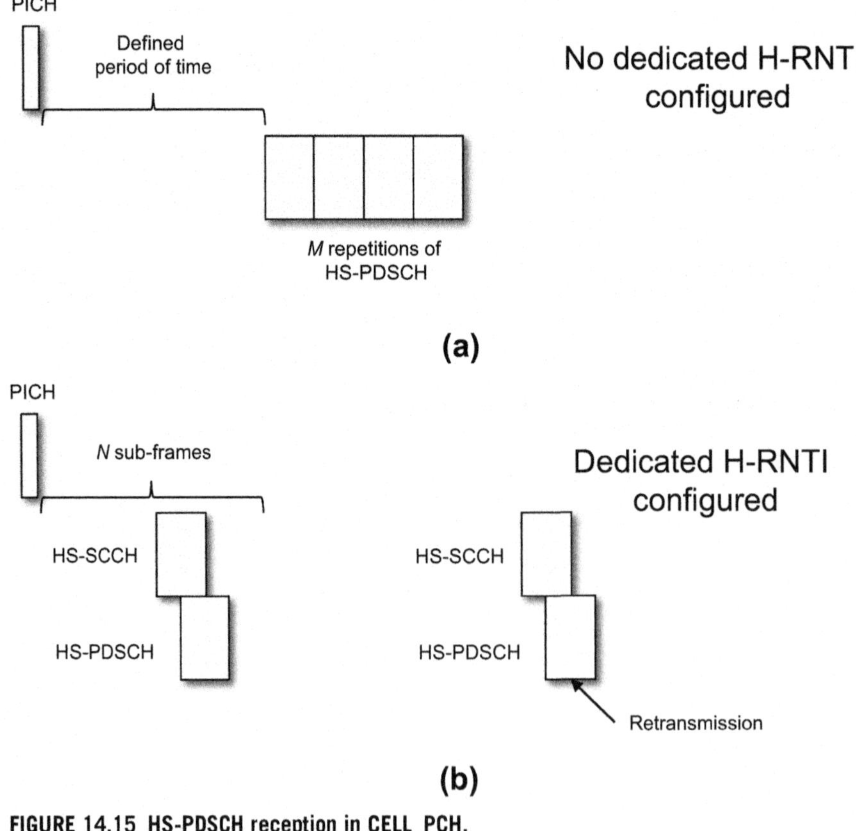


**FIGURE 14.15 HS-PDSCH reception in CELL_PCH.**

the network). If an HS-SCCH is detected, the UE attempts to decode it and also moves to CELL_FACH state.

For enhanced CELL_PCH state operation, there is no HS-DPCCH uplink feedback in any release of the 3GPP specifications.

#### 14.2.1.2 RRC protocol operation

If a UE is configured to operate Enhanced CELL_FACH in downlink (i.e., HS-PDSCH in CELL_FACH), then the UE does not monitor S-CCPCH, unless the MBMS transport channel MTCH is mapped to S-CCPCH. Thus, any broadcast channel parameters that are mapped to S-CCPCH must also be transmitted using HS-PDSCH.

In traditional CELL_DCH operation, the H-RNTI identifier is used to identify the UE, such that it can identify when it is scheduled in an HS-SCCH message. For Enhanced CELL_FACH operation, three types of H-RNTI identifiers may be configured:

- The broadcast H-RNTI may be common to all UEs configured with Enhanced CELL_FACH. It is used for scheduling HS-PDSCH transmissions that contain BCH SIBs.
- The common H-RNTI may be common to all UEs configured with Enhanced CELL_FACH.
- The UE may be configured with a dedicated H-RNTI.

UEs configured with Enhanced CELL_FACH must always decode transmissions scheduled with the broadcast H-RNTI and in addition either the common H-RNTI or, if configured, the dedicated H-RNTI.

Enhanced CELL_FACH MAC operates only flexible L2 (i.e., MAC-i). L2 hybrid-ARQ operation is much the same as is the case with CELL_DCH HSDPA operation. However, in this case, if a different H-RNTI is used for a particular hybrid-ARQ process compared with the previous transmission on the same hybrid-ARQ process and following a NACK, the hybrid-ARQ buffer is flushed and the PDU is assumed to be new data.

A common configuration for Enhanced CELL_FACH is broadcast in system information. For specific UEs, the Enhanced CELL_FACH parameter settings can be set to UE-specific values by UE-specific signaling. The UE capabilities in respect of Enhanced CELL_FACH are included in RRC connection request and cell update messages.

L2 operation in CELL_PCH mode is somewhat similar to CELL_FACH mode. However, when in CELL_PCH mode, the HS-PDSCH scheduled to the UE can only contain the PCCH paging channel. The UE may monitor only the common H-RNTI or both the common and dedicated H-RNTI. As is noted in Section 14.2.1.1, the behavior in respect to hybrid-ARQ retransmissions is somewhat different in CELL_PCH, in which retransmissions, if any, are made in consecutive sub-frames. The need for scheduling of hybrid-ARQ retransmissions is dependent on whether the common H-RNTI or dedicated H-RNTI was used for scheduling.

## 14.2.2 UPLINK ENHANCEMENTS: E-DCH OPERATION IN CELL_FACH

The downlink HSDPA operation was standardized in release 7, with some small improvements and corrections in release 8. During release 8, 3GPP extended uplink operation in CELL_FACH.

In the uplink, the UE may transmit CCCH, DCCH, or small amounts of user data in DTCH (see Figure 14.16a). These are mapped to the RACH transport channel. Release 99 RACH operation is described in Chapter 6 and consists of two stages: preamble ramping, in which the UE announces to the network that it has something to transmit and the message part, as depicted in Figure 14.17. The amount of data that can be carried in RACH is limited by the maximum RACH message size.

In designing an improved random access scheme, 3GPP attempted to reuse much of the HSUPA functionality introduced in release 6. Fundamentally, the operation in Enhanced CELL_FACH consists of a modification to the release 99 preamble ramping and network acknowledgment protocol followed by a restricted E-DCH operation. CCCH, DCCH, or DTCH may be mapped to E-DCH in CELL_FACH state, as illustrated in Figure 14.16b, and similarly to CELL_DCH operation, multiple MAC-d flows (that is, multiple logical channels) may be multiplexed onto the E-DCH transport channel.

### *14.2.2.1 Physical layer operation*

The physical layer operation consists of two stages: preamble ramping and then E-DCH transmission as depicted in Figure 14.18. When the UE has data to transmit, the MAC initiates a preamble ramping operation at the physical layer. The physical layer selects a RACH slot and signature sequence according to the *Access Service Class* and begins transmitting preambles with increasing Tx power. The preamble selection and transmission procedure is exactly the same as in the case of the release 99 RACH procedure, as described in Chapter 6.

If the network detects a preamble, it may respond using the AICH channel. For Enhanced CELL_FACH operation, an *extended* AICH (E-AICH) channel has been

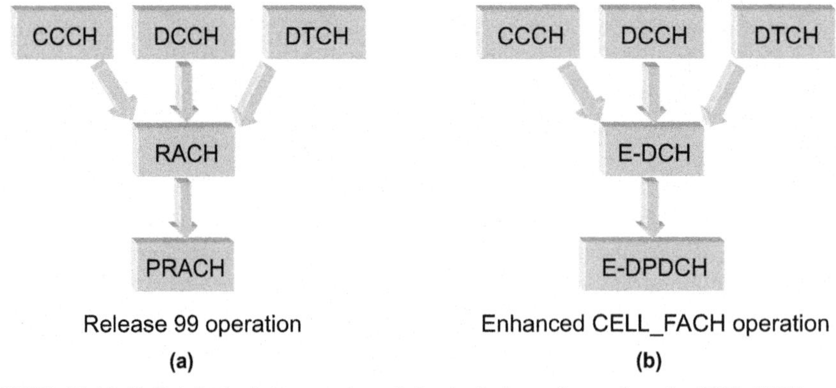


**FIGURE 14.16 Uplink logical, transport, and physical channel mappings for CELL_FACH.**

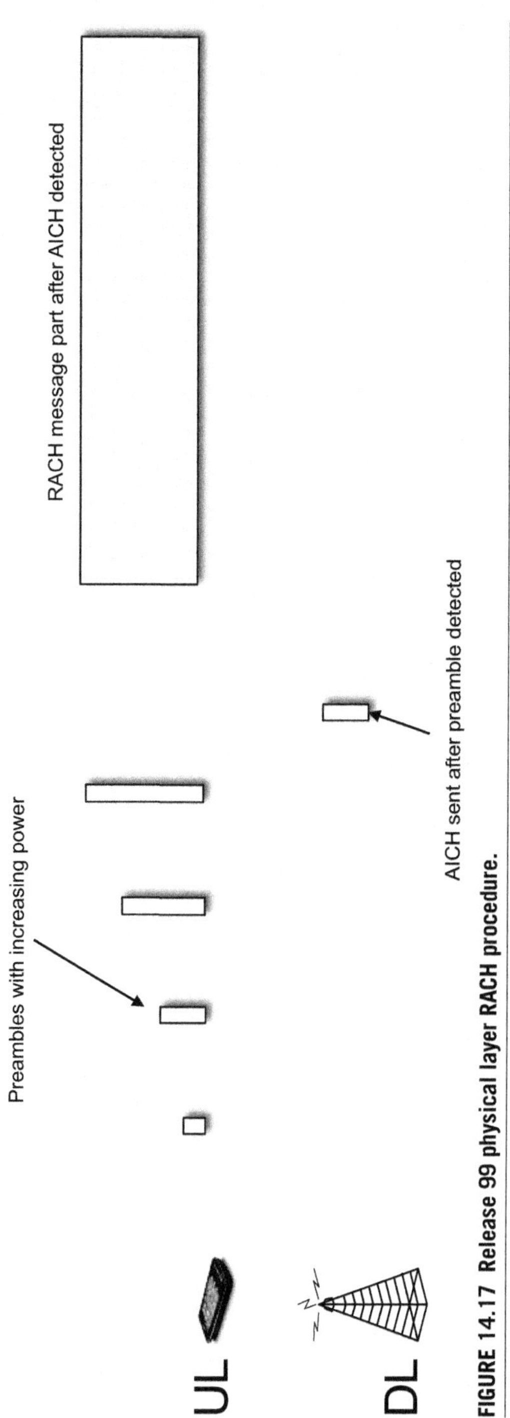


**FIGURE 14.17** **Release 99 physical layer RACH procedure.**

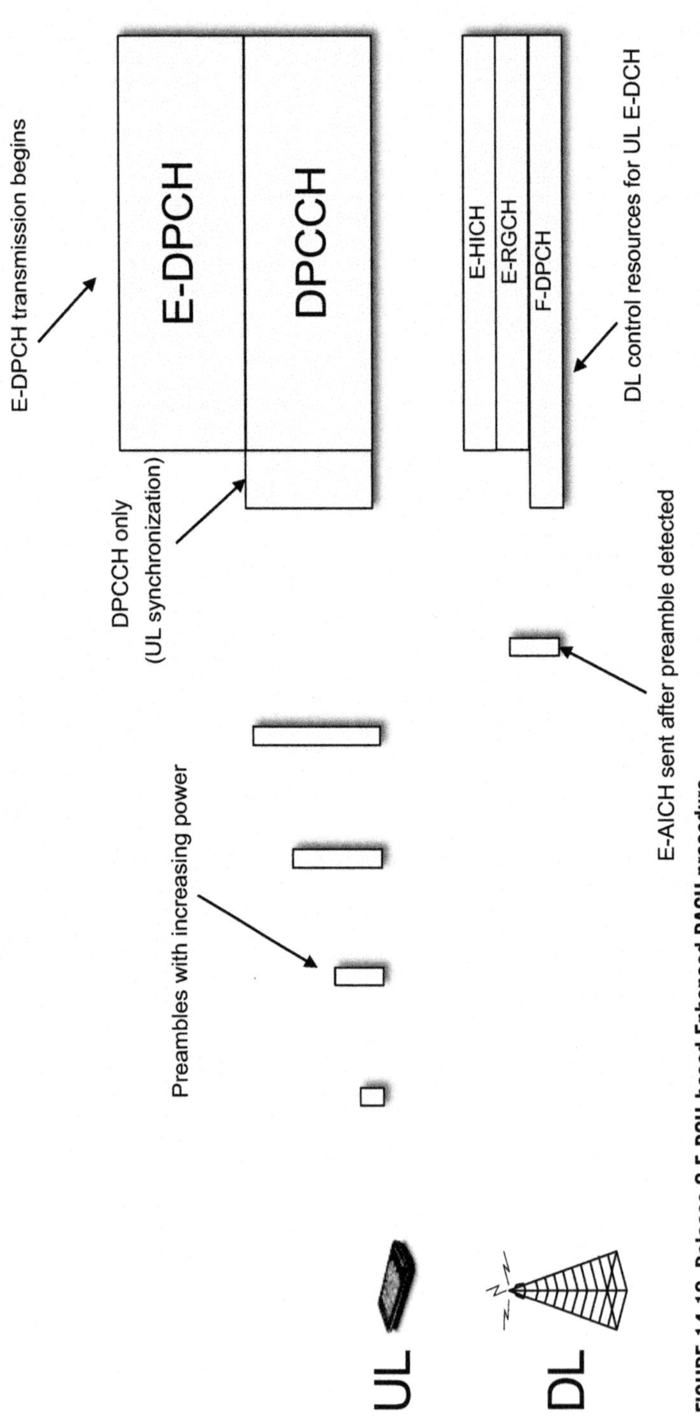


**FIGURE 14.18** Release 8 E-DCH-based Enhanced RACH procedure.

introduced. In the release 99 AICH channel, a signature is transmitted that maps to the preamble that has been detected and can take one of two values, corresponding to ACK (in which case the UE can continue with the RACH message part) or NACK (in which case the UE must either start a new preamble ramping sequence or report to higher-layers that the RACH access failed). For the Enhanced CELL_FACH operation, the E-AICH may transmit one or two E-AI signature sequences simultaneously. In case two E-AI sequences are transmitted, the UE must be able to detect both sequences.

The first sequence corresponds to the preamble sequence that has been detected and can indicate one of two values. The first value indicates "ACK," and in this case, the UE can proceed in setting up the E-DCH channel. In order to operate E-DCH, the UE needs to know which set of downlink common control resources it should monitor. The downlink common control resource set is inferred from the index of the particular signature sequence that has been selected, which in turn relates to the randomly selected preamble index. Figure 14.19 shows the operation when an ACK is received on the first E-AICH sequence.

Alternatively, the first sequence may indicate its second value, in which case the UE must detect which second sequence index has been transmitted and what second sequence value has been applied. One of the secondary sequences can be used to indicate the value "NACK." This is illustrated in Figure 14.20a. If the UE detects this sequence and the "NACK" value, then it must restart the preamble ramping procedure. For the rest of the sequences, the combination of sequence and value indicates a set of downlink control resources that the UE should monitor for operating E-DCH. This is illustrated in Figure 14.20b. In this case, the signature used on the second E-AI sequence together with the sequence index references one of the available sets of downlink control resources.

The E-AICH channel provides the network with flexibility in allocating downlink control resources for E-DCH operation. If the network would have to rely on the control resources being implied from the AICH sequence (as is the case when ACK is sent), then a problem would arise if, for example, a UE would make a random access attempt and select a preamble corresponding to an AICH sequence corresponding to an E-DCH control resource set that is already in use for another UE. In such a case, without the ability to redirect the UE to use a different set of E-DCH resources using the E-AICH, the network would need to reject the UE's random access attempt, despite the potential existence of other unused control resources. Such issues could severely limit the capacity of the extended RACH protocol. Figure 14.21 depicts the potential problem that could arise if only one signature sequence would be used.

Once the UE has received an ACK, or has received a secondary E-AI indicating an alternative set of downlink resources, it can then begin the uplink synchronization procedure for E-DCH transmission. The downlink control resource set consists of an indication of codes that are temporarily used for indicating F-DPCH, E-RGCH, E-AGCH, and E-HICH to the UE. These downlink resources are used for controlling the uplink transmission.

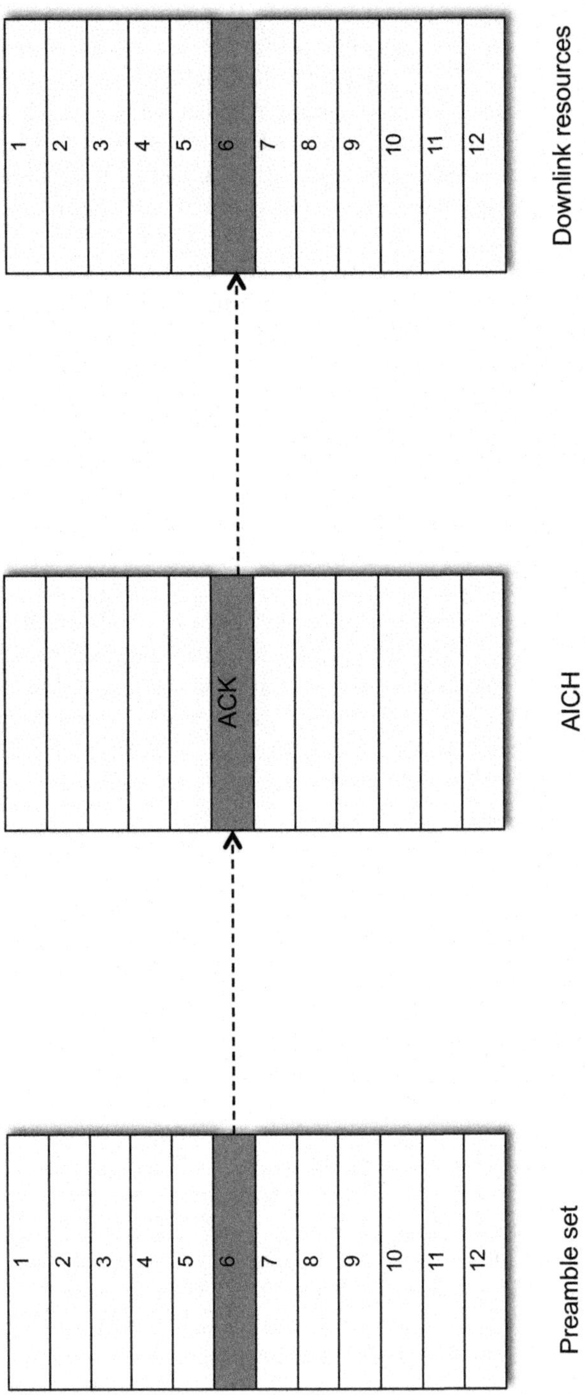

**FIGURE 14.19** Mapping between preambles, AICH sequences, and downlink control resource groups.

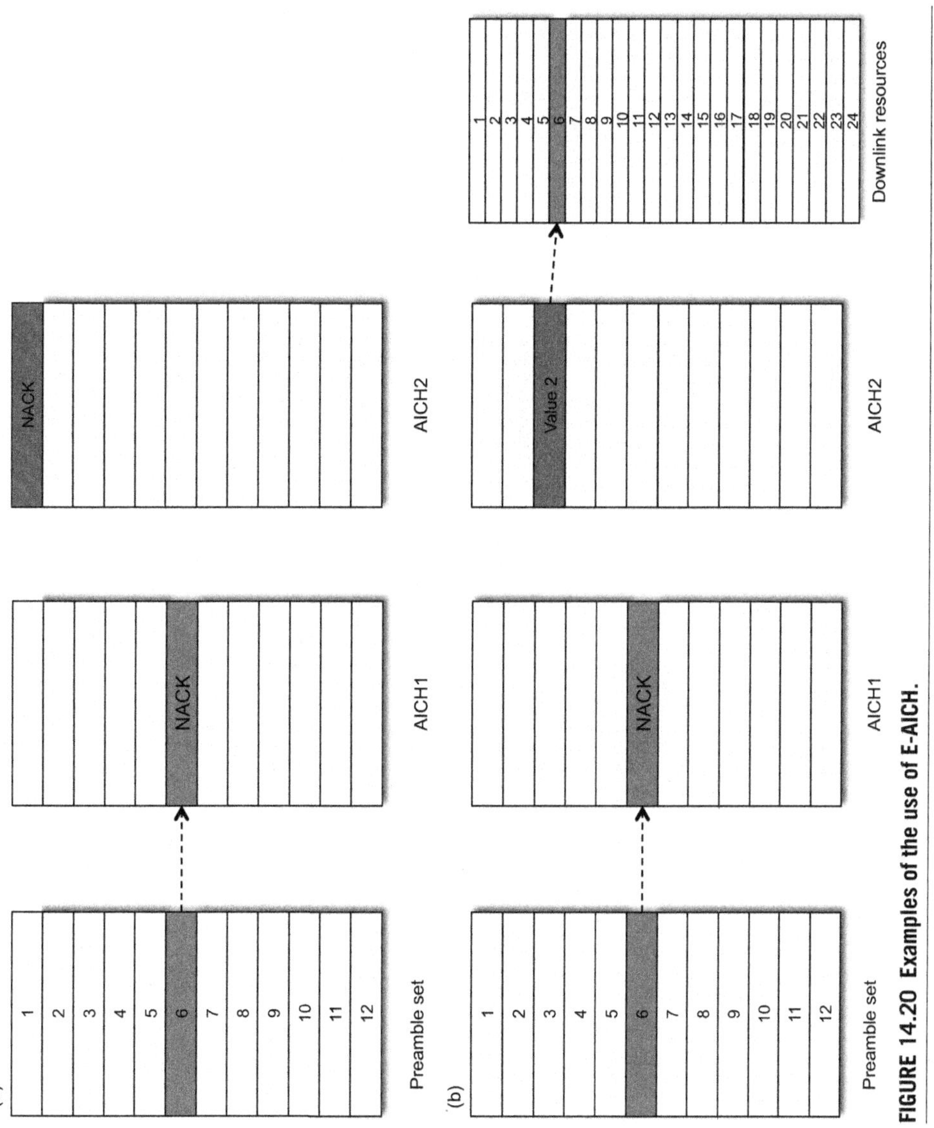


**FIGURE 14.20** Examples of the use of E-AICH.

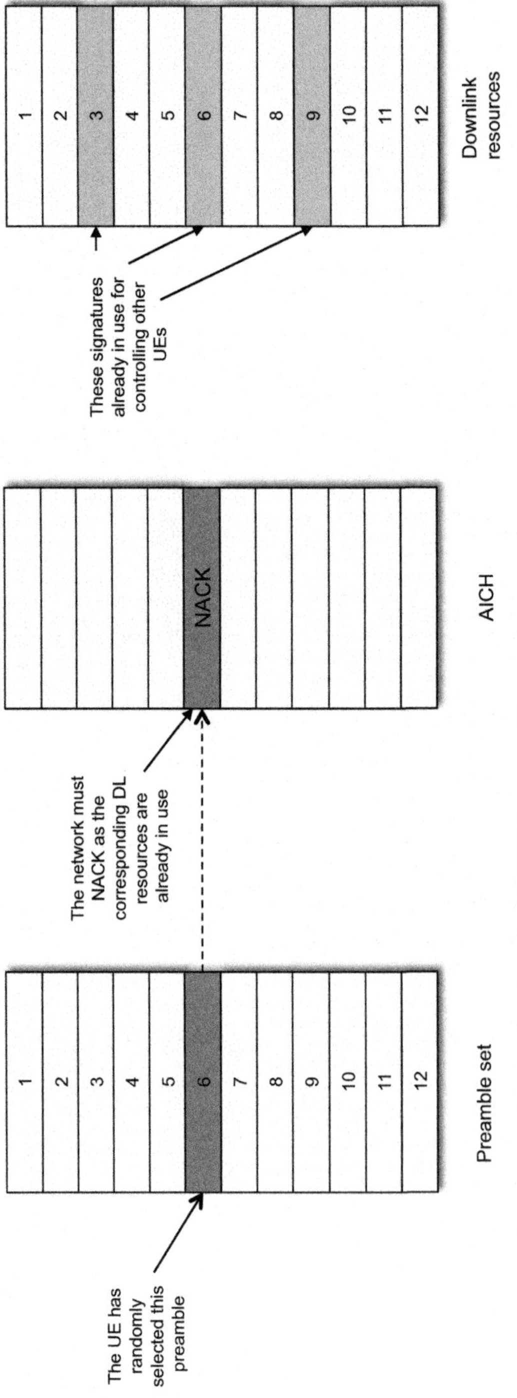


**FIGURE 14.21 Problem of resource blocking when E-AICH is not used.**

Following completion of preamble ramping, to start uplink E-DCH transmission, the UE must initiate a so-called synchronization procedure. The procedure known in the 3GPP specifications as "synchronization procedure AA" is used. In essence, this synchronization procedure consists of

- First, the network transmits F-DPCH, and the UE establishes frame and chip synchronization.
- Then, the UE transmits DPCCH in uplink for 40 ms, starting with a transmit power that is related to the power of the last transmitted preamble and then following the F-DPCH TPC commands.
- Finally, the UE estimates the F-DPCH quality during the 40 ms. If the quality falls below a threshold, the synchronization is judged to have failed. Otherwise, the synchronization is judged to have been effective. At this stage, the UE is transmitting DPCCH and is under correct power control.

Following successful synchronization, the UE can start E-DCH transmission. The E-DCH scheduling and transmission procedure operates in a similar manner as CELL_DCH, using the common downlink control resources. The network is able to configure either the 10 ms TTI or the 2 ms TTI for E-DCH operation in CELL_FACH using cell-specific signaling in the broadcast channel.

#### *14.2.2.2 Scheduling and RRC operation*

An initial grant level is set by higher-layers for the UE to use for starting transmissions. The differences between E-DCH operation in CELL_FACH and E-DCH operation in CELL_DCH from a UE perspective are as follows:

- In CELL_FACH in releases 8-10, there is no active set and soft and softer handover are not operated. Thus, the UE does not receive F-DPCH or non-serving E-RGCH from neighbor cells. The neighbor cells have no means to control inter-cell interference arising from a UE that is operating in the CELL_FACH state. However, typically, transmissions in the CELL_FACH state, and the associated interference, are short-lived. Note that in release 11, one of the additional enhancements introduced to the E-DCH transmission in the CELL_FACH state was reception of non-serving E-RGCH from neighbor cells; see Section 14.2.3 for a more complete description.
- There is no secondary E-AGCH E-RNTI in CELL_FACH, unlike CELL_DCH operation.
- E-DCH in the CELL_FACH state can only operate with a single transmit branch (i.e., no Tx diversity) and QPSK transmission.

From a network perspective, some restrictions apply to scheduler operation in CELL_FACH in order to simplify operation. For releases 8-10, restrictions include the following:

- The network cannot use the non-serving E-RGCH to control interference from CELL_FACH UEs in neighbor cells.

- Individual hybrid-ARQ processes cannot be deactivated (prior to release 11). However, it is possible for the network to deactivate all hybrid-ARQ processes; this ceases the E-DCH transmission.
- If the UE has no grant, it may not make non-scheduled transmissions during CELL_FACH operation.

Because of the random nature of RACH access, RACH and the E-DCH in CELL_FACH is a contention-based operation and thus it is entirely possible that two or more UEs may attempt RACH access using the same preamble resources. In such a case, the network may be able to detect a preamble that is transmitted from more than one UE and respond with AICH or E-AICH, as described in Section 14.2.2.1. All UEs that have transmitted the preamble may then detect the AI and assume that the downlink resources are assigned to them and that they can start E-DCH transmission. To cover such situations, a *contention resolution* mechanism is included in the 3GPP specifications to resolve the contention. The contention resolution mechanism consists of two phases: first, the "contention resolution" phase, in which more than one UE may be transmitting, and second, the "contention-free" phase, in which it is certain that only one UE transmits.

During the contention resolution phase depicted in Figure 14.22, when a UE first starts transmitting on E-DCH, if it has a dedicated E-RNTI, it includes the E-RNTI in the MAC header for each transmission. As soon as the network successfully receives a MAC-PDU with a UE E-RNTI, the network sends an E-AGCH

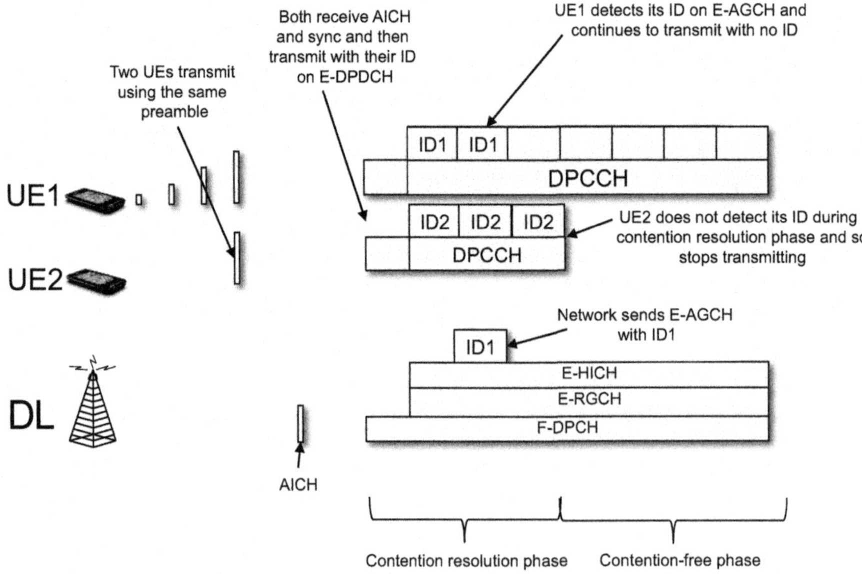


**FIGURE 14.22 Contention resolution for the Enhanced CELL_FACH uplink operation.**

transmission using the successfully decided E-RNTI. The UE that is addressed can then continue to transmit and does not further include E-RNTI in the MAC header. Other UEs will not receive an E-AGCH, as either their PDUs have not successfully been received or, if they have been received; the network has not sent a further E-AGCH with their own E-RNTI. At the end of the contention resolution phase, any UE that has not received an E-AGCH containing its E-RNTI must stop transmission. At this stage, a maximum of one UE will be transmitting and the contention-free period begins.

The above-described contention resolution mechanism applies to DCCH and DTCH transmissions. For CCCH transmissions, there is no contention resolution mechanism. For CCCH transmissions, the UE that sends a PDU can be identified either from its E-RNTI in the MAC header or from the RRC message.

In release 8, once the contention-free stage begins, then if it is configured to do so by the network, the UE starts to transmit HS-DPCCH.

When a UE is scheduled using E-AGCH, then it must be identified using an E-RNTI identifier. For operation in CELL_FACH, two types of E-RNTI may be used:

- The network may have configured the UE with a dedicated E-RNTI; in this case this should be used.
- If the UE does not have a dedicated E-RNTI, then a common E-RNTI may be used. The UE may only transmit CCCH if it does not have a dedicated E-RNTI.

The UE may transmit control data using CCCH or DCCH or user-plane data using DTCH. To transmit DCCH or DTCH, the UE must have been configured with a dedicated E-RNTI.

If the UE transmits CCCH, then the network does not control the UE using E-AGCH or E-RGCH. However, a maximum time duration is set for CCCH transmissions; after the time limit is reached, the UE must stop transmission and the downlink control resources are released.

Once the UE has transmitted all of its uplink data, it does not necessarily release the downlink control resources and the right to transmit E-DCH immediately. However, immediately after the last PDU of data, the UE sends a MAC status report to the network indicating that its buffer status is empty; this informs the network that the UE has no further data to transmit and allows for the network to cease the E-DCH resource allocation using an E-AGCH command deactivating all hybrid-ARQ processes.

After the DTX cycle commences, the UE is expected to continue to transmit DPCCH until a timer expires. Following expiry of the timer, the UE periodically transmits DPCCH. The duration of the timer, the interval between DPCCH transmissions, and the length of DPCCH transmissions are configurable by the network. The timing of uplink DPCCH transmissions and the downlink DRX cycle are closely related such that F-DPCH power control commands associated with the uplink DPCCH transmissions can be received during the times in which the UE must monitor HS-SCCH in downlink.

Eventually, the UE must release the downlink control resources and stop transmitting E-DCH. There are several mechanisms by which this may happen:

- The downlink resources may be released after a (network configurable) timer expires following transmission of the last data PDU.
- The downlink resources may be released after the network sends an E-AGCH command deactivating all hybrid-ARQ processes.
- For CCCH transmissions, the duration of the transmission is fixed and the downlink resources are released after the time available for CCCH transmission has expired.
- Any UE that has not received its E-RNTI after the contention resolution phase must release the downlink control resources.

#### 14.2.2.3 E-DCH operation in idle mode

In addition to CELL_FACH, the UE is able to initiate RACH transmissions while it is in idle mode. This is necessary because the UE needs to be able to communicate with the network and send an RRC connection request in order to be able to transit to connected mode. While in idle mode, the UE does not have a dedicated E-RNTI and thus can only send CCCH. Idle-mode operation of E-DCH is the same as CELL_FACH operation.

### 14.2.3 RELEASE 11 ENHANCEMENTS TO ENHANCED CELL_FACH

As smartphone growth and the associated app ecosystem have taken off, the characteristics of traffic and its impact on UE performance and network behavior have become better understood. In release 11, it became apparent that the Enhanced CELL_FACH behaviors introduced in the previous releases could be further optimized, and thus 3GPP undertook a Work Item to improve the specifications in a manner that would enable further enhancements of E-DCH and HSDPA operation in CELL_FACH. The Work Item consisted of a basket of specification changes, whose aims fell into one of four categories:

- supporting a larger population of UEs remaining in CELL_FACH state;
- improving further the potential for extending UE battery life;
- improving the efficiency of resource scheduling, considering that the network operates users in both CELL_DCH and CELL_FACH; and
- reduction of latency.

It is interesting to note that peak-rate improvements was not included in the release 11 enhancements; in release 11, for example, HSDPA MIMO operation continues to not be configurable in CELL_FACH, while E-DCH operation in uplink remains constrained to QPSK. However, improving peak rates in CELL_FACH is not of importance, because the intention for CELL_FACH is to handle short bursts of data while the CELL_DCH state should continue to be used for high-performance and high-speed transfer of larger amounts of data.

#### *14.2.3.1 Supporting a larger population of UEs and improved mobility in the CELL_FACH state*

Supporting improved mobility is complementary to increasing the amount of supportable UEs, because improved mobility reduces the need to switch UEs into other states for mobility reasons. Features introduced into release 11 Enhanced CELL_FACH operation that fall into a general category of supporting a larger population of UEs and improved mobility include the following.

*Fallback to release 99 operation*

In release 8 E-DCH in CELL_FACH operation, UEs that are able and configured to use E-DCH in CELL_FACH may not use the release 99 RACH procedure. Nevertheless, the cell still needs to operate release 99 RACH in order to support legacy users. When release 8 or later UEs make a RACH access and are granted E-DCH resources, then the E-DCH resources are blocked for the duration of the E-DCH transmission, plus the configured continuation time after the last MAC PDU is transmitted (see Section 14.2.2.2). For very small PDUs, and in particular PDUs containing DCCH or CCCH, the benefits of utilizing E-DCH compared to the release 99 RACH become small. Hence, in circumstances in which there is a lot of E-DCH activity and resources are likely to become blocked, it is advantageous to be able to redirect some attempts at using E-DCH to instead use the Rel-99 RACH channel, which does not require E-DCH resources to be allocated. Such an ability to redirect carries two advantages. First, if the common E-DCH resources are overloaded, some CELL_FACH traffic can be off-loaded to the release 99 RACH, enabling a larger total number of users to be supported. Second, the continuation timer can be optimized to consider only larger packet sizes.

Thus, in release 11, a mechanism was introduced to enable the network to redirect a UE to use release 99 RACH in place of E-DCH. The decision to redirect the UE is fully under the control of the network, and only signaling traffic can be redirected. A redirection is indicated to a UE in the following manner:

- First, the possibility of redirection is configured by means of common RRC signaling.
- Then, if a specific access attempt by a specific UE is to be redirected, the redirection is indicated by means of sending the UE a pre-configured NACK using E-AI.

Once a UE configured with redirection receives a NACK on E-AI, if the UE data is from DCCH or CCCH, the UE starts a release 99 RACH access with no backoff. If the UE data is from DTCH, then the UE interprets the E-AI NACK as a NACK and after backoff restarts an E-DCH access attempt. Legacy UEs and UEs that are configured without redirection simply interpret the E-AI as a NACK.

*Priority-based reselection and reselection to E-UTRA*

Prior to release 11, UEs in CELL_FACH could be configured to perform intra- or inter-frequency measurements and reselection using measurement occasions or DRX (see Section 14.2.1). However, the reselection ability in CELL_FACH excluded two deployment cases. The first was reselection to E-UTRA, in particular E-UTRA hotspots, and the second was efficient reselection to UTRA hotspots on other carriers. The inability to perform such reselections implies that idle mode or CELL_PCH/URA_PCH must be used in such deployments, which limits the amount of users that can be held in CELL_FACH. Reselection to E-UTRA is a fairly straightforward replication of the idle-mode E-UTRA reselection method. The priority-based reselection mechanism governs the frequency with which the UE measures on other carriers and *radio access technologies* (RAT) for reselection candidates and also which carriers are searched, and it is also the same as for idle mode, consisting of the following principles:

- If the quality (in terms of Rx Ec/No) and Rx signal level for the serving cell are above preconfigured thresholds, then the UE searches so-called higher priority carriers or RATs once per minute. "Higher priority" carriers or RATs are carriers/RATs that have been used for hotspot deployment. The UE is thus periodically searching to check whether it is near to a hotspot.
- If the quality or the Rx signal level for the serving cell falls below preconfigured thresholds, then the UE searches on all carriers/RATs that it has been configured to search. In this situation, the serving cell signal level or Rx quality is degrading and it is possible that the UE is approaching the cell border. Thus, the UE must search on carriers or RATs that are not provided as hotspots but rather for providing continuous coverage; these are equal or lower priority carriers. Still, the UE can be close to hotspots; hence, higher priority carriers and RATs are still searched.

The areas intended to be covered with low- and high-priority measurements are illustrated in Figure 14.23. Improving mobility in this manner enables UEs to be held in the CELL_FACH state for a longer period of time, because it is not necessary to move them to other states for mobility purposes.

*Network controlled measurement reporting*

For the purpose of traffic steering, it is possible for the network to reject an RRC connection request or release an RRC connection and direct the UE to another RAT. In the release 8 specifications, when the UE makes an RRC connection request, it sends the results of neighbor cell measurements that have triggered a reporting criterion on RACH. This information can be used by the network to redirect to other frequency layers. However, in release 8, the UE will not include measurements made on E-UTRA on RACH. Thus, if the network attempts to redirect the UE to E-UTRA, it has to do it blindly without knowing if the UE is in proximity to an E-UTRA cell.

With increasing uptake of E-UTRA, it is preferable for the network to obtain information relating to E-UTRA for UEs in CELL_FACH (that is, E-UTRA RSRP or RSRQ). Thus, two mechanisms for E-UTRA reporting were introduced in release 11.

The first of these is inclusion of E-UTRA measurement reports in the measurement results that are sent on RACH. Configuration for E-UTRA reporting, in addition to other measurement reports that are sent on RACH, is included in SIB 19. The SIB also configures which measurement reports should be prioritized in case more measurements are triggered to report than can fit in the RACH payload.

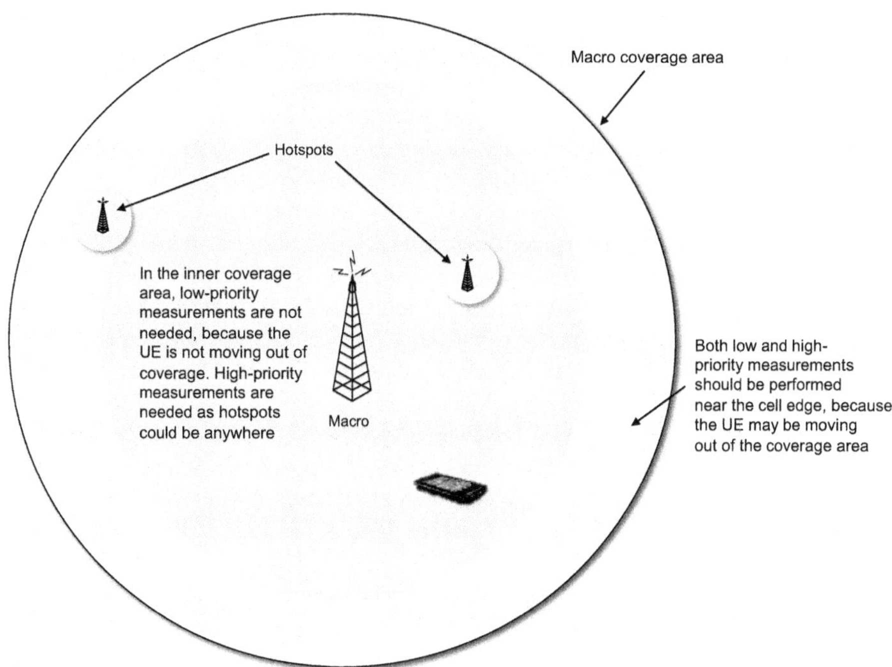


**FIGURE 14.23 Use of hierarchical measurements.**

A second reporting mechanism is a UE-specific configuration of a measurement using a MEASUREMENT CONTROL message. Two types of E-UTRA measurements can be configured. An event-triggered configuration causes the UE to report the measurement if its value exceeds a given threshold. The report is a once-off and after being triggered, the UE does not report again. A second possibility is to configure periodic reporting.

Reporting measurements in RACH and dedicated measurement configurations using a MEASUREMENT CONTROL are mutually exclusive; if a UE is configured to report an event-based or periodic measurement, then it does not send measured results on RACH. Furthermore, if configured to report measurements, a UE does not perform cell reselection; hence, its mobility is under the control of the network.

#### *14.2.3.2 Improving further the potential for extending UE battery life*

Keeping the UE modem active in either the transmit or receive directions causes a current drain on the battery and reduces the battery life. For this reason, it is desirable to deactivate the modem either in the transmit or receive directions during any slots in which the UE does not need to be transmitting or receiving information from the network. Release 99 included DTX and DRX in idle mode and CELL_PCH, and this was supplemented by CPC in release 7 and Enhanced CELL_FACH DRX in release 8.

In the CELL_FACH state, the transmitter can be deactivated whenever the UE has no E-DCH resources allocated. When there is a need for a new uplink transmission, the RACH procedure is used to re-establish uplink synchronization. The configured length of the downlink DRX cycle determines the responsiveness of the UE when new data arrives in the downlink.

The necessary length and on on-time the DRX duty cycle depend on the expected traffic characteristics. Where traffic is expected to be infrequent, a long DRX cycle can be deployed; for responsiveness to frequent small burst, a shorter DRX/DTX cycle must be configured. In the release 8 Enhanced CELL_FACH specifications, it was envisaged that UEs would be held in the CELL_FACH state when they had relatively frequent but small packets, and in order to ensure responsiveness, it was assumed that relatively short DRX cycles would be configured. The maximum DRX cycle length was fixed at 320 ms.

Some smartphone traffic types, however, may experience periods of relative inactivity followed by renewed activity. In these cases, it is power inefficient to hold such UEs in release 8 Enhanced CELL_FACH with DRX, and the UEs should be moved to CELL_PCH or idle mode. Moving between the states is undesirable from an RRC and latency perspective.

The decision was therefore made in release 11 to enable a longer DRX cycle in Enhanced CELL_FACH, such that UE modem power consumption for release 11 UEs could be roughly similar in CELL_FACH and idle mode. This increased DRX cycle length was introduced by means of a so-called *second DRX cycle*.

The second DRX cycle can be activated in one of two configurations. In the first configuration, the second DRX cycle is directly activated once an inactivity timer expires following the last downlink or uplink transmission. When this configuration is used, the first DRX cycle is not defined, and the second DRX cycle is thus a straightforward extension of the release 8 DRX behavior.

The second configuration utilizes both first and second DRX cycles and is illustrated in Figure 14.24. After the last downlink or uplink transmission, a timer is initiated and once this timer expires, the UE enters the first DRX cycle. At this point, a second timer is started. If there is no further downlink or uplink activity until the second timer expires, then the UE enters the second DRX cycle. It is envisaged that in this configuration, the first DRX cycle is relatively short and the second DRX cycle long. This enables a stage transition of the UE to a lower responsiveness.

With either the first or the second configuration, as soon as there is downlink scheduling or uplink data, the UE exits the DRX cycle and begins transmission/reception.

It should be noted that the fast dormancy functionality described in Chapter 7 is still available to UEs that are configured in Enhanced CELL_FACH with second DRX; it may still be the case that the idle-mode DRX cycle is longer. If the second DRX cycle is equal to or longer than the idle-mode DRX cycle, the UE may only send one fast dormancy request; if the second DRX cycle is shorter than the idle-mode cycle, then the UE may send multiple requests.

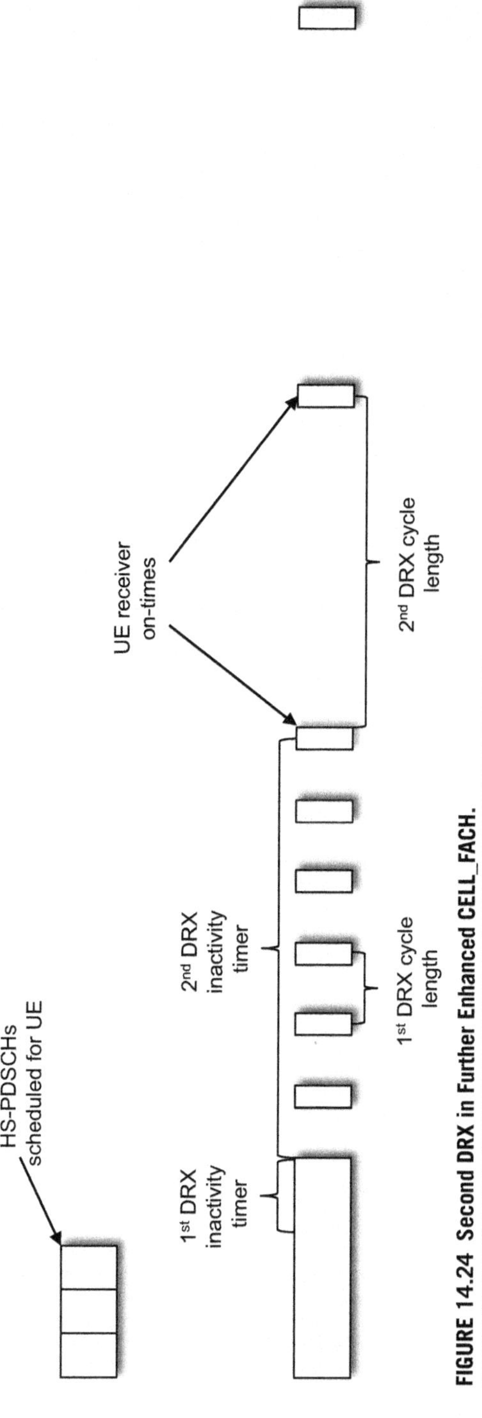


**FIGURE 14.24 Second DRX in Further Enhanced CELL_FACH.**

#### 14.2.3.3 Improving the efficiency of resource scheduling

*2/10 ms TTI length selection*

In the release 8 Enhanced CELL_FACH specifications, either a 2 ms TTI or a 10 ms TTI may be configured for CELL_FACH E-DCH users on a cellwide basis using SIB (i.e., broadcast) signaling. Ten-millisecond TTI is needed in cells in which a proportion of the cell area is uplink coverage limited (see Figure 14.25). In such cells, with the release 8 specifications, 10 ms TTI must be configured for all Enhanced CELL_FACH users by means of SIB.

There are significant advantages to using a 2 ms TTI in CELL_FACH. These include a reduction in latency, improved link-level performance, the ability to include per hybrid-ARQ process scheduling, the ability to coordinate scheduling with 2 ms TTI CELL_DCH users, and greater granularity in scheduling, all of which can lead to an improvement in user experience, cell capacity, or both. Thus, it is desirable to configure 2 ms TTI for as many users as possible.

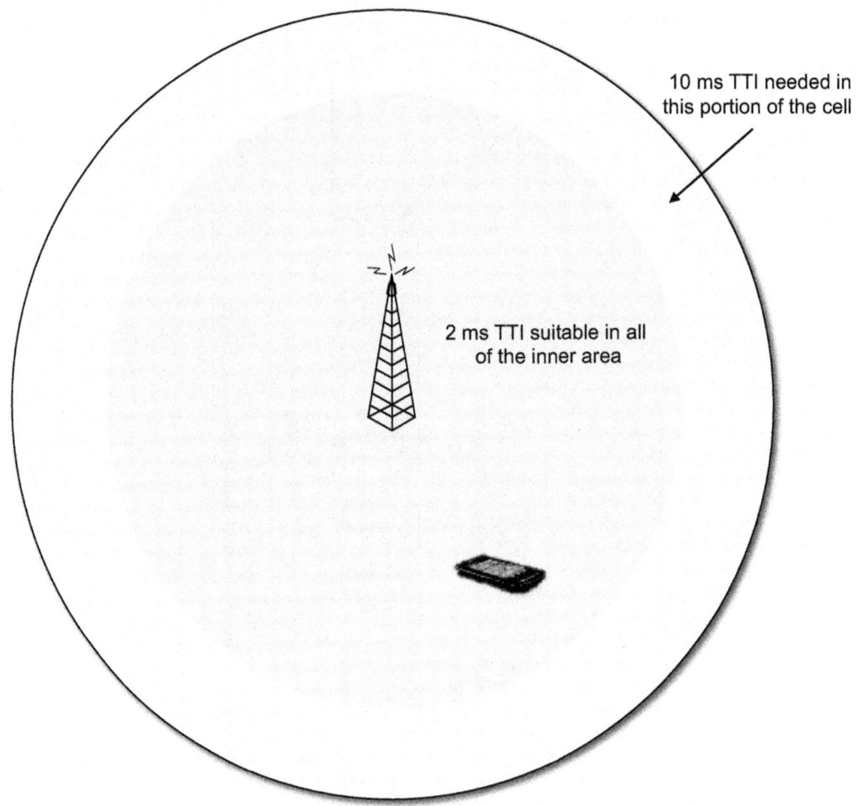


**FIGURE 14.25 Distribution of 2 ms and 10 ms TTI requirement.**

The release 11 improvements to Enhanced CELL_FACH addressed a means to increase the amount of UEs that can use a 2 ms TTI for E-DCH access for cells in which a 10 ms TTI needs to be configured in order to provide coverage to cell-edge UEs.

In order to use a 2 ms TTI, a UE has to have an uplink link budget that is not power limited. Unfortunately, when the UE makes a random access in CELL_FACH in order to request E-DCH resources, the network is unaware whether the UE will be power limited. Hence, it is necessary for the UE to determine whether it is power limited and indicate to the network if it is able to use a 2 ms TTI.

The UE determines whether it is power limited during the PRACH power ramping stage. Each time a preamble is sent, the UE calculates the uplink power headroom using the following formula (see Figure 14.26):

$$Headroom = \{\min(Maximum\, allowed\, uplink\, Tx\, power,\\ P\_MAX) - Preamble\_Power + deltaPp\text{-}e\}.$$

In the above, *Maximum allowed uplink Tx power* is a configured maximum UE power, and *P_MAX* is the UE's maximum Tx power. *Preamble_Power* is the power that will be applied for the preamble and *deltaPp-e* is a network-configurable offset

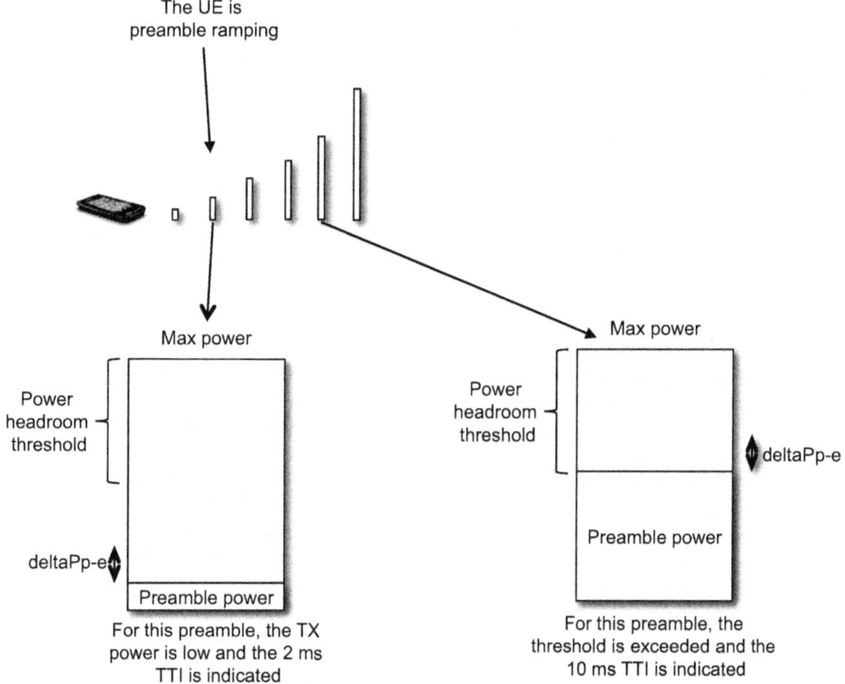


**FIGURE 14.26 UE procedure for 2/10 ms TTI selection.**

between the preamble power and the initial DPCCH transmit power that will be applied if the UE receives an ACK with E-DCH resources. The uplink power headroom is then compared with a threshold that is configurable by the network via SIB. If the power is above the threshold, a 2 ms TTI is indicated; if it is below the threshold, then a 10 ms TTI is selected.

Thus, in effect the formula is emulating the power headroom estimation that is made during regular E-DCH transmission, and then the threshold that is used for deciding on the required TTI length can be configured by the network by considering the amount of power that is needed for transmitting the minimum E-DCH TFC that should be supported; if the headroom is below the power required for the minimum E-TFC, then the 10 ms TTI should be selected.

After determining which TTI length is preferred, the UE must indicate the preference to the network. The preference is indicated by means of selecting a preamble from a predefined set. Preamble partitioning for release 11 Enhanced CELL_FACH is described in Section 14.2.3.5.

The network allocates E-DCH control resources for E-DCH transmission. The control resources are partitioned into two types. One type of control resource is always used for controlling 10 ms TTI E-DCH transmissions. The other type may be used for controlling either 2 ms or 10 ms TTI transmissions. As with release 8 Enhanced CELL_FACH, the network may ACK the UE with the default E-DCH resources for the used preamble, or redirect the UE to use different E-DCH control resources. The TTI length that the UE uses depends on the preferred TTI length that the UE indicated to the network and the control resources assigned by the network:

- If the UE is a release 8 UE, then regardless of which E-DCH control resources are assigned to the UE, the UE will use the SIB configured TTI length.
- If the UE is a release 11 UE and indicates a preference for the 10 ms TTI, then the UE will apply a 10 ms TTI regardless of the assigned E-DCH control resources.
- If the UE is a release 11 UE and indicates a preference for the 2 ms TTI then
  - If the network assigns the default E-DCH resources in the manner depicted in Figure 14.19 for the preamble or E-DCH resources that can be used for controlling 2 ms or 10 ms TTI (e.g., resources 13-24 in Figure 14.27), the UE will apply the 2 ms TTI.
  - If the network assigns E-DCH resources that are only used for controlling the 10 ms TTI, the UE must apply the 10 ms TTI. This is illustrated in Figure 14.27. The UE has requested a 2 ms TTI, but is directed to 10 ms TTI resources, and so must use a 10 ms TTI.

Thus, if the UE indicates a preference for a 2 ms TTI, the network is able to override this preference by means of the allocation of E-DCH resources indicated by the E-AICH. However, if the UE indicates a preference for a 10 ms TTI, then the network is unable to override this preference.

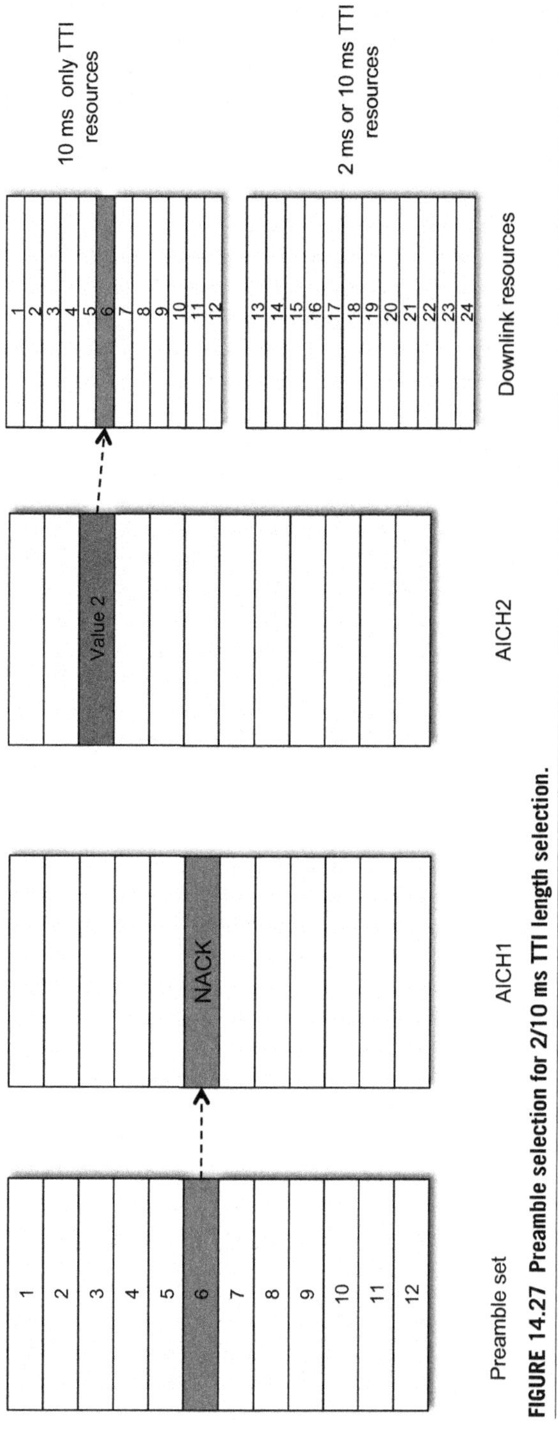


**FIGURE 14.27** Preamble selection for 2/10 ms TTI length selection.

*Per hybrid-ARQ process scheduling*

When operating in CELL_DCH, it is possible for the E-DCH scheduler to activate a limited set of hybrid-ARQ processes, while the remaining ones are blocked for uplink transmission. This functionality enables, among other things, TDM scheduling of uplink resources, which improves uplink orthogonality. In the release 8 specifications, activation and deactivation of individual hybrid-ARQ processes is not supported in Enhanced CELL_FACH.

Because of the introduction of 2/10 ms TTI length selection, which enables a wider deployment of Enhanced CELL_FACH together with 2 ms TTI and CELL_FACH/CELL_DCH user timing alignment, which enables TDM scheduling, the benefits of being able to do hybrid-ARQ process activation and deactivation are greater in release 11. Thus, hybrid-ARQ process activation and deactivation in CELL_FACH was introduced into the release 11 specifications.

A default set of activated hybrid-ARQ processes are indicated by means of SIB signaling and applies at the start of E-DCH transmission. During E-DCH transmission, the E-AGCH can be used to activate or deactivate individual hybrid-ARQ processes using the same mechanism as release 6 CELL_DCH scheduling.

One difference between Enhanced CELL_FACH and CELL_DCH is in the transmission of scheduling information. In CELL_DCH, MAC PDUs containing SI can only be transmitted in activated hybrid-ARQ processes. In CELL_FACH, however, SI is transmitted in all hybrid-ARQ processes, regardless of activation status. The reason behind this is that because of the lack of softer handover in CELL_FACH, the SI may be less reliable and transmission in all processes improves reliability. Other than for transmission of SI, the UE may not use deactivated hybrid-ARQ processes.

*Time alignment between CELL_DCH and CELL_FACH E-DCH transmissions*

The intention of introducing 2 ms TTI length selection and per hybrid-ARQ process scheduling is to enable TDM scheduling in the uplink. In order to enable full TDM scheduling, E-DCH transmissions from CELL_FACH users and CELL_DCH users must be time-aligned.

In the release 8 specifications, the start time of E-DCH transmission is tied to the access slot that was used by the preamble. After the preamble is transmitted, the time between the preamble and AICH is either three or five slots. The time between the start of the AICH and the start of the DPCCH is then calculated using the following formula, and is thus around four slots, as illustrated in Figure 14.28

$$\tau_{\text{a-m}} = 10240 + 256 \times S_{offset} + \tau_0 \text{ chips}, \tag{14.1}$$

where $S_{offset}$ denotes a symbol offset configured by higher-layers, {0, …, 9}, and $\tau_0 = 1024$ chips defining the downlink-to-uplink frame timing difference.

Access slots are two slots in length. Thus, the timing of E-DCH transmission will depend on the timing of the access slot that was selected for the preamble, and E-DCH transmissions from different users will differ by multiples of two slots. Many users will not be aligned with the TTI length of three slots.

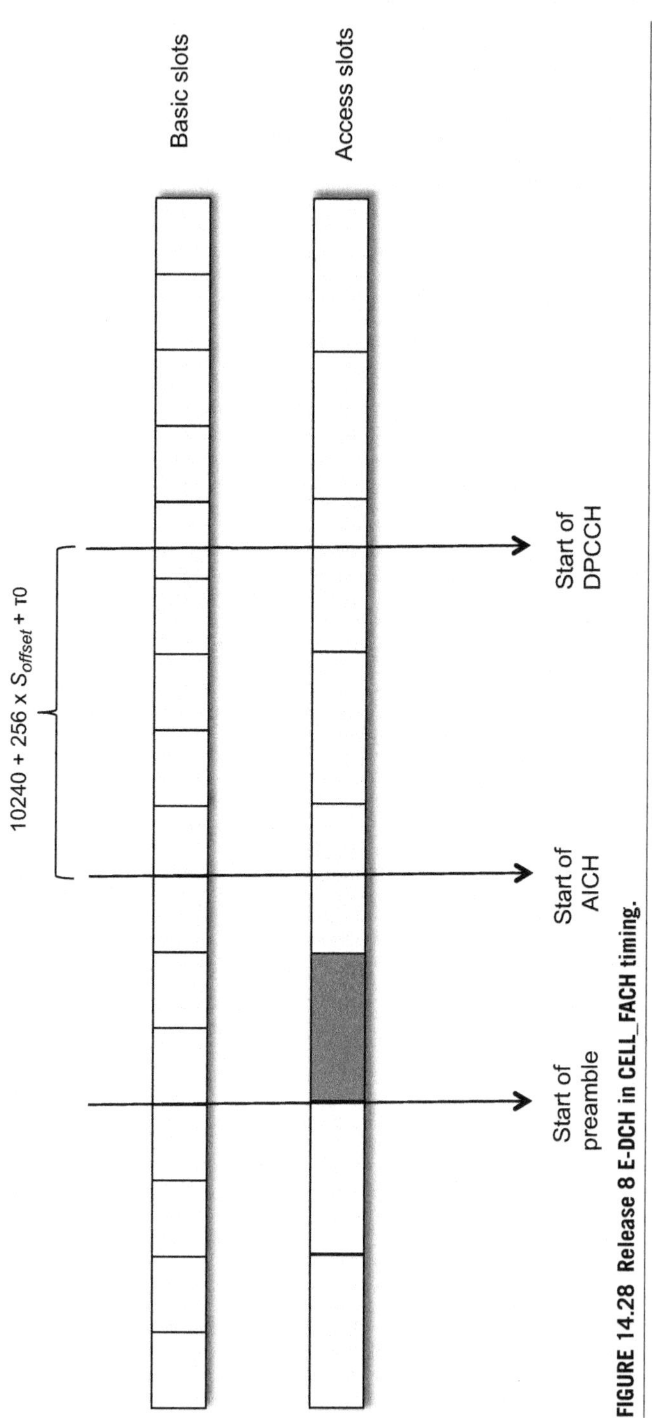


**FIGURE 14.28** Release 8 E-DCH in CELL_FACH timing.

TTI alignment was enabled in release 11 by means of introducing an additional parameter, $T_{offset}$, that adjusts the timing of the E-DCH transmission such that it always begins on a multiple of three slots. Thus, the time difference between the start of the E-DCH differs from the E-AICH according to the following formula

$$\tau_{\text{a-m}} = 10240 + 2560 \times T_{offset} + 256 \times C_{offset} + \tau_0 \text{ chips,} \tag{14.2}$$

where $C_{offset}$ is a cell-specific symbol offset configured by higher-layers, {0, …, 29}, and

$$T_{offset} = \begin{cases} 0, \text{ if the preamble was transmitted in access slot } \{0, 3, ..., 12\} \\ \quad \text{using an even preamble signature } \{0, 2, ..., 14\} \\ 1, \text{ if the preamble was transmitted in access slot } \{1, 4, ..., 13\} \\ 2, \text{ if the preamble was transmitted in access slot } \{2, 5, ..., 14\} \\ 3, \text{ if the preamble was transmitted in access slot } \{0, 3, ..., 12\} \\ \text{using an odd preamble signature } \{1, 3, ..., 15\} \end{cases}$$

The operation of the timing adjustment is shown in Figure 14.29. In the figure, $C_{offset}$ and $\tau_0$ are set to zero for clarity. In Figure 14.29a, access slot 0 is used. In this case, the E-DCH transmission starts in slot 7. In Figure 14.29b, access slot 1 is used. This causes the E-DCH transmission to start in slot 10. In Figure 14.29c, access slot 2 is used. In this case, E-DCH transmission starts in slot 13. The adjustment in the timing calculation leads to the E-DCH timing always being placed with a three-slot (one TTI) separation. With appropriate setting of $C_{offset}$, this enables CELL_FACH and CELL_DCH users to be aligned in uplink.

*Non-serving common E-RGCH in CELL_FACH state*

Soft and softer handover cannot be configured for UEs operating in the CELL_FACH state. With CELL_DCH E-DCH operation, neighboring cells have two means to control interference from UEs transmitting in neighbor cells: the DPCCH level can be reduced using TPC commands when the instantaneous coupling loss between the UE and neighboring cell is lower than that between the UE and serving cell, and the non-serving E-RGCH can be used to reduce the E-DPDCH Tx power. In CELL_FACH, no such mechanisms are available. In release 8, it was envisaged that data transmissions would be short-lived and thus interference effects would be momentary.

In release 11, 3GPP concluded that it could be useful for neighbor cells to have some means of controlling interference from CELL_FACH users in uplink by means of CELL_FACH UEs listening to the E-RGCH from neighbor cells when they are near to a cell border.

Before starting an uplink transmission, a UE must decide whether and which E-RGCHs it should listen to. The UE makes a decision once prior to the start of E-DCH transmission and then continues to monitor the same set of E-RGCHs until the E-DCH resources are released. No reassessment of the decision on which

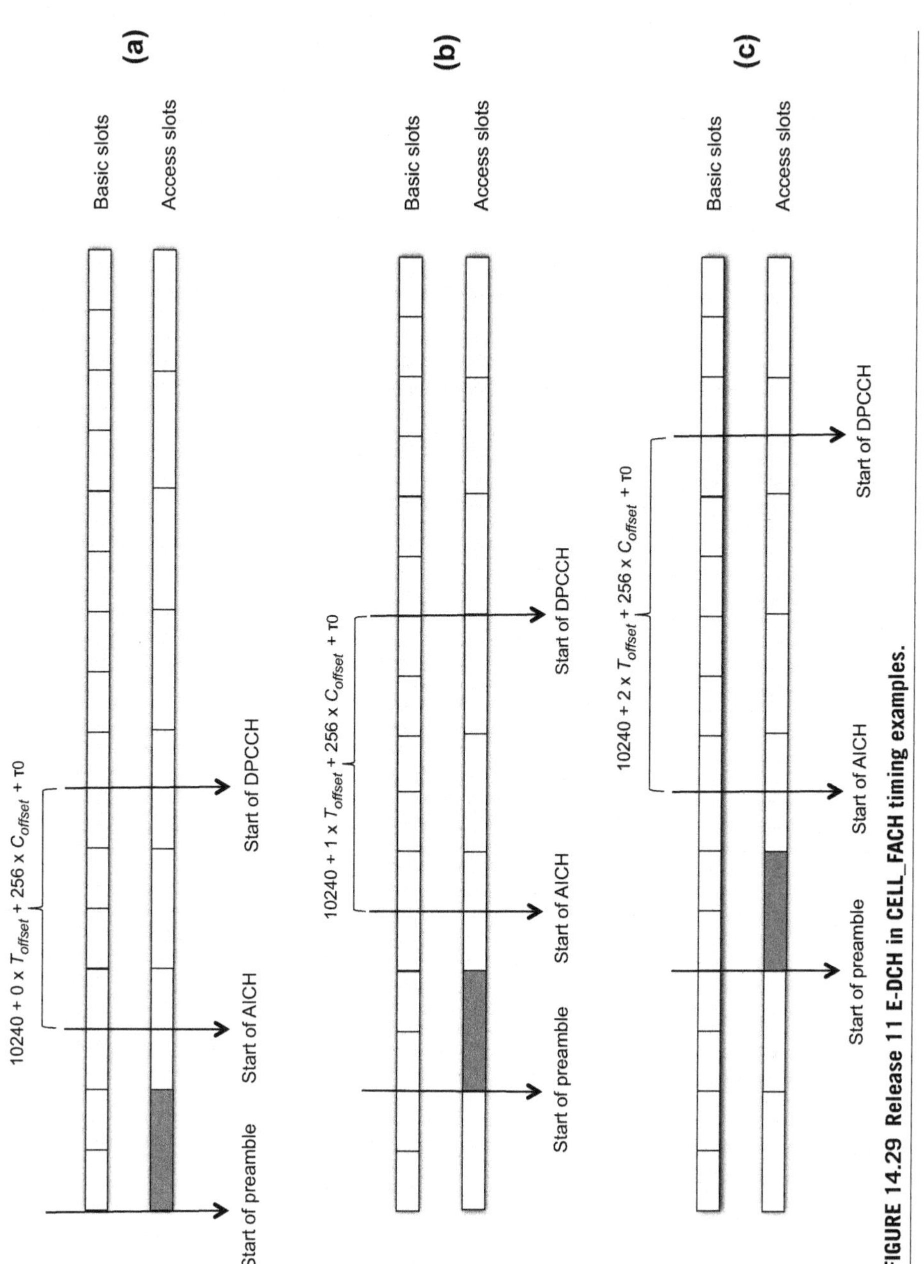


**FIGURE 14.29 Release 11 E-DCH in CELL_FACH timing examples.**

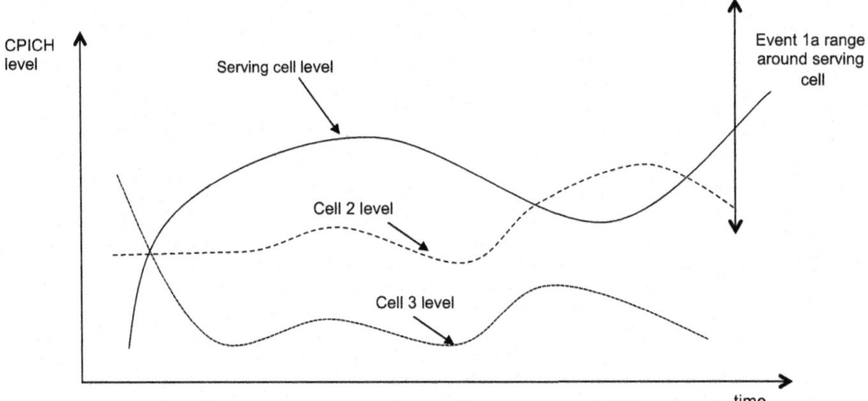


**FIGURE 14.30 Example of cells meeting and not meeting event 1a criterion.**

E-RGCHs to monitor is made during the E-DCH transmission, because it is expected that the E-DCH transmission will be relatively short in length.

The decision on which cells' E-RGCH to monitor is based on CPICH Ec/Io measurements. Cells that fulfill the event 1a criterion are candidates for E-RGCH monitoring. The event 1a criterion is fulfilled when the CPICH Ec/No for a cell is within $\pm R_a$ dB of the Ec/No of the best measured cell, where $R_a$ is configurable (see Figure 14.30). The UE monitors a maximum of three E-RGCHs; these are the best three (or fewer) E-RGCHs for the three cells with highest CPICH Ec/Io that meet the event 1a criterion.

Once E-RGCH transmission starts, if a UE detects that the E-RGCH is set in a cell that it is monitoring, it must reduce its E-DPDCH Tx power if the E-DPDCH Tx power is below a signaled minimum grant level.

*Standalone HS-DPCCH transmission*

The HS-DPCCH control channel carries CQI and ACK/NACK information relating to downlink HS-PDSCH scheduling and transmissions. In release 7 HSDPA in CELL_FACH operation, no HS-DPCCH is transmitted in uplink and the only information available to the scheduler for making scheduling decisions is via RRM measurement reporting. In release 8, HS-DPCCH can be transmitted in uplink if E-DCH is triggered due to uplink data.

In many cases, triggering HS-DPCCH when uplink data arises is sufficient because downlink activity leads quickly to uplink activity; for example, TCP ACKs, or else downlink activity, are very short, and hence there is not much benefit to hybrid-ARQ and CQI feedback. However, in release 11 it was recognized that there are some circumstances in which it is preferable to be able to activate uplink HS-DPCCH reporting independently of uplink data activity in CELL_FACH in order to make HSDPA scheduling more efficient. Thus, the ability to activate an HS-DPCCH transmission using HS-SCCH orders was standardized. Standalone HS-DPCCH transmission can also be activated from CELL_PCH state in release 11, provided that a dedicated H-RNTI is available for the UE in question so that an HS-SCCH order can be sent.

When the UE receives an HS-SCCH order indicating that it should start HS-DPCCH transmission, it initiates a random access with preamble ramping. The RACH access procedure operates in the same manner as is the case when the UE has uplink data to transmit on E-DCH. Upon receiving an ACK and E-DCH resources, the UE sends an SI report including its E-RNTI for the purpose of contention resolution. HS-DPCCH transmission begins immediately, even before the contention resolution phase is completed.

A timeout value may be configured, such that after receiving the last TTI of downlink data, the UE will release the E-DCH resources and stop transmitting once a timeout value, $T_{bhs}$, is reached. If during the uplink transmission the UE has uplink data to transmit, then the UE starts its inactivity timer after the last downlink or uplink transmission and releases the E-DCH resources after whichever the greater of the uplink inactivity timer $T_b$ and the downlink inactivity timer $T_{bhs}$ expires.

#### 14.2.3.4 Reduction of latency

*PRACH access time reduction*

Prior to starting preamble ramping at the start of an RACH access, the UE must select an access slot. Access slots are divided into two groups: access slots 0–7 and access slots 8–15. Prior to release 11, a UE is required to randomly select an access slot from the *next* access slot set relative the current one, as indicated in Figure 14.31a. This leads to an average delay of 7.5 slots and potentially a larger peak delay before preamble ramping can start.

In release 11, the UE is able to select a sequence from any of the next seven access slots, rather than needing to wait until the next access slot set, as illustrated in Figure 14.31b. This has the effect of reducing the average delay to four slots.

It should be noted that it is not visible to the network whether the UE has implemented the access time reduction or not, and the access time reduction is not captured in any RAN5 testing.

#### 14.2.3.5 UE capabilities for release 11 and signaling of UE support for features during preamble ramping

The release 11 enhancements to CELL_FACH are mainly optional and are bundled into groups of features that the UE can indicate support for. The feature bundles are as follows:

1. fallback to release 99 PRACH;
2. support for concurrent 2 ms and 10 ms TTI deployment;
3. support for TTI alignment between CELL_DCH and CELL_FACH users and support for per hybrid-ARQ process grants; and
4. support for E-UTRA measurements in CELL_FACH.

The UE can inform the network about its capabilities to support each of the feature bundles via UE capability signaling.

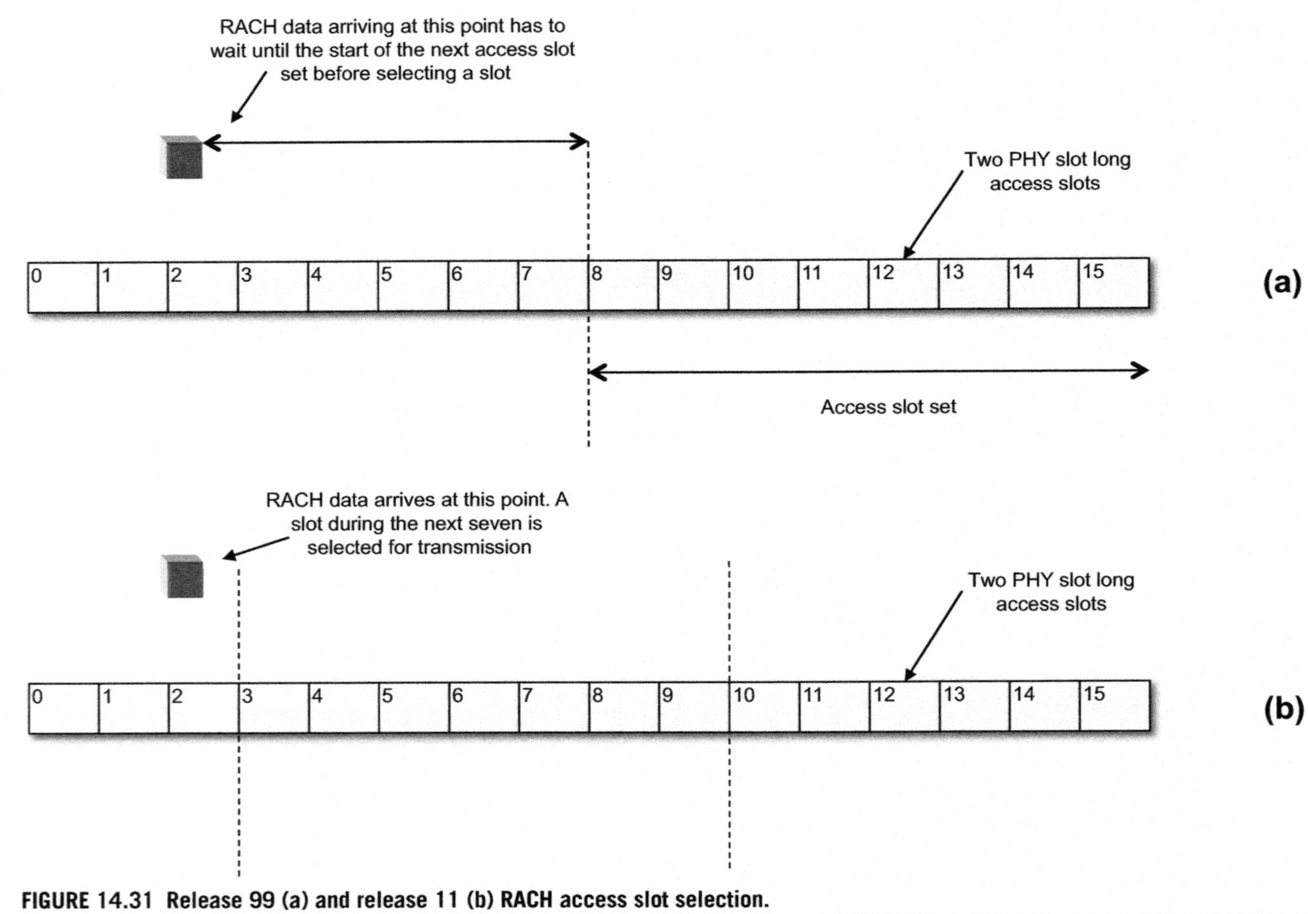


**FIGURE 14.31** Release 99 (a) and release 11 (b) RACH access slot selection.

For some of the features though, it is necessary for the network to know during the preamble ramping cycle whether the UE that has initiated the preamble ramping supports the feature, in order to react appropriately. UEs indicate support of a feature by means of preamble selection.

Preambles are grouped into five groups. The number of groups that are required depends on which features are configured in the cell. The preamble groups are as follows:

- The legacy release 99 preamble group is for UEs that support Rel-99 CELL_FACH.
- The legacy EUL preamble group for release 8 is used by UEs that do not support Further Enhanced CELL_FACH.
- The first extension preamble group can be used for two purposes:
  - For UEs that support concurrent 10 ms and 2 ms TTI selection, this extension group is used to indicate a preference for 10 ms TTI.
  - For UEs that do not support concurrent 10 ms and 2 ms TTI selection, or if the network has not configured 2/10 ms TTI selection, if the network has configured a 10 ms TTI in the cell then this preamble group is used by UEs that support the mandatory release 11 Enhanced CELL_FACH features.

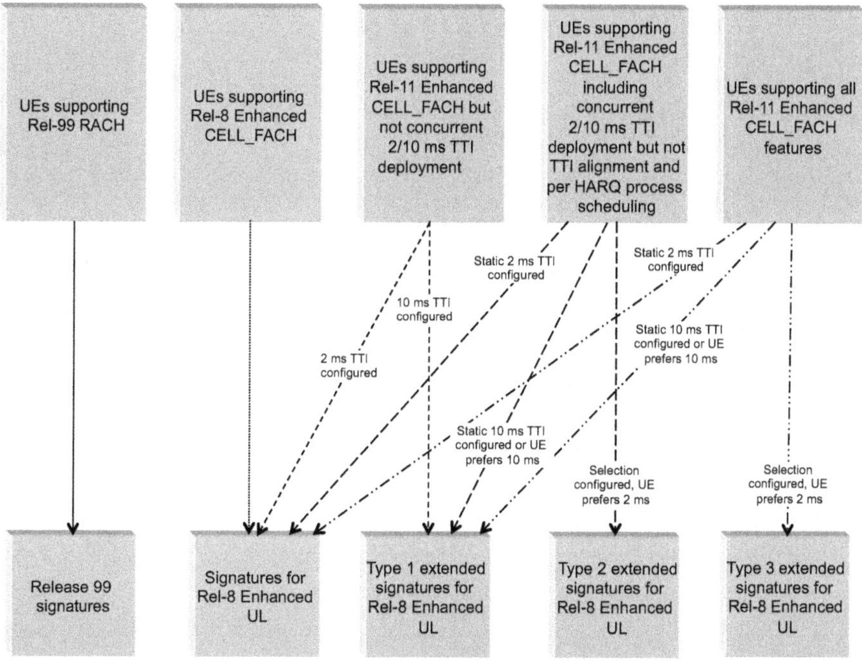


**FIGURE 14.32 Mapping of preamble portioning to UE capabilities.**

- The second extension preamble group can also be used for other purposes:
  - For UEs that support 2 ms and 10 ms concurrent TTI, when TTI length selection is configured, this second group can be used to indicate that the 2 ms TTI length is preferred:
    - If TTI timing alignment is not configured by the network, then this preamble group is always used for indicating 2 ms TTI.
    - UEs that do not support TTI alignment and per hybrid-ARQ process scheduling always use this preamble group for indicating a preference for 2 ms TTI.
- The third extension preamble group is used by UEs that support TTI length alignment and per hybrid-ARQ processing, when these are configured by the network. It is used to indicate both that a 2 ms TTI is preferred and that the UE can support TTI length alignment and per hybrid-ARQ process scheduling.

Figure 14.32 illustrates the potential preamble group selections by different groups of UEs and the feature bundling of UEs.

CHAPTER

# Multimedia broadcast multicast services

# 15

## CHAPTER OUTLINE



In the past, cellular systems have mostly focused on transmission of data intended for a single user and not on broadcast services. Traditional broadcast networks, exemplified by the radio and TV broadcasting networks, have on the other hand focused on covering very large areas and have offered no or limited possibilities for transmission of data intended for a single user. *Multimedia Broadcast and Multicast Services* (MBMS), introduced for WCDMA in release 6, supports multicast/broadcast services in a cellular system, thereby combining multicast and unicast transmissions within a single network.

With MBMS, the same content is transmitted to multiple users located in a specific area, the *MBMS service area*, in a unidirectional fashion. The MBMS service area typically covers multiple cells, although it can be made as small as a single cell.

Broadcast and multicast describe different, although closely related, scenarios:

- In *broadcast*, a point-to-multipoint radio resource is set up in each cell being part of the MBMS broadcast area, and all users subscribing to the broadcast service simultaneously receive the same transmitted signal. No tracking of users' movement in the radio access network is performed, and users can receive the content without notifying the network. Mobile TV is an example of a service that could be provided through MBMS broadcast.
- In *multicast*, users request to join a multicast group prior to receiving any data. The user movements are tracked and the radio resources are configured to match the number of users in the cell. Each cell in the MBMS multicast area may be configured for point-to-point or point-to-multipoint transmission. In sparsely


HSPA Evolution: The Fundamentals for Mobile Broadband. DOI: 10.1016/B978-0-08-099969-2.00015-6

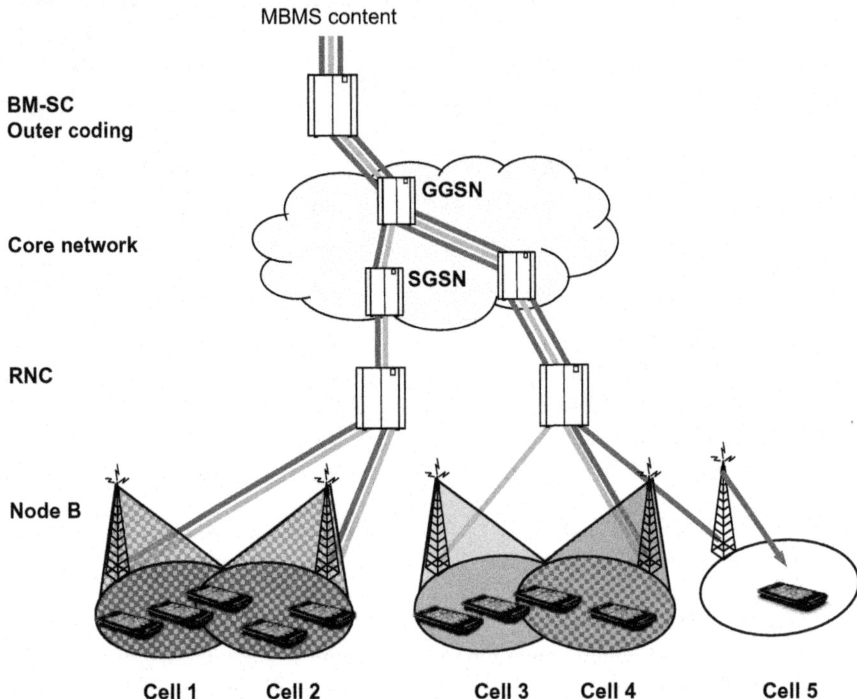


**FIGURE 15.1 Example of MBMS services. Different services are provided in different areas using broadcast in cells 1–4. In cell 5, unicast is used as there is only a single user subscribing to the MBMS service.**

populated cells with only one or a few users subscribing to the MBMS service, point-to-point transmission may be appropriate, while in cells with a larger number of users, point-to-multipoint transmission is better suited. Multicast therefore allows the network to optimize the transmission type in each cell.

To a large extent, MBMS affects mainly the nodes above the radio-access network. A new node, the *Broadcast Multicast Service Center* (BM-SC), illustrated in Figure 15.1, is introduced. The BM-SC is responsible for authorization and authentication of content provider, charging, and the overall configuration of the data flow through the core network. It is also responsible for application-level coding, as discussed below.

As the focus of this book is on the radio-access network, the procedures for MBMS will only be briefly described. In Figure 15.2, typical phases during an MBMS session are illustrated. First, the service is announced. In case of broadcast, there are no further actions required by the user; the user simply "tunes" to the channel of interest. In case of multicast, a request to join the session has to be sent to become a member of the corresponding MBMS service group and, as such, receive the data. Before the MBMS transmission can start, the BM-SC sends a session-start

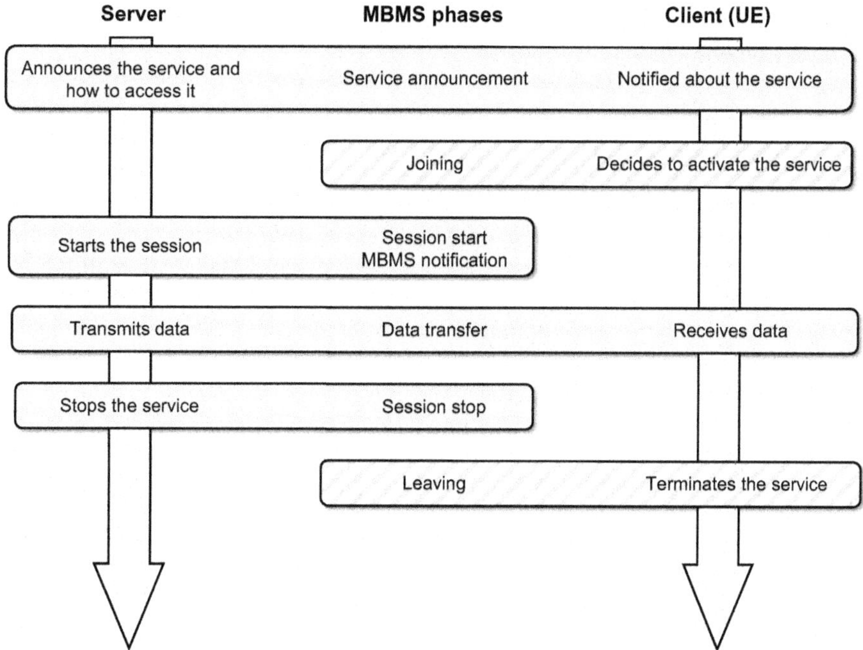


**FIGURE 15.2 Example of typical phases during an MBMS session. The dashed phases are only used in case of multicast and not for broadcast.**

request to the core network, which allocates the necessary internal resources and requests the appropriate radio resources from the radio-access network. All UEs of the corresponding MBMS service group are also notified that content delivery from the service will start. Data will then be transmitted from the content server to the end users. When the data transmission stops, the server will send a session-stop notification. Also, users who want to leave an MBMS multicast service can request to be removed from the MBMS service group.

One of the main benefits brought by MBMS is the resource saving in the network as a single stream of data may serve multiple users. This is illustrated in Figure 15.1, where three different services are offered in different areas. From the BM-SC, data streams are fed to each of the Node Bs involved in providing the MBMS services. As seen in the figure, the data stream intended for multiple users is not split until necessary. For example, there is only a single stream of data sent to all the users in cell 3. This is in contrast to UTRAN prior to release 6, where one stream per user has to be configured throughout both the core network and the radio access network.

In the following, the principles behind MBMS in the radio access network and their introduction into UTRAN will be discussed. The focus is on point-to-multipoint transmission as this requires some new features in the radio interface. Point-to-point transmission uses either dedicated channels or HS-DSCH and is, from a radio-interface perspective, not different from any other transmission.

A description of MBMS from a specification perspective is found in [1] and the references therein.

## 15.1 RELEASE 6 MBMS OVERVIEW

As discussed above, one of the main benefits with MBMS is resource saving in the network as multiple users can share a single stream of data. This is valid also from a radio-interface perspective, where a single transmitted signal may serve multiple users. Obviously, point-to-multipoint transmission puts very different requirements on the radio interface than point-to-point unicast. User-specific adaptation of the radio parameters, such as channel-dependent scheduling or rate control, cannot be used as the signal is intended for multiple users. The transmission parameters such as power must be set taking the worst case user into account as this determines the coverage for the service. Frequent feedback from the users, for example, in the form of CQI reports or hybrid-ARQ status reports, would also consume a large amount of the uplink capacity in cells where a large number of users simultaneously receive the same content. Imagine, for example, a sports arena with thousands of spectators watching their home team playing, all of them simultaneously wanting to receive results from games in other locations whose outcome might affect their home team. Clearly, user-specific feedback would consume a considerable amount of capacity in this case.

From the above discussion, it is clear that MBMS services are power limited and maximizing the diversity without relying on feedback from the users is of key importance. The two main techniques for providing the diversity for MBMS services are

1. *Macro-diversity* by combining transmissions from multiple cells.
2. *Time-diversity* against fast fading through a long 80 ms TTI and application-level coding.

Fortunately, MBMS services are not delay sensitive and the use of a long TTI is not a problem from the end-user perspective. Additional means for providing diversity can also be applied in the network, for example open-loop transmit diversity. Receive diversity in the UE also improves the performance, but as the 3GPP UE requirements for release 6 are set assuming single-antenna UEs, it is hard to exploit this type of diversity in the planning of MBMS coverage. Also, note that application-level coding provides additional benefits not directly related to diversity, as discussed below.

### 15.1.1 MACRO-DIVERSITY

Combining transmissions of the same content from multiple cells (macro-diversity) provides a significant diversity gain [2], in the order of 4–6 dB reduction in transmission power compared to single-cell reception, as illustrated in Figure 15.3. Two combining strategies are supported for MBMS, *soft combining* and *selection combining*, and the principles of both are illustrated in Figure 15.4.

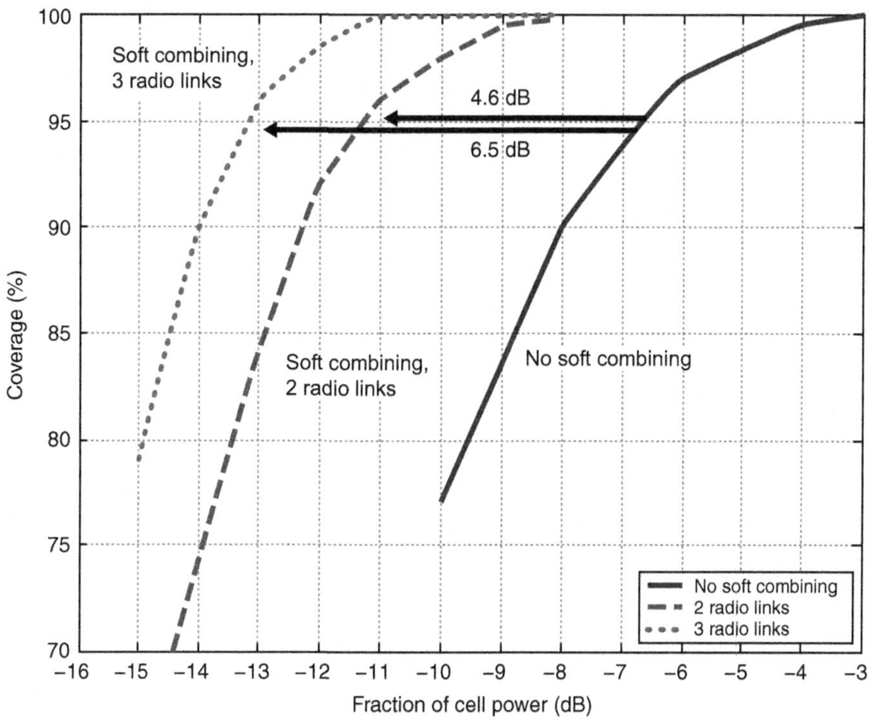


**FIGURE 15.3 The gain with soft combining and multicell reception in terms of coverage versus power for 64 kbit/s MBMS service (vehicular A, 3 km/h, 80 ms TTI, single receive antenna, no transmit diversity, 1% BLER).**

Soft combining, as the term indicates, combines the soft bits received from the different radio links prior to (Turbo) decoding. In principle, the UE descrambles and RAKE combines the transmission from each cell individually, followed by soft combining of the different radio links. Note that, in contrast to unicast, this macro-diversity gain comes "for free" in the sense that the signal in the neighboring cell is anyway present. Therefore, it is better to exploit this signal rather than treat it as interference. However, as WCDMA uses cell-specific scrambling of all data

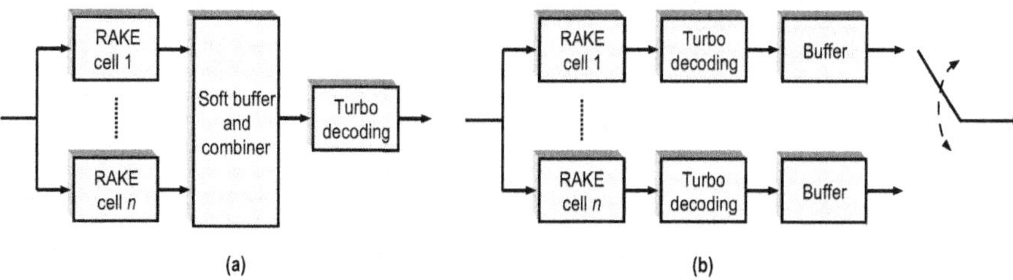


**FIGURE 15.4 Illustration of the principles for (a) soft combining and (b) selection combining.**

transmissions, the soft combining needs to be performed by the appropriate UE processing. This processing is also responsible for suppressing the interference caused by (non-MBMS) transmission activity in the neighboring cells. To perform soft combining, the physical channels to be combined should be identical. For MBMS, this implies the same physical channel content and structure should be used on the radio links that are soft combined.

Selection combining, on the other hand, decodes the signal received from each cell individually and for each TTI selects one (if any) of the correctly decoded data blocks for further processing by higher-layers. From a performance perspective, soft combining is preferable as it provides not only diversity gains but also a power gain as the received power from multiple cells is exploited. Relative to selection combining, the gain is in the order of 2–3 dB [2].

The reason for supporting two different combining strategies is to handle different levels of asynchronism in the network. For soft combining, the soft bits from each radio link have to be buffered until the whole TTI is received from all involved radio links and the soft combining can start, while for selection combining, each radio link is decoded separately and it is sufficient to buffer the decoded information bits from each link. Hence, for a large degree of asynchronism, selection combining requires less buffering in the UE at the cost of an increase in Turbo decoding processing and loss of performance. The UE is informed about the level of synchronism and can, based upon this information and its internal implementation, decide to use any combination scheme as long as it fulfills the minimum performance requirements mandated by the specifications. With similar buffering requirements as for a 3.6 Mbit/s HSDPA UE, which is the basis for the definition of the UE MBMS requirements, soft combining is possible provided the transmissions from the different cells are synchronized within approximately 80 ms, which is likely to be realistic in most situations.

As mentioned above, the UE capabilities are set assuming similar buffering requirements as for a 3.6 Mbit/s HSDPA UE. This results in certain limitations in the number of radio links a UE is required to be able to soft combine for different TTI values and different data rates. This is illustrated in Table 15.1, from which it is

**Table 15.1** Requirements on UE Processing for MBMS Reception [5]

| | **Soft Combining** | | **Selection Combining** | |
|---|---|---|---|---|
| **Data Rate (on MTCH)** | **Maximum Number of RLs** | **TTI** | **Maximum Number of RLs** | **TTI** |
| 256 kbit/s | 3 | 40 | 2 | 40 |
| | ≤2 | 80 | 1 | 80 |
| 128 kbit/s | ≤3 | 80 | 3 | 40 |
| | | | 2 | 80 |
| | | | 1 | 80 |
| ≤64 kbit/s | ≤3 | 80 | | |

also seen that all MBMS-capable UEs can support data rates up to 256 kbit/s. It is worth noting that in release 6 there is a single MBMS UE capability—either the UE supports MBMS or not. As network planning has to be done assuming a certain set of UE capabilities in terms of soft combining, etc., exceeding these capabilities cannot be exploited by the operator. The end user may of course benefit from a more advanced UE, for example, through the possibility for receiving multiple services simultaneously.

### 15.1.2 APPLICATION-LEVEL CODING

Many end-user applications require very low error probabilities, in the order of $10^{-6}$. Providing these low error probabilities on the transport channel level can powerwise be quite costly. In point-to-point communications, some form of (hybrid) ARQ mechanism is therefore used to retransmit erroneous packets. HSDPA, for example, uses both a hybrid-ARQ mechanism (see Chapter 8) and RLC retransmissions. In addition, the TCP protocol itself also performs retransmissions to provide virtually error-free packet delivery. However, as previously discussed, broadcast typically cannot rely on feedback, and, consequently, alternative strategies need to be used. For MBMS, application-level forward error-correcting coding is used to address this problem. The application-level coding resides in the BM-SC and is thus not part of the radio-access network but is nevertheless highly relevant for a discussion of the radio-access-network design for support of MBMS. With application-level coding, the system can operate at a transport-channel block-error rate in the order of 1–10% instead of fractions of a percent, which significantly lowers the transmit power requirement. As the application-level coding resides in the BM-SC, it is also effective against occasional packet losses in the transport network, for example, due to temporary overload conditions.

Systematic Raptor codes [3] have been selected for the application-level coding in MBMS [4], operating on packets of constant size (48–512 bytes). Raptor codes belong to a class of *Fountain codes*, and as many encoding packets as needed can be generated on the fly from the source data. For the decoder to be able to reconstruct the information, it only needs to receive sufficiently many coded packets. It does not matter which coded packets it received, in what order they are received, or if certain packets were lost (Figure 15.5).

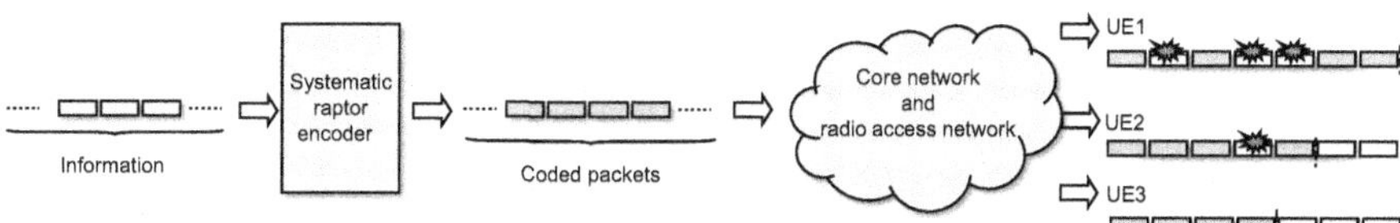


**FIGURE 15.5 Illustration of application-level coding. Depending on their different ratio conditions, the number of coded packets required for the UEs to be able to reconstruct the original information differs.**

In addition, to provide additional protection against packet losses and to reduce the required transmission power, the use of application-level coding also simplifies the procedures for UE measurements. For HSDPA, the scheduler can avoid scheduling data to a given UE in certain time intervals. This allows the UE to use the receiver for measurement purposes, for example, to tune to a different frequency and possibly also to a different radio access technology. In a broadcast setting, scheduling measurement gaps is cumbersome as different UEs may have different requirements on the frequency and length of the measurement gaps. Furthermore, the UEs need to be informed when the measurement gaps occur. Hence, a different strategy for measurements is adopted in MBMS. The UE measurements are done autonomously, which could imply that a UE sometimes misses (part of) a coded transport block on the physical channel. In some situations, the inner Turbo code is still able to decode the transport channel data, but if this is not the case, the outer application-level code will ensure that no information is lost.

## 15.2 DETAILS OF RELEASE 6 MBMS

One requirement in the design of MBMS was to reuse existing channels to the maximum extent possible. Therefore, the FACH transport channel and the S-CCPCH physical channel are reused without any changes. To carry the relevant MBMS data and signaling, three new logical channels are added to release 6:

1. *MBMS Traffic Channel* (MTCH), carrying application data.
2. *MBMS Control Channel* (MCCH), carrying control signaling.
3. *MBMS Scheduling Channel* (MSCH), carrying scheduling information to support discontinuous reception in the UE.

All these channels use FACH as the transport channel type and the S-CCPCH as the physical channel type. In addition to the three new logical channels, one new physical channel is introduced to support MBMS – the *MBMS Indicator Channel* (MICH), used to notify the UE about an upcoming change in MCCH contents.

### 15.2.1 MTCH

The MTCH is the logical channel used to carry the application data in case of point-to-multipoint transmission (for point-to-point transmission, DTCH, mapped to DCH or HS-DSCH, is used). One MTCH is configured for each MBMS service and each MTCH is mapped to one FACH transport channel. The S-CCPCH is the physical channel used to carry one (or several) FACH transport channels.

The RLC for MTCH is configured to use unacknowledged mode as no RLC status reports can be used in point-to-multipoint transmissions. To support selective combining (discussed in Section 15.1.1), the RLC has been enhanced with support for in-sequence delivery using the RLC PDU sequence numbers and the same type of mechanism as employed in MAC-hs (see Chapter 8). This enables the UE to do

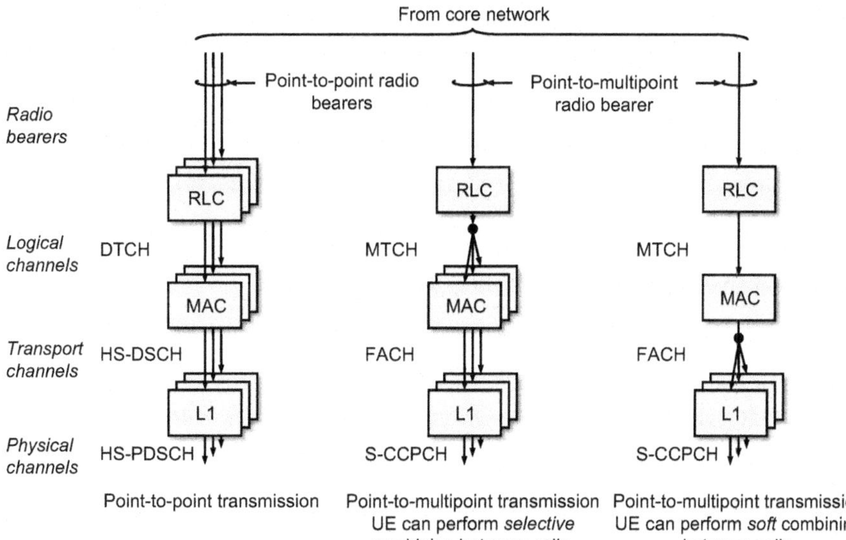


**FIGURE 15.6 Illustration of data flow through RLC, MAC, and L1 in the network side for different transmission scenarios.**

reordering up to a depth set by the RLC PDU sequence number space in case of selection combining.

In Figure 15.6, an example of the flow of application data through RLC, MAC, and physical layer is illustrated. The leftmost part of the figure illustrates the case of point-to-point transmission, while the middle and rightmost parts illustrate the case of point-to-multipoint transmission using the MTCH. In the middle part, one RLC entity is used with multiple MAC entities. This illustrates a typical situation where selection combining is used, where multiple cells are loosely time aligned and the same data may be transmitted several TTIs apart in the different cells. Finally, the rightmost part of the figure illustrates a typical case where soft combining can be used. A single RLC and MAC entity is used for transmission in multiple cells. To allow for soft combining, transmissions from the different cells need to be aligned within 80.67 ms (assuming 80 ms TTI).

## 15.2.2 MCCH AND MICH

The MCCH is a logical channel type used to convey control signaling necessary for MTCH reception. One MCCH is used in each MBMS-capable cell and it can carry control information for multiple MTCHs. The MCCH is mapped to FACH (note, a different FACH than used for MTCH), which in turn is transmitted on an S-CCPCH physical channel. The same S-CCPCH as for the MTCH may be used, but if soft combining is allowed for MTCH, different S-CCPCHs for MTCH and MCCH should be used. The reason for using separate S-CCPCHs in this case is that no

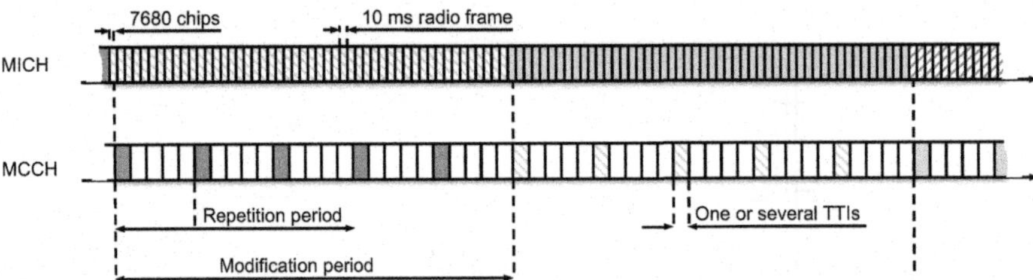


**FIGURE 15.7 MCCH transmission schedule. Different shades indicate (potentially) different MCCH content, for example, different combinations of services.**

selection or soft combining is used for the MCCH, and the UE receives the MCCH from a single cell only. The RLC is operated in unacknowledged mode for MCCH. Where to find the MCCH is announced on the BCCH (the BCCH is the logical channel used to broadcast system configuration information).

Transmission on the MCCH follows a fixed schedule, as illustrated in Figure 15.7. The MCCH information is transmitted using a variable number of consecutive TTIs. In each *modification period*, the critical information remains unchanged[1] and is periodically transmitted based on a *repetition period*. This is useful to support mobility between cells; a UE entering a new cell or a UE which missed the first transmission does not have to wait until the start of a new modification period to receive the MCCH information.

The MCCH information includes information about the services offered in the modification period and how the MTCHs in the cell are multiplexed. It also contains information about the MTCH configuration in the neighboring cells to support soft or selective combining of multiple transmissions. Finally, it may also contain information to control the feedback from the UEs in case counting is used.

Counting is a mechanism where UEs connect to the network to indicate whether they are interested in a particular service or not and is useful to determine the best transmission mechanism for a given service. For example, if only a small number of users in a cell are interested in a particular service, point-to-point transmission may be preferable over point-to-multipoint transmission. To avoid the system being heavily loaded in the uplink as a result of counting responses, only a fraction of the UEs transmit the counting information to the network. The MCCH counting information controls the probability with which a UE connects to the network to transmit counting information. Counting can thus provide the operator with valuable feedback on where and when a particular service is popular, a benefit typically not available in traditional broadcast networks.

To reduce UE power consumption and avoid having the UE constantly receiving the MCCH, a new physical channel, the *MBMS Indicator Channel* (MICH), is

[1]The MBMS access information may change during a modification period, while the other MCCH information is considered as critical and only may change at the start of a modification period.

introduced to support MBMS. Its purpose is to inform UEs about upcoming changes in the critical MCCH information and the structure is identical to the paging indicator channel. In each 10 ms radio frame, 18, 36, 72, or 144 MBMS indicators can be transmitted, where an indicator is a single bit, transmitted using on–off keying and related to a specific group of services.

By exploiting the presence of the MICH, UEs can sleep and briefly wake up at predefined time intervals to check whether an MBMS indicator is transmitted. If the UE detects an MBMS indicator for a service of interest, it reads the MCCH to find the relevant control information, for example, when the service will be transmitted on the MTCH. If no relevant MBMS indicator is detected, the UE may sleep until the next MICH occasion.

### 15.2.3 MSCH

The purpose of the MSCH is to enable UEs to perform discontinuous reception of the MTCH. Its content informs the UE in which TTIs a specific service will be transmitted. One MSCH is transmitted in each S-CCPCH carrying the MTCH and the MSCH content is relevant for a certain service and a certain S-CCPCH.

## 15.3 RELEASE 7 MBSFN

### 15.3.1 SFN UNDERLYING PRINCIPLE

In a traditional HSPA network, each cell applies a scrambling code and serves users within the cell with common and dedicated channels. In the downlink, non-serving cells are experienced as interference. MBMS services, as described in Section 15.2, may cover a larger area than a single cell. When this is the case, several cells transmit the same MBMS data, and UE throughput can be improved by means of soft and selective combining of multiple cell data, as described in Section 15.1.1.

With soft and selective combining, however, when despreading and demodulating each cell's MTCH, the surrounding cells still continue to be experienced as interference. Soft combining can provide a gain in received softbit power but cannot decrease the interference level. This can be illustrated by considering a simple example of three cells, as depicted in Figure 15.8. It is assumed that the received power from each of the cells is $S_1$, $S_2$, and $S_3$, respectively, that the proportion of the cell power spent on MTCH is $\alpha$, and that the combined interference from all other cells is $I$. When decoding MTCH in each individual cell, if there is no channel fading then the SINR achieved in cell 1, cell 2, and cell 3 is

$$\frac{\alpha S_1}{S_2+S_3+I}, \frac{\alpha S_2}{S_1+S_3+I}, \frac{\alpha S_3}{S_1+S_2+I}. \tag{15.1}$$

If MRC combining is applied on the softbits, the combined SINR is the sum of the individual SINRs.

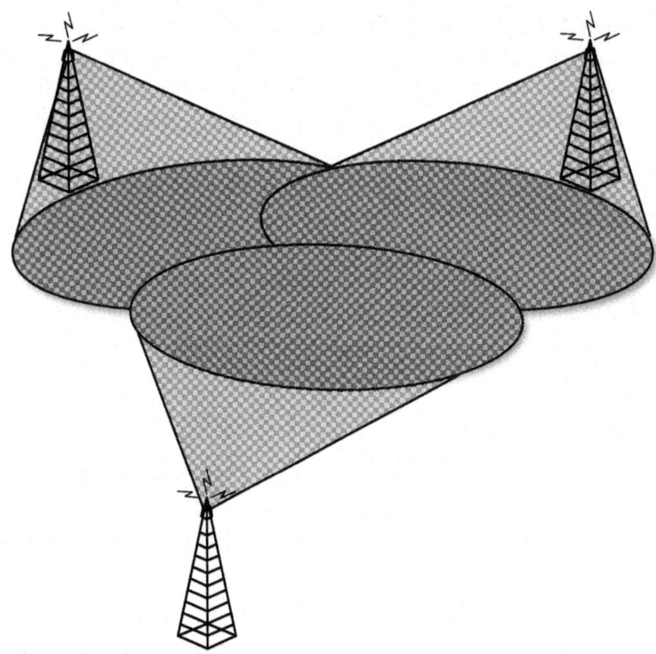

**FIGURE 15.8 Three-cell MBSFN cluster.**

If all of the information to be transmitted from a cell is exactly the same, then it is more efficient to transmit the information using the same scrambling code in each cell. In this case, the cells over which the transmission takes place are known as a *Single Frequency Network* (SFN). The UE experiences all of the cells that are part of an SFN as if they would be a single cell, since they have the same scrambling code and exactly the same content. In an SFN, in the previous scenario and in AWGN conditions, the SINR becomes

$$\alpha \frac{(S_1 + S_2 + S_3)}{I}. \tag{15.2}$$

The AWGN SINR in this case is obviously superior to the soft combining SINR. However, in order to achieve SFN gains in a real environment, several implementation-related issues must be overcome:

- In order to operate the UEs with a reasonable receiver complexity, the individual Node Bs must be tightly synchronized, such that they effectively operate as a distributed antenna.
- Because of the distributed placement of Node Bs, the UEs will experience a larger delay spread than is the case for conventional single cell transmissions. Equalization of this larger delay spread will increase UE complexity.
- All information transmitted within the SFN area, that is, all data and control information for all channels, must be exactly the same and encoded, spread, and modulated in exactly the same manner.

### 15.3.2 MBSFN IN THE 3GPP SPECIFICATIONS

SFN operation was standardized by 3GPP in release 7. To operate SFN, it is necessary to transmit exactly the same signal from all of the involved nodes. SFN is not, therefore, appropriate for normal cellular operation, since it does not allow for the use of scrambling codes or differing cell identities. SFN was standardized for operation on a separate, downlink-only standalone carrier dedicated for MBMS operation.

In FDD, the carrier transmits only MBSFN in every slot and has no associated uplink. In TDD, it is possible to configure the downlink-only MBSFN serving in a subset of slots on a TDD carrier, while configuring a regular TDD cell to use other uplink and downlink slots. To support the larger amount of delay spread, an MBSFN-specific TDD slot format was specified.

The FDD standalone carrier carries P-CCPCH (carrying a limited BCH), S-CCPCH (carrying FACH that carries MTCH), P-CPICH, and SCH only. Optionally, an MICH may also be transmitted. The BCH carries a limited number of SIBs that contain required information for the carrier.

The carrier can only carry unicast information, and no counting procedure is available. MBSFN reception is, however, completely independent of the UE's operation on the regular cell and is independent of the UE's RRC state. Furthermore, receiving MBSFN may not impact the UE's operation on the regular carrier.

MBSFN cells are arranged into MBSFN clusters; all cells within a cluster transmit the same information in SFN. Between clusters, the services may differ. UEs perform reselection between clusters. MBSFN cells within a cluster must be tightly synchronized in order to provide the SFN service.

## 15.4 RELEASE 8 INTEGRATED MOBILE BROADCAST

The release 7 MBSFN solution created differing slot formats for the FDD and TDD modes for MBSFN operation. In release 8, the solution was further enhanced to become a single unified solution known as *Integrated Mobile Broadcast* (IMB). IMB offers the combined advantages of reducing development efforts by means of a single slot format for FDD and TDD and greater opportunity for UE DRX in the FDD mode, because of the use of a shorter TTI.

The IMB solution is similar to release 7 MBSFN in that it is intended for downlink-only and broadcast-only operation. The carrier carries a P-CPICH, SCH, P-CCPCH, S-CCPCH, a new S-CCCPH with a so-called frame type 2, a new TDM pilot channel known as T-CPICH, and MICH. A limited set of eight scrambling codes are available for MBSFN clusters; this is in particular necessary because the T-CPICH sequence depends on the scrambling code.

The P-CCPCH carries a limited BCH, as in release 7. The conventional S-CCPCH carries FACH, carrying MCCH. The new frame type 2 S-CCPCH carries a second FACH to which MTCH is mapped.

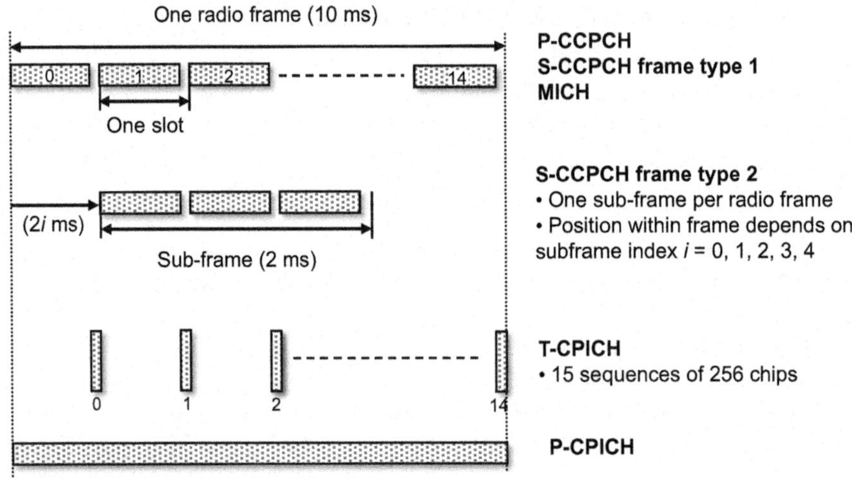


**FIGURE 15.9 IMB channel structure.**

The frame type 2 S-CCPCH differs from the release 99 S-CCPCH in five aspects:

- The TTI is 2 ms (compared to multiples of 10 ms for S-CCPCH).
- The spreading factor is 16 (it is configurable for S-CCPCH).
- Multicode transmission of 15 SF16 codes is applied.
- The first 256 chips in each slot are DTX.
- The modulation can be either QPSK or 16-QAM.

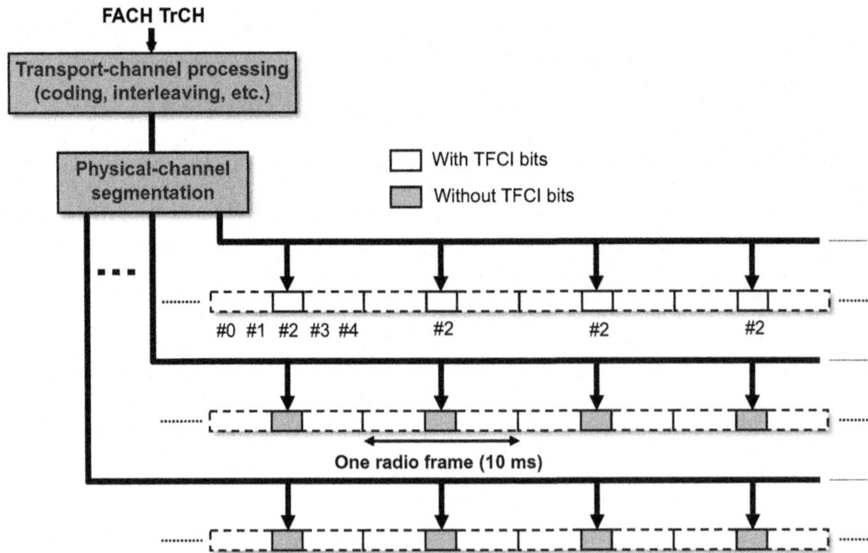


**FIGURE 15.10 IMB physical channel mapping procedure.**

The 2 ms TTI length for the frame type 2 S-CCPCH enables DRX for a UE not listening to MTCH continuously. The UE also does not need to continuously receive MCCH on the first S-CCPCH. The first type 2 S-CCPCH code carries a 5-bit TFCI; the other 14 codes do not carry TFCI.

The T-CPICH is transmitted using the 15 codes allocated to the frame type 2 S-CCPCH during the first 256 chips of each slot. The T-CPICH can be used for enhanced channel estimation. The actual sequence used for the T-CPICH depends on the scrambling code; it is designed such that the first 128 chips of T-CPICH and P-CPICH when combined are approximately the same as the last 128 chips after scrambling. This effectively provides a cyclic prefix and allows for delay spreads of up to 33 μs in the MBSFN cluster.

Figure 15.9 depicts the physical channels for IMB and Figure 15.10 shows the physical channel mapping procedure.

## REFERENCES

[1] 3rd Generation Partnership Project; Technical Specification Group Radio Access Network; Introduction of the Multimedia Broadcast Multicast Service (MBMS) in the Radio Access Network (RAN); Stage 2 (Release 6), 3GPP, 3GPP TS 25.346.

[2] 3rd Generation Partnership Project; Technical Specification Group Radio Access Network; S-CCPCH Performance for MBMS (Release 6), 3GPP, 3GPP TR 25.803.

[3] A. Shokrollahi, Raptor codes, IEEE Trans. Info. Theory 52 (6) (2006) 2551–2567.

[4] 3rd Generation Partnership Project; Technical Specification Group Services and System Aspects; Multimedia Broadcast/multicast; Protocols and codecs, 3GPP TS 26.346.

[5] 3rd Generation Partnership Project; Technical Specification Group Radio Access Network; UE Radio Access Capabilities, 3GPP, 3GPP TS 25.306.

## FURTHER READING

P.J. Czerepinski, T.M. Chapman, J. Krause, Coverage and planning aspects of MBMS in UTRAN, Fifth IEE International Conference on 3G Mobile Communication Technologies (3G 2004).

CHAPTER

# 16 Spectrum and RF characteristics

## CHAPTER OUTLINE




HSPA Evolution: The Fundamentals for Mobile Broadband. DOI: 10.1016/B978-0-08-099969-2.00016-8

This chapter describes the RF characteristics of the *UMTS Terrestrial Radio Access* (UTRA), which consists of both WCDMA and its evolution into HSPA. The RF characteristics of UTRA are determined by a set of RF requirements that base stations and UEs must meet. These are strongly related to the frequency bands that UTRA is able to operate in and the global and regional regulation of how that spectrum can be used. There is also a large set of performance requirements for the baseband, setting the minimum performance for the different physical channels.

There are a number of frequency bands identified for mobile use and specifically for IMT. Many of these bands were first defined for operation with UTRA but are now also defined for LTE and shared with LTE deployments.

The RF requirements impact the RF implementation in terms of filters, amplifiers, and all other RF components that are used to transmit and receive the signal. In later releases, the addition of multi-carrier techniques further impacts the implementation.

## 16.1 SPECTRUM FOR UTRA

UTRA is designed to be deployed both in existing IMT bands and in future bands that may be identified. The possibility of operating a radio-access technology in different frequency bands is, in itself, nothing new. For example, 2G, 3G, and 4G UEs are multiband capable, covering bands used in the different regions of the world to provide global roaming. From a radio-access functionality perspective, this has no or limited impact and the UTRA physical-layer specifications [1–4] do not assume any specific frequency band. What may differ, in terms of specification, between different bands are mainly the more specific RF requirements, such as the allowed maximum transmit power, requirements/limits on *out-of-band* (OOB) emission, and so on. One reason for this is that external constraints, imposed by regulatory bodies, may differ between different frequency bands. A paired frequency range where UTRA is designed to operate is called an *Operating band* but is often just referred to as a "band."

### 16.1.1 SPECTRUM DEFINED FOR IMT SYSTEMS BY THE ITU-R

The global designations of spectrum for different services and applications are done within the ITU-R. The *World Administrative Radio Congress* WARC-92 identified the bands 1885–2025 and 2110–2200 MHz as intended for implementation of IMT-2000. Of these 230 MHz of 3G spectrum, $2 \times 30$ MHz were intended for the satellite component of IMT-2000 and the rest for the terrestrial component. Parts of the bands were used during the 1990s for deployment of 2G cellular systems, especially in the Americas. The first deployments of 3G in 2001–2002 in Japan and Europe were done in this band allocation, and for that reason it is often referred to as the IMT-2000 "core band."

Additional spectrum for IMT-2000 was identified at the World Radiocommunication Conference WRC-2000, where it was considered that an additional need for 160 MHz of spectrum for IMT-2000 was forecasted by the ITU-R. The identification included the bands used for 2G mobile systems at 806–960 and 1710–1885 MHz, and "new" 3G spectrum in the bands at 2500–2690 MHz. The identification of bands previously assigned for 2G was also recognition of the evolution of existing 2G mobile systems into 3G. Additional spectrum was identified at WRC'07 for IMT, encompassing both IMT-2000 and IMT-Advanced. The bands added were 450–470, 698–806, 2300–2400, and 3400–3600 MHz, but the applicability of the bands varies on a regional and national basis. At WRC'12, there were no additional spectrum allocations identified for IMT, but the issue was put on the agenda for WRC'15. It was also determined to study the use of the band 694-790 MHz for mobile services in Region 1 (Europe, Middle East, and Africa).

The somewhat diverging arrangement between regions of the frequency bands assigned to IMT means that there is not one single band that can be used for 3G and 4G roaming worldwide. Large efforts have, however, been put into defining a minimum set of bands that can be used to provide roaming. In this way, multiband devices can provide efficient worldwide roaming for 3G and 4G devices.

### 16.1.2 FREQUENCY BANDS FOR UTRA

3GPP makes spectrum identified by ITU available for UTRA by means of defining UTRA bands and corresponding requirements. The frequency bands in which UTRA FDD can operate are paired bands, with one band designated for uplink and one for downlink. There are also unpaired bands for UTRA TDD operation, but those are not discussed in this book.

Release 11 of the 3GPP specifications for UTRA FDD includes 20 paired frequency bands. The paired bands are numbered with Roman numerals [5] for UTRA. Note that the same frequency bands when used for LTE have the same numbers as the paired bands for UTRA FDD but are labeled with Arabic numerals and that numbering scheme is also used for MSR (multistandard) base stations. Since the number of bands used for LTE is larger than for UTRA FDD, there are some "gaps" in the sequence of band numbers for UTRA. All bands for UTRA FDD are summarized in Table 16.1 and in Figures 16.1 and 16.2, which also show the corresponding frequency allocation defined by the ITU.

Some of the frequency bands are partly or fully overlapping. In most cases, this is explained by regional differences in how the bands defined by the ITU are implemented. At the same time, a high degree of commonality between the bands is desired to enable global roaming. A set of bands was first specified as bands for UTRA, with each band originating in global, regional, and local spectrum developments.

*Band I* is the first band that was defined for UTRA in release 99 of the 3GPP specifications, also known as the 2 GHz "core band." *Band II* was added later for operation in the US PCS1900 band and *Band III* for 3G operation in the GSM1800 band. *Band IV* was introduced as a new band for the Americas following the addition of the 3G bands at WRC-2000. Its downlink overlaps completely with the downlink of Band I, which facilitates roaming and eases the design of dual Band I + IV UEs. *Band X* is a later extension of Band IV from 2 × 45 to 2 × 60 MHz.

**Table 16.1** Paired Frequency Bands Defined by 3GPP for UTRA

| Band (UTRA) | Band (MSR BS) | Uplink range (MHz) | Downlink range (MHz) | Main region(s) |
|---|---|---|---|---|
| I | 1 | 1920–1980 | 2110–2170 | Europe, Asia |
| II | 2 | 1850–1910 | 1930–1990 | Americas (Asia) |
| III | 3 | 1710 –1785 | 1805 –1880 | Europe, Asia (Americas) |
| IV | 4 | 1710–1755 | 2110–2155 | Americas |
| V | 5 | 824–849 | 869–894 | Americas, Asia |
| VI | 6 | 830–840 | 875–885 | Japan |
| VII | 7 | 2500–2570 | 2620–2690 | Europe, Asia |
| VIII | 8 | 880–915 | 925–960 | Europe, Asia |
| IX | 9 | 1749.9–1784.9 | 1844.9–1879.9 | Japan |
| X | 10 | 1710 –1770 | 2110–2170 | Americas |
| XI | 11 | 1427.9–1447.9 | 1475.9–1495.9 | Japan |
| XII | 12 | 698–716 | 728–746 | United States |
| XIII | 13 | 777–787 | 746–756 | United States |
| XIV | 14 | 788–798 | 758–768 | United States |
| XIX | 19 | 830–845 | 875–890 | Japan |
| XX | 20 | 832–862 | 791–821 | Europe |
| XXI | 21 | 1447.9–1462.9 | 1495.9–1510.9 | Japan |
| XXII | 22 | 3410–3490 | 3510–3590 | Europe |
| XXV | 25 | 1850–1915 | 1930–1995 | Americas |
| XXVI | 26 | 814–849 | 859–894 | Americas |

*Band IX* overlaps with Band III but is intended only for Japan. The specifications are drafted in such a way that implementation of roaming dual Band III + IX UEs is possible. The 1500 MHz frequency band is also identified in 3GPP for Japan as *Bands XI* and *XXI*. It is allocated globally to mobile service on a co-primary basis and was previously used for 2G in Japan.

With WRC-2000, the band 2500–2690 MHz was identified for IMT-2000 and it is identified as *Band VII* in 3GPP for FDD operation, together with an unpaired band identified for "center gap" of the FDD allocation. The band has a slightly different arrangement in North America, where a U.S.-specific unpaired band is defined.

WRC-2000 also identified the frequency range 806–960 MHz for IMT-2000, complemented by the frequency range 698–806 MHz in WRC'07. As shown in Figure 16.2, *several bands* are defined for FDD operation in this range. *Band XIII* uses the same band plan as GSM900. *Bands V*, *VI*, *XIX*, and *XXVI* overlap but are intended for different regions. Band V is based on the U.S. cellular band, while Bands VI and XIX are restricted to Japan in the specifications. 2G systems in Japan had a very specific band plan and Band VI was a way of partly aligning the Japanese spectrum plan in the 810–960 MHz range to that in other parts of the world, while Band XIX is a later extension of that allocation.

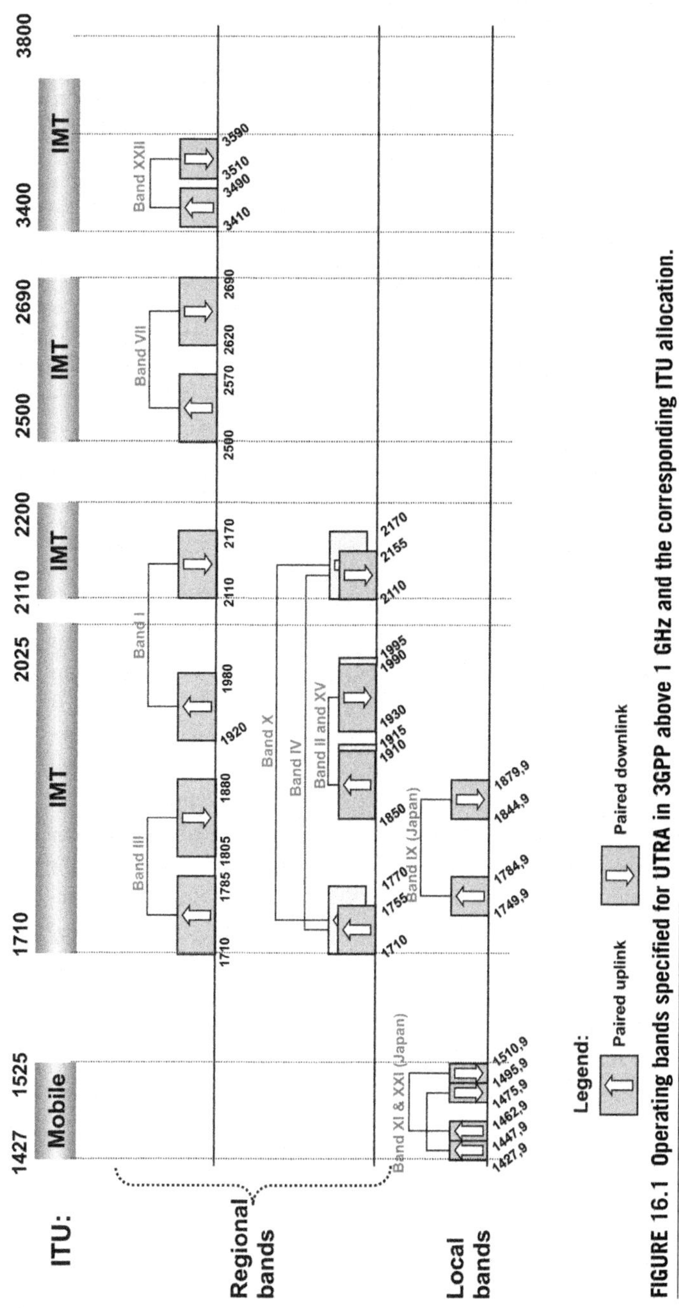


**FIGURE 16.1** Operating bands specified for UTRA in 3GPP above 1 GHz and the corresponding ITU allocation.

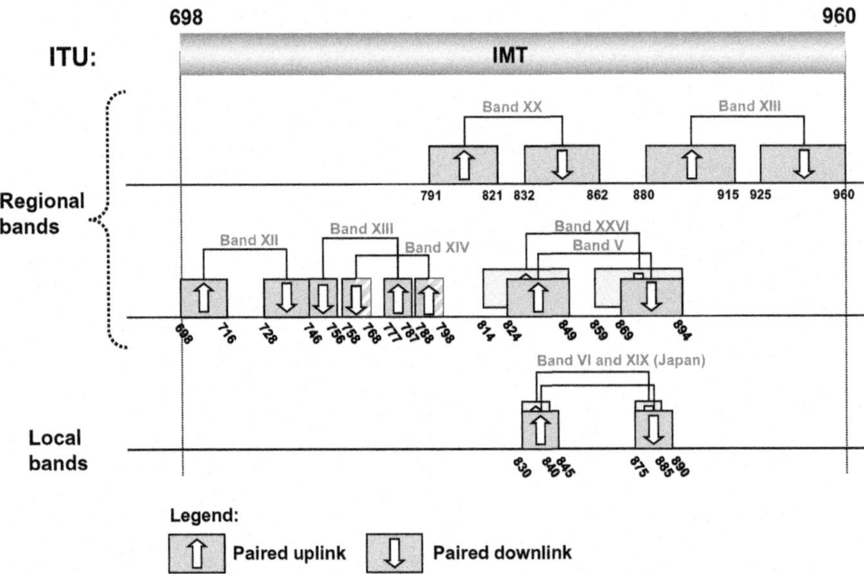


**FIGURE 16.2 Operating bands specified for UTRA in 3GPP below 1 GHz and the corresponding ITU allocation.**

An extensive study was performed in 3GPP to create an extension of Band V (850 MHz), which is one of the bands with the most widespread deployment globally. The extension adds additional frequency ranges below the present Band V and is done with two new operating bands, of which one is defined for UTRA. Band XXVI is the "Upper Extending 850 MHz" band, which encompasses the Band V range, adding 2 × 10 MHz to create an extended 2 × 35 MHz band for UTRA.

*Bands XII, XIII*, and *XIV* make up the first set of bands defined for what is called the *digital dividend*—that is, for spectrum previously used for broadcasting. This spectrum is partly migrated to be used by other wireless technologies, since TV broadcasting is migrating from analog to more spectrum-efficient digital technologies. Another regional band for the digital dividend is *Band XX* that is defined in Europe.

The paired *Band XXII* is specified for UTRA FDD in the frequency range 3.4–3.6 GHz [6]. In Europe, a majority of countries already license this band for both Fixed Wireless Access and mobile use and there is a European spectrum decision for 3.4–3.8 GHz with "flexible usage modes" for deployment of fixed, nomadic, and mobile networks. In Japan, not only 3.4–3.6 GHz but also 3.6–4.2 GHz will be available to terrestrial mobile services in the future. The band 3.4–3.6 GHz has also been licensed for wireless access in Latin America.

### 16.1.3 NEW FREQUENCY BANDS

Additional frequency bands are continuously specified for UTRA. WRC'07 identified additional frequency bands for IMT, which encompasses both IMT-2000 and

IMT-Advanced. Several of the bands defined by WRC'07 are already available for UTRA as described above, or will become available partly or fully for deployment on a global basis:

- *450–470 MHz* was identified for IMT globally. It is already allocated to mobile service globally, but it is only 20 MHz wide and has a number of different arrangements.
- *698–806 MHz* was allocated to mobile service and identified IMT to some extent in all regions. Together with the band at 806–960 MHz identified at WRC-2000, it forms a wide frequency range from 698 to 960 MHz that is partly identified to IMT in all regions, with some variations. A number of UTRA bands are already defined in this frequency range.
- *2300–2400 MHz* was identified for IMT on a worldwide basis in all three regions. It is defined today for LTE but not for UTRA.
- *3400–3600 MHz* was allocated to the mobile service on a primary basis in Europe and Asia and partly in some countries in the Americas. There is also satellite use in the bands today. Part of it is defined as UTRA Band XXII.

Additional bands for IMT were not on the agenda for WRC'12 but are on the agenda for WRC'15. For the frequency ranges below 1 GHz identified at WRC-07, 3GPP has already specified several UTRA operating bands, as shown in Figure 16.2. The bands with the widest use are Bands V and XIII, while most of the other bands have regional or more limited use. With the identification of bands down to 698 MHz for IMT use and the switch-over from analog to digital TV broadcasting, Bands XII, XIII, and XIV are defined in the United States and Band XX in Europe for the digital dividend.

## 16.2 FLEXIBLE SPECTRUM USE

Most of the frequency bands identified above for deployment of UTRA are existing IMT bands, and some bands also have other systems deployed, including LTE and GSM. Bands are also in some regions defined in a "technology neutral" manner, which means that coexistence between different technologies is a necessity.

The fundamental UTRA requirement to operate in different frequency bands does not, in itself, impose any specific requirements on the radio-interface design. There are, however, implications for the RF requirements and how those are defined, in order to support the following:

- *Coexistence between operators in the same geographical area in the band.* These other operators may deploy UTRA or other technologies, such as LTE or GSM/EDGE. There may also be non-IMT technologies in some cases. Such coexistence requirements are to a large extent developed within 3GPP, but there may also be regional requirements defined by regulatory bodies in some frequency bands.
- *Co-location of base station equipment between operators.* There are in many cases limitations to where base station equipment can be deployed. Often, sites must be shared between operators or an operator will deploy multiple

technologies in one site. This puts additional requirements on both base station receivers and transmitters.

- *Coexistence with services in adjacent frequency bands and across country borders*. The use of the RF spectrum is regulated through complex international agreements, involving many interests. There will therefore be requirements for coordination between operators in different countries and for coexistence with services in adjacent frequency bands. Most of these are defined in different regulatory bodies. Sometimes the regulators request that 3GPP includes such coexistence limits in the 3GPP specifications.
- *Release-independent frequency-band principles*. Frequency bands are defined regionally and new bands are added continuously. This means that every new release of 3GPP specifications will have new bands added. Through the "release independence" principle, it is possible to design UEs based on an early release of 3GPP specifications that support a frequency band added in a later release.
- *Aggregation of spectrum allocations*. Operators of UTRA systems have quite diverse spectrum allocations, which may fit multiple RF carriers. The allocation may even be non-contiguous, consisting of multiple blocks spread out in a band. For these scenarios, the UTRA specifications support *multi-carrier operation*, where multiple carriers in contiguous or non-contiguous blocks within a band, or in multiple bands, can be combined to create larger transmission bandwidths.

Figure 16.3 demonstrates the WCDMA pulse shape in the frequency domain; see also Figure 2.6 describing the theoretical *Root-Raised cosine* (RRC) pulse shape in detail. The actual transmitted spectrum emissions will also depend on the transmitter

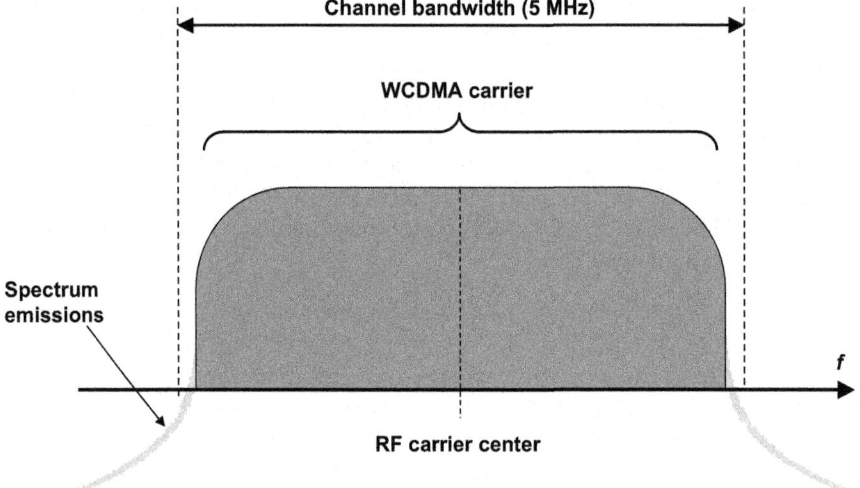


**FIGURE 16.3 The WCDMA RRC pulse shape in the frequency domain.**

RF chain and other components, as is illustrated in principle in Figure 16.3. The emissions outside the channel bandwidth are called *unwanted emissions* and the requirements for those are discussed further below.

## 16.3 MULTI-CARRIER OPERATION

Multi-carrier HSDPA was first introduced in 3GPP release 8, with *Dual-Cell HSDPA* (DC-HSDPA) as a means to increase the transmission bandwidth and consequently the maximum achievable data rate. In subsequent releases, further steps for multi-carrier support in the downlink were introduced, with multi-carrier for up to eight carriers, multiple bands, and also in non-contiguous spectrum added. Multi-carrier HSUPA was standardized in release 9. Detailed descriptions of the different multi-carrier schemes are found in Chapter 12, while the RF impacts are presented here.

Multi-carrier transmission has impact on the spectrum use and the RF characteristics for the base station and the UE. The impact is minor on the base station side, while there are more impacted requirements for the UE, in particular for dual-cell HSUPA.

From a spectrum point of view, it would in theory be possible to introduce the multi-carrier schemes in any band and dual band schemes in any band combination. Because of implementation restrictions and the need to set band-specific RF requirements, the multi-carrier schemes are however restricted to certain bands and band combinations. A list of the bands and band combinations for the different multi-carrier HSDPA schemes is presented in Table 12.9, including a mapping of each feature to the respective 3GPP releases.

### 16.3.1 MULTI-CARRIER DOWNLINK SCHEMES IN A SINGLE BAND

Adjacent multi-carrier operation is defined for operation in contiguous spectrum. In this case, the nominal carrier spacing is 5 MHz, but other values of the channel spacing are feasible. The set of RF requirements and their fundamental definitions for each carrier are fundamentally the same as those for single carrier operation or for operation of multiple independent carriers. There are, however, some new requirements, especially for the UE receiver.

For the base station, transmission of any of the multi-carrier schemes DC-HSDPA, 4C-HSDPA, or 8C-HSDPA will from an RF point of view look like any transmission of two, four, or eight carriers. All RF requirements are identical to what they are for transmission of two, four, or eight independent RF carriers, with one notable exception. Since the carriers in DC-HSDPA, 4C-HSDPA, or 8C-HSDPA will be received simultaneously by a single UE, there is a special timing requirement between the carriers. This requirement is expressed as a maximum time alignment error between the carriers. There are also some additional demodulation performance requirements for the HS-DPCCH, since the control signaling is redesigned for multi-carrier transmissions (as described in Chapter 12).

For the UE, receiving a multi-carrier signal has more far-reaching impacts, since this was not an inherent property of the downlink receiver design before the multi-carrier schemes were introduced. All of the RF characteristics for the receiver, except receiver spurious emissions, have additional requirements defined for each of the downlink multi-carrier schemes, since it is essential to characterize the UE receiver thoroughly for all possible receive signals. The different RF characteristics for the receiver are summarized in Section 16.5.2 and further described in the sections following. Note that the spurious emissions requirement at the receiver is a regulatory limit and does not depend on the number of carriers transmitted. There are also major implications for the HSDPA demodulation performance requirements for the UE, with additional requirements defined for all multi-carrier schemes.

### 16.3.2 DUAL-BAND DOWNLINK SCHEMES

If an operator has spectrum available in multiple bands, there are several multi-carrier schemes defined, making it possible to transmit multiple carriers across the two bands. The following options are defined in the 3GPP specifications:

- DB-DC-HSDPA, with one carrier in each band.
- Dual-Band 4C-HSDPA, with up to four carriers distributed over two bands, with one to three (adjacent) carriers in each band.

The band and carrier combinations are restricted to certain bands, as further detailed in Table 12.9. Additional band combinations and options may be defined in future releases. 8C-HSDPA is presently only defined for single-band contiguous operation.

From an RF characteristics point of view, the situation is very similar to the one for the single-band multi-carrier schemes. For the base stations, requirements are in general only defined for one antenna connector per band, implying that a multi-band transmission will have individual RF characteristics defined per band, which are identical to single-band operation. Again, the exception is an additional time-alignment requirement between the carriers in both bands. Note that there are special requirements defined for multiband base stations, where a common RF is used for multiple bands. Those requirements are also applicable to the multiband schemes in that case. Further details of multiband base stations are given in Section 16.11.

For the UE, there are additional receiver RF requirements for each multiband/multi-carrier scheme. Also here, the receiver spurious emissions requirement is unchanged.

### 16.3.3 MULTI-CARRIER UPLINK SCHEME

There is presently one uplink multi-carrier scheme defined, referred to as dual-cell HSUPA (DC-HSUPA). The feature is not restricted to any specific band, and new requirements are defined in a band agnostic way.

From an RF characteristic point of view, DC-HSUPA has major impact for all UE transmitter requirements, which are all redefined for transmission of two adjacent carriers with 5 MHz spacing, in principle forming a 10 MHz transmission. For the UE receiver, there are also modified DC-HSUPA requirements for the receiver susceptibility to in-band interfering signals (blocking and receiver intermodulation).

For the base station, there is no impact on the RF requirements, since it is assumed that a base station can anyway receive two adjacent carriers. The only difference with DC-HSUPA is that the two RF carriers received come from the same UE, and they are therefore demodulated jointly in the baseband. There are thus new baseband demodulation performance requirements for receiving DC-HSUPA.

### 16.3.4 MULTI-CARRIER IN NON-CONTIGUOUS SPECTRUM

Some spectrum allocations used for HSPA deployments consist of fragmented parts of spectrum for different reasons. The spectrum may be recycled 2G spectrum, where the original licensed spectrum was "interleaved" between operators. This was quite common for original GSM deployments, for implementation reasons (the original combiner filters used were not easily tuned when spectrum allocations were expanded). In some regions, operators have also purchased spectrum licenses on auctions and have for different reasons ended up with multiple allocations in the same band that are not adjacent.

For deployment of non-contiguous spectrum allocations, there are a few implications:

- If the full spectrum allocation in a band is to be operated with a single base station, the base station has to be capable of operation in non-contiguous spectrum.
- If a larger transmission bandwidth is to be used than what is available in each of the spectrum fragments, both the UE and the base station have to be capable of handling non-contiguous carriers in that band.

Non-contiguous operation is defined for the downlink multi-carrier scheme NC-4C-HSDPA. Note that in general terms, the capability for the base station to operate in non-contiguous spectrum is not directly coupled to multi-carrier transmission as such. From an RF point of view, what will be required by the base stations is to receive and transmit carriers over an RF bandwidth that is split in two (or more) separate sub-blocks with a sub-block gap in between, as shown in Figure 16.4. The spectrum in the sub-block gap can be deployed by any other operator, which means that the RF requirements for the base station in the sub-block gap will be based on coexistence for uncoordinated operation. This has implications for some of the base station RF requirements within an operating band.

If the non-contiguous spectrum is operated with multi-carrier transmission, the RF requirements for the base station will be fundamentally the same as in general for non-contiguous spectrum. The main difference is that a few RF requirements have additional definitions for how they apply in the sub-block gap. For the UE, however, there are additional implications and limitations to handle the simultaneously received carriers. There are thus specific NC-4C-HSDPA requirements defined for all receiver RF characteristics, except for receiver spurious emissions, which is a regulatory requirement and independent of RF configuration.

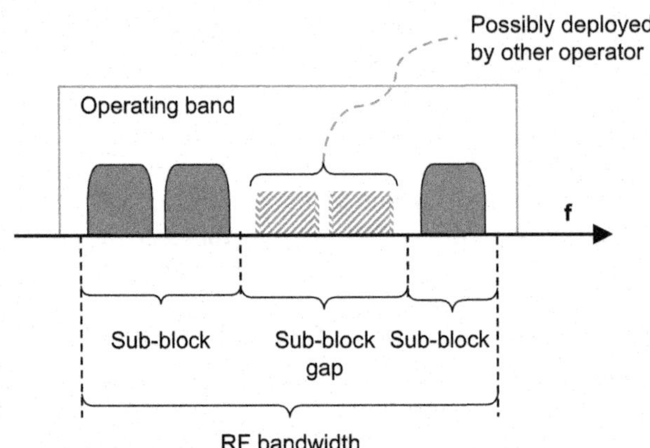


**FIGURE 16.4 Example of non-contiguous spectrum operation, illustrating the definitions of *RF bandwidth*, *sub-block*, and *sub-block gap*.**

## 16.4 MULTISTANDARD RADIO BASE STATIONS

Traditionally the RF specifications have been developed separately for the different 3GPP radio-access technologies GSM/EDGE, UTRA, and E-UTRA (LTE). The rapid evolution of mobile radio and the need to deploy new technologies alongside the legacy deployments has, however, lead to implementation of different *Radio-Access Technologies* (RAT) at the same sites, often sharing antennas and other parts of the installation. A natural further step is then to also share the base station equipment between multiple RATs. This requires multi-RAT base stations.

The evolution to multi-RAT base stations is also fostered by the evolution of technology. While multiple RATs have traditionally shared parts of the site installation, such as antennas, feeders, backhaul, or power, the advance of both digital baseband and RF technologies enables a much tighter integration. A base station consisting of two separate implementations of both baseband and RF, together with a passive combiner/splitter before the antenna, could in theory be considered a multi-RAT base station. 3GPP has, however, made a narrower, but more forward-looking, definition.

In a *Multistandard Radio* (MSR) base station, both the receiver and the transmitter are capable of simultaneously processing multiple carriers of different RATs in common *active* RF components. The reason for this stricter definition is that the true potential of multi-RAT base stations, and the challenge in terms of implementation complexity, comes from having a common RF. This principle is illustrated in Figure 16.5 with an example base station capable of both UTRA and LTE. Much of the UTRA and LTE baseband functionality may be separate in the base station but possibly implemented in the same hardware. The RF must, however, be implemented in the same active components, as shown in the figure.

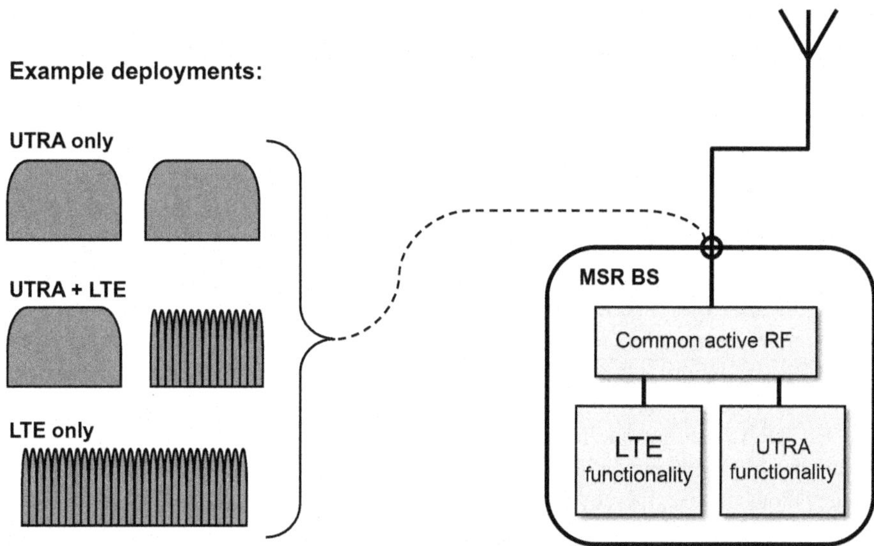


**FIGURE 16.5 Example of deployments of UTRA and LTE using an MSR base station.**

The main advantages of an MSR base station implementation are twofold:

- Migration between RATs in a deployment, for example from GSM/EDGE to UTRA or LTE, or from UTRA to LTE, is possible using the same base station hardware.
- A single base station designed as an MSR base station can be deployed in various environments for single-RAT operation for each RAT supported, as well as for multi-RAT operation where that is required by the deployment scenario. This is also in line with the recent technology trends seen in the market, with fewer and more generic base station designs. Having fewer varieties of base station is an advantage both for the base station vendor and for the operator, since a single solution can be developed and implemented for a variety of scenarios.

The single-RAT 3GPP radio access standards, with requirements defined independently per RAT, do not support such migration scenarios with an implementation where common base station RF hardware is shared between multiple access technologies, and hence a separate set of requirements for multistandard radio equipment is needed.

An implication of a common RF for multiple RATs is that carriers are no longer received and transmitted independently of each other. For this reason, a common RF specification must be used to specify the MSR base station. From 3GPP release 9, there is a set of MSR base station specifications for the core RF requirements [7] and for test requirements [8]. Those specifications support GSM/EDGE,[1] UTRA and E-UTRA, and all combinations thereof. To support all possible RAT combinations,

[1]The MSR specifications are not applicable to single-RAT operation of GSM/EDGE.

the MSR specifications have many generic requirements applicable regardless of RAT combination, together with specific single-access-technology-specific requirements to secure the integrity of the systems in single-RAT operation.

The MSR concept has a substantial impact for many requirements, while others remain completely unchanged. A fundamental concept introduced for MSR base stations is *RF bandwidth*, which is defined as the total bandwidth over the set of carriers transmitted and received. Many receiver and transmitter requirements for GSM/EDGE and UTRA are specified relative to the carrier center and for LTE in relation to the channel edges. For an MSR base station, they are instead specified relative to the *RF bandwidth edges*, in a way similar to carrier aggregation in LTE. In the same way as for carrier aggregation, a parameter $F_{\text{offset}}$ is also introduced to define the location of the RF bandwidth edges relative to the edge carriers. For GSM/EDGE carriers, $F_{\text{offset}}$ is set to 200 kHz, while it is in general half the channel bandwidth for UTRA and E-UTRA. By introducing the RF bandwidth concept and introducing generic limits, the requirements for MSR shift from being carrier centric toward being frequency block centric, thereby embracing technology neutrality by being independent of the access technology or operational mode.

While E-UTRA and UTRA carriers have quite similar RF properties in terms of bandwidth and power spectral density, GSM/EDGE carriers are quite different. The operating bands for which MSR base stations are defined are therefore divided into three *Band Categories* (BCs):

- BC1 – All paired bands where UTRA FDD and E-UTRA FDD can be deployed.
- BC2 – All paired bands where in addition to UTRA FDD and E-UTRA FDD, GSM/EDGE can also be deployed.
- BC3 – All unpaired bands where UTRA TDD and E-UTRA TDD can be deployed.

Since the carriers of different RATs are not transmitted and received independently, it is necessary to perform parts of the testing with carriers of multiple RATs being activated. This is done through a set of multi-RAT *Test Configurations* defined in [8], specifically tailored to stress transmitter and receiver properties. These test configurations are of particular importance for the unwanted emission requirements for the transmitter and for testing of the receiver susceptibility to interfering signals (blocking, etc.). An advantage of the multi-RAT test configurations is that the RF performance of multiple RATs can be tested simultaneously, thereby avoiding repetition of test cases for each RAT. This is of particular importance for the very time-consuming tests of requirements over the complete frequency range outside the operating band.

The requirement with the largest impact from MSR is the spectrum mask, or the *operating band unwanted emissions* requirement, as it is called. The spectrum mask requirement for MSR base stations is applicable for multi-RAT operation where the carriers at the RF bandwidth edges are either GSM/EDGE, UTRA, or E-UTRA carriers of different channel bandwidths. The mask is generic and applicable to all cases and covers the complete operating band of the base station. There is an exception for the 150 kHz closest to the RF bandwidth edge, where the mask is aligned with the GSM/EDGE modulation spectrum for the case when a GSM/EDGE carrier or a 1.4/3 MHz E-UTRA carrier is transmitted adjacent to the edge.

An important aspect of MSR is the declaration by the base station vendor of the supported RF bandwidth, power levels, multi-carrier capability, etc. All testing is based on the capability of the base station through a declaration of the supported *Capability Set* (CS), which defines all supported single RAT and multi-RAT combinations. There are currently six capability sets, CS1 to CS6, defined in the MSR test specification [8], allowing full flexibility for implementing and deploying base stations compliant with the MSR specification. These capability sets are listed in Table 16.2 together with the band categories where the capability set is applicable and the RAT configurations that are supported by the BS. Note the difference between the capability of a base station (as declared by the manufacturer) and the configuration in which a BS is operating. CS1 and CS2 define capabilities for base stations that are only single-RAT capable and make it possible to apply the MSR base station specification for such base stations, instead of the corresponding single-RAT UTRA or E-UTRA specifications. There is no capability set defined for base stations that are *only* single-RAT GSM capable, since that is a type of BS that is solely covered by the single-RAT GSM/EDGE specifications.

For a large part of the base station RF requirements, multi-RAT testing is not necessary and the actual test limits are unchanged for the MSR base station. In these cases, both the requirements and the test cases are simply incorporated through direct references to the corresponding single-RAT specifications.

Multi-carrier operation as described above in Section 16.3 is also applicable to MSR base stations. Since the MSR specification has most of the concepts and definitions in place for defining multi-carrier RF requirements, whether aggregated or not, the differences for the MSR requirements compared to non-aggregated carriers are very minor.

**Table 16.2** Capability Sets (CSx) Defined for MSR Base Stations and the Corresponding RAT Configurations

| Capability Set CSx Supported by a Base Station | Applicable Band Categories | Supported RAT Configurations |
|---|---|---|
| CS1 | BC1, BC2, or BC3 | Single-RAT: UTRA |
| CS2 | BC1, BC2, or BC3 | Single-RAT: E-UTRA |
| CS3 | BC1, BC2, or BC3 | Single-RAT: UTRA or E-UTRA<br>Multi-RAT: UTRA + E-UTRA |
| CS4 | BC2 | Single-RAT: GSM or UTRA<br>Multi-RAT: GSM + UTRA |
| CS5 | BC2 | Single-RAT: GSM or E-UTRA<br>Multi-RAT: GSM + E-UTRA |
| CS6 | BC2 | Single-RAT: GSM, UTRA, or E-UTRA<br>Multi-RAT: GSM + UTRA, GSM + E-UTRA, UTRA + E-UTRA, or GSM + UTRA + E-UTRA |

## 16.5 OVERVIEW OF RF REQUIREMENTS FOR UTRA

The RF requirements define the receiver and transmitter RF characteristics of a base station or UE. The base station is the physical node that transmits and receives RF signals on one or more antenna connectors. Note that a base station is not the same thing as a Node B, which is the corresponding logical node in the UTRA Radio-Access Network. The terminal is denoted UE in all RF specifications.

The set of RF requirements defined for UTRA is fundamentally the same as those defined for LTE or any other radio system. Some requirements are also based on regulatory requirements and are more concerned with the frequency band of operation and/or the place where the system is deployed, than with the type of system.

The types of transmitter requirements defined for the UE are very similar to what is defined for the base station, and the definitions of the requirements are often similar. The output power levels, however, are considerably lower for a UE, while the restrictions on the UE implementation are much higher. There is tight pressure on cost and complexity for all telecommunications equipment, but this is much more pronounced for UEs because of the scale of the total market, being close to *two billion* devices per year. In cases where there are differences in how requirements are defined between UE and base station, they are treated separately in this chapter.

The RF requirements for the base station are specified in [9] and for the UE in [5]. The RF requirements are divided into transmitter and receiver characteristics. There are also *performance characteristics* for base station and UE that define the receiver baseband performance for all physical channels under different propagation conditions. These are not strictly RF requirements, though the performance will also depend on the RF to some extent.

Each RF requirement has a corresponding test defined in the UTRA test specifications for the base station [10] and the UE [11]. These specifications define the test setup, test procedure, test signals, test tolerances, etc. needed to show compliance with the RF and performance requirements.

### 16.5.1 TRANSMITTER CHARACTERISTICS

The transmitter characteristics define RF requirements for the wanted signal transmitted from the UE and base station but also for the unavoidable unwanted emissions outside the transmitted carrier(s). The requirements are fundamentally specified in three parts:

- *Output power level* requirements set limits for the maximum allowed transmitted power, for the dynamic variation of the power level, and in some cases for the transmitter OFF state.
- *Transmitted signal quality* requirements define the "purity" of the transmitted signal and also the relation between multiple transmitter branches.
- *Unwanted emissions* requirements set limits to all emissions outside the transmitted carrier(s) and are tightly coupled to regulatory requirements and coexistence with other systems.

**Table 16.3** Overview of UTRA Transmitter Characteristics

| | Base Station Requirement | UE Requirement |
|---|---|---|
| Output power level | Maximum output power<br>Output power dynamics | Transmit power<br>Output power dynamics<br>Power control<br>Transmit ON/OFF power |
| Transmitted signal quality | Frequency error<br>Transmit modulation quality<br>Time alignment between transmitter branches | Frequency error<br>Transmit modulation quality<br>Time alignment error<br>In-band emissions |
| Unwanted emissions | Spectrum emissions mask<br>Adjacent channel leakage ratio (ACLR and CACLR)<br>Spurious emissions<br>Occupied bandwidth<br>Transmitter intermodulation | Spectrum emissions mask<br>Adjacent channel leakage ratio (ACLR)<br>Spurious emissions<br>Occupied bandwidth<br>Transmit intermodulation |

A list of the UE and base station transmitter characteristics arranged according to the three parts defined above is shown in Table 16.3. A more detailed description of the requirements can be found later in this chapter.

### 16.5.2 RECEIVER CHARACTERISTICS

The set of receiver requirements for UTRA is quite similar to what is defined for other systems such as LTE. The receiver characteristics are fundamentally specified in three parts:

- *Sensitivity and dynamic range* requirements for receiving the wanted signal.
- *Receiver susceptibility to interfering signals* defines receivers' susceptibility to different types of interfering signals at different frequency offsets.
- *Unwanted emissions* limits are also defined for the receiver.

A list of the UE and base station receiver characteristics arranged according to the three parts defined above is shown in Table 16.4. A more detailed description of each requirement can be found later in this chapter.

### 16.5.3 REGIONAL REQUIREMENTS

There are a number of regional variations to the RF requirements and their application. The variations originate in different regional and local regulations of spectrum and its use. The most obvious regional variation is the different frequency bands and

**Table 16.4** Overview of UTRA Receiver Characteristics

| | Base Station Requirement | UE Requirement |
|---|---|---|
| Sensitivity and dynamic range | Reference sensitivity | Reference sensitivity power level |
| | Dynamic range | Maximum input level |
| Receiver susceptibility to interfering signals | Blocking (in-band and out-of-band) | Out-of-band blocking |
| | Narrowband blocking | Narrowband blocking |
| | | Spurious response |
| | Adjacent channel selectivity (ACS) | Adjacent channel selectivity (ACS) |
| | Intermodulation characteristics | Intermodulation characteristics |
| Unwanted emissions from the receiver | Receiver spurious emissions | Receiver spurious emissions |

their use, as discussed above. Many of the regional RF requirements are also tied to specific frequency bands.

When there is a regional requirement on, for example, spurious emissions, this requirement should be reflected in the 3GPP specifications. For the base station, it is entered as an optional requirement and is marked as "regional." For the UE, the same procedure is not possible, since a UE may roam between different regions and will therefore have to fulfill all regional requirements that are tied to an operating band in the regions where the band is used.

### 16.5.4 BASE STATION CLASSES

In the base station specifications, there is one set of RF requirements that is generic, applicable to what is called "general purpose" base stations. This is the original set of UTRA requirements developed in 3GPP release 99. It has no restrictions on base station output power and can be used for any deployment scenario. When the RF requirements were derived, however, the scenarios used were macro scenarios [12]. For this reason, in release 5 additional base station classes were introduced that were intended for micro-cell and pico-cell scenarios. An additional class for femto-cell scenarios was added in release 9. It is also clarified that the original set of "general purpose" RF parameters are applicable for macro-cell scenarios. The terms *macro*, *micro*, *pico*, and *femto* are not used in 3GPP to identify the base station classes; instead, the following terminology is used:

- *Wide Area base stations*. This type of base station is intended for macro-cell scenarios, characterized with a minimum coupling loss between base station and UE of 70 dB.

- *Medium-Range base stations*. This type of base station is intended for micro-cell scenarios, characterized with a minimum coupling loss between base station and UE of 53 dB. Typical deployments are outdoor below-rooftop installations, giving both outdoor hot spot coverage and outdoor-to-indoor coverage through walls.
- *Local Area base stations*. This type of base station is intended for pico-cell scenarios, characterized with a minimum coupling loss between base station and UE of 45 dB. Typical deployments are indoor offices and indoor/outdoor hotspots, with the base station mounted on walls or ceilings.
- *Home base stations*. This type of base station is intended for femto-cell scenarios, which are not explicitly defined. Minimum coupling loss between base station and UE of 45 dB is also assumed here. Home base stations can be used both for open access and in closed subscriber groups.

The Local Area, Medium-Range, and Home base station classes have modifications to a number of requirements compared to Wide Area base stations, mainly as a result of the assumption of a lower minimum coupling loss:

- Maximum base station power is limited to 38 dBm output power for Medium-Range base stations, 24 dBm output power for Local Area base stations, and to 20 dBm for Home base stations. This power is defined per antenna and carrier, except for home base stations, where the power over all antennas (up to four) is counted. There is no maximum base station power defined for Wide Area base stations.
- Home base stations have an additional requirement for protecting systems operating on adjacent channels. The reason is that a UE connected to a base station belonging to another operator on the adjacent channel may be in close proximity to the Home base station. To avoid an interference situation where the adjacent UE is blocked, the Home base station must make measurements on the adjacent channel to detect adjacent base station operations. If an adjacent base station transmission is detected under certain conditions, the maximum allowed Home base station output power is reduced in proportion to how weak the adjacent base station signal is, in order to avoid interference to the adjacent base station.
- Unwanted emission limits for protecting Home base station operation (from other Home base stations) are lower, since a stricter through-the-wall indoor interference scenario is assumed. Limits for co-location for Medium-Range and Local Area are, however, less strict, corresponding to the relaxed reference sensitivity for the base station.
- Receiver reference sensitivity limits are higher (more relaxed) for Medium-Range, Local Area, and Home base stations. Receiver dynamic range is also adjusted accordingly.
- All Medium-Range, Local Area, and Home base station limits for receiver susceptibility to interfering signals are adjusted to take the higher receiver sensitivity limit and the lower assumed minimum coupling loss (base station-to-UE) into account.

## 16.6 OUTPUT POWER LEVEL REQUIREMENTS

### 16.6.1 BASE STATION OUTPUT POWER AND DYNAMIC RANGE

There is no general maximum output power requirement for base stations. The output power capability is declared by the manufacturer. As mentioned in the discussion of base station classes above, there is, however, a maximum declared output power limit of 38 dBm for Medium-Range base stations, 24 dBm for Local Area base stations, and of 20 dBm for Home base stations. In addition to this, there is a tolerance specified, defining how much the actual maximum power may deviate from the power level declared by the manufacturer.

The base station also has a specification of the power control dynamic range, defining the power range over which it should be possible to configure. There is also an accuracy requirement for the downlink inner-loop power control steps. Finally, there is a dynamic range requirement for the total base station power.

A separate requirement is set on the P-CPICH power and S-CPICH power accuracy relative to the P-CPICH power.

### 16.6.2 UE OUTPUT POWER, ACCURACY, AND DYNAMIC RANGE

The UE output power level is defined in three steps:

*UE power class* defines a *nominal* maximum output power for QPSK modulation. It may be different in different operating bands, but the main UE power class is today set at 23 dBm for all bands.

*Maximum Power Reduction (MPR)* defines an allowed reduction of maximum power level for certain physical channels and/or combination of uplink features such as E-DCH for uplink transmit diversity, DC-HSUPA, and uplink MIMO.

*Allowed adjustments of tolerances in UE maximum power* is used as an alternative way to MPR, in order to allow the UE to transmit at lower power for some features, including DB-DC-HSDPA and 4C-HSDPA. For UEs that in addition to UTRA support LTE for the same operating bands and also support LTE carrier aggregation (a form of LTE "multi-carrier" operation) in those bands, there is additional allowance to adjust the tolerances in minimum power.

There are requirements for the UE's ability to set the power of individual codes relative to the total transmitted power, called a *UE relative code domain power accuracy*. The requirement applies for each carrier in case of DC-HSUPA and for each antenna in case of uplink MIMO or transmit diversity.

The UE has a definition of the *transmitter Off power level*, applicable to conditions when the UE is not allowed to transmit. There is also a general On/Off time mask specified, plus specific time masks for change of *Transport Format Combination* (TFC) and for power setting in uplink compressed mode.

There is a special time mask defined for *Out-of-synchronization handling of output power*. The time mask specifies how fast the UE must shut its power off in case the downlink signal is lost and how fast it must turn the power back on again when the downlink signal is recovered. The reason for the requirement is to make sure that the UE does not create harmful interference in the uplink because of lost power control ("rogue UE" behavior) in case the downlink power control commands do not reach the UE.

The UE transmit power control is specified through requirements for the absolute power tolerance for the *Open loop power control* setting. For the *Inner-loop power control*, there are requirements for the *Transmitter power control range* for individual power control steps and for the *Transmitter aggregate power control range* that is accumulated after multiple identical groups of TPC commands.

## 16.7 TRANSMITTED SIGNAL QUALITY

The requirements for transmitted signal quality specify how much the transmitted base station or UE/base station signal deviates from an "ideal" modulated signal in the signal and the frequency domains. Impairments on the transmitted signal are introduced by the transmitter RF parts, with the nonlinear properties of the power amplifier being a major contributor. The signal quality is assessed for base station and UE through requirements on *EVM* and *frequency error*. An additional UE requirement is UE in-band emissions (for multi-carrier transmission).

### 16.7.1 EVM AND FREQUENCY ERROR

While the theoretical definitions of the signal quality measures are quite straightforward, the actual assessment is a very elaborate procedure, described in great detail in the 3GPP specification. The reason is that it becomes a multi-dimensional optimization problem, where the best match for the timing, the frequency, and the signal constellation are found.

The *Error Vector Magnitude* (EVM) is a measure of the error in the *Root-Raised Cosine* (RRC) modulated output signal consisting of a pre-determined number of transmitted codes, relative to an ideal reference signal, taken as the root mean square of the error vectors over the signal samples. It is expressed as a percentage value in relation to the power of the ideal signal. The EVM fundamentally defines the maximum SINR that can be achieved at the receiver, if there are no additional impairments to the signal between transmitter and receiver. The evaluation assumes an ideal receiver. The frequency offset resulting from the EVM evaluation is averaged and used as a measure of the *frequency error* of the transmitted signal.

The requirements on *Peak Code Domain error* are calculated with a similar principle, where the same error signal as determined for EVM is projected onto every code in the code domain, and the mean power of each projection relative to the mean power of the composite reference signal is determined. The Peak code domain error is the peak error ratio across all codes.

There is also a *Relative Code Domain Error* defined for the UE transmitter and for 64-QAM modulation for the BS transmitter. The requirement specifies the average code domain error taken across all active code channels.

Finally, for the UE there are requirements relating to the maximum allowable *phase discontinuity* between slot and HS-DPCCH boundaries when changing signal properties due to, for example, TX power-level changes.

### 16.7.2 UE IN-BAND EMISSIONS

*In-band emissions* are emissions within assigned carriers in case of multi-carrier transmission (DC-HSUPA). The requirement limits how much a UE can leak into a carrier set to transmit at minimum power from the adjacent carrier set at maximum power. This requirement is needed for multi-carrier transmissions, since the EVM requirement only defines the signal quality for each individual carrier and not for the combination of multiple carriers.

### 16.7.3 TIME ALIGNMENT

Several UTRA features require the base station to transmit from two or more antennas, such as transmitter diversity and MIMO. For multi-carrier transmissions, the carriers may also be transmitted from different antennas. In order for the UE to properly receive the signals from multiple antennas, the timing relation between any two transmitter branches is specified in terms of a maximum time alignment error between transmitter branches. The maximum allowed error depends on the feature or combination of features in the transmitter branches.

For the UE, transmit diversity and MIMO also require transmission from multiple antennas. There is also dual-cell transmission from the UE for DC-HSUPA. In the same way as for the similar BS features, these UE features have requirements for the time alignment error between transmitter branches or cells.

## 16.8 UNWANTED EMISSIONS REQUIREMENTS

Unwanted emissions from the transmitter are divided into *out-of-band (OOB) emissions* and *spurious emissions* in ITU-R recommendations [13]. OOB emissions are defined as emissions on a frequency close to the RF carrier, which results from the modulation process. Spurious emissions are emissions outside the RF carrier that may be reduced without affecting the corresponding transmission of information. Examples of spurious emissions are harmonic emissions, intermodulation products, and frequency conversion products. The frequency range where OOB emissions are normally defined is called the *OOB domain*, whereas spurious emissions limits are normally defined in the *spurious domain*.

ITU-R also defines the boundary between the OOB and spurious domains at a frequency separation from the carrier center of 2.5 times the necessary bandwidth, which corresponds to 12.5 MHz for UTRA. This division of the requirements is easily applied for systems such as UTRA that have a fixed channel bandwidth. It does, however, become more complex for MSR base stations, where multiple carriers of different channel bandwidths are transmitted. The approach taken for defining the boundary in 3GPP is slightly different for UTRA and MSR base stations, respectively.

With the recommended boundary between OOB emissions and spurious emissions set at 2.5 times the channel bandwidth, third- and fifth-order intermodulation products from the carrier will fall inside the OOB domain, which will cover a frequency range of twice the channel bandwidth on each side of the carrier. For the OOB domain, two overlapping requirements are defined for both base station and UE: *Spectrum Emissions Mask* (SEM) and *Adjacent Channel Leakage Ratio* (ACLR). The details of these are further explained below. For MSR base stations, the SEM is replaced by the operating band *unwanted emissions mask* (UEM).

## 16.8.1 IMPLEMENTATION ASPECTS

The spectrum of an ideal RRC filtered WCDMA signal, as described in Chapter 3, is zero outside of the allocated transmission bandwidth. In reality, this cannot be achieved, however, since the transmitter is always to some extent nonlinear and does therefore not have ideal RF properties.

The nonlinear characteristic of the *Power Amplifier* (PA) used to amplify the RF signal is the main source of intermodulation products outside the channel bandwidth. Power back-off to give a more linear operation of the PA can be used but at the cost of a lower power efficiency. The power back-off should therefore be kept to a minimum. For this reason, additional linearization schemes can be employed. These are especially important for the base station, where there are fewer restrictions on implementation complexity and use of advanced linearization schemes is an essential part of controlling spectrum emissions. Examples of such techniques are feed-forward, feedback, predistortion, and postdistortion.

There are also special limits defined to meet a specific regulation set by the FCC [14] for the operating bands used in the United States and by the ECC for some European bands. These are specified as separate limits in addition to the unwanted emission limits.

## 16.8.2 LIMITS FOR OUT-OF-BAND SPECTRUM EMISSIONS

The spectrum emissions mask defines the permissible out-of-band spectrum emissions outside the necessary bandwidth. The spectrum mask for UTRA UEs and BS are defined in a similar way, but the BS mask has a broader application, as further discussed below. For MSR base stations, where not only UTRA but also LTE and GSM transmissions may occur, a different concept is used called *Operating Band Unwanted Emissions.*

### 16.8.2.1 Spectrum emissions masks

The spectrum mask for UEs extends 12.5 MHz from the carrier, corresponding to the out-of-band domain for a 5 MHz carrier, as illustrated in Figure 16.6. Outside of the frequency range of the mask, the spurious emissions limits apply. The mask has some

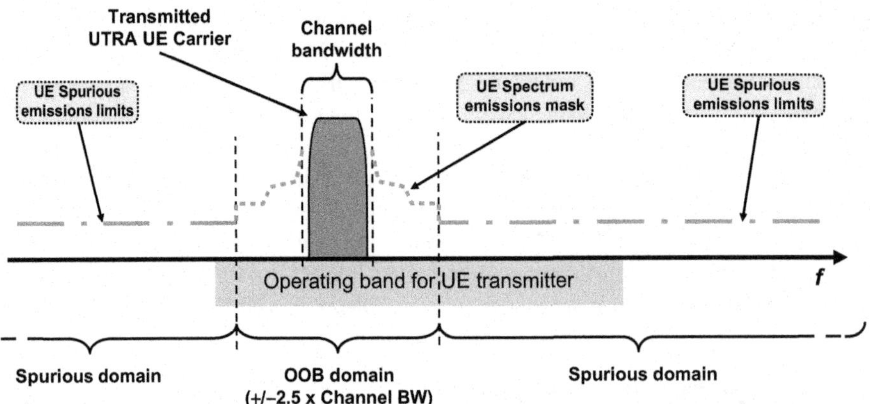


**FIGURE 16.6 Frequency ranges for spectrum emissions mask and spurious emissions applicable to UTRA UE.**

variations for some operating bands, where there are specific regional requirements to meet in the out-of-band domain. There is a special mask for DC-HSUPA which extends out to 20 MHz from the center frequency of the two carriers. The extended mask is still within the out-of-band domain, which from a regulatory point of view is considered to be twice as large because of the double total transmission bandwidth. The UE spectrum mask is relative to the output power for all power levels.

For the UTRA base station, the mask extends out to the transmitter operating band edge, or 12.5 MHz from the carrier, as illustrated in Figure 16.7. This means that for the base station, the mask extends into the spurious domain and thus overlaps with the spurious emissions requirement. With two limits for the same frequency range, the stricter one will apply. The UTRA base station mask is defined relative to

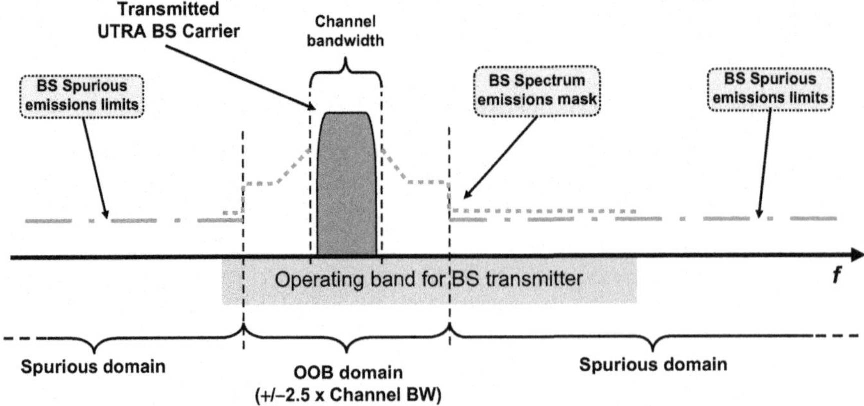


**FIGURE 16.7 Frequency ranges for spectrum emissions mask and spurious emissions applicable to UTRA base stations.**

the base station output power for the most common output power ranges of the Wide Area (macro) and Medium-Range (micro) base station classes. For the highest power levels, the mask is however limited to a fixed level. The same is true for the power levels of Local Area (pico) and Home BS (femto) base stations, where a fixed mask level applies.

#### 16.8.2.2 MSR base station operating band unwanted emission limits

For the MSR base station, the problem of the variation of the boundary between OOB and spurious domain with the varying channel bandwidth for the different RATs is handled by not defining an explicit boundary. The solution is a unified concept of *operating band unwanted emissions mask* (UEM) for the MSR base station instead of the spectrum mask usually defined for OOB emissions. The operating band unwanted emissions requirement applies over the whole base station transmitter operating band, plus an additional 10 MHz on each side, as shown in Figure 16.8. All requirements outside of that range are set by the regulatory spurious emissions limits, based on the ITU-R recommendations [13]. As seen in the figure, a large part of the operating band unwanted emissions are defined over a frequency range that for smaller channel bandwidths can be both in spurious and OOB domains. This means that the limits for the frequency ranges that may be in the spurious domain also have to align with the regulatory limits from the ITU-R. The shape of the mask is generic for all channel bandwidths from 5 to 20 MHz, with a mask that consequently has to align with the ITU-R limits starting 10 MHz from the channel edges. The operating band unwanted emissions are defined with a 100 kHz measurement bandwidth.

### 16.8.3 ADJACENT CHANNEL LEAKAGE RATIO

In addition to a spectrum emissions mask, the OOB emissions are defined by an *Adjacent Channel Leakage Ratio* (ACLR) requirement. The ACLR concept is very

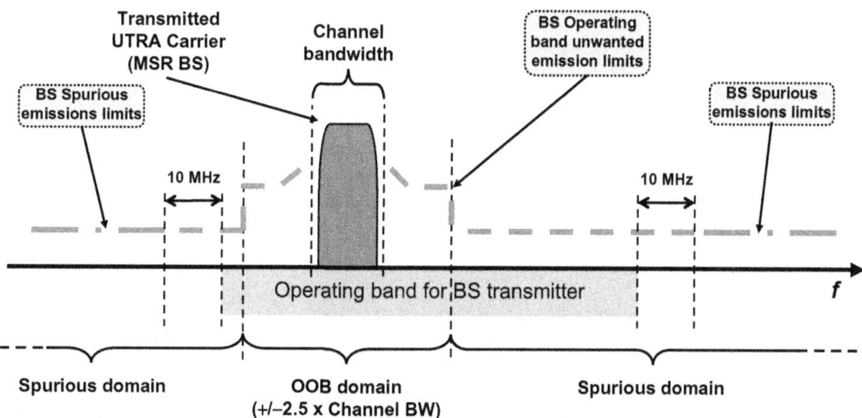


**FIGURE 16.8 Frequency ranges for operating band unwanted emissions and spurious emissions applicable to MSR base station.**

useful for analysis of coexistence between two systems that operate on adjacent frequencies. The ACLR defines the ratio of the power transmitted within the assigned channel bandwidth to the power of the unwanted emissions transmitted on an adjacent channel. There is a corresponding receiver requirement called *Adjacent Channel Selectivity* (ACS), which defines a receiver's ability to suppress a signal on an adjacent channel.

The definitions of ACLR and ACS are illustrated in Figure 16.9 for a wanted and an interfering signal received in adjacent channels. The interfering signal's leakage of unwanted emissions at the wanted signal receiver is given by the ACLR and the ability of the receiver of the wanted signal to suppress the interfering signal in the adjacent channel is defined by the ACS. The two parameters when combined define the total leakage between two transmissions on adjacent channels. That ratio is called the *Adjacent Channel Interference Ratio* (ACIR) and is defined as the ratio of the power transmitted on one channel to the total interference received by a receiver on the adjacent channel, due to both transmitter (ACLR) and receiver (ACS) imperfections.

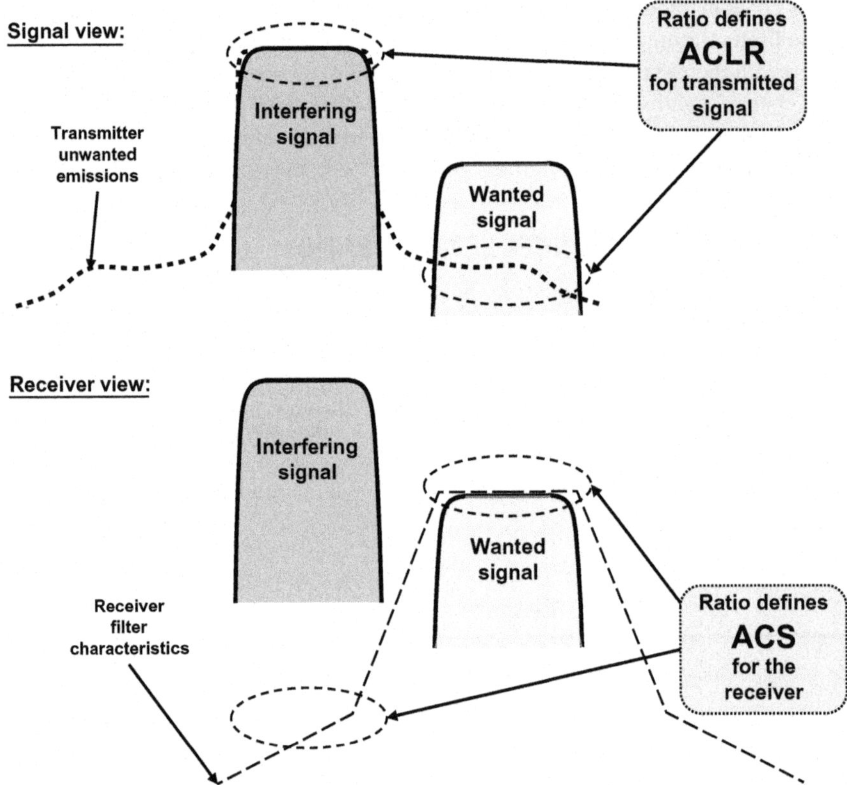


FIGURE 16.9 Illustration of ACLR and ACS, with example characteristics for an "aggressor" interferer and a receiver for a "victim" wanted signal.

This relation between the adjacent channel parameters is [12]:

$$\mathrm{ACIR} = \frac{1}{\frac{1}{\mathrm{ACLR}} + \frac{1}{\mathrm{ACS}}} \tag{16.1}$$

ACLR and ACS can be defined with different channel bandwidths for the two adjacent channels. The equation above will also apply for different channel bandwidths but only if the same two channel bandwidths are used for defining all three parameters ACIR, ACLR, and ACS used in the equation.

The ACLR limits for UTRA UE and base station are derived based on extensive analysis [12] of UTRA coexistence with UTRA or other systems on adjacent carriers.

In case of multi-carrier operation for a base station, the ACLR (as other RF requirements) apply as for any multi-carrier transmission, where the ACLR requirement will be defined for the carriers on the edges of the RF bandwidth. In case of non-contiguous multi-carrier operation where the sub-block gap is so small that the ACLR requirements at the edges of the gap will "overlap," a special *Cumulative ACLR* requirement (CACLR) is defined for the gap. For CACLR, contributions from carriers on both sides of the sub-block gap are accounted for in the CACLR limit.

### 16.8.4 SPURIOUS EMISSIONS

The limits for base station spurious emissions are taken from international recommendations [13] and are defined in the complete spurious domain. As noted above, the frequency range of the requirement overlaps partly with the UTRA spectrum emissions mask inside the operating band. There are also additional regional or optional limits for protection of other systems that UTRA may coexist with or even be co-located with. Examples of other systems considered in those additional spurious emissions requirements are GSM, LTE, UTRA TDD, CDMA2000, and PHS.

UE spurious emissions limits are defined for all frequency ranges outside the frequency range covered by the SEM. The limits are generally based on international regulations [13], but there are also additional requirements for coexistence with other bands when the mobile is roaming.

In addition, there are base station and UE emission limits defined for the receiver. Since receiver emissions are dominated by the transmitted signal, the receiver spurious emissions limits are only applicable when the transmitter is Off, and also when the transmitter is On for a base station that has a separate receiver antenna connector.

### 16.8.5 OCCUPIED BANDWIDTH

Occupied bandwidth is a regulatory requirement that is specified for equipment in some regions, such as Japan and the United States. It is originally defined by the ITU-R as a maximum bandwidth, outside of which emissions do not exceed a certain percentage of the total emissions. The occupied bandwidth is for UTRA equal to the channel bandwidth (5 MHz), outside of which a maximum of 1% of the emissions are allowed (0.5% on each side).

### 16.8.6 TRANSMITTER INTERMODULATION

An additional implementation aspect of an RF transmitter is the possibility of intermodulation between the transmitted signal and another strong signal transmitted in the proximity of the base station or UE. For this reason there is a requirement for *transmitter intermodulation*.

For the base station, the requirement is based on a stationary scenario with a co-located other base station transmitter, with its transmitted signal appearing at the antenna connector of the base station being specified but attenuated by 30 dB. Since it is a stationary scenario, there are no additional unwanted emissions allowed, implying that all unwanted emission limits also have to be met with the interferer present.

For the UE, there is a similar requirement based on a scenario with another UE transmitted signal appearing at the antenna connector of the UE being specified but attenuated by 40 dB. The requirement specifies the minimum attenuation of the resulting intermodulation product below the transmitted signal.

## 16.9 SENSITIVITY AND DYNAMIC RANGE

The primary purpose of the *reference sensitivity requirement* is to verify the receiver *Noise Figure*, which is a measure of how much the receiver's RF signal chain degrades the SNR of the received signal. The reference sensitivity is defined as the receiver input level where the physical channel BER must be less than 0.001.

For the UE, reference sensitivity is defined not only for DPCH with a physical *Bit Error Ratio* (BER) limit but also for HSDPA with a *Block Error Ratio* (BLER) limit and with separate requirements for the different multi-carrier schemes.

The intention of the *dynamic range requirement* is to ensure that the receiver can also operate at received signal levels considerably higher than the reference sensitivity. For the base station, the requirement is specified with a physical channel BER requirement in the same way as for reference sensitivity. The scenario assumed is with presence of increased interference and corresponding higher wanted signal levels, thereby testing the effects of different receiver impairments. The dynamic range requirement for the UE is specified as a *maximum signal level* at which the BER requirement is met. It is specified at an increased interference level to match the higher signal level and also with separate throughput-based requirements for HSDPA and the different downlink multi-carrier schemes.

## 16.10 RECEIVER SUSCEPTIBILITY TO INTERFERING SIGNALS

There is a set of requirements for base station and UE, defining the receiver's ability to receive a wanted signal in the presence of a stronger interfering signal. The reason for the multiple requirements is that, depending on the frequency offset of the interferer from the wanted signal, the interference scenario may look very different and different types of receiver impairments will affect the performance. The intention

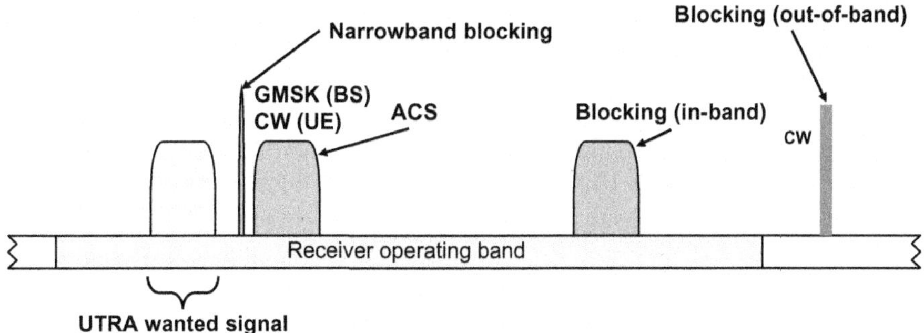


**FIGURE 16.10 Base station and UE requirements for receiver susceptibility to interfering signals in terms of blocking, ACS, and narrowband blocking.**

of the different combinations of interfering signals is to model as far as possible the range of possible scenarios with interfering signals of different bandwidths that may be encountered inside and outside the base station and UE receiver operating band.

While the types of requirements are very similar between base station and UE, the signal levels are different, since the interference scenarios for the base station and UE are very different.

The following requirements are defined for UTRA base station and UE, starting from interferers with large frequency separation and going close in (see also Figure 16.10).

- *Blocking*. This corresponds to the scenario with strong interfering signals received outside the operating band (out-of-band blocking) or inside the operating band (in-band blocking) but not adjacent to the wanted signal. In-band blocking includes interferers in the first 20 MHz outside the operating band for the base station and the first 15 MHz for the UE. The scenarios are modeled with a *Continuous Wave* (CW) signal for the out-of-band case and a WCDMA signal for the in-band case. There are additional (optional) base station blocking requirements for the scenario when the base station is co-located with another base station in a different operating band. For the UE, a fixed number of *exceptions* are allowed from the out-of-band blocking requirement for each assigned frequency channel and at the respective *spurious response frequencies*. At those frequencies, the UE must comply with the more relaxed spurious response requirement.
- *Narrowband blocking*. The scenario is an adjacent strong narrowband interferer, which in the requirement is modeled as a GMSK (GSM) signal for the base station and a CW signal for the UE.
- *Adjacent channel selectivity*. The ACS scenario is a strong signal in the channel adjacent to the wanted signal and is closely related to the corresponding ACLR requirement (see also the discussion in Section 16.8.3). The adjacent interferer is a WCDMA signal. For the UE, the ACS is specified for two cases with a lower and a higher signal level. For MSR base stations,

there is no specific ACS requirement defined. It is instead replaced by the narrowband blocking requirement, which covers the adjacent channel properties fully.

- *Receiver intermodulation*. The scenario is *two* interfering signals near to the wanted signal, where the interferers are one CW and one WCDMA signal (not shown in Figure 16.10). The purpose of the requirement is to test receiver linearity. The interferers are placed in frequency in such a way that the main intermodulation product falls inside the received wanted signal. There is also a *narrowband intermodulation* requirement for both the base station and the UE, where a CW signal is very close to the wanted signal and the other interferer is a GMSK (GSM) signal.

For all requirements, the wanted signal uses the same reference channel as in the corresponding reference sensitivity requirement. With the interference added, the same BER or a BLER limit is to be met as for the reference sensitivity, but at a "de-sensitized" higher wanted signal level.

For the UE, all requirements above are in addition specified for the different downlink multi-carrier options, in order to ensure that the receiver susceptibility to interference is maintained with reception of multi-carrier signals.

## 16.11 MULTIBAND-CAPABLE BASE STATIONS

The 3GPP specifications have been continuously developed to support larger RF bandwidths for transmission and reception through multi-carrier and multi-RAT operation over contiguous and non-contiguous spectrum allocations. This has been made possible with the evolution of RF technology supporting larger bandwidths for both transmitters and receivers. The next step in RF technology for base stations is to support simultaneous transmission and/or reception in multiple bands through a common radio. A multiband base station could cover multiple bands over a frequency range of a few hundred MHz.

The RF requirements for such multiband-capable base stations are defined in the MSR specifications of 3GPP release 11. The specification supports multiband operation both with a single RAT UTRA or E-UTRA and with multiple RATs, also called *Multiband Multistandard Radio* (MB-MSR) base stations. The specifications cover all combinations of RATs, except pure single-RAT GSM operation across the supported bands.

There are several scenarios envisioned for multiband base station implementation and deployment. The possibilities for the multiband capability are

- Multiband transmitter + multiband receiver
- Multiband transmitter + single-band receiver
- Single-band transmitter + multiband receiver

The first case is demonstrated in Figure 16.11, which shows an example base station with a common RF implementation of both transmitter and receiver for two

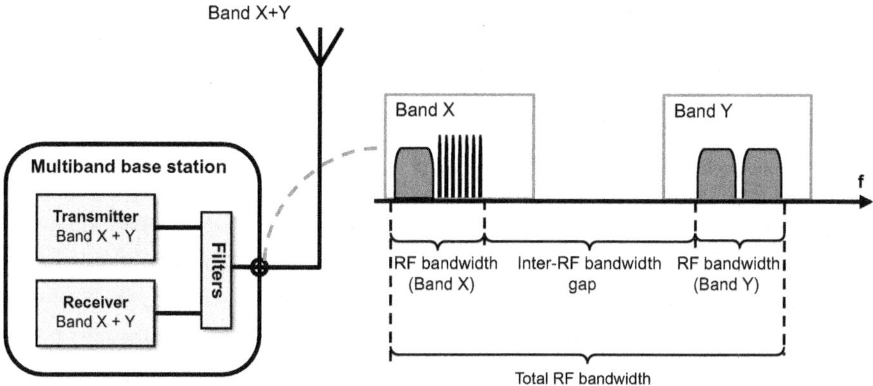


**FIGURE 16.11 Example of multiband base station with multiband transmitter and receiver for two bands with one common antenna connector.**

operating bands X and Y. Through a duplex filter, the transmitter and receiver are connected to a common antenna connector and a common antenna. The example is also a multi-RAT capable MB-MSR base station, with WCDMA + GSM configured in Band X and WCDMA configured in Band Y. Note that the figure has only one diagram showing the frequency range for the two bands, which could either be the receiver or transmitter frequencies.

Figure 16.11 also illustrates some new parameters that are defined for a multiband base station.

- *RF bandwidth* has the same definition as for a multistandard base station but is defined individually for each band.
- *Inter-RF bandwidth gap* is the gap between the RF bandwidths in the two bands. Note that the inter-RF bandwidth gap may span a frequency range where other mobile operators can be deployed in bands X and Y, as well as a frequency range between the two bands that may be used for other services.
- *Total RF bandwidth* is the full bandwidth supported by the base station to cover the multiple carriers in both bands.

In principle, a multiband base station can be capable of operating in more than two bands. The requirements and testing developed for the new type of base stations in 3GPP release 11 will, however, in general only cover a two-band capability.

While having only a single antenna connector and a common feeder that connects to a common antenna is desirable to reduce the amount of equipment needed in a site, it is not always possible. It may also be desirable to have separate antenna connectors, feeders, and antennas for each band. An example of a multiband base station with separate connectors for two operating bands X and Y is shown in Figure 16.12. Note that while the antenna connectors are separate for the two bands, the RF implementation for transmitter and receiver are in this case common for the bands. The RF for the two bands is separated into individual paths for Band X and Band Y before

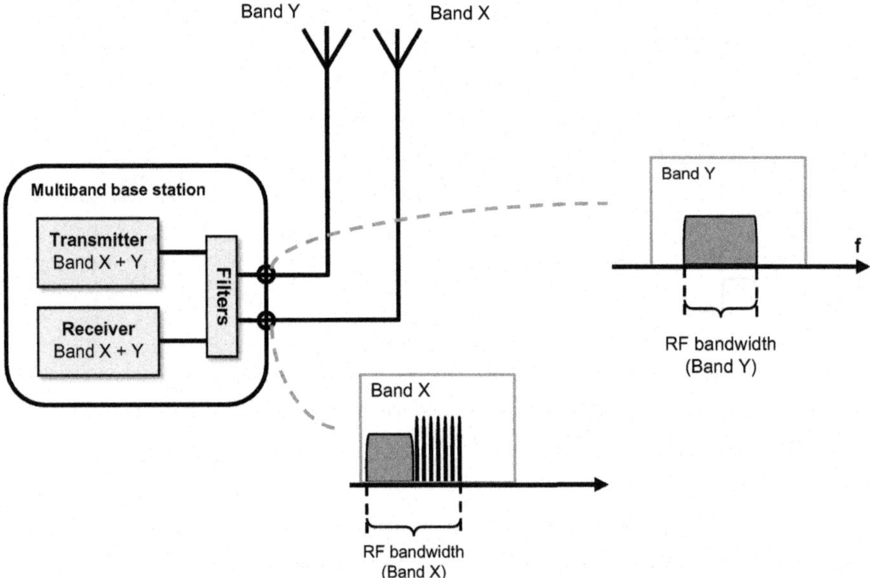


**FIGURE 16.12 Multiband base station with multiband transmitter and receiver for two bands with separate antenna connectors for each band.**

the antenna connectors through the use of a filter. As for multiband base stations with a common antenna connector for the bands, it is also possible to have either the transmitter or receiver be a single-band implementation, while the other is multiband.

Further possibilities are base station implementations with separate antenna connectors for receiver and transmitter, in order to give better isolation between the receiver and transmitter paths. This may be desirable for a multiband base station, considering the large total RF bandwidths, which will in fact also overlap between receiver and transmitter.

For a multiband base station, with a possible capability to operate with multiple RATs and several alternative implementations with common or separate antenna connectors for the bands and/or for the transmitter and receiver, the declaration of the base station capability becomes quite complex. What requirements will apply to such a base station and how they are tested will also depend on these declared capabilities.

Most RF requirements for a multiband base station remain the same as for a single-band implementation. There are some notable exceptions, however:

- *Transmitter spurious emissions*: For UTRA base stations, the requirement excludes the frequency range closer than 12.5 MHz from the carrier center and for MSR base stations, the requirements exclude frequencies in the operating band plus an additional 10 MHz on each side of the operating band, since those frequency ranges are covered by the spectrum mask and UEM limits respectively. For a multiband base station, the exclusions apply to the transmissions in both operating bands, and only the spectrum mask/UEM limits apply in those frequency ranges. This is called "joint exclusion band."

- *Spectrum mask (SEM) and operating band unwanted emissions mask (UEM)*: For multiband operation, when the inter-RF bandwidth gap is less than 20 MHz, the SEM for UTRA base stations and UEM limit for MSR base stations apply as a cumulative limit with contributions counted from both bands, in a way similar to operation in non-contiguous spectrum.
- *ACLR*: For multiband operation, when the inter-RF bandwidth gap is less than 20 MHz, the *Cumulative ACLR* (CACLR) will apply with contributions counted from both bands, in a way similar to operation in non-contiguous spectrum.
- *Transmitter intermodulation*: For a multiband base station, when the inter-RF bandwidth gap is less than 15 MHz, the requirement only applies for the case when the interfering signals fit within the gap.
- *Blocking requirement*: For multiband base station, the in-band blocking limits apply for the in-band frequency ranges of *both* operating bands. This can be seen as a "joint exclusion," similar to the one for spurious emissions. The blocking and receiver intermodulation requirements also apply inside the inter-RF bandwidth gap.
- *Receiver spurious emissions*: For a multiband base station, a "joint exclusion band" similar to the one for transmitter spurious emissions will apply, covering both operating bands plus 10 MHz on each side.

In the case where the two operating bands are mapped on separate antenna connectors, as shown in Figure 16.12, the above exceptions for transmitter/receiver spurious emissions, SEM/UEM, ACLR, and transmitter intermodulation do not apply. Those limits will instead be the same as for single-band operation for each antenna connector. In addition, if such a multiband base station with separate antenna connectors per band is operated in only one band with the other band (and other antenna connector) inactive, the base station will from a requirement point of view be seen as a single-band base station. In this case, all requirements will apply as single-band requirements.

## REFERENCES


[1] 3GPP, 3rd Generation Partnership Project; Technical Specification Group Radio Access Network; Physical Channels and Mapping of Transport Channels onto Physical Channels (FDD), 3GPP TS 25.211.

[2] 3GPP, 3rd Generation Partnership Project; Technical Specification Group Radio Access Network; Multiplexing and Channel Coding (FDD), 3GPP TS 25.212.

[3] 3GPP, 3rd Generation Partnership Project; Technical Specification Group Radio Access Network; Spreading and Modulation (FDD), 3GPP TS 25.213.

[4] 3GPP, 3rd Generation Partnership Project; Technical Specification Group Radio Access Network; Physical Layer Procedures (FDD), 3GPP TS 25.214.

[5] 3GPP, 3rd Generation Partnership Project; Technical Specification Group Radio Access Network; User Equipment (UE) Radio Transmission and Reception, 3GPP TS 25.101.

[6] 3GPP, 3rd Generation Partnership Project; Technical Specification Group Radio Access Network; UMTS-LTE 3500 MHz Work Item Technical Report (Release 10), 3GPP TR 37.801.
[7] 3GPP, E-UTRA, UTRA and GSM/EDGE; Multi-Standard Radio (MSR) Base Station (BS) Radio Transmission and Reception, 3GPP TR 37.104.
[8] 3GPP, E-UTRA, UTRA and GSM/EDGE; Multi-Standard Radio (MSR) Base Station (BS) Conformance Testing, 3GPP TR 37.141.
[9] 3rd Generation Partnership Project; Technical Specification Group Radio Access Network; Base Station (BS) Radio Transmission And Reception (FDD), 3GPP TS 25.104.
[10] 3rd Generation Partnership Project; Technical Specification Group Radio Access Network; Base Station (BS) Conformance Testing (FDD), 3GPP TS 25.141.
[11] 3rd Generation Partnership Project; Technical Specification Group Radio Access Network; User Equipment (UE) Conformance Specification; Radio Transmission and Reception (FDD); 3GPP TS 34.121.
[12] 3rd Generation Partnership Project; Technical Specification Group Radio Access Network; Radio Frequency (RF) System Scenarios, 3GPP TS 25.942.
[13] ITU-R, Unwanted Emissions in the Spurious Domain, Recommendation ITU-R SM.329-12, September 2012.
[14] FCC, Title 47 of the Code of Federal Regulations (CFR), Federal Communications Commission.

CHAPTER

# Heterogeneous networks 17

## CHAPTER OUTLINE



As discussed in Chapter 1, the concept of cellular mobile radio has existed since the first generation of analog mobile radio systems in 1947. During the first couple of generations of mobile radio, services were uniform (mainly voice) and radio equipment was expensive and required significant maintenance. Users were sufficiently few in number and services expensive enough that a large concentration of users in one place and time was unlikely. These factors implied that a grid of wide area base station sites planned to provide a reasonable level of coverage was an effective and economic means of providing service.

In some areas, such as dense city centers, capacity became a more significant concern during the growth phase of 2G and 3G deployments and it became more necessary to densify the deployment of macro sites. Still, the type of site was fairly uniform and propagation and interference to some extent predictable.

In the modern era of mobile broadband, complex mobile devices have become widely available and expectation levels have risen of a ubiquitous provision of high-quality, high-speed services at a low cost. At the same time, base station and chipset technologies have emerged that have enabled a new generation of low-power base stations that can provide benefits such as capacity, increased user throughputs, and/or improved coverage over a small area. As part of an overall solution to be able to cost-effectively meet demand and expectations for mobile broadband, operators are increasingly turning to more complex network deployment solutions that consist of

HSPA Evolution: The Fundamentals for Mobile Broadband. DOI: 10.1016/B978-0-08-099969-2.00017-X a mixture of traditional macro base stations, small low-power base stations, and base stations of intermediate size. Coverage and interference planning may be performed in the traditional manner for some types of base stations, while others may be deployed in a less predictable manner. A name commonly given to a network built with a variety of different base station types is "heterogeneous network."

Heterogeneous deployments are not a technology component; deployment of differing types of nodes has been a possibility from the first release. As the name suggests, heterogeneous networks may look quite different from one another, may be deployed for a variety of reasons, and may enable different types of opportunities. The emergence of new types of deployments may lead to new and different problems that can be mitigated or solved using technology and standards solutions. Thus, progressive releases have added enhancements that facilitate new types of heterogeneous network deployments. It is of interest, at a general level, to review and consider the reasons behind deployment of heterogeneous network nodes, different types of nodes and their characteristics, and some of the most common issues that need to be addressed in heterogeneous deployments. 3GPP specifically addressed heterogeneous networks in release 12, as discussed in Chapter 19.

## 17.1 MOTIVATION FOR DEPLOYMENT OF HETEROGENEOUS NETWORKS

The motivation behind the design of a particular network is very much related to the specific geographic, demographic, market, and economic situation of an operator. To provide higher user data rates and/or higher capacity, an operator will need to either improve the capabilities of its existing network or densify the network (or both). Densification can be by means of adding further macro nodes to a homogeneous macro network. However, this is not always possible because of practical limitations on sites, cost, and backhaul. Some general situations that motivate deployment of low-power base stations in addition to an overlaying macro network are outlined here as a background to consideration of heterogeneous network issues.

### 17.1.1 GENERAL OFF-LOADING

Situations can arise in which the macro network is overloaded and it is not feasible to deploy further macro nodes. An example could be a city center area with a large traffic volume, in which the density of macro sites is large and practical sites are no longer available. Reasons for a lack of practical sites can include geography, the radio environment, mounting constraints, and regulations. In such cases, non-macro *low-power nodes* (LPNs; different variants of LPNs are defined in Section 17.2) are intended to off-load traffic from the macro network, which is of benefit both to the users that are off-loaded to the low-power node and the remaining macro users who enjoy additional macro capacity. Strategies for deploying the low-power nodes in such situations may be opportunistic (where there are suitable sites available), driven by the location of traffic hotspots, or even random.

### 17.1.2 ENHANCEMENT OF COVERAGE

In some locations, coverage may be limited, or only low data rates available, in particular for the uplink. One example may be indoors, in which wall penetration losses attenuate the signal from the macro network. In such cases, improved user throughputs may be provided by deployment of low-power nodes, and the low-power nodes are themselves less likely to provide interference toward the macro network because of the isolation that leads to the poor macro coverage.

### 17.1.3 EXTREME HOTSPOT COVERAGE

Some locations are likely to attract extraordinarily high levels of traffic. Examples are large railway stations, stadiums, and popular city center squares. If served from the macro network, these hotspots may reduce macro capacity and indirectly reduce data rates available at other locations that are distant from the hotspot. Such increased macro activity could even increase macro inter-cell interference and thus reduce the data rates available in other macro-cells.

### 17.1.4 IN-HOME COVERAGE

The advent of so-called "Femto-cells" with low capacity and low transmit power enables provision of such cells within users' homes. In such cases, the backhaul is via the fixed Internet connection to the home. The advantage to the end-user is the potential to provide good in-home coverage, high data rates, seamless mobility with the surrounding network, and the potential for charging schemes that take into account the user's home location. For the network, potential benefits include off-loading, potentially reduced mobility issues, and backhaul provided via the user's Internet connection. On the other hand, unlike other types of cells, such cells are often placed by users in an uncoordinated manner, which can risk causing negative interference effects.

Deployment and ownership models began to change with the introduction of home Node Bs. One development that is of particular interest is that of the so-called *Closed Subscriber Group* (CSG). A CSG is a node that is not generally open to all of the subscribers to a network, but only a limited number of users, typically the ones approved by the owner of the node. A common scenario for deploying a CSG is as a home Node B. Typically, such a node is restricted to, for example, family members.

## 17.2 TYPES OF NODES IN HETEROGENEOUS NETWORKS

From a technology perspective, a node could potentially be built in any size, shape, and form that fulfills a market demand. To be declared 3GPP compliant, however, base stations must meet a set of RAN4 *Radio Frequency* (RF) conformance requirements that are described in [1] and which cover multiple aspects of behavior of the RF system such as transmit power level, receiver sensitivity, emissions levels, receiver blocking levels, and so on. Prior to release 6, the same set of RF requirements needed to be applied to all types of base stations. The transmit power of base stations could be declared, but other parameters needed to be met as absolute values. The original RF requirements were designed for macro base stations and were in some cases overly stringent for smaller base stations, which could lead to unnecessary cost, power consumption, form factor, and weight, and in other cases not stringent enough considering the potentially closer proximity of UEs to smaller base stations. For example, a small base station does not need to meet as stringent a reference sensitivity requirement as a macro base station, since it will be deployed in a small area and will not need to receive from power-limited UEs that are several kilometers distant. Another example is the receiver blocking level, which may be larger for a smaller base station, because an interfering UE on another carrier may be located nearby but transmitting at high power to a distant macro node.

To overcome the issue that the requirements on different types of base stations can look different, 3GPP created the concept of base station classes. Four types of base station classes have been defined:

- *Wide area* base stations are traditional macro base stations that are expected to cover larger urban and rural cells. It is expected that the antennas will be located on masts or at roof top level.
- *Medium-range* base stations are micro base stations. These are generally outdoor base stations and are likely to be located below rooftop level and to provide additional capacity to the macro network.
- *Local area* base stations are small base stations that are typically deployed in hotspot areas with a high traffic density. They may be indoor (for example, shopping malls) or outdoor and are typically wall or ceiling mounted.
- *Home* base stations are femto base stations intended for home use.

The base station vendor declares which types of requirements a base station conforms to and hence which class a base station belongs to. In this way, the specifications allow for several different types of base stations to be built for different heterogeneous network scenarios.

## 17.3 ISSUES TO CONSIDER WITH HETEROGENEOUS NETWORKS

Deployment of heterogeneous network technology presents challenges that must be addressed by means of network configuration, standardized behavior, or technology.

### 17.3.1 CELL RANGE EXPANSION

Small nodes are often deployed with the purpose of off-loading traffic from the surrounding macro network. Off-loading traffic provides higher throughputs to the users that are off-loaded, while enabling other macro users to enjoy a larger share of the available macro resources.

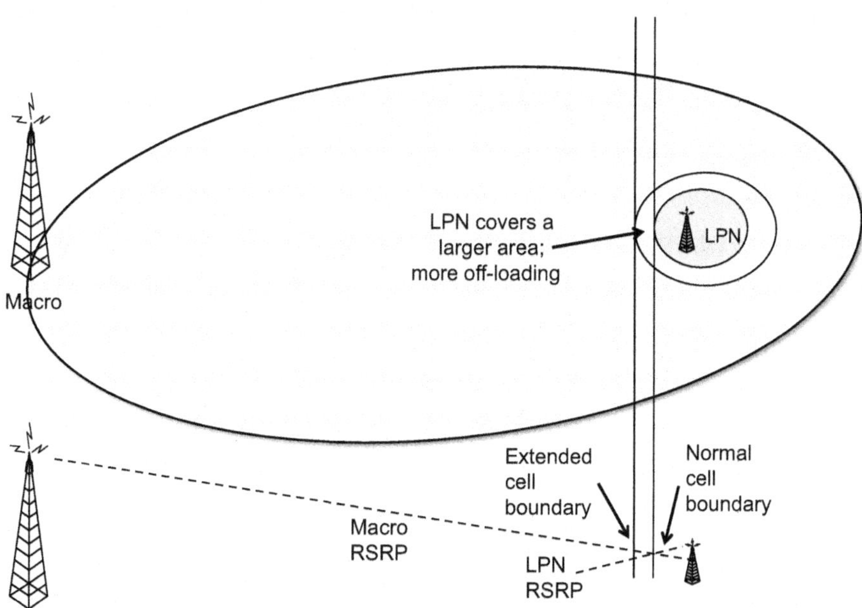


**FIGURE 17.1 Cell range expansion to increase capacity. RSRP stands for Reference Signal Received Power.**

Traditionally, in connected mode, a UE will be handed over to the cell with the strongest received common pilot signal, that is, the best downlink. In heterogeneous networks, however, it may not always be the case that being served from the cell with the best downlink is optimal from a system and user performance perspective. A simple deployment is depicted in Figure 17.1 consisting of a macro-cell containing a small cell node that is intended for off-loading. The area within which the small cell common pilot level is greater than the macro common pilot level is depicted in the inner circle around the small cell node. However, there is a wider area around the small cell node within which the small cell downlink common pilot level is no more than 6 dB worse than the macro level. If UEs in this area are served from the small cell node, they will experience a lower per TTI throughput than if they would be served from the macro. However, the small cell node shares its resources among fewer UEs than the macro node, and hence the UEs can potentially be served more frequently and experience an improved user throughput. Furthermore, serving the UEs from the small cell would off-load the macro and improve the experience of other macro users.

For heterogeneous networks, it may be desirable to be able to serve users outside of the normal cell range of the small cell. In the downlink, users that are located in a region around the small cell for which SINR is low experience reduced per TTI throughput but may be scheduled more frequently than if they were served from an overloaded macro with better SINR. Also, it may be the case that it is preferable for a user to be served from a small cell to which it has a low uplink pathloss, despite the user having poor SINR and being beyond the usual downlink cell boundary of the small cell. The technique of serving users in this manner is known as cell range expansion. Operating UEs in a cell range expansion zone can necessitate consideration of appropriate thresholds for operating handover and the overhead of control signaling to UEs experiencing poor SINR.

### 17.3.2 UPLINK/DOWNLINK LINK IMBALANCE

In a traditional macro network, all of the nodes transmit with the same power level. HSPA typically deploys soft handover, that is, maintains links with a number of cells in the so-called active set. In the uplink, the signal transmitted by the UE is received at all of the cells. The UE effectively follows power control commands from the cell with the lowest instantaneous pathloss, where pathloss is defined as the average attenuation of the uplink signal from the UE to a receiving cell. In the downlink, the UE receives dedicated channels from all of the cells in the active set, but one cell, usually the cell with the best DL SINR, serves the UE with HS-PDSCH.

The difference in pathloss between the cell with the lowest pathloss and other cells in the active set is known as the link imbalance. If the cells have the same transmit power in the downlink, then the average link imbalance is also equal to the difference in average downlink received signal strengths. The lower the average link imbalance, the greater is the gain from SHO for the uplink and the downlink DCH.

Handover decisions and decisions on which cells should be admitted to a UE's active set are made on the basis of averaged measurements of common pilot signals from different cells that are made by the UE and reported to the network. When cells transmit with equal pilot power, measurements of receive pilot power are equivalent to measurements of average link imbalance. Cells are typically added to the active set when their received pilot power is no more than 3-6dB different to the pilot power from the strongest cell. With a macro-only deployment (or more generally, a deployment with the same pilot power at all nodes), the UL pathloss will be similar to the downlink; hence, deductions about relative DL pathloss to each cell will also be valid for the uplink.

Where low-power nodes are deployed, however, the transmit power on the common pilots is greater for macro nodes than for small cell nodes. Thus, at the point at which the measured pilot strength is similar for both macro and small nodes, the pathloss will be lower to the small node, as illustrated in Figure 17.2. The region between the uplink cell boundary and the downlink cell boundary is typically referred to as the imbalance area.

Consider a UE that is inside of the UL cell boundary, as illustrated in Figure 17.2, but is not in soft handover and is only connected to the macro node. This region is referred to as the so-called *strong mismatch zone*. The UE will transmit with high power and, because of the lower pathloss to the small node, risks causing excessive interference to the uplink of the small node, as illustrated in Figure 17.3. This issue can be solved by either handing the UE over to the small node or operating soft handover. In the latter case, the UE will follow uplink power control commands from the small node and will not cause excessive interference to the small node.

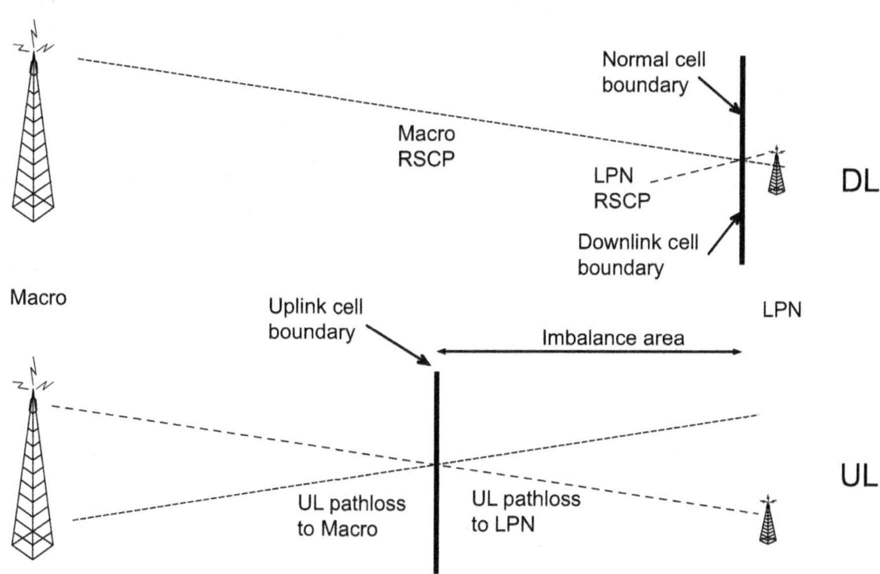


**FIGURE 17.2 Differing positions for downlink and uplink boundaries for a small cell. RSCP stands for received signal code power.**

However, the UE uplink will not reach the macro node (as illustrated in Figure 17.4), which is a potential problem if the uplink is carrying control information that should be received at the macro node as may, for example, be the case if the macro node is the serving node and must receive HS-DPCCH.

## 17.3.3 MOBILITY AND RADIO RESOURCE MANAGEMENT (RRM)

Mobility refers to the procedures involved in relocating UEs between serving cells, while RRM refers to the management of UE traffic between cells and frequency layers such that the available capacity from the network is utilized optimally.

**FIGURE 17.3 UE transmitting to a macro and causing excessive interference to a small cell.**

FIGURE 17.4 A UL transmission controlled from the small cell does not reach the macro.

A heterogeneous network may be deployed on multiple carriers with a mixture of macro and low-power nodes. In order to obtain sufficient information for mobility decisions, the UE needs to constantly search on all carriers for all small cells, since it does not know if it is in proximity to a small cell. When the UE is near to the macro-cell border, it may also need to measure adjacent macro-cell levels on one or more carriers in order to handover at the right time before coverage is lost. Constant searching of multiple carriers requires the UE receiver to be turned on and tuned to the appropriate carriers, which causes additional current drain in the UE and reduces battery life. Furthermore, when connected and receiving data, retuning of the UE receiver for measurement purposes may cause data loss. Thus, a compromise must be sought with regard to UE measurement frequency and optimization of radio resource management.

Because of the larger number of nodes, it can be expected that the number of mobility events in heterogeneous networks (for example, handovers) will be larger, and corresponding levels of Layer 3 (L3) processing must be provided within the network. L3 processing in this context refers to the RNC processing and RRC signaling that is required for managing mobility by means of handovers.

### 17.3.4 SELF-OPTIMIZATION

Some types of heterogeneous networks may be deployed in a planned fashion, for example, deployment of micro nodes within a macro network. However, in other circumstances, the network may be unplanned. An example of this is deployment of home nodes, which may be consumer driven. In both cases, and in particular where networks are unplanned, it is desirable for the network to be able to take steps to optimize. For example, interference needs to be managed, potentially by means of adjusting nodes' transmit power levels. Also, when nodes are deployed in an unplanned manner, the network needs to gain intelligence about the network topology and in particular which nodes are located in proximity to one another in order to make mobility decisions correctly. Thus, the network and to some extent the standardization needs to address means of enabling the network to self-optimize.

### 17.3.5 INTERFERENCE MANAGEMENT FOR UNCOORDINATED HOME NODE Bs

To users in the surrounding network that are not part of the closed subscriber group, CSG home Node Bs appear as interference. Non-CSG users in the vicinity may experience reduced performance because of this interference, and in the worst case even no coverage. In addition to users on the same channel, users on adjacent channels potentially subscribed to different operators may experience strong adjacent channel interference if they are in close proximity to a home Node B. Unlike the case with planned cells, an operator is unable to control where such adjacent channel interference issues may occur.

## REFERENCE

[1] 3rd Generation Partnership Project; Technical Specification Group Radio Access Network; Base Station (BS) radio transmission and reception (FDD), 3GPP, 3GPP TS 25.104.

## FURTHER READING

S. Landström, A. Furuskär, K. Johansson, L. Falconetti, F. Kronestedt, Heterogeneous networks—increasing cellular capacity, Ericsson Rev. (January 2011).

PART

# Performance and future outlook

CHAPTER

# HSPA system performance 18

## CHAPTER OUTLINE



HSPA Evolution: The Fundamentals for Mobile Broadband. DOI: 10.1016/B978-0-08-099969-2.00018-1

The performance of a mobile broadband communications system such as HSPA is driven by a number of factors, such as a mixture of traffic behaviors, connectivity states, user population and positions, propagation environment effects, algorithmic effects in, for example, the baseband or scheduling, operator policies, inter-cell and inter-UE interference, signaling error levels, reporting accuracies, and user behavior. These factors do not act in isolation but interact in a non-trivial manner. A proportion of the factors that influence the network performance relates to human behavior. Thus, the performance of a HSPA deployment is never analytically predictable. Nonetheless, statistical analysis and simulation of system behavior is needed for planning and monitoring the deployment and benchmarking improvements.

Link level simulations of HSPA performance provide information about the efficiency of the air interface. Since improving individual links can in many cases lead to increased performance for the whole system, link-level evaluations are useful in evaluating and improving HSPA. Since release 11 HSPA air interface is already highly efficient, only a few efficiency improvements can be obtained by targeting link-level performance. Thus, improvements to HSPA generally target system improvements rather than link improvement. It should be noted that link-level performance does not provide information about user experience, since user experience depends both on link efficiency and the proportion of time and resources that are dedicated to the link for the user in question. The amount of resources available for a particular user depends, in turn, on how the total resources of the system are spent, that is, on system behavior.

In order to properly characterize HSPA performance in terms of capacity and user experience, it is necessary to model system behavior to some extent. In this context, the term "system" refers to the interaction of all factors that impact capacity and user experience, including traffic, propagation, algorithmic behavior, and the impact of multiple users and cells. An accurate system model would, however, be both highly complex and scenario dependent. Typically, a simplified system model that provides a statistical picture of HSPA performance in a statistically average environment is employed for research and standardization. Some stakeholders, such as operators and network vendors, may also utilize more complex models that relate more directly to scenarios that they intend their network and equipment to cover.

# 18.1 TYPICAL SYSTEM MODELING ASSUMPTIONS

A general statistical system model for HSPA needs to make a number of assumptions.

## 18.1.1 TRAFFIC CHARACTERISTICS

Traffic in mobile networks can arise from a variety of sources, including among other things voice calls, file downloads, video streaming, web browsing, online chatting, online gaming, e-mail polling, and enterprise VPN. A typical network will contain a mixture of traffic types, and the mixture will vary depending on the location and the time of day. Since the traffic mixture is dependent on human behavior, there is not an analytical means to calculate traffic characteristics.

For modeling HSPA behavior, there are three common types of traffic that might be assumed:

*Full Buffer* traffic is an extremely simplified scenario in which it is assumed that there is a fixed number of users in each cell whose buffers for upload/download are constantly full. The scenario tends to lead to full TTI utilization in all cells for serving the traffic. Full buffer is a highly unrealistic scenario, even for a highly loaded network. More realistically, with real types of traffic, an operator will not load a network beyond the point at which, at peak load, the mean TTI utilization is around 50%; otherwise, the network would be unable to cope with instantaneous peaks in TTI utilization while maintaining QoS. Furthermore, a scenario where users request an unlimited amount of data in every cell is not realistic.

Nonetheless, full buffer–based simulations can be useful in some scenarios. In particular, when evaluating features that are intended to provide link-level benefits, full-buffer simulations can relate link-level improvements (which are usually dependent on C/I) to a cell-level C/I distribution. Furthermore, full-buffer simulations are computationally simple and can be useful for making approximate estimations of HSPA performance.

*Bursty* traffic is often represented using a statistical model. Bursts are considered to be somewhat variable in two aspects, *burst size* and *inter-arrival time*, as illustrated in Figure 18.1. The pdf of the burst size may be modeled using a *Log-normal* distribution:

$$f_x = \frac{1}{\sqrt{2\pi}\sigma x} e^{\frac{-(\ln(x)-\mu)^2}{2\sigma^2}} \quad x > 0, \tag{18.1}$$

where $\mu$ is the mean burst size and $\sigma$ the standard deviation of the burst size. Typically, the burst size is constrained to be within a certain range—referred to as a *truncated Log-normal* distribution. The pdf for the inter-arrival time is often calculated using an exponential distribution:

$$f_x = \lambda e^{-\lambda x} \; x \geq 0, \tag{18.2}$$

where $\lambda$ is the mean inter-arrival time.

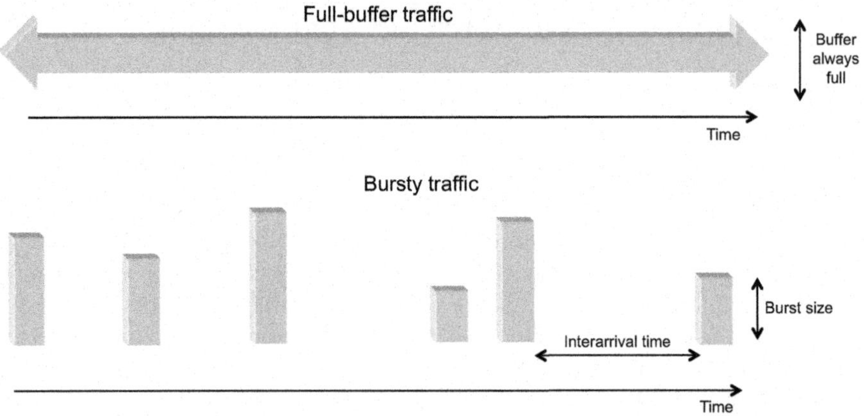


**FIGURE 18.1 Full buffer and bursty traffic characteristics.**

The parameters of the distribution depend on the type of bursty traffic to be modeled. For file download, the mean burst size will be several megabytes and the inter-arrival times large. For web browsing, file sizes may be a few tens or hundreds of kilobytes, with much shorter inter-arrival times.

A further aspect of the model is how different bursts are modeled. In some models, bursts are modeled as coming from a fixed number of users in the cell, as shown in Figures 18.2a and 18.3a. Thus one user will generate a succession of bursts. In other models, each arriving burst is modeled as a new user, as shown in Figures 18.2b and 18.3b.

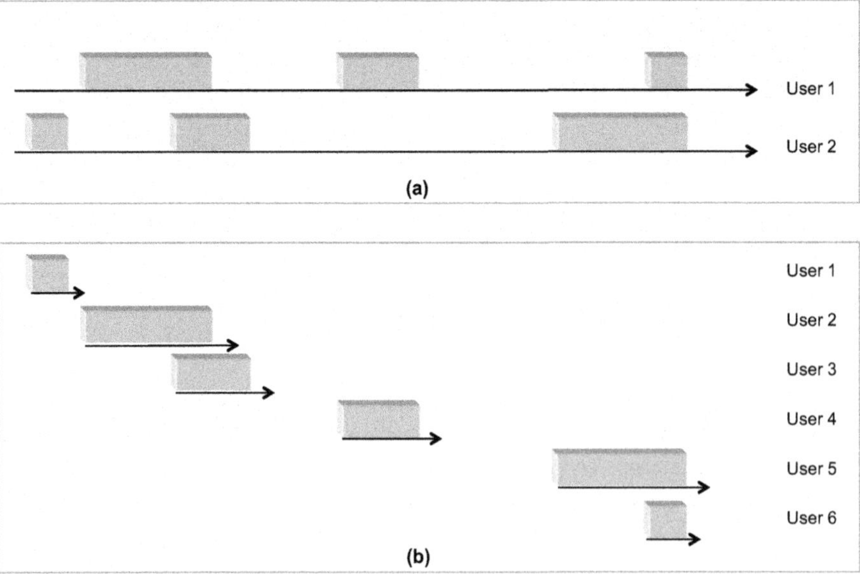


**FIGURE 18.2 Traffic models with multipacket users and single-packet users.**

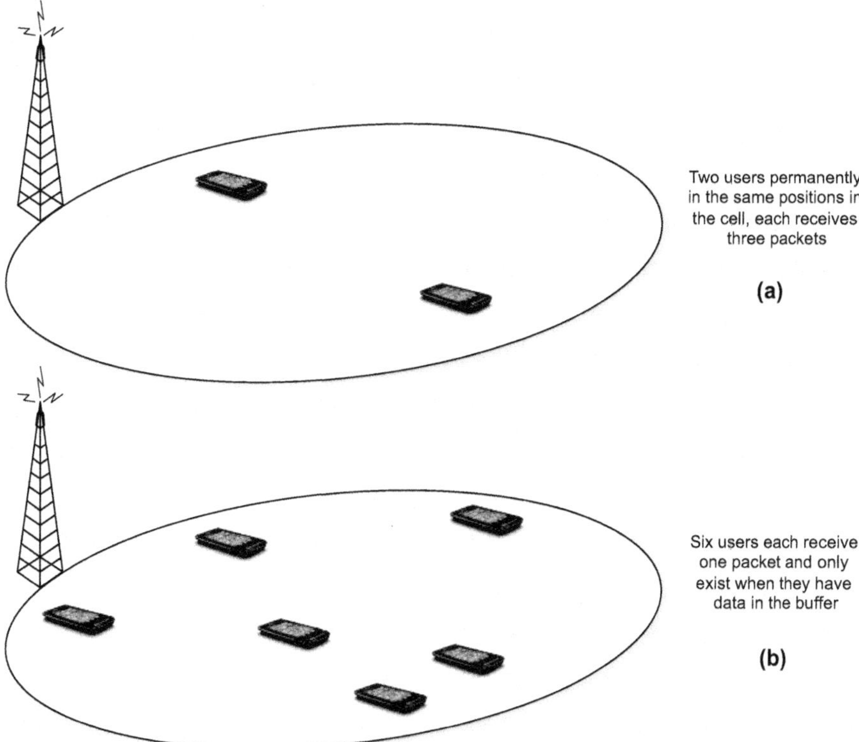


**FIGURE 18.3 User positions with multipacket users and single-packet users.**

The type of model that is appropriate will depend upon the type of traffic being modeled; for example, it is likely that a single web browsing user may generate a succession of bursts, whereas file downloads with a large inter-arrival time may arrive from different users.

A figure of interest for a traffic model is the total *offered load*. The total offered load is the total amount of bits arising from the model during a fixed period of time (regardless of whether the traffic is served or not) divided by the length of the period of time. The offered load is approximately equal to the mean burst size divided by the mean inter-arrival time in the model described above.

*Voice* is one of the key services for mobile networks and voice performance is of interest. Voice, in particular VoIP, arrives in fixed size packets every 20 ms. To model network jitter, a variation of around 20 ms inter-arrival time may be added to voice modeling in some circumstances, as illustrated in Figure 18.4.

The full-buffer, bursty, and voice models described above are well-proven means of evaluating system performance, in particular, in 3GPP. Alternatives that may be used for increasing the realism of the simulations include traffic data obtained from real networks.

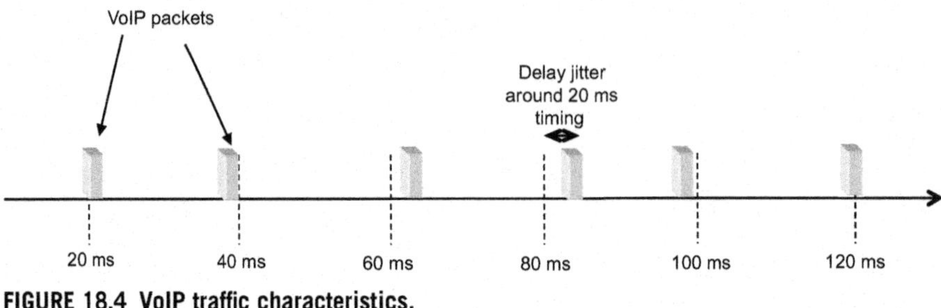


**FIGURE 18.4 VoIP traffic characteristics.**

### 18.1.2 NETWORK DEPLOYMENT CHARACTERISTICS

A real network deployment will depend on factors such as local terrain, population distribution, and the availability of cell sites. For system performance simulation, deployment models that are statistically representative and are conceptually and computationally simple are often adopted. In some circumstances, however, it is desirable to incorporate real terrain and propagation data and even 2D and 3D maps in order to increase the realism of the simulation. In this section, some simple models are discussed.

*Hexagonal grid* is a conceptually simple and widely used approximation to a macro network. Three-sector sites are placed on a grid such that each cell can be represented by a hexagon (Figure 18.5). The geometrically simple shape allows for the concept of "wrap around," in which the interference from cells is replicated around the edges of the modeled deployment in order to avoid edge effects. Terrain and propagation are typically modeled as uniform.

The *Manhattan grid* model is intended to represent an average "city" and is an early heterogeneous network model. The model consists of a uniform set of buildings overlaid onto a hexagonal macro network. Users may be restricted to be outdoors only or users inside buildings may experience an additional pathloss representative of wall penetration loss. In addition to macro nodes, directional small nodes may be placed at building corners to point along streets (Figure 18.6).

In a *simplified heterogeneous network*, low-power nodes are dropped randomly within a hexagonal grid macro deployment. A minimum distance between low-power nodes and between the macro nodes and low-power nodes is enforced (Figure 18.7). The propagation model between a UE and a low-power node differs from the propagation between a UE and a macro node. The number of low-power nodes dropped in each macro-cell area is a parameter of the model.

### 18.1.3 PROPAGATION CHARACTERISTICS

The propagation environment is modeled using a statistical model of average characteristics. Generally, a propagation model consists of three components: a pathloss-dependent component, shadow fading, and fast fading. Traditional propagation models are *two-dimensional* (2D) (Figure 18.8a), in that they consider only the azimuth

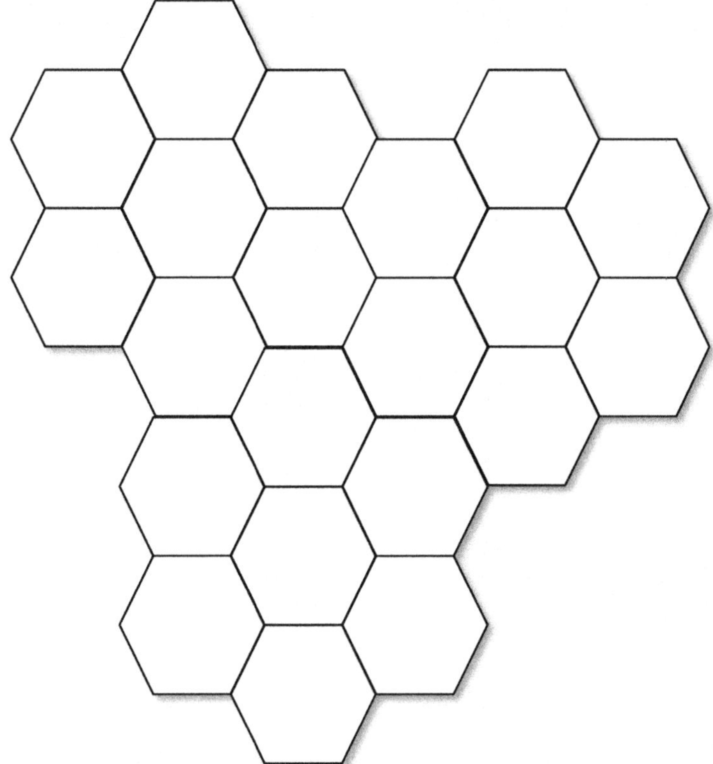

**FIGURE 18.5 Hexagonal grid layout.**

angle and the distance of the UE from the base station. However, in recent years, to properly model the impact of antenna downtilt, *three-dimensional* (3D) models have become more popular that consider both the azimuth and elevation angles of the UE (Figure 18.8b).

The pathloss-dependent component depends on the distance of the UE from the base station. A typical pathloss-dependent model for a macro-cellular environment is based on [1]:

$$L = 40(1 - 4\times10^{-3}\Delta h_b)\log_{10}(R) - 18\log_{10}(\Delta h_b) + 21\log_{10}(f) + 80\,dB, \tag{18.3}$$

where $\Delta h_b$ is the difference in height in meter between the UE and the base station, $f$ is the carrier frequency, and $R$ is the distance in kilometers between the UE and the base station. For a carrier frequency of 2 GHz and a height difference of 15 m, this formula is simplified to

$$L = 128.1 + 37.6\log_{10}(R). \tag{18.4}$$

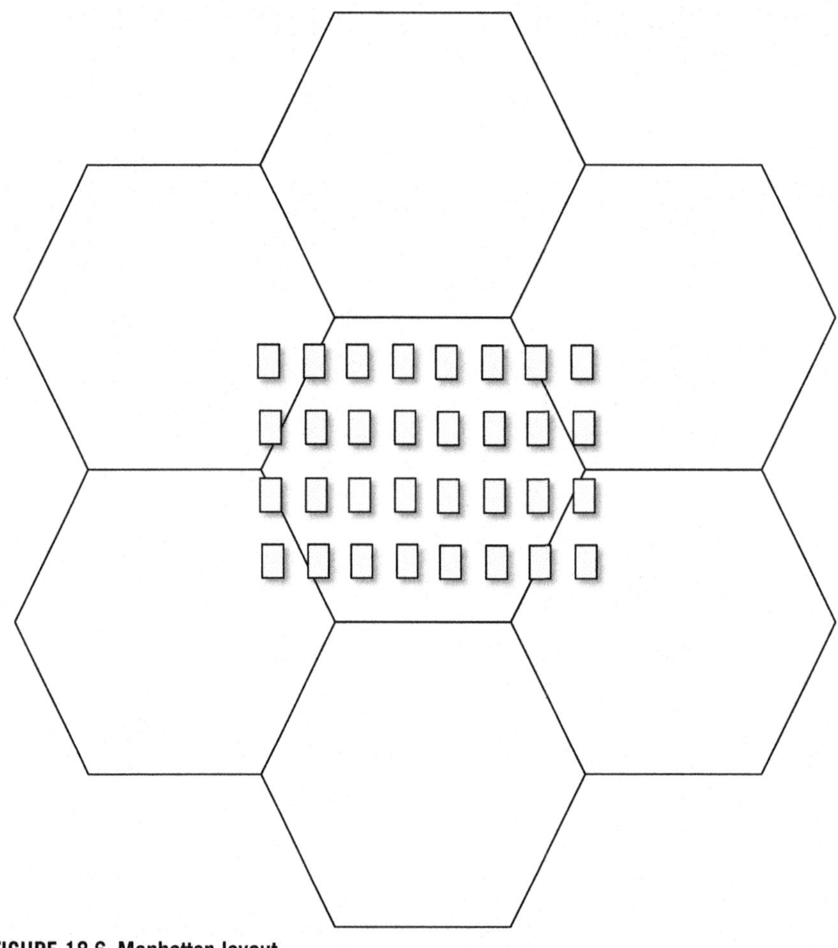

**FIGURE 18.6 Manhattan layout.**

In a heterogeneous network environment and an urban Manhattan grid, the LPN model may include a *Line of Sight* (LoS) component in addition to a non-LoS component [2].

In addition to the pathloss component, an antenna gain component needs to be added. For three-sector macro sites, a simple 2D antenna model is shown in Figure 18.9. The antenna gain depends on the angle of arrival of the UE signal with respect to the antenna boresight. For 3D modeling, both the azimuth and elevation angles with respect to boresight need to be considered.

An important aspect that is not considered in these simple antenna models is antenna pattern sidelobes. In particular in the horizontal domain, sidelobes may in some circumstances contribute to inter-cell interference in the downlink.

On top of the distance-dependent pathloss, shadow fading is added. Shadow fading is not time variant and is assumed to be log-normal distributed, typically with a

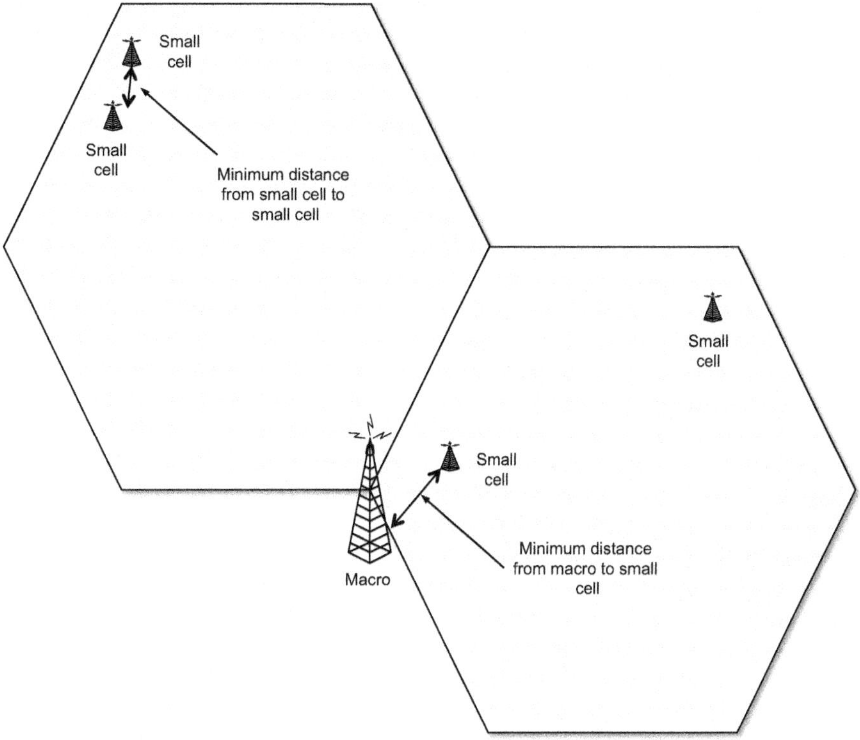


**FIGURE 18.7 Generic heterogeneous network layout.**

variance of 8–10 dB. Furthermore, shadow fading is often assumed to have 100% correlation between sectors of the same site and 50% correlation between sites.

The combination of antenna gain, pathloss, and shadow fading gives the long-term path characteristics to the UE. In addition to the long-term characteristics, short-term fast fading must also be calculated. Often a multipath Rayleigh fading channel model is used to calculate the fast fading. The multipath channel consists of a profile of path delays and mean gain factors. Fading on each path is modeled statistically independently. Around each path is added a Doppler spectrum. Typically, standard multipath models are used, such as the ITU Pedestrian A, Vehicular A, or the Pedestrian B model [1]. The instantaneous multipath fading profile is usually calculated once per TTI.

For advanced MIMO schemes, a more complex model of the channel may be needed, including a spatial channel model [3,4].

### 18.1.4 MODELING OF BASE STATION SCHEDULING

Scheduling decisions, as described in Chapter 5, are usually modeled once per TTI.

For the downlink HSDPA, a couple of standard models are used for scheduling: *round robin* and *proportional fair.* With *round robin*, users who have data in their

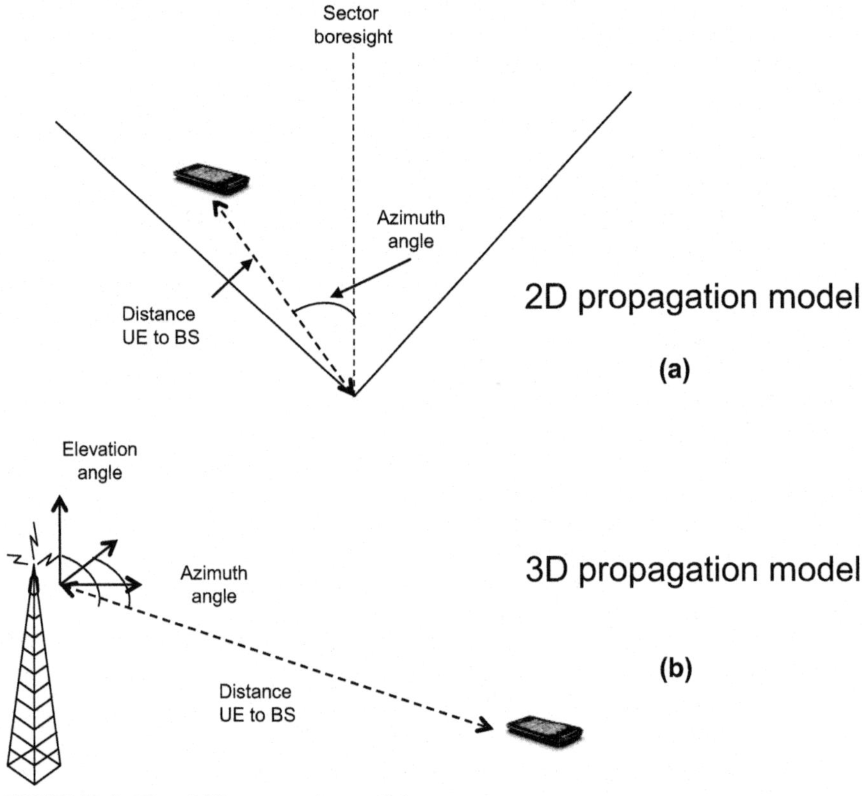


**FIGURE 18.8 2D and 3D propagation models.**

buffer are scheduled one after the other. With proportional fair, a throughput metric is calculated for each user *i*, and the user with the highest metric $PF_i$ is scheduled. The throughput metric is depicted in (18.5). $R(i)$ is the throughput supported by the UE in the current TTI and $\overline{R}(i)$ is a rolling average of the throughput to the UE during previous scheduling occasions. As depicted in (18.6), $\overline{R}(i)$ is calculated using a leaky filter. With large values of α, the scheduler tends to quickly forget previous scheduling, so the users with the higher instantaneous throughput tend to be scheduled, improving cell capacity. With low α, the scheduler memory is long and the weighting leads to users being scheduled on a more "fair" basis, independent of their channel conditions. Cell capacity can be traded off against fairness by adjusting α.

$$PF_i = \frac{R(i)}{\overline{R}(i)}, \tag{18.5}$$

$$\overline{R}(i) = \alpha R(i) + (1-\alpha)\overline{R}(i). \tag{18.6}$$

By adjusting the proportional fair coefficient $\alpha$, the scheduler can be varied between being fair in the sense of giving all users equal air-time and being optimal from

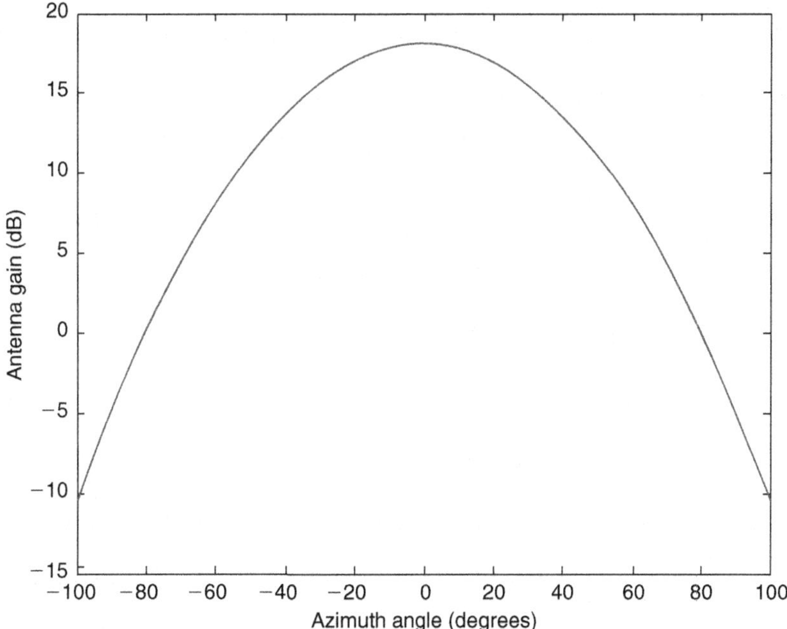


**FIGURE 18.9 2D antenna pattern.**

a system point of view by scheduling users with good channel conditions that can achieve high throughputs. The parameter $\alpha$ is selected in order to maximize system throughput while ensuring an acceptable performance for cell-edge users that experience a high carrier-to-interference ratio. Scheduling is described in more detail in Chapter 5.

The scheduling metric may be calculated from a simulated CQI reporting based on users' channel conditions.

The uplink scheduling requires more careful consideration. There are two principle approaches to scheduling the uplink: *Time Division Multiplexing* (TDM), in which the scheduler aims to schedule one user in each 2 ms TTI (for example, by means of hybrid-ARQ process restriction) and *Code Division Multiplexing* (CDM), in which all users are scheduled simultaneously and the uplink RoT is shared.

### 18.1.5 MODELING OF ADAPTIVE MODULATION AND CODING AND HYBRID-ARQ

For HSDPA, *Adaptive Modulation and Coding* (AMC) and hybrid-ARQ must be accounted for in each TTI in which a user is scheduled. In a very simplified simulation, each time a user is scheduled for a TTI, a mean achievable throughput can be calculated for that user given the pathloss and instantaneous fading profile. Alternatively, and more commonly, hybrid-ARQ can be explicitly modeled. For the first transmission, a modulation and coding format is selected based on the simulated CQI reporting. Then for the TTI in which the transmission takes place, the probability of a correct reception can be calculated based on the link-level Eb/No and the Eb/No achieved in any previous transmissions. A decision can be made randomly based on the calculated probability on whether a retransmission should be made or not.

For HSUPA, the process is similar although a decision on modulation and coding format at the Node B scheduler is not needed because of the action of power control in maintaining Rx Eb/No (however, a decision is still made on E-TFC in the UE scheduler based on grants, power, and buffer status). The need for a retransmission of a TTI can be calculated in a similar manner.

### 18.1.6 RECEIVER AND LINK-LEVEL PERFORMANCE

After a user is scheduled in a particular TTI, it is necessary to calculate the receiver Eb/No. Calculating the Eb/No involves considering the characteristics of the received signal, the interference, the noise, and the receiver algorithms. In the downlink, the wanted signal comes from the serving cell, while interfering signals come from neighbor cells. In the uplink, the wanted signal comes from the UE for which the Eb/No is to be calculated, while the interference comes from all other UEs in the system, including UEs in neighbor cells.

For the wanted signal, the combination of Tx power, pathloss, and multipath fading enables a channel impulse response, $\boldsymbol{h}$, to be known. In some simulations, a channel estimate, $\hat{\boldsymbol{h}}$, may be calculated. For the interfering cells, the interference is approximated to AWGN; that is, $\Sigma|h_i|^2 = \mathrm{I}$, although for interference-mitigating receivers, the impulse response of some of the strongest interferers may be taken into account. The thermal noise must be calculated, taking into account a suitable noise figure for the receiver. The calculated channel estimates, interference, and thermal noise can then be used to calculate a receiver Eb/No. The Eb/No will depend on the type of receiver, for example, RAKE, LMMSE, or interference mitigating (see Chapter 3), and thus the post-receiver Eb/No must be modeled by the simulator based on the channel state and, if applicable, channel estimate.

After the Eb/No is calculated, the probability of success is determined by using a look-up table. The look-up table is based on link-level simulations capturing the performance of the decoding chain.

### 18.1.7 HANDOVER AND MOBILITY EVENTS

Some simulators may include modeling of the movement of UEs during the simulation, including recalculation of pathloss and modeling of handover events. Such simulators are, however, not commonly used for the calculation of HSPA performance metrics, as described in this chapter. Semi-static system simulators do not model movement of UEs, although they do model the impact of Doppler spread in the multipath fading channel.

## 18.2 SYSTEM SIMULATION PERFORMANCE METRICS

To assess the performance of HSPA, a number of metrics are considered. The metrics are in some cases contradictory and can be traded off against one another, and so metrics should be considered together to gain a true picture of the performance of an HSPA system.

### 18.2.1 CELL SERVED THROUGHPUT

The cell served throughput is the average amount of information that is transferred to/from the cell. For the system to be stable, the cell served throughput must be equal to the offered load; if the served throughput is lower than the offered load, then the system is not delivering all of the traffic that is offered to it (Figure 18.10).

### 18.2.2 PER-TTI USER THROUGHPUT

The per-TTI user throughput is the throughput experienced in each TTI by users when they are scheduled. The information may be plotted as a *cumulative distribution function* (CDF). Interesting points in the CDF may be examined such as the mean per-TTI throughput or the 5th percentile (often considered to represent the "cell-edge" performance). Per-TTI user throughput gives information about the link performance of the system, but it does not give any information about user experience, since it does not consider in how many TTIs a user is scheduled. Users

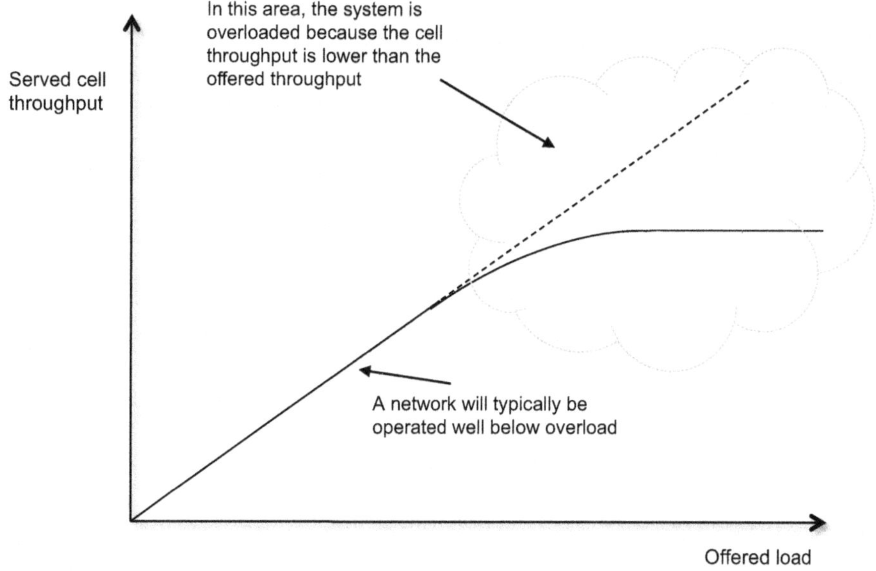


**FIGURE 18.10 Offered and served load characteristics.**

who are very infrequently scheduled will not experience a high data rate, even if they are able to transfer a large number of bits during the TTIs in which they are scheduled.

### 18.2.3 MEAN USER THROUGHPUT

The mean user throughput is, for each user, the mean throughput experienced over all TTIs per second. Since there are many users, a CDF of mean user throughput may be plotted and examined at the mean or 5th percentile. Mean user throughput is a useful metric for full buffer simulations. However, for bursty traffic, mean user throughput is still not a sufficient metric of user experience. With bursty traffic, user experience is determined by how long a packet takes to transfer between transmitter and receiver, and delays to packet transfer are not captured with mean user throughput.

### 18.2.4 BURST THROUGHPUT

For bursty traffic, the key aspect of user experience is how much time it takes to transfer bursts after they arise. The time taken to transfer bursts depends on scheduling delay, the proportion of TTIs in which the burst is scheduled, and the throughput. All of this can be captured with the so-called *burst throughput*. The burst throughput of a burst is the size of the burst divided by the time difference between the point in time at which the last TTI containing the burst is successfully received and the time at which the burst first arrives in the buffer for transmission. Since there are many bursts during a simulation, a CDF of burst throughput may be constructed. Once again, the mean and 5th percentile points on the CDF are often of interest.

Figure 18.11 illustrates the different user throughput metrics.

Burst throughput is typically traded off against offered load and served throughput (Figure 18.12). Increasing offered load implies that resources must be shared among more users and the burst throughput decreases.

An interesting secondary metric is to define a minimum user experience for the system (for example, by deciding a minimum 5th percentile packet throughput) and

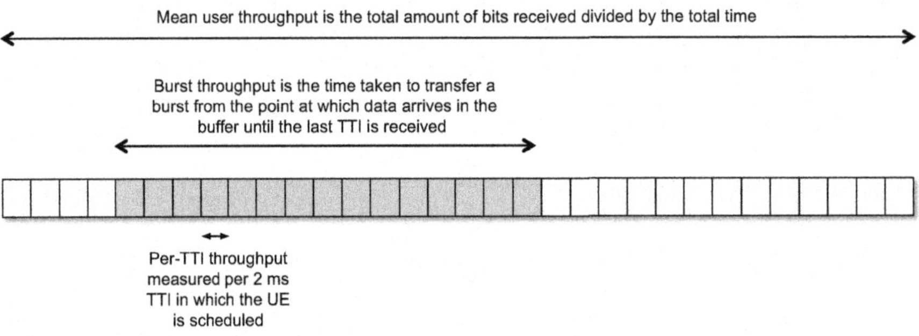


**FIGURE 18.11 TTI, mean user, and burst throughput metrics.**

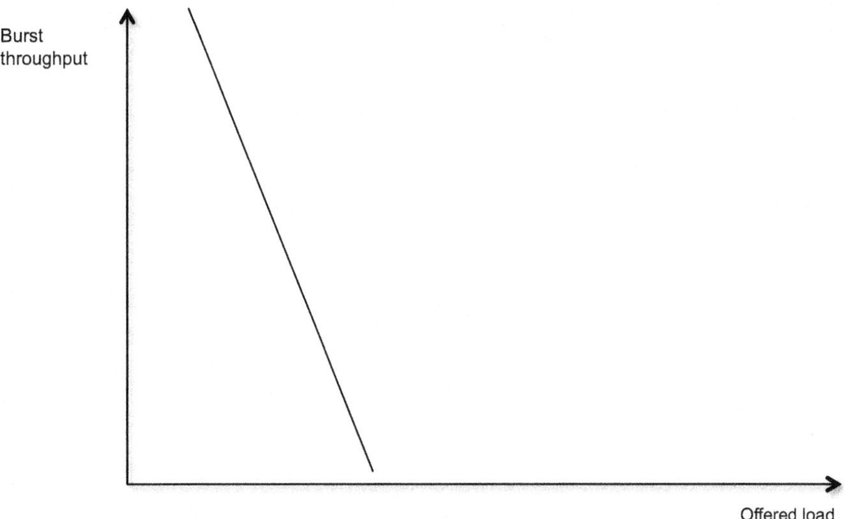


FIGURE 18.12 Illustration of relationship between offered load and burst throughput.

examining the offered load level at which this minimum packet throughput/user experience is achieved. This will be the maximum offered load that the system can handle while still meeting its packet throughput *Key Performance Indicator* (KPI) and hence is a packet call capacity for the system. Reducing the throughput expectation will obviously increase the packet call capacity, and vice versa.

Another metric of secondary interest is the TTI utilization, which is the proportion of TTIs in which any user is scheduled. Generally, with packet traffic, when operating at full capacity the TTI utilization will still be much less than 100%. This is because at 100% TTI utilization, scheduling delays would become excessive and packet throughput would suffer.

Other secondary metrics that may be of interest include, among other things, packet latency (that is, the transfer times for packets), mean amounts of hybrid-ARQ retransmissions and hybrid-ARQ round trip times, inter-cell interference levels, and uplink RoT.

## 18.3 HSPA PERFORMANCE DRIVERS

The performance of a HSPA network in terms of the metrics outlined in Section 18.2 is driven by a large number of factors relating to the design of the UE, Node B, RNC, and core network, backhaul, traffic, and the radio propagation environment. Depending on the system requirements, a number of factors can be optimized for system performance, although in some cases these factors are contradictory (for example, increased control signaling to improve scheduling and link adaptation will increase control overhead), and other factors will add additional cost to the network or UE.

### 18.3.1 DOWNLINK PERFORMANCE DRIVERS

Important performance indicators for the downlink are the achievable burst throughput (which relates to user experience) and capacity. Coverage is generally not as significant an issue in the downlink as in the uplink since the downlink is usually interference rather than noise limited, and base stations have significantly larger transmit power than mobile UEs. However, reaching indoor users with reasonable data rates can imply situations in which receive power to noise is a limiting factor.

Bursty-type traffic is typical in the downlink of mobile broadband systems. Bursty traffic has the characteristic that bursts arising near the cell center can be served relatively quickly while bursts arising near to the cell-edge take longer to serve. Assuming that the probability of bursts arising is geographically evenly distributed, then the unequal times to serve bursts will lead to a skew in the distribution of active users toward the cell-edge. Thus, cell-edge performance can have a strong influence on the achievable capacity, and improving cell-edge performance can have a disproportionate influence on capacity.

Furthermore, improving link-level performance and shortening burst times reduce the amount of TTIs in which the Node B must serve HS-DPSCH traffic for a given load level, which will decrease the amount of inter-cell interference caused by the Node B. Thus, in addition to the direct benefit of increasing data rates, improving the link-level performance also reduces inter-cell interference, which indirectly increases link-level performance in other cells. Hence, increasing link-level performance can yield a gain when considering the system performance of a HSPA network with bursty traffic that is higher than the gain observed on individual links.

#### *18.3.1.1 Cell size and number of cells*

Apart from reaching indoor users, in general the downlink is limited by inter-cell interference, and the instantaneous data rates that can be achieved in scheduled TTIs are not strongly dependent on inter-cell distance. In high-traffic scenarios, however, capacity in the downlink (that is, the amount of users that can be supported simultaneously) can be a performance-limiting factor. Capacity can be increased by either improving per-cell downlink capacity or increasing the number of cells. The network can be adapted to the geographical distribution of the traffic by, for example, decreasing the inter-site distance in densely populated areas or by placing additional cells in hotspot areas, such as railway stations.

Capacity can also be increased by increasing the number of sectors served by each site. Although not common, deployment of six sector sites in place of the more usual three sectors can increase capacity in some circumstances.

#### *18.3.1.2 Receiver performance*

Improving the performance of UE receivers increases the link efficiency and directly increases user experience and capacity. Means of improving UE receiver performance include more advanced equalization algorithms, such as LMMSE, increasing the number of receive antennas at the UE (for most handheld UEs, however, two antennas is a limit) and, for UEs with two or more antennas, applying interference mitigation in the receiver, as described in Chapter 3. As described above, improving link performance for an individual link will yield an even larger benefit to a whole HSPA system.

Improvements in the UE RF components, which lead to a decrease in the noise factor, have the potential to improve the available data rates in scenarios in which the UE is noise limited (for example, indoor).

#### *18.3.1.3 Transmitter performance*

A means to increase the link performance by improving the Node B transmitter is to add more antennas. The release 11 HSPA specifications support two- and four-branch MIMO, which can be used for beamforming and for spatial multiplexing. Deploying an increased number of transmit antennas in an HSPA network can cause issues relating to the existing UE base that must be carefully considered. Many UEs do not support two-branch MIMO, and four-branch MIMO is relatively new in the specifications. Deploying two antenna branches requires additional pilot overhead. For UEs that are MIMO capable, the MIMO gain overcomes the potential throughput loss that could be caused by the additional pilot overhead. However, for UEs that are not MIMO capable, the additional pilot is pure overhead and such UEs will experience worse throughput when MIMO is deployed, causing reduced user experience for these users and potentially a system capacity reduction.

#### *18.3.1.4 Interference management*

In most deployments, downlink performance is limited by inter-cell interference. Inter-cell interference can be reduced by applying so-called downtilt to base station antennas, as illustrated in Figure 18.13. Tilting the antennas downward reduces the footprint of the main lobe into the neighbor cell and reduces inter-cell interference. However, if the downtilt is too large, parts of the intended cell coverage area will also experience reduced receive power and throughput will be lost. Setting the correct

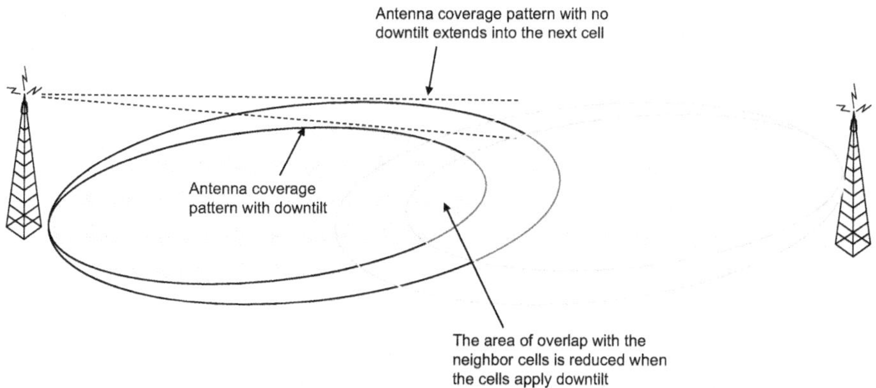


**FIGURE 18.13 The use of downtilt to reduce inter-cell interference.**

downtilt is a deployment-specific optimization that must be determined carefully. Furthermore, there is a conflict between the need of the downlink to reduce cell overlap and the uplink in which cell overlap can increase soft handover gain. Typically, the same antenna unit serves both downlink and uplink. In the future, *Advanced Antenna Systems* (AAS) that are capable of electronic beamforming for setting downtilt will be capable of adjusting downtilt for downlink and uplink independently.

#### 18.3.1.5 Latency

Many types of applications that make interaction via a mobile broadband network utilize the TCP protocol. The TCP protocol was originally designed for the fixed Internet and incorporates mechanisms, such as slow start, that are designed to avoid traffic congestion in the Internet. When operated across a mobile network, the TCP protocol is sensitive to round trip delay, in particular during the initial slow start phase of an interaction. If round trip time is significant, then the source data rate will be throttled and the packet call throughput to the UE will become limited. Some types of interaction are in particular likely to generate a lot of short lived TCP connections, such as browsing web pages, and with poor round trip latency, user experience will be deteriorated.

Other types of application, such as online gaming, require short ping times for user experience. Speech calls tolerate a latency of 50–60 ms over the air interface within their total latency budget.

Latency can be traded off with other system performance aspects, such as over-the-air user throughput and system capacity, by means of, for example, QoS-sensitive scheduling or adjusting the average amount of hybrid-ARQ retransmissions. Depending on the service type, the latency and system performance can be optimized appropriately.

#### 18.3.1.6 Radio resource management (RRM), scheduling, and mobility management

In many networks, some UEs will be in the coverage of several cells, which may have different levels of loading. These cells may be on the same carrier (for example, near cell borders or in heterogeneous networks) or on different carriers or even different *Radio Access Technologies* (RATs). The network radio resource management should consider the radio characteristics of UEs and the overall traffic situation and use handover and reselection measurements to ensure that traffic is steered optimally between available carriers and cell resources and that heavily loaded cells are offloaded.

#### 18.3.1.7 Multi-carrier

Multi-carrier operation (DC-HSDPA and MC-HSDPA) can increase user throughput and also provide system gains.

### 18.3.2 UPLINK PERFORMANCE DRIVERS

Unlike the downlink, the uplink operates fast closed-loop power control. In the absence of UE power limitations, the position of a UE in a cell would in principle not influence the achievable data rate. However, cell-edge UEs are of course more likely to experience power limitations and also cause a larger amount of inter-cell interference.

#### 18.3.2.1 Rise over Thermal (RoT)

The HSPA uplink is interference limited. Users are not orthogonal and thus interfere with one another. In order to maintain stability of the power control loops and coverage to the cell-edge, it is necessary to limit the maximum total received power experienced at the Node B receivers. The limit on interference is expressed as *Rise over Thermal* (RoT). The scheduler manages interference by assigning Tx power and implicitly data rates to users. The RoT limit restricts the uplink capacity. The maximum RoT limit that the scheduler can use is driven by deployment-related factors, such as the cell size and isolation between cells.

#### 18.3.2.2 Cell size and the number of cells

Because of the limited transmit power of the UE, supporting high data rates may not be possible for cells in excess of 1 km in the uplink (although coverage with medium rates will still be feasible). For urban cells with an inter-site distance of less than 500 m, the uplink data rates are more likely to be limited by the available RoT budget than UE transmit power, and thus similar considerations apply to cell size and the number of cells as in the downlink in such circumstances.

#### 18.3.2.3 Receiver performance

Since the uplink is not orthogonal and achieves reduced SINR compared with the downlink, the direct gain of applying LMMSE type of receivers is not as significant as in the downlink. LMMSE equalization is in principle required where high data rates and 16-QAM are supported; for lower data rates, more simple receivers such as RAKE are adequate. However, uplink performance can be significantly improved if receivers are introduced that are capable of performing interference cancellation between the signals received from non-orthogonal users, in effect increasing the orthogonality of the uplink.

There exists greater potential for increasing the number of base station receive antennas in the uplink than there is for UE Rx antennas in the downlink. Increasing the number of receive chains that can be coherently combined in baseband leads to substantial gains in uplink throughput.

#### 18.3.2.4 Transmitter performance

The HSPA release 11 specifications support two-antenna closed-loop transmit diversity and uplink two-branch MIMO from the UE. Because of the action of closed-loop power control, uplink Tx diversity does not influence much the Rx SINR and thus the proportion of RoT required for supporting UEs. However, Tx diversity can reduce the transmit power required, thus enabling power-limited UEs to achieve higher data rates and reducing inter-cell interference. Reducing the level of inter-cell interference then increases the proportion of RoT that can be allocated for data transmission and the uplink throughput and capacity.

#### 18.3.2.5 CDM and TDM scheduling

The fundamentally performance-limiting factor in the HSPA uplink is the non-orthogonality of users transmitting in the same cell. As discussed above, orthogonality can be improved by means of an interference canceling receiver. A complementary approach is to time domain multiplex users or groups of users using hybrid-ARQ process restrictions rather than scheduling users simultaneously. Time multiplexing users provides orthogonality between users but may also increase variance in the power control loops as a result of more rapidly changing inter-cell interference conditions as different users in different positions transmit.

#### 18.3.2.6 Power control and scheduling accuracy

Uplink operation in HSPA is a process of continuously managing overall interference by means of inner- and outer-loop power control and scheduling. Inner-loop power control aims to maintain the DPCCH receive level from each UE around the SINR target set by the outer-loop power control. If the interference level fluctuates rapidly, then the inner-loop power control will more frequently miss the target, causing either an excessive use of the RoT budget in the case of overshooting or additional hybrid-ARQ retransmissions or even lost transmissions in the case of undershooting. Frequent over- or undershooting of the SINR target by the inner-loop power control will cause the outer-loop power control to raise the SINR target in order to keep the number of lost transmissions to the quality target, which will in turn increase the long-term RoT usage of the UE. Frequent variations of the interference level will cause the scheduler to become less able to accurately manage the RoT, which will in turn exacerbate the interference fluctuations.

Thus uplink capacity and performance are improved by improving the stability of the interference management in the uplink. Since the interference level is impacted by the inner- and outer-power control loops and scheduling of all UEs in the system across multiple cells, there are multiple control loops involved in uplink interference management, and optimizing performance is a non-trivial task.

#### 18.3.2.7 Latency

Similarly to the downlink, TCP-based processes in the uplink are sensitive to round-trip time latency. The uplink latency impacts the downlink as well as the uplink, since the TCP performance depends on round-trip time latency.

#### 18.3.2.8 Radio resource management (RRM), scheduling, and mobility management

Similar considerations apply as in the case of the downlink. In the uplink, management of the soft handover active set is also required.

#### 18.3.2.9 Dual-cell HSUPA

Similarly to the downlink, operating two carriers can increase user throughput and provide system gains.

## 18.4 HSPA PERFORMANCE FIGURES

During 2008–2010, the ITU developed a set of requirements to be fulfilled by radio access technologies in order to fulfill the IMT-Advanced criteria and effectively be credible 4G technologies. Although no formal submission was made to ITU for HSPA, HSPA release 11 has been evaluated against the IMT-A criteria and passes the minimum requirements.

The IMT-A criteria are evaluated in four different types of propagation conditions: urban macro, urban micro, indoor, and rural. Urban macro-cells are characterized by a grid of three sector sites with an ISD of 500 m, aiming to provide continuous coverage to pedestrians and vehicular users. Urban micro-cells are characterized by a reduced site-to-site distance and are interference-limited cells that aim to provide capacity where there is an increased user and traffic density. Indoor cells are ceiling mounted low-power nodes that aim to provide high user throughput where there is a very high density of users and traffic. Rural cells are characterized with a higher inter-site distance and higher vehicular speeds.

The evaluation considered supported peak spectral efficiency, average user throughput, cell-edge user throughput, and control- and user-plane latency.

### 18.4.1 PEAK SPECTRAL EFFICIENCY

Peak spectral efficiency is the theoretical maximum throughput per Hertz of bandwidth assuming that the radio conditions are suitable to support the throughput. The peak throughput is a straightforward calculation of the maximum number of bits that can be packed into a TTI, normalized by time and bandwidth. The peak spectral efficiency is unlikely to be attained under real radio conditions and will not be sustained but is useful as a guide to the boundary conditions of the radio technology.

For release 11 HSPA, peak spectral efficiency is calculated from the main features as follows. Note that multi-carrier operation does not influence the peak spectral efficiency, since the spectral efficiency is bandwidth normalized.

### 18.4.2 DOWNLINK

15 (maximum number of HSDPA codes) $\times$ 3 $\times$ 2560/16 (number of symbols in an SF16 code per 3-slot TTI) $\times$ 6 (64-QAM) $\times$ 4 (4-stream MIMO) $\times$ 500 (500 TTIs per second)/$5\times10^6$ (5 MHz) = 17.28 bps/Hz.

### 18.4.3 UPLINK

2560 $\times$ 3 (chips per 3-slot TTI) $\times$ 1.5 (maximum code tree utilization, 2 $\times$ SF2 + 2 $\times$ SF4) $\times$ 3 (8-PAM) $\times$ 2 (dual-stream MIMO) $\times$ 500 (500 TTIs per second)/$5\times10^6$ (5 MHz) = 6.9 bps/Hz.

These numbers exceed the IMT-A requirement of 15 bps/Hz in downlink and 6.75 bps/Hz in uplink.

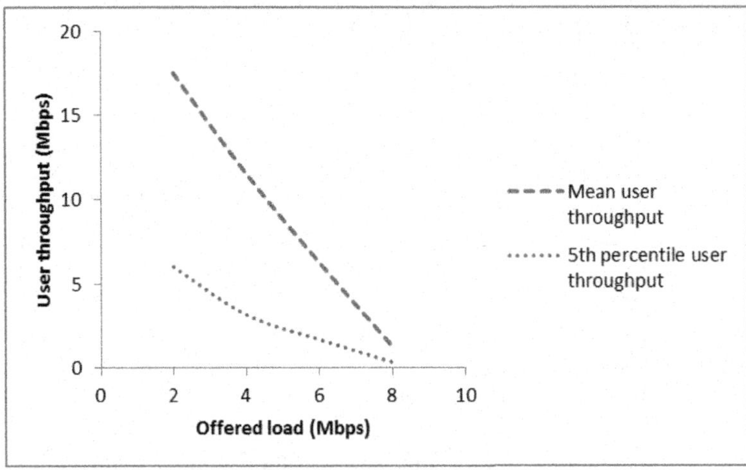


FIGURE 18.14 Mean and 5th percentile burst throughput for 1 × 2 HSDPA.

### 18.4.4 MEAN BURST THROUGHPUT

Mean burst throughput depends on the offered load level. Figure 18.14 shows the mean and the 5th percentile downlink burst throughput for single-carrier HSDPA with a 1 × 2 configuration (that is, one transmit and two receive antennas). An operator may define a minimum user throughput level to be met in the network. In this example, if a minimum (that is, 5th percentile) user throughput of 0.5 Mbps was to be set, then the capacity of the carrier is around 7.5 Mbps of total bursty traffic.

Downlink performance may be improved by means of multi-carrier and/or MIMO. Figure 18.15 depicts the user throughput achievable with multi-carrier. Increasing

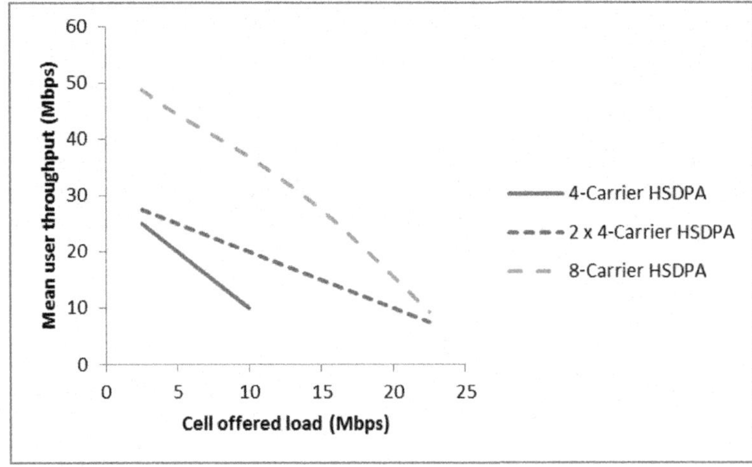


FIGURE 18.15 Multi-carrier HSDPA performance.

the number of carriers from four to eight increases the achievable user throughput at a given offered load level by a factor greater than the additional bandwidth. The reason for this is additional multi-carrier scheduling and trunking gain and also the fact that carrying bursts faster reduces the number of scheduler TTIs and hence the inter-cell interference, and a lower interference level increases link performance.

Comparing Figure 18.15 along the *x* axis, it can be seen that the capacity when multi-carrier (that is, eight-carrier aggregation at the UE, as opposed to transmitting eight carriers as two groups of four) is applied increases by a factor greater than the number of carriers. If a base station was to transmit multiple carriers without UE carrier aggregation, such that RRM had to allocate UEs to carriers by means of handover, then the capacity increase gained by transmitting *N* carriers would be limited to *N* and no individual users would experience increased throughput.

The benefits of additional transmit and receive antennas are shown in Figure 18.16. It can be seen that with two receiver antennas, introducing 2 × 2 MIMO significantly increases system capacity, provided that all of the UEs are 2 × 2 MIMO capable (which is an underlying assumption in this figure). Increasing the number of transmit antennas to four while maintaining two receive antennas gives a small gain.

Increasing the number of receive antennas to four yields a significant gain in user throughput and system capacity, although again it must be noted that the size of this gain is dependent on the proportion of four-antenna MIMO-capable UEs, and in these curves it is assumed that all UEs possess four receive antennas and are MIMO capable. With four receive antennas, increasing the number of transmit antennas from two to four yields a relatively larger gain increase than is the case when the UEs have two receive antennas. The larger gain arises because with four Rx antennas increased spatial multiplexing, with up to four parallel streams, can be achieved.

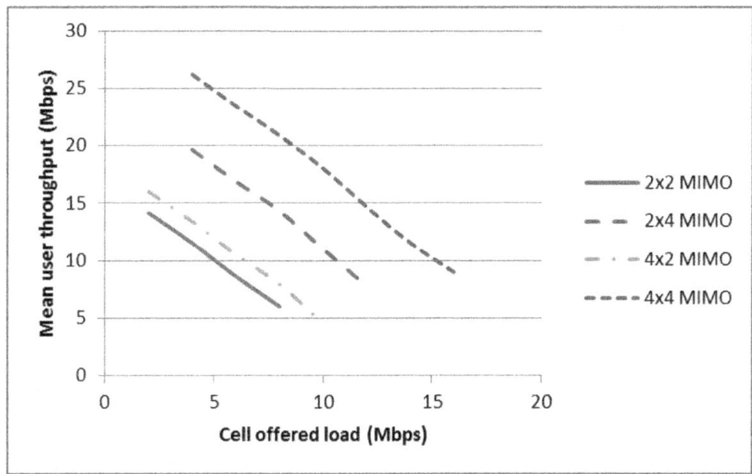


**FIGURE 18.16 HSDPA MIMO performance.**

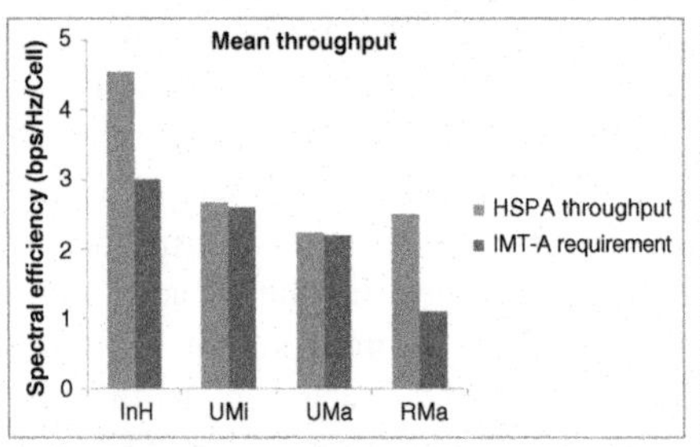


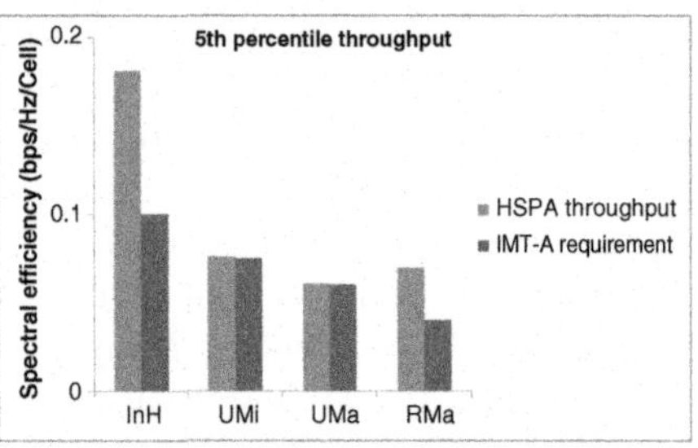


**FIGURE 18.17 IMT-A HSDPA performance.**

InH – Indoor Hotspot; UMi – Urban Micro; UMa – Urban Macro; RMa – Rural Macro

Figure 18.17 depicts the mean and cell-edge spectral efficiency compared to the IMT-A requirement in each of the scenarios investigated for IMT-A. Spectral efficiency in this case is the packet throughput normalized by the system bandwidth. The left figure compares mean throughput and the right figure 5th percentile throughput. In these figures, release 10 HSDPA is applied; that is, 2 × 2 MIMO + 64-QAM, four-carrier HSDPA. It can be observed that already with release 10, HSDPA exceeds 4G spectral efficiency requirements.

HSUPA achievable user throughputs in a macro environment versus offered load are shown in Figure 18.18.

Figure 18.19 compares HSUPA mean and cell-edge user throughput with the IMT-A requirement. The uplink in this case is utilizing 16-QAM higher-order modulation and eight receive antennas. As with the downlink, it can be observed that release 10 HSUPA already significantly exceeds the requirements for 4G data rates.

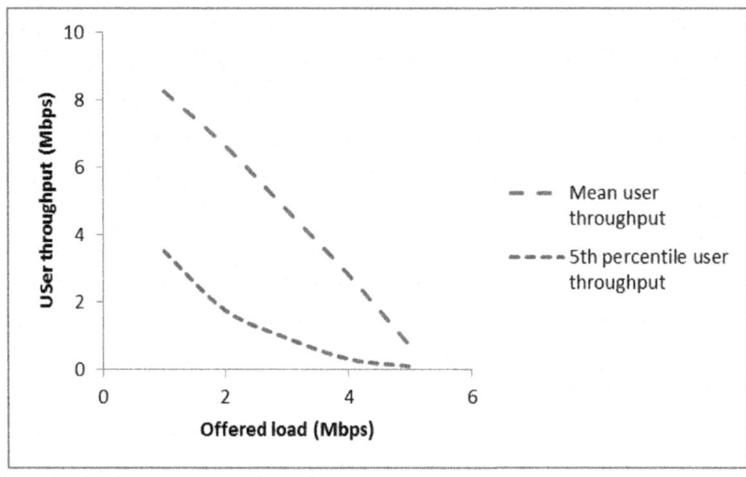


**FIGURE 18.18 HSUPA system performance.**

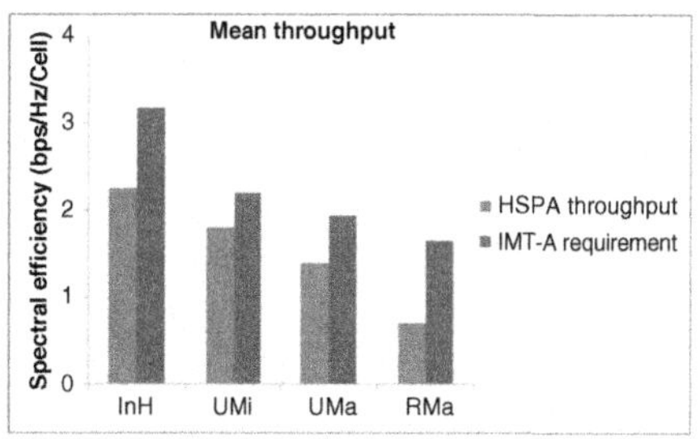


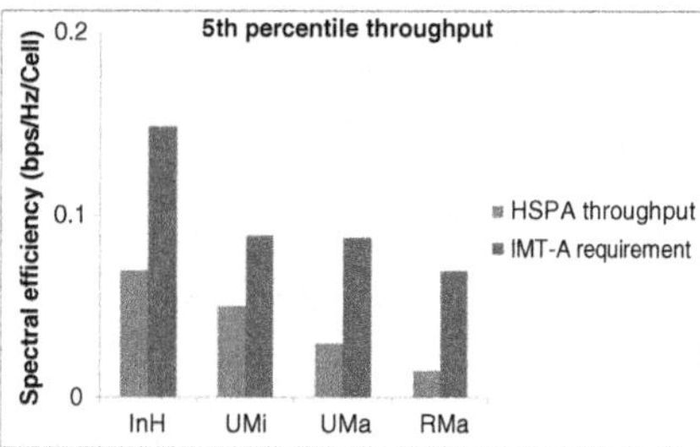


**FIGURE 18.19 HSUPA cell-edge and 5th percentile throughput.**

InH – Indoor Hotspot; UMi – Urban Micro; UMa – Urban Macro; RMa – Rural Macro

### 18.4.5 LATENCY

IMT-A also sets requirements on user- and control-plane latency. User-plane latency is the lowest time required for transferring a PDU across the physical layer. The lowest time is achieved when the hybrid-ARQ operating point is set such that no hybrid-ARQ retransmissions are required. In this case, HSPA in both uplink and downlink satisfies the IMT-A requirement of 10 ms.

Control-plane latency is the time required for call setup. IMT-A requires control plane latency to be less than 100 ms; the HSPA RAN satisfies this requirement.

## REFERENCES

[1] 3GPP TR 101.112; Selection Procedures for the Choice of Radio Transmission Technologies of the UMTS; UMTS 30.03 Version 3.2.0.

[2] Digital Mobile Radio Towards Future Generation Systems; COST 231 Final Report; CEST.

[3] 3GPP TR 25.996; Spatial Channel Model for Multiple Input Multiple Output (MIMO) Simulations.

[4] G. Calcev, D. Chizhik, B. Goransson, et al. A wideband spatial channel model for system-wide simulations, IEEE Vehicular Technol (March 2007). (Volume 56, Issue: 2). Page(s): 389–403

## FURTHER READING

M. Jurvansuu, J. Prokkola, M. Hanski, P. Perala, HSDPA performance in live networks. IEEE ICC 07, pp. 467–471.

M. Maternia, M. Januszewski, K. Ranta-Aho, J. Wigard, A. Bohdanowicz, M. Marzynski, M. Wcislo, On performance of long term HSPA evolution: towards meeting IMT-Advanced requirements. IEEE ICC 2012, pp. 6040–6044.

J. Peisa, H. Ekstrom, H. Hannu, S. Parkvall, End-to-end performance of WCDMA enhanced uplink. IEEE VTC 05, pp. 1432–1436, Vol. 3.

CHAPTER

# Release 12 and Beyond

# 19

## CHAPTER OUTLINE



HSPA continues to evolve and will do so for many years to come. To address the requirements of changing user behavior, new techniques are developed and become standardized, thereby providing operators flexibility, capacity, and coverage needed for the connected world of the future.


**HSPA Evolution: The Fundamentals for Mobile Broadband. DOI: 10.1016/B978-0-08-099969-2.00019-3**

Since release 99, the 3GPP WCDMA standard has undergone continuous development. Today, WCDMA/HSPA is a best-in-class voice solution with exceptional voice accessibility and retainability. In addition, HSPA provides an excellent access technology for mobile broadband, as it delivers high data rates and high cell-edge throughput – all of which enable good user experience across the entire network. The same driving forces to further enhance performance and capabilities continue to exist, but the targets move further ahead as the technology state of the art improves.

This chapter provides an overview of some of the main features standardized during 3GPP release 12 and also describes areas considered for release 13. In addition to these features, some other aspects of 3GPP work, such as home Node B and development of enablers for base station technologies (for example, AAS and MSR), will also play a role in the evolution of HSPA. Because much of this release 13 work is still being discussed in 3GPP at the time of writing, the provided overview should be seen as indicative.

## 19.1 HETEROGENEOUS NETWORKS

As discussed in Chapter 17, deployment of heterogeneous networks has attracted significant attention in recent years as a means to increase capacity, coverage, and end-user performance. It should, however, be emphasized that a heterogeneous network is a deployment strategy and not a new technology component. Deployment of heterogeneous networks has been possible from the very first releases of WCDMA/HSPA. For example, functionality offered by soft-handover facilitates efficient deployments of heterogeneous networks in WCDMA/HSPA. Nevertheless, the growth of heterogeneous networks deployments has revealed some new challenges that need to be addressed to improve the efficiency of heterogeneous networks operation. Two main areas that attracted a lot of attention during release 12 discussions [1]-[3] are uplink/downlink imbalance issues and mobility enhancements. In release 12, two *work items*[1] (WIs), both addressing improvements in so-called co-channel heterogeneous network deployments, were completed. One targeted mobility enhancements [1] and the other addressed various improvements related to uplink/downlink imbalance problems [2]. The co-channel deployment scenario, also referred to as separate cell, is one in which *low-power nodes* (LPNs) are deployed within the macro-cell coverage area, where the macro and the LPNs share the same frequency but have different cell IDs such that each LPN creates its own separate cell (see Figure 19.1).

As described in Chapter 17, one of the main challenges in a co-channel deployment is caused by the uplink/downlink imbalance. The transmit power of the common pilots is greater for macro nodes than for LPNs. Hence, the equal downlink receive power border and the equal uplink pathloss border do not coincide, giving rise to the so-called *imbalance region* (see Figure 19.2). The uplink/downlink imbalance can cause two main uplink problems: one relates to excessive interference

[1]See Chapter 1 for a discussion about the ways of working in 3GPP.

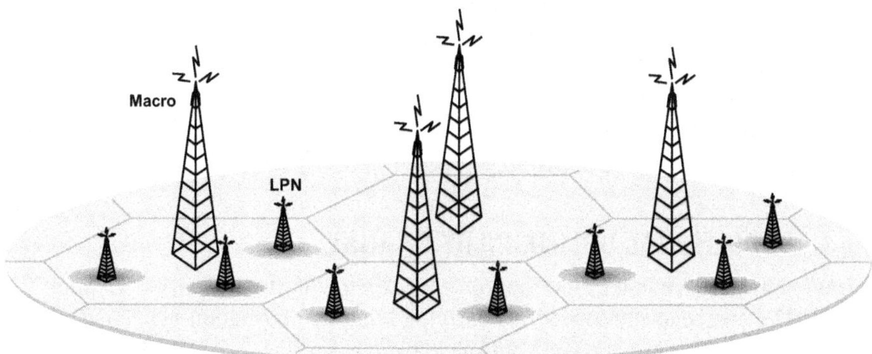


**FIGURE 19.1 So-called co-channel type of heterogeneous networks deployment, where each node creates its own cell.**

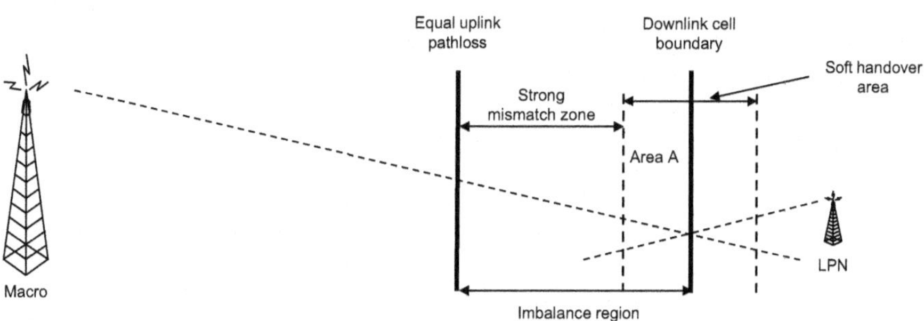


**FIGURE 19.2 Imbalance region in a heterogeneous network deployment.**

being created toward an LPN, and the second relates to insufficient receive SINR at a macro node:

- A user in the so-called *strong mismatch zone* (see Figure 19.2) has the macro node as its serving cell, a smaller pathloss toward the LPN, and is not yet in soft handover. Hence, this UE will create excessive interference toward the LPN, without the LPN having the possibility to affect the transmit power of the UE. It should be noted, however, that with proper network settings, problems associated with the strong mismatch zone can be mitigated.
- The second problem occurs when a user in Area A in Figure 19.2 has the macro as its serving cell, a much smaller pathloss toward the LPN, and is in soft handover with both the macro and the LPN in its active set. In this scenario, the UE is essentially power controlled by the LPN and will transmit with a relatively low power because of the low pathloss toward the LPN. The uplink signal quality in the serving macro cell might then be too weak for successful decoding of uplink carried control information. The serving cell needs to receive downlink-related hybrid-ARQ ACK/NACK information and

CQI/PCI information carried on the HS-DPCCH, as well as uplink scheduling information. Scheduling information can be either out-band via E-DPCCH or in-band via E-DPDCH.

The next subsections will describe the release 12 standardized solutions for addressing the problems discussed above.

### 19.1.1 ADDITIONAL UPLINK PILOT CHANNEL

A new physical channel, the uplink *Dedicated Physical Control Channel 2* (DPCCH2), has been introduced for carrying control information generated at L1. The DPCCH2 is solely power controlled by the serving HS-DSCH cell. When DPCCH2 is configured, then HS-DPCCH uses DPCCH2 as its reference, that is, the power of HS-DPCCH is set relative to DPCCH2 and pilot symbols in DPCCH2 are used for channel estimation and coherent detection of HS-DPCCH at the serving HS-DSCH cell. Hence, two independent fast power control loops are maintained, and unlike other UL physical channels, the power of DPCCH2 and HS-DPCCH are set "independently" of the DPCCH in this case.

The support of DPCCH2 is optional for the UE, and if enabled there can only be one DPCCH2 configured and only on the primary uplink frequency. The DPCCH2 slot format consists of eight pilot bits to support channel estimation for coherent detection, and two TPC commands. Spreading of the DPCCH2 for the cases with or without DPDCH configured is shown in Figure 19.3. Since the DPCCH2 is power controlled, it follows that $\beta_{c2}$ is always set to 1.0. The DPCCH2 is only transmitted in slots where DPCCH is transmitted. Thus, if DPCCH is DTXed, then so is DPCCH2.

If DPCCH2 is configured, then the HS-DPCCH(s) can only exist together with an uplink DPCCH2. The intention is to ensure that downlink scheduling information (CQI/PCI/rank and hybrid-ARQ) is reliably received at the serving cell. However, uplink scheduling information (happy bit, buffer status, etc.) is also required in the serving cell. One solution to this would be to configure E-DCH decoupling (see Section 19.1.2). Another solution would be to increase the power offset used for E-DPCCH and use DPCCH2 for channel estimation and coherent detection. This would increase the likelihood to reliably receive the happy bit in the serving cell.

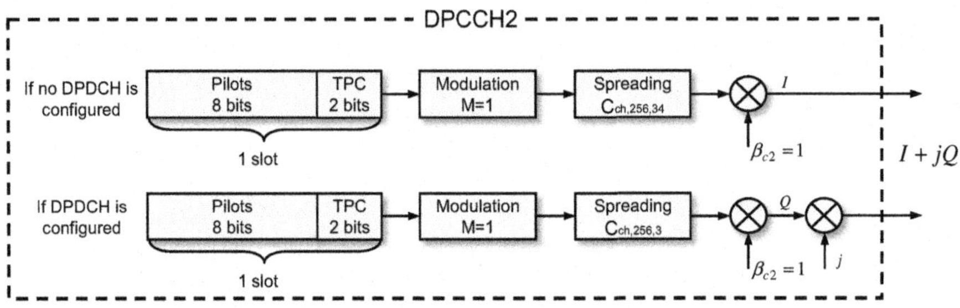


**FIGURE 19.3 Spreading for DPCCH2.**

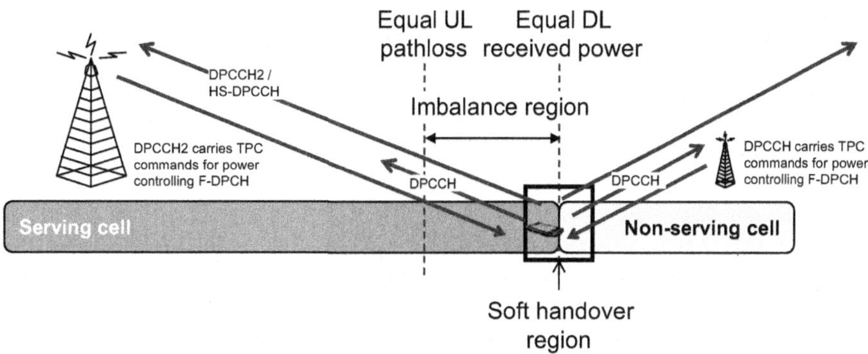


**FIGURE 19.4 DPCCH2 operation.**

To facilitate this operation, the E-DPCCH beta value range has been increased for DPCCH2 operation. One thing to notice is that DPCCH and DPCCH2 experience the same radio channel. Hence, it seems reasonable that any or both of these channels can be used for channel estimation and coherent detection of any of the other channels. In many respects this is true, but some observations need to be taken into account. First of all, in scenarios with severe imbalance (see Figure 19.4), the DPCCH will have much lower transmit power than DPCCH2. Hence, the quality of DPCCH may be inadequate. Also, since DPCCH and DPCCH2 are independently power controlled there will be a mismatch if, for example, DPCCH2 is used for channel estimation and coherent detection of a physical channel that uses DPCCH as its reference channel. This mismatch matters mainly if time-averaging is used for channel estimation, and higher-order modulation is employed by the physical channel (for example, E-DPDCH).

The initial uplink DPCCH2 transmit power is given by the DPCCH power plus an additional offset that is signaled by higher layers. After initialization, the HS-DSCH serving cell in the active set monitors the DPCCH2 quality and provides TPC commands to continuously adjust the uplink DPCCH2 transmit power in steps of 1 dB. The TPC commands used to control the DPCCH2 transmit power are carried by an additional F-DPCH resource. The slot format for the F-DPCHes associated with DPCCH (if applicable) and DPCCH2, respectively, must be different, while the timing is the same.

The main purpose of the DPCCH2 is to ensure reliable reception of HS-DPCCH in the serving HS-DSCH cell. Another benefit of having two independent radio-link loops between the serving HS-DSCH cell and the UE is that two DL power control commands can be issued by the UE to independently control the DL quality of the serving HS-DSCH cell and another cell in the active set. This is attractive in scenarios with link imbalances, for example, when the DL pathloss is significantly less between the serving HS-DSCH cell and the UE compared to between the UE and other cells in the active set. Hence, when DPCCH2 is configured, the power control loop associated with DPCCH2 adjusts the F-DPCH power at the serving HS-DSCH cell, and the power control loop associated with DPCCH adjusts the F-DPCH power at the so-called *designated non-serving HS-DSCH cell*, which is set by higher-layer signaling. That is, the UE generates TPC commands to control the transmit power from the serving HS-DSCH cell and send them in the TPC field of the uplink DPCCH2, and in addition, the UE generates TPC commands to control the transmit power from the designated non-serving HS-DSCH cell and send them in the TPC field of the uplink DPCCH. If E-DCH decoupling is configured, then the designated non-serving HS-DSCH cell must be the E-DCH serving cell.

For the serving HS-DSCH cell, both the F-DPCH associated with DPCCH2 and the F-DPCH associated with DPCCH are transmitted with the same power dictated by the TPC commands received on DPCCH2. One exception is if DPCH is configured, in which the downlink power of F-DPCH associated with DPCCH2 for the serving HS-DSCH cell is adjusted based on TPC commands from the DPCCH2, and the downlink power of DPCCH/DPDCH is adjusted based on the TPC commands received on the DPCCH.

Unless DPCH is configured, a single DL quality target by means of a downlink TPC command error rate target for the F-DPCHs is set by higher-layer signaling.

Power balancing is typically employed to ensure that the DL transmission powers from different cells in the active set do not deviate too much. However, with DPCCH2 configured two independent power control loops between links with potentially large imbalances are provided. Hence, when DPCCH2 is configured, power balancing between the serving HS-DSCH cell and the non-serving HS-DSCH cells should be avoided, while power balancing between the different non-serving HS-DSCH cells may still be beneficial.

### 19.1.2 E-DCH DECOUPLING

The introduction of the *E-DCH decoupling* feature provides a separation of the downlink and uplink serving cells. The intention is to keep the LPN as the serving E-DCH (uplink) cell at the same time as the macro is the serving HS-DSCH (downlink) cell for users in the imbalance zone. This provides an uplink performance gain attributed to macro off-loading and because the *rise-over-thermal* (RoT) headroom in the LPN can be utilized more efficiently than the macro headroom because of lower required uplink transmission power. If E-DCH decoupling is applied, the problem of reliably receiving uplink control information in the serving E-DCH cell is alleviated, while the corresponding downlink problem (receiving HS-DPCCH in the serving HS-DSCH cell) still exists. Also, the cost for sending downlink-related information from the LPN, for example E-AGCH, increases because of potentially strong downlink macro interference.

E-DCH decoupling is an optional release 12 feature allowing decoupling of the E-DCH and HS-DSCH serving cells. From a specifications point of view, the impact of E-DCH decoupling is rather limited. A new UE capability indicates whether the UE supports E-DCH decoupling. Initiation and termination of E-DCH decoupling is handled by higher-layer signaling, and existing signaling procedures for serving cell change are reused. E-DCH decoupling is only applicable when E-DCH and HS-DSCH serving cells are configured.

The L1 definition of the E-DCH serving cell is the cell from which the UE receives absolute grants from the Node-B scheduler. Hence, with E-DCH decoupling configured, the E-AGCH is only sent from the E-DCH serving cell and not from the HS-DSCH serving cell.

E-DCH decoupling also affects the operation of UL MIMO. More specifically, the E-ROCH is sent from the E-DCH serving cell, and higher layers indicate to the UE which cells in the active set should transmit the F-TPICH, with the restriction that either only the E-DCH serving cell or the HS-DSCH serving cell transmits the F-TPICH or all cells from the serving radio link set transmit the F-TPICH.

Another L1 impact of decoupling the E-DCH and HS-DSCH serving cells is that some features will require coordination between these cells via Iub signaling. For example, the E-DCH serving cell needs to inform the HS-DSCH serving cell if an HS-SCCH order should be issued based on a decision in the EUL scheduler. This applies, more specifically, to orders for CLTD activation state changes and CPC orders for activating/deactivating DRX and DTX. Furthermore, communication between the serving cells is needed to ensure that existing multi-carrier configuration rules are satisfied. That is, if all secondary HS-DSCH serving cells are deactivated, then the secondary E-DCH frequency also needs to be deactivated.

In scenarios where E-DCH decoupling is beneficial, it typically holds that the downlink quality from the serving E-DCH cell is worse than the quality of the downlink from the serving HS-DSCH cell. Hence, in this case, downlink synchronization is based on the quality of the TPC fields of the F-DPCH frame received from the serving (or secondary serving) E-DCH cell instead of the serving HS-DSCH cell (see Figure 19.5). Special considerations apply when DPCCH2 is configured.

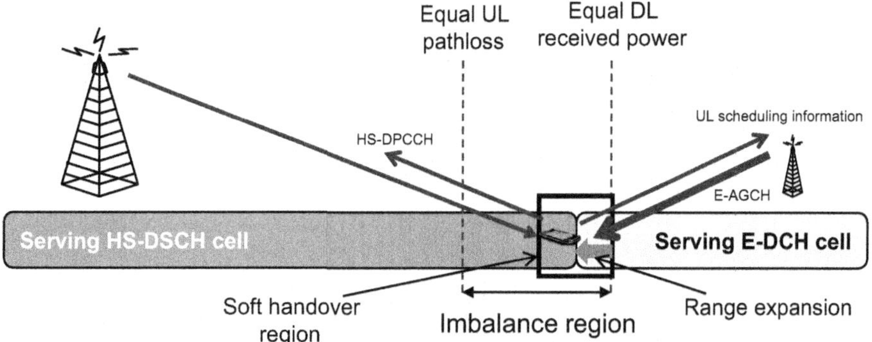


**FIGURE 19.5 E-DCH decoupling with CIO-based range expansion. The downlink from the serving E-DCH cell needs to overcome strong DL interference from the serving HS-DSCH cell. The uplink toward the serving HS-DSCH cell is potentially weak, affecting HS-DPCCH reception quality.**

### 19.1.3 RADIO LINKS WITHOUT DPCH/F-DPCH

Release 12 introduces the possibility to set up radio links without an associated DPCH or F-DPCH. More specifically, the network can configure a subset of the non-serving E-DCH radio links in the UE's active set to operate without an F-DPCH or DPCH. This means that the UE is only receiving E-HICH and optionally E-RGCH from these non-serving cells. When dual-cell E-DCH operation is configured, a radio link without DPCH/F-DPCH can be configured on both the primary and the secondary downlink frequencies. A new UE capability bit is used to indicate whether the feature is supported.

One problem with not sending the F-DPCH/DPCH is that the UE needs to know when to listen to E-HICH and possibly E-RGCH. Since the timing of these channels is related to the F-DPCH/DPCH timing, the (virtual) offset between P-CCPCH and the F-DPCH/DPCH needs to be signaled to the UE.

Clearly, the UE does not monitor radio links for which DPCH/F-DPCH is not transmitted, which needs to be taken into consideration in synch and TPC combining procedures.

The possibility to set up radio links without an F-DPCH/DPCH is an enabler for several modes of operation, such as

- *Early soft handover* (also referred to as *extended E-HICH*). The network configures an early Event 1a, that is, a low threshold for when to add a new cell to the UEs active set is used. This cell will then be configured without an F-DPCH/DPCH and the UE can enjoy macro diversity gains without being unreliably power controlled by the early added cell. A typical scenario where this can be beneficial is the strong mismatch zone where the UL is good toward a cell while the DL is weak.
- *Common grant (common RGCH).* As discussed before, one problem for the network with UEs in the strong mismatch zone is that these UEs can cause excessive interference toward LPNs, and the LPNs cannot affect the interference (grant of these UEs). Similar to that for early soft handover, by adding a cell early to a UE's active set, this cell can reduce the interference level by issuing relative grants using the E-RGCH, while still guaranteeing that the power control is unaffected.
- *Inner-loop power control restriction.* One approach to ensure that UL/DL control information is reliably received in the serving cell would be to let the serving cell alone power-control the UE. To guarantee robust operation of such an approach, the UE should preferably not receive TPC commands from non-serving cells.

Figure 19.6 provides one example of how the network could be operated to leverage the release 12 standardized features discussed above.

### 19.1.4 MOBILITY ENHANCEMENTS

Heterogeneous networks, in particular with massive deployments of small cells, lead to new mobility-related challenges compared to macro-only scenarios. For example, discovery and management of a large number of cells are needed and more frequent

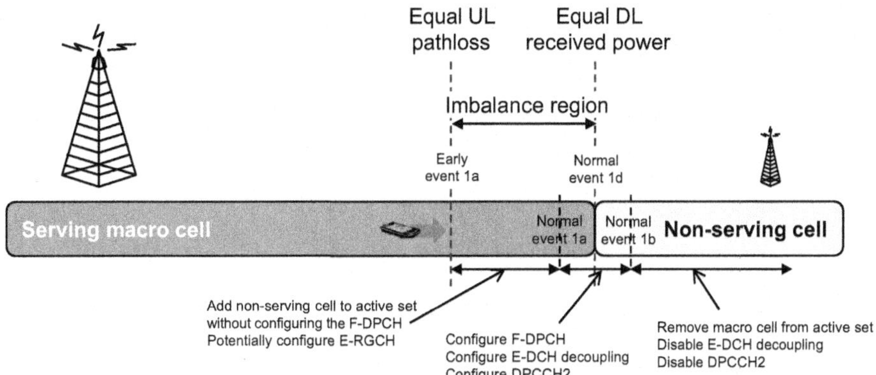


**FIGURE 19.6 A heterogeneous network operation using release 12 standardized features.**

handovers or cell reselections occur. The work item on mobility enhancements in heterogeneous networks [1] consisted of three main topics: *small cell discovery and identification*, *high speed UEs*, and *massive deployment of small cells*. Small cell discovery based on inter-frequency measurements in cases when LPNs are deployed on a different frequency than the macro cells becomes costly in terms of UE battery power, and proximity detection schemes were investigated. High-speed UEs (above 60 km/h) may pose a challenge for dense small cell deployments, where fast-moving UEs can experience frequent handovers and intense mobility signaling. Massive deployment of small cells introduces a number of potential problems, such as *Primary Scrambling Code* (PSC) confusion, and limitations in the *Neighbor Cell List* (NCL).

The following with respect to enhanced mobility handling was standardized in release 12:

- *Massive deployment of small cells.* The size of the inter-frequency NCL is extended from max 32 to max 64 for CELL_DCH, CELL_FACH, CELL_PCH, and idle states, with the intention to enhance the UE's ability to detect cells in scenarios with massive deployment of small cells. The intra-frequency NCL is, however, not affected.
  The intra-frequency NCL extension does not introduce any new UE capabilities or impact any RAN4 specifications, but there is related work ongoing in the WI [8] that will further impact the NCL. It was further concluded that potential problems with PSC confusion can be sufficiently mitigated with existing mechanisms, such as cell planning and *Automatic Neighbor Relation* (ANR) functionality.
- *Mobility enhancements.* Several mobility enhancements were discussed during release 12, but in the end it was concluded that the existing standard provides sufficient tools to handle mobility issues efficiently. The only standardized enhancements were an optional enhancement of the eSCC mechanism for Event 1c and a decision to make the support of the extended measurement IDs up to 32 mandatory.
- *Further mobility enhancements.* It was agreed to introduce a new Event 2g for inter-frequency measurements as an optional feature.

## 19.2 HSUPA ENHANCEMENTS

A work item, referred to as further *enhanced uplink* (EUL) enhancements, targeting technical solutions for increasing the uplink capacity, coverage, and end-user experience was started in release 12. In the following subsections, a short description of the four main areas within this WI is given.

### 19.2.1 HSUPA COVERAGE IMPROVEMENTS

The E-DCH 2 ms TTI offers higher peak rate and lower latency compared to the 10 ms TTI. For various *mobile broadband* (MBB) applications, the HSUPA 2 ms TTI is therefore essential for a good end-user experience and is generally preferred over the 10 ms TTI. The 10 ms TTI provides, however, better cell-edge coverage because of the longer frame duration. In real networks, it is therefore necessary to switch between the 2 and 10 ms TTIs in order to maximize coverage and end-user performance. There are, however, some issues with the TTI switch mechanism; the triggers are non-optimal and the switching procedures are slow, resulting in relatively large safety margins and thereby additional loss of 2 ms TTI coverage. In release 12, various enhancements were proposed, including both improvements to the coverage measurements used for deciding when to switch the TTI length and mechanisms to perform faster switching.

A new L1 HS-SCCH order has been introduced to inform the UE to perform a TTI switch (either from 10 ms TTI to 2 ms TTI or vice versa). It was also agreed to introduce a network-configured activation time that determines when the UE shall be able to start transmitting E-DCH data using the new E-DCH TTI length. The activation time is measured from the beginning of the first E-DCH radio frame after the end of the HS-SCCH subframe containing the order. The UE reconfiguration process begins after the activation time and lasts for 20 ms, after which the UE starts transmitting E-DCH data using the new TTI length. The E-DCH is not transmitted during the reconfiguration process, while other transmissions, that is, uplink DPCCH/DPDCH/HS-DPCCH and downlink HS-DSCH/DPCH, are unaffected by the TTI switching procedure. Figure 19.7 illustrates the L1-initiated TTI switching procedure.

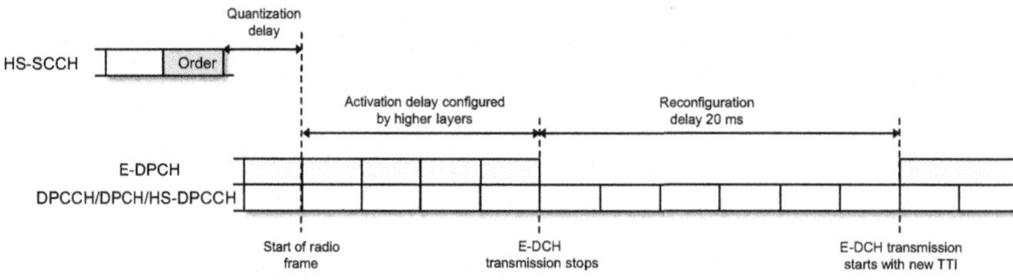


**FIGURE 19.7 L1-initiated TTI switch.**

### 19.2.2 IMPROVEMENTS TO CURRENT ACCESS CONTROL MECHANISM

Smartphone users want to be able to rapidly access the network whenever they need to. Maintaining a device in a connected-mode state, such as CELL_FACH, CELL_PCH, or URA_PCH, for as long as possible is one way of achieving this—access from these states is much faster than from the IDLE state. In recent releases, connected-mode states have been made more efficient from a battery and resource point of view, making it feasible to maintain inactive devices in these states for longer.

As the number of smartphone users increases, networks need flexible mechanisms to maintain high system throughput even during periods of extremely heavy load. Allowing the network to limit the number of concurrently active users as well as the number of random accesses is one such mechanism. Improvements in release 12 include enabling admission decisions that are differentiated on traffic type and enhanced control over the number of users simultaneously accessing the network.

### 19.2.3 INTRODUCTION OF ENHANCEMENTS FOR ENABLING HIGH USER BIT RATES IN SINGLE- AND MULTI-CARRIER UPLINK MIXED-TRAFFIC SCENARIOS

In release 9, dual-cell HSUPA operation was standardized, allowing two carriers, a primary and a secondary, to be assigned to a user. The user's traffic could thereby be flexibly allocated between the carriers, and the achievable peak rate was doubled. In release 12, various single and dual UL carrier enhancements for enabling high user bit rates in mixed-traffic scenarios have been introduced.

The maximum allowed uplink interference level in a cell, also known as the maximum *rise-over-thermal* (RoT), is an important quantity relating to the peak rate that can be achieved in the cell. The RoT level that an operator can allow is determined by a number of factors, such as the density of the network, the capability of the network to handle interference (for example, with advanced interference suppression or cancelation techniques), and the capabilities of the devices in the network, including both smartphones and legacy feature phones. Typically, macro cells are dimensioned with an average RoT of around 7 dB, enabling uplink data rates in the order of 5 Mbps, while also securing voice and data coverage for cell-edge users. Higher data rates, such as 11 Mbps (available since release 7) and 34 Mbps (available since release 11) require RoT levels greater than 10 and 20 dB, respectively.

To increase network energy efficiency and UE battery life, the cost in system and UE resources to stay in a connected state should be as low as possible. It would also be desirable to increase the ability to deploy carriers flexibly according to operators' traffic and deployment scenarios. In this context, one of the enhancements considered in release 12 was optimization of the secondary uplink carrier for fast and flexible handling of multiple high rate data users, through faster and more efficient granting and lower cost per bit. The design target was to allow multiple bursty data users in a cell to transmit at the highest rates on the secondary carrier without causing any uplink interference to each other or to legacy users.

Additional enhancements included improved RoT management and longer DTX/DRX cycles. Enabling the maximum RoT on a user's enhanced secondary carrier to be configured to support any available uplink data rate, while the maximum RoT on a user's primary carrier is configured to secure cell-edge coverage for signaling, random access, and legacy (voice) users[1], will improve RoT management.

3GPP agreed to include the following features as part of release 12:

- *DTX enhancements.* Three sub-features are encompassed under DTX enhancements:
    - *DTX enhancements on the secondary carrier*. In release 9 DC-HSUPA, the primary and the secondary uplink frequencies operate with identical DTX parameters (that is, DTX cycle lengths, inactivity thresholds, etc.). In release 12, the possibility to decouple the DTX parameters between the primary and the secondary carriers has been introduced. The intention is to provide more efficient support for different traffic characteristics on each of the carriers. Moreover, in order to further reduce the uplink interference on the secondary carrier, the set of configurable values corresponding to *DTX cycle 2* has been extended up to 1280 subframes (corresponding to 2.56 seconds). This provides the opportunity to operate the secondary uplink frequency virtually in an ideal gating mode[2]. Note, though, that the DTX cycle extension is only applicable to the secondary carrier.
    - *Improved power control after a DTX gap*. In CPC, when there is a gap in transmission due to DTX, the power of the uplink DPCCH after the transmission gap is set to the value used in the last slot before the transmission gap. This is a good solution as long as the DTX gap length is shorter than the coherence time of the wireless channel (as an example, the coherence time of a UE moving at 3 km/h is roughly 45 ms). If, on the other hand, the DTX gap is long relative to the coherence time, then other solutions might perform better. One example is the release 12 DC-HSUPA scenario where the secondary carrier is configured with long DTX cycles while the primary carrier has smaller DTX cycles. In this case, it is better to infer information from the primary carrier when setting the initial power of the secondary carrier after a DTX gap. Even though the fast fading typically is independent between the different carriers, there is often correlation between the long-term channel characteristics. Also, some differences between the carriers, for example, RoT characteristics and carrier frequency differences, can be taken into account by means of network configurations. More specifically, for DC-HSUPA in release 12, the initial power of the secondary carrier can be set based on a filtered version of the power of the primary carrier, where the time constant of the filter is set by the network. A gap-length threshold is also introduced in order to switch between the release 7 solution and the averaging filter approach. This preserves the advantages of the legacy solution for gaps smaller than the coherence time. Moreover, an offset parameter has also been incorporated as a means

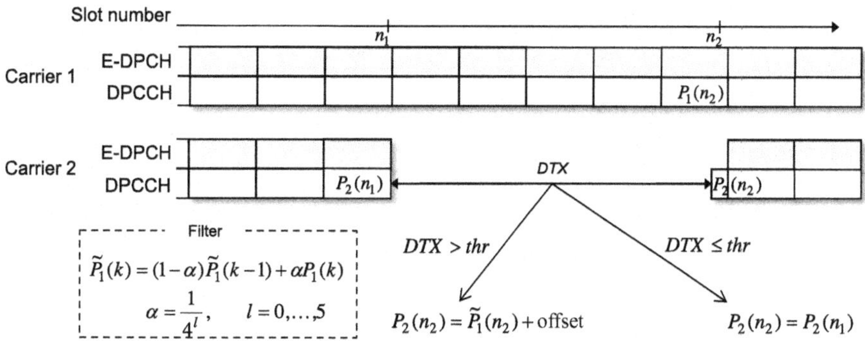


**FIGURE 19.8 Process of setting the power of the secondary carrier after a transmission gap.**

to compensate for differences between the carriers, for example, diverse traffic behaviors. The initial power setting of the secondary carrier after a transmission gap is illustrated in Figure 19.8.

- *Removal of post-verification period of synchronization procedure A*. The post-verification period of the synchronization procedure A allows the UE to start its transmission before the downlink dedicated channel is considered established by higher layers. A potential failure of the post-verification period is, however, likely to occur when transmissions are highly bursty, resulting in long inactivity periods—the reason being that the criteria for reporting in-sync will fail whenever the number of TTIs of the initial transmission is shorter than the minimum amount of time (here 40 ms) used in phase I of the downlink synchronization status evaluation procedure. In release 12, it is therefore possible to initiate the synchronization procedure A directly from phase II of the downlink synchronization status evaluation procedure, having as initial conditions an automatic report of in-sync and the ability to be able to transmit E-DCH data immediately. The above solution preserves the benefits brought by the post-verification process but at the same time avoids a potential failure that may occur due to potential bursty data traffic.

- *DRX enhancements*. An *Inactivity Threshold for UE DRX cycle 2* has been introduced in release 12 along with *UE DRX cycle 2* to allow further UE battery savings. The second DRX cycle is composed of a set of configurable values that extend up to 80 sub-frames (before release 12, the maximum value of the *DRX cycle* was 20 sub-frames). The DRX enhancements not only bring benefits to single-carrier operation but also to multi-carrier transmission. Instead of entering the *UE DRX cycle 2*, secondary DL carriers are autonomously deactivated once the configurable value of the *Inactivity Threshold for UE DRX cycle 2* is reached, and reactivation of secondary DL carriers is performed in the conventional way by means of HS-SCCH orders. When DC-HSUPA is configured and the secondary E-DCH carrier is activated, the secondary HS-DSCH carrier associated with the secondary E-DCH carrier cannot

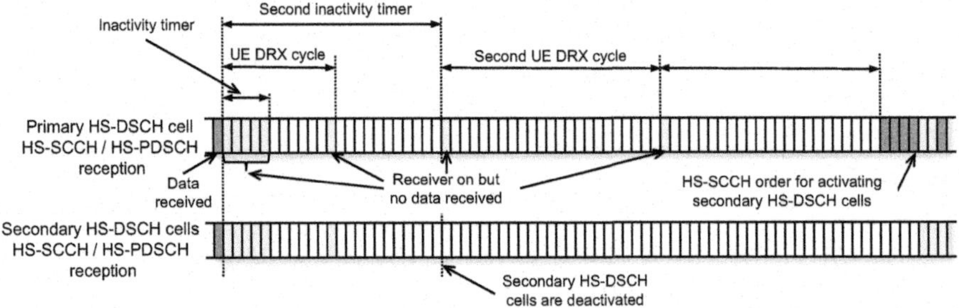


**FIGURE 19.9 Enhanced DRX mechanism.**

be deactivated and will instead use the secondary DRX cycle. Figure 19.9 illustrates the enhanced DRX mechanism.

- *Implicit Grant Handling*. To enhance the *time division multiplexing* (TDM) operation, a scheduling algorithm referred to as *Grant Detection* has been introduced. In this case, all hybrid-ARQ processes are used and the UE is allowed to transmit E-DCH data if and only if the UE-specific CRC of the E-AGCH carrying the grant message is successfully decoded (that is, the CRC checks with its E-RNTI); otherwise the UE cannot transmit any data. If there is an ongoing transmission, the UE must stop it immediately in order to avoid interfering with the newly granted UE. This TDM scheduling algorithm eliminates the gaps between data transmissions of different granted UEs and reduces significantly the E-AGCH load since no signaling for deactivation is needed. This operation is particularly useful for bursty traffic scenarios.

## 19.2.4 UPLINK CONTROL CHANNEL OVERHEAD REDUCTION FOR MULTI-RAB OPERATION

In release 7, the introduction of CPC allowed a reduction of the uplink control channel overhead by enabling discontinuous transmission of the DPCCH channel when a low amount or no data are transmitted. A reduction in control channel overhead results in increased capacity for data on the uplink. In release 12, one of the considered scenarios is when power-limited UEs in multi-RAB (that is, simultaneously transmitting multiple services such as voice and data) do not have sufficient uplink transmit power to keep a desired level of uplink services. The UE would then benefit from reducing the CQI reporting if there are no active downlink data transmissions. The limited uplink transmit power can thereby be more efficiently used for the DPCCH and DPDCH, which, in turn, leads to improved coverage. A second, longer CQI feedback cycle has therefore been introduced to which the UE will revert after a configured number of HS-SCCH TTIs of no transmission to the UE. Normal CQI feedback cycle operation resumes again when an HS-SCCH with consistent control channel information is detected by the UE. The second CQI feedback cycle can also

**Table 19.1** Summary of capabilities related to HSUPA enhancements

| Feature | Support | Capability signaling | Feature dependences |
|---|---|---|---|
| Access control in connected mode | Optional | Yes | |
| Enhanced EUL TTI switching | Optional | Yes | |
| Enhanced downlink DRX | Optional | Yes | |
| Implicit grant handling | Optional | Yes | |
| Enhanced secondary uplink carrier operation (DTX enhancements, power control enhancements, postverification Sync A) | Optional | Yes | Requires implicit grant handling |
| HS-DPCCH overhead reduction for multi-RAB operation | Optional | Yes | |
| Filtered UPH reporting | Optional | Yes | |

be configured to be equal to the normal CQI feedback cycle. To improve coverage, when the UE hits the maximum power limit and no other channels but DPCCH, DPDCH, and HS-DPCCH have available power to transmit, the UE transmits CQI = 0 in subframes where CQI is required to be transmitted. The UE can thereby indicate to the Node B that it is power limited and that the Node B could stop scheduling data on the downlink. Because of power limitation, power scaling must also be carried out. Slots containing CQI are seen as less important than those containing hybrid-ARQ ACK; hence, slots corresponding to CQI are scaled while slots corresponding to hybrid-ARQ ACK are not. DPCCH, DPDCH, and HS-DPCCH are scaled equally for HS-DPCCH subframes where hybrid-ARQ ACK is transmitted. The new scaling behavior and criteria for deriving CQI = 0 are configurable by higher layer signaling.

A number of UE capabilities are introduced to indicate whether the features described above are supported; see Table 19.1 for a summary.

## 19.3 DCH ENHANCEMENTS

The *Dedicated Channel* (DCH) is still useful for some types of services, in particular voice calls. To enhance the link efficiency of DCH traffic, a work item was opened in release 12 on *DCH Enhancements* [5]. During the preceding study item phase, it was shown [6] that optimizing the DCH efficiency may not only lead to higher capacity for *circuit-switched* (CS) traffic but also improved capacity for *packet-switched* (PS) traffic as more resources become available for HSPA. Some of the optimizations also provided data throughput gains for scenarios involving a mixture of voice (CS voice over dedicated DCH channels) and data (PS data over HSPA channels) transfer. However, the most interesting aspect of DCH Enhancements may be its potential to reduce the UE battery consumption. The UE battery consumption during voice calls is to a large extent dependent on the power consumption from the RF parts in the UE, and normally these are switched on continuously during a speech call. By using discontinuous transmission and reception, the RF parts can be switched on/off dynamically, thereby significantly reducing the time the RF parts are active and thus also decreasing the battery consumption. The main concepts employed in DCH Enhancements are outlined below.

### 19.3.1 MODES OF OPERATION

In DCH Enhancements, two different modes of operation have been defined, the so-called basic and full modes of operation. A UE capable of supporting DCH Enhancements can support either only the basic mode, or both the basic and full mode.

A number of different subfeatures are active in basic and full operation modes; see Table 19.2. These subfeatures are described in the following sections.

### 19.3.2 UPLINK DPDCH DYNAMIC 10/20 MS TRANSMISSION

In voice calls using the AMR speech codec, the codec is operating on a 20 ms basis. This means that blocks of bits are passed from the speech codec every 20 ms, and these blocks of bits are transmitted on DCHs using a 20 ms TTI. In parallel to the radio bearers carrying the speech data, an associated *signaling radio bearer* (SRB) with 40 ms TTI is also configured. This SRB is used to send RRC messages between UE and RNC.

Prior to the introduction of DCH Enhancements, transmission of the speech transport blocks and SRB transport blocks would cause the UE transmitter to be continuously active since the transport blocks would be spread over the full TTI periods, and

**Table 19.2** Subfeatures active in basic and full operation modes of DCH Enhancements

| Subfeature | Active in basic mode | Active in full mode |
|---|---|---|
| Uplink DPDCH dynamic 10/20 ms transmission | Yes | Yes |
| Uplink DPCCH slot format #5 | Yes | Yes |
| Pilot-free downlink DPCH slot formats | Yes | Yes |
| Pseudo-flexible rate matching | Yes | Yes |
| Downlink frame early termination mode 0 | Yes | No |
| Downlink frame early termination mode 1 | No | Yes |
| Transport channel concatenation | No | Yes |

the UE would also transmit a *transport format combination indicator* (TFCI) spread over the DPCCH radio frames to tell the Node B what transport format combination it should attempt to decode.

With DCH Enhancements, the uplink may be operated discontinuously in a 10 ms transmission mode, where during the first of the two radio frames in a 20 ms speech TTI, the speech transport blocks are transmitted and during the second radio frame there is no transmission. That is, the data (both speech and SRB) is compressed in time so that the instantaneous data rate during a radio frame is doubled, and every second radio frame there is no transmission.

A drawback with 10 ms transmission is that since the instantaneous data rate is doubled, the required instantaneous transmit power is doubled as well. This means that the cell coverage is reduced by approximately 3 dB, which in some scenarios is a very significant coverage reduction that would lead to additional dropped calls and worse speech quality. To avoid this drawback, the UE is allowed to dynamically switch between the new 10 ms transmission mode and the old 20 ms transmission mode. The transmission mode selection is done by the UE based on its available transmission power headroom, but the dynamics of this switch can be controlled with parameters under network control.

### 19.3.3 PILOT-FREE DOWNLINK DPCH SLOT FORMATS

Prior to introduction of DCH Enhancements, all downlink DPCH slot formats contain pilot bits. These pilot bits are typically only used by the UE to estimate the downlink SNR, which in turn is the basis for which TPC commands the UE generates and sends to the Node B. However, it is fully possible to estimate the downlink SNR from the TPC bits instead of pilot bits. This is, for example, how it is done for the F-DPCH, which does not contain any pilot bits. This overhead optimization has been done in DCH Enhancements, with the introduction of two new DPCH slot formats: slot formats #17 and #18. The benefit is twofold: removing the pilot bits saves downlink power and interference since these bits do not need to be transmitted any more, but there is also a benefit since the vacated bits now can be used for coded data bits (speech and SRB) which gives slightly lower channel coding rates and better channel coding performance. Overall, this results in a link performance benefit.

### 19.3.4 PSEUDO-FLEXIBLE RATE MATCHING

To reduce the downlink overhead, blind transport format detection is typically used on downlink. This means that no TFCI is transmitted on downlink, and instead the UE has to try to blindly decode different transport formats. Hopefully the decoding results in a correct CRC, which the UE then assumes indicates that the guess of the actual transport format was correct and further that the data was received correctly. To aid this brute-force transport format detection, so-called fixed position rate matching has earlier been used in downlink, where the different transport channels are allocated fixed spaces within the DPCH. Then the number of bits actually transmitted in such a fixed space is determined by the transport-block size in the current TTI.

Fixed position rate matching means that a transport channel A is allocated space on the DPCH for its highest rate, and this space cannot be used by any other transport channel B, even if transport channel A is currently not operating at a maximum rate. Instead, the unused bits within transport channel A's allocation in the TTI will be DTXed.

Pseudo-flexible rate matching introduced with DCH Enhancements allows for transport channels to share the available DPCH bits more flexibly. Since the SRB is typically only transmitted a small percentage of the time, it is good to be able to not reserve DPCH bits for SRB and instead use all the DPCH bits for transmission of speech, since with more available physical channel bits the channel coding rate is lower and channel coding performance is better. This leads to a link efficiency benefit, at the expense of a slightly more complicated decoding process in the UE.

### 19.3.5 DOWNLINK FRAME EARLY TERMINATION

It may not always be necessary to receive all the slots within a TTI to perform a successful decoding. In particular, when a received block passes the *cyclic redundancy check* (CRC), the transmitter may be instructed via quick feedback signaling to stop its DPCH transmission since these additional slots would only cause additional interference without improving the reception quality further. Several decoding attempts can be made, where if the decoding was unsuccessful, additional coded bits are received to increase the probability that the next decoding attempt will be successful. This is referred to as *Frame Early Termination* (FET) and is used in downlink with DCH Enhancements.

Two different modes of operation have been specified for DCH Enhancements, Mode 0 and Mode 1, where Mode 0 is a simplified version of Mode 1.

In FET Mode 1, multiple FET attempts are made by the UE, and the UE signals a FET ACK/NACK to the Node B on the uplink DPCCH. The UE will start evaluating if the decoded data passes the CRC criteria after 10 slots of received data and if needed continue with decoding attempts every second slot until the full 20 ms TTI has been received. To be able to aid the UE's outer loop power control, both an early and a late BLER target is signaled to the UE. For example, one could instruct the UE to send TPC commands to the Node B with the intent that after 15 slots the probability of a block error is 10% and that after 30 slots it shall be 1%. Using these two BLER targets, the UE can make sure to meet the quality of service target needed for the speech service, while enabling frame early terminations benefits.

FET Mode 0 is quite different from Mode 1 but can still be seen as a significantly simplified frame early termination technique. In this operation mode, the transmission is made with the intent of being decodable after 10 ms, and the UE makes only one decoding attempt after 10 ms and is instructed to let its outer loop power control aim at a single BLER target at that point in time. Hence, the 20 ms speech frames are coded as if they are to be transmitted over 20 ms (two radio frames), but transmission stops right after the first radio frame. This enables the UE to switch off its receiver chain during the second radio frame and thereby save battery power. There is no FET ACK/NACK signaling in Mode 0; the transmitter assumes that the UE's decoding after 10 ms is successful.

### 19.3.6 UPLINK DPCCH SLOT FORMAT #5

A new DPCCH slot format #5 is used when DCH Enhancements is configured. In this case, over a 20 ms compression interval consisting of 30 slots, the bits in the TFCI field of the slot format carry TFCI information in the first 10 slots. In FET Mode 0 the TFCI field in the next 20 slots is not used, while in FET Mode 1 the TFCI field carries the ACK/NACK indicator for the downlink frame early termination.

### 19.3.7 TRANSPORT CHANNEL CONCATENATION

AMR-coded speech employs unequal error protection, where the speech codec bit classes (A, B, and possibly C) are protected differently, depending on how susceptible the human ear is to errors on the different bits. Class A bits are the most important and are protected by CRC as it is better to discard the speech frame containing erroneous Class A bits when reconstructing the speech, since the bit errors could cause strange sounds. To support unequal error protection, AMR-coded speech is carried using two or three transport channels. These transport channels then have different quality of service requirements, for example, subflow A can have a target BLER of 1% and include a CRC for the receiver's outer loop power control to evaluate if this is met, while subflows B and C can have BER requirements of 0.1% and 0.5% respectively and include no CRC.

When FET Mode 1 is configured, the decoding may lead to a successfully passed CRC already after a subset of transmitted slots. If this would be evaluated on one of the speech transport channels only, there would be no guarantee that the quality of the other transport channels is good enough at that point in time. To avoid problems with this, and to not have to include CRC also for the other two transport channels carrying speech, transport channel concatenation is performed of the speech transport channels before further processing in the physical layer in Node B. Hence, all speech data is concatenated into one transport block and it is the CRC of this transport block that is the basis for the ACK/NACK indicator signaled for the frame early termination.

## 19.4 BCH ENHANCEMENTS

The amount of information that can be carried on the *broadcast channel* (BCH) is determined by the P-CCPCH capacity (which is fixed) and by the *System Information Block* (SIB) repetition periods (which are limited by mobility considerations and practical latency when a UE must acquire all system information due to, for example, CS fallback, reselection, or redirection). As HSPA has become increasingly feature rich, the need for transferring information relating to an expanding range of features on the BCH has increased. The exact amount of information that must be transmitted on the BCH depends on which features are implemented in a network and the network configuration. However, a study in 3GPP predicted that in some networks, the amount of SIBs is sufficiently large that the BCH could approach its capacity limit. A BCH capacity shortage would potentially block deployment of new features, and thus 3GPP decided to start a release 12 work item [7] with the aim of developing a secondary broadcast channel.

The broadcast channel designed in release 99 carries general system information that UEs use for configuration purposes. The BCCH logical channel is mapped to the BCH transport channel, which is mapped to the P-CCPCH physical channel. The P-CCPCH physical channel has a fixed transport format and a TTI length of 20 ms and is able to transfer 246 bits/TTI using rate ½ convolutional encoding and a 16-bit CRC. The P-CCPCH is time multiplexed with the synchronization channel. Broadcast information is encoded into a so-called *Master Information Block* (MIB), scheduling blocks, and a number of SIBs. The MIB contains information about the subsequent scheduling blocks and/or SIBs, and the SIBs themselves contain system information. See Figure 19.10 for an illustration.

In order to extend the capacity for broadcast information in release 12, a second broadcast channel may be configured and mapped to a special S-CCPCH, which is transmitted in addition to the P-CCPCH. The first broadcast channel and P-CCPCH remain unchanged in order that they can be read by legacy UEs. Needless to say, legacy users will be unable to read the secondary BCH; thus, any features relating to parameters carried on the secondary BCH will not be accessible to legacy UEs.

The S-CCPCH that carriers a secondary broadcast channel is identical in physical layer parameterization to the P-CCPCH (apart from the fact that it is mapped to a different spreading code), including the first 256 chips of each slot being DTX. However, unlike the P-CCPCH, which must transmit the *System Frame Number* (SFN) in every TTI, the S-CCPCH carrying a second broadcast channel may be DTX when there are no SIBs scheduled on the S-CCPCH.

The UE may be configured to receive two further S-CCPCHs: one that carries the *Paging Channel* (PCH) and a second carrying FACH or CTCH occasions for idle mode users. However, the pre–release 12 requirement that a UE must only

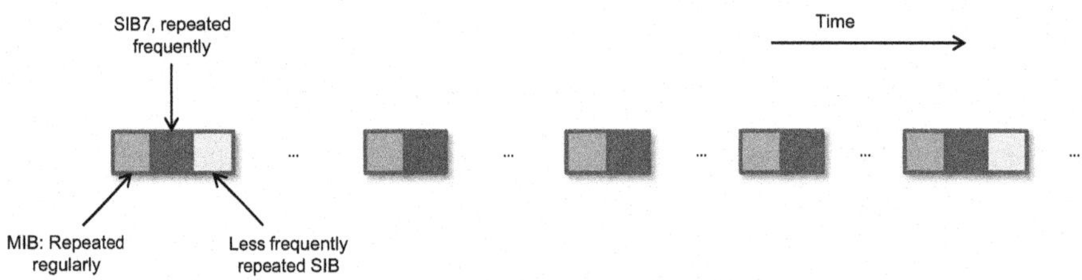


**FIGURE 19.10 Simplified example of BCH scheduling involving MIB and two SIBs.**

be able to receive at most two S-CCPCHs simultaneously is retained. For CELL_FACH users, it is unnecessary to read PCH. For users in CELL_PCH or idle mode, the UE in general reads the S-CCPCHs carrying PCH and the secondary broadcast where necessary, but during a CTCH occasion, if the UE also receives data on PCH it is not required to read the secondary broadcast channel in addition to CTCH and PCH.

The presence of the secondary broadcast channel is indicated in the MIB on the primary broadcast channel. The MIB indicates the spreading code used for the S-CCPCH for the secondary broadcast channel, the repetition period, and the number of segments. Forty milliseconds after the MIB, the S-CCPCH containing the secondary broadcast channel transmits a so-called SIB block, which indicates further information on SIB scheduling on the secondary BCH. It is not necessary to schedule SIBs in every TTI on the secondary BCH.

The secondary BCH may be used to carry copies of any SIB (either pre–release 12 or post–release 12 SIBs) that is also carried on the primary BCH. It may also carry unique SIBs for release 12 and beyond features. Of course, if an SIB for a release 12+ feature is only transmitted on the secondary BCH, then it is not available for legacy users.

The UE is expected to be able to read information from both broadcast channels simultaneously, and thus the time required to read complete system information is unaffected.

## 19.5 LOOKING TO THE FUTURE - RELEASE 13

Work on release 13 commenced in 3GPP in the fall of 2014. Areas that have emerged as potential HSPA evolution topics in release 13 include *small data transmission enhancements for UMTS*, *downlink enhancements*, *network-assisted interference cancelation* (NAIC), *data compression enhancements,* increases in the scope of *advanced receiver technologies* and the possibility of operating a *dual-cell uplink in two bands*.

### 19.5.1 SMALL DATA TRANSMISSION ENHANCEMENTS FOR UMTS

Small data packet size (typically for machine-to-machine communications) applications are expected to be a large growth area, with the potential for billions of connected devices. In many cases, these applications pose different service requirements compared to conventional traffic types and for an important segment of applications, the requirements on power consumption and coverage versus data rate and latency may vary.

WCDMA has a coverage footprint that is suitable for operators to provide small data-oriented services, and the devices are already available at a cost level suitable for the machine-oriented communication market. However, the current generation of WCDMA specifications has been optimized for mobile broadband traffic, and to ensure that new-device markets are properly served, the specifications should enable requirements on coverage in challenging device locations, low device power consumption, and machine type data rates to be met.

3GPP has initiated a study, the objective of which is to identify any potential problems and system bottlenecks and also develop technical solutions for improved support for small data applications, delay-tolerant applications, and massive deployment of devices over HSPA-based transport. A key constraint is to ensure coexistence with existing HSPA traffic.

### 19.5.2 DOWNLINK ENHANCEMENTS

It was noted toward the end of release 12 that there may be some benefits in improving the downlink TPC signaling for scenarios where large numbers of users must be supported.

In particular, for some users with low mobility, a TPC update rate of once per slot is not strictly necessary. In these cases, the TPC update rate could be reduced by a factor $N$ (for example, 5), and TPC bits could be repeated in the $N$ consecutive slots between update periods. The UE would then be able to use soft combining of the $N$ repeated TPC bits to improve reliability. The improved reliability of the repeated TPC bits could enable the base station to reduce the power allocated for TPC and instead use the power for data transfer. Of course, the power control update rate would reduce and the power control latency would increase, and the trade-off between soft combining gain and power control algorithm loss will need to be evaluated. Furthermore, 3GPP will need to verify that there are scenarios in which the overall power overhead of F-DPCH is sufficiently high that reducing the overhead will yield system gains.

Release 13 will study the potential of this and other potential downlink enhancements. These include mechanisms to enhance downlink signaling performance on overhead and latency, especially for the case of RRC state transition and parameter updating as well as mechanisms to enhance SRB coverage over HSPA.

### 19.5.3 NETWORK ASSISTED INTERFERENCE CANCELATION

*Other-Cell Interference* (OCI) is the dominant limitation for HSPA DL performance in many typical scenarios. Cell-edge regions in traditional deployments suffer from high-rate coverage limitations due to interference from other cells, so that the own-cell geometry may end up around or below 0 dB. In heterogeneous networks, a UE located in the *Range Expansion* (RE) region and served by an LPN may encounter geometries down to −10 dB as a result of strong macro interference. This inhibits DL data, DL control, and UL grant reception by an LPN UE in the RE region.

The classic approach to OCI mitigation is to apply *Interference Rejection Combining* (IRC), using multi-antenna receivers. This has been shown to work well in scenarios with one dominant interferer, but a dual-antenna RX at the UE lacks sufficient spatial degrees of freedom to combat interference from several significant interference sources.

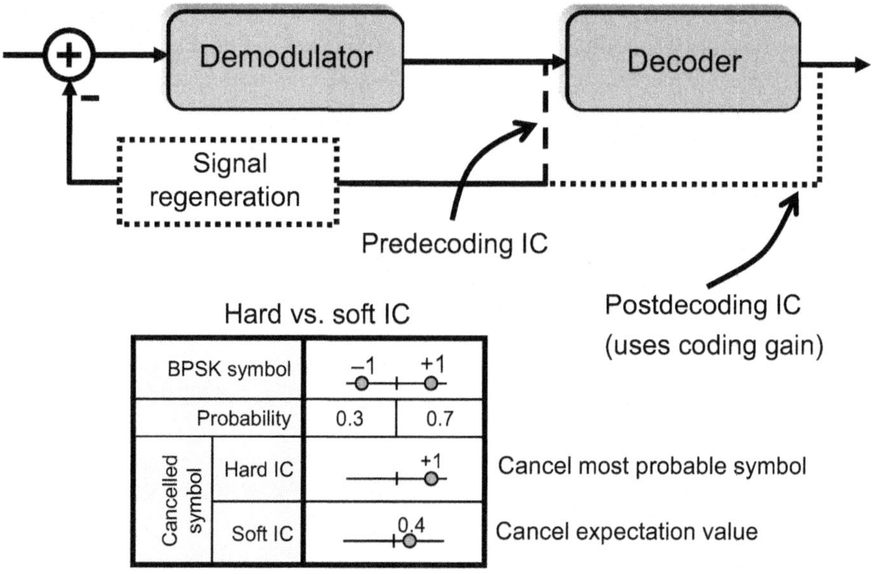


**FIGURE 19.11 Post- and pre-decoding interference cancelation.**

Another approach to achieve improved interference mitigation is to employ *OCI cancelation* (OCIC). The idea is that an interfering signal can be canceled after detecting the interfering symbols at the victim UE. An interfering symbol may be detected without the help of the *forward-error-correcting* (FEC) decoder. This is referred to as pre-decoding *interference cancelation* (IC). FEC code provides additional structured redundancy, which can be exploited to better detect the interfering signal - referred to as post-decoding IC (see also Figure 19.11). However, the victim UE needs to know the parameters of the FEC to be able to exploit the FEC code structure. This motivates the idea of *Network Assisted Interference Cancelation* (NAIC).

A key question for NAIC is whether network assistance (that is, the network sending the UE information about the signal to be canceled in order that the UE can operate more efficient cancelation algorithms) is beneficial, and if so, what information should be signaled from the network to the UE and when the information should be signaled.

## 19.5.4 DATA COMPRESSION

It has been observed that UL data transmissions often exhibit a considerable amount of redundancy across IP packets. Reasons for this can include the sending of TCP/IP ACKs relating to downlink transmissions and commonality in TCP/IP headers. During release 12, the potential gains in uplink RoT and capacity that could be achieved by compressing this redundant data across multiple packets were studied.

UL compression is proposed to be supported in RLC in release 13. RRC signaling would be introduced, but the compression algorithm itself will not be specified (that is, the means to detect and remove duplicate data will be implementation specific). In order to provide a reasonable expectation of compression performance, new performance requirements relating the compression ratios achieved by new implementations may be introduced.

### 19.5.5 INCREASED SCOPE FOR ENHANCED UE RECEIVERS

Performance requirements for dual antenna and interference mitigating UE receivers were introduced into the release 8 specifications in the form of so-called Type 3 (RX diversity) and Type 2i/Type 3i (interference mitigation) requirements. The release 8 requirements are applicable to the reception of HSDPA, since this is the dominant channel for data reception.

It is still the case, however, that some services (such as voice and signaling) may be configured to be carried over dedicated channels. Currently, advanced receiver requirements are not applied to dedicated channels. Introducing requirements enables UEs that can improve their DCH reception to be predictably deployed, which can in turn reduce the overhead for DCH services and increase the downlink capacity.

### 19.5.6 DUAL-BAND DUAL-CELL HSUPA

Dual-cell (DC) HSUPA, as described in Chapter 12, was introduced into release 9 of the specifications. DC-HSUPA in release 9 must be operated on contiguous carriers in the same band. This restriction enables DC-HSUPA UEs to operate with a single power amplifier.

However, operators may often have spectrum available in two bands. A common example of this is the so-called core band (Band I) and the 900 MHz band (Band VIII). Since release 9, UE PA technologies have improved and it has become feasible to build a dual PA UE that is able to transmit simultaneously in two bands. To enable such operation, release 13 is expected to extend DC-HSUPA to enable dual band transmission (with one carrier in each band).

Most of the work to support DB-DC-HSUPA is likely to be in the form of new UE radio requirements. Some minor issues might, however, need to be considered for the air interface procedures (such as handling power control and maximum power constraints with two PAs) and RRC signaling procedures.

### 19.5.7 MULTI-FLOW ENHANCEMENTS

There may be some scope for increasing the number of options for configuring multi-flow in release 13, in particular, the possibility for configuring multi-flow together with a larger number of multi-carrier features. At the time of writing, consideration is still ongoing on the usefulness of such enhancements.

## REFERENCES


[1] RP-131348, 'New WI proposal: UMTS Mobility enhancements for Heterogeneous Networks', Huawei, HiSilicon, 3GPP TSG RAN Meeting #61, Porto, Portugal, 3rd – 6th September 2013.

[2] RP-140463, 'Revised WID: UMTS Heterogeneous Networks enhancements, Huawei, 3GPP TSG RAN Meeting #63, Fukuoka, Japan, 3rd – 6th March 2014.

[3] '3rd Generation Partnership Project; Technical Specification Group Radio Access Network; Study on UMTS heterogeneous networks', 3GPP, 3GPP TR 25.800.

[4] 3GPP Technical Specifications 25.331, 'Radio Resource Control (RRC); Protocol specification'.

[5] RP-131357 'WID on DCH Enhancements'.

[6] TR 25.702 'TR on Study on Dedicated Channel (DCH) enhancements for UMTS'.

[7] RP-140131 'Revised WID: Enhanced Broadcast of System Information', Ericsson, 3GPP TSG RAN Meeting #63, Fukuoka, Japan, 3rd – 6th March 2014.

[8] RP-132061 'WID on Increasing the minimum number of carriers for UE monitoring in UTRA and E-UTRA'.

# List of acronyms

| | |
|---|---|
| 3G | 3rd Generation |
| 3GPP | 3rd Generation Partnership Project |
| 4G | 4th Generation |
| AAS | Adaptive/Advanced Antenna System |
| ACIR | Adjacent Channel Interference Ratio |
| ACK | Acknowledgment (in ARQ protocols) |
| ACLR | Adjacent Channel Leakage Ratio |
| ACS | Adjacent Channel Selectivity |
| ACTS | Advanced Communications Technology and Services |
| AI | Acquisition Indicator |
| AICH | Acquisition Indicator Channel |
| AM | Acknowledged Mode (RLC configuration) |
| AMC | Adaptive Modulation and Coding |
| AMPR | Additional Maximum Power Reduction |
| AMPS | Advanced Mobile Phone System |
| AMR-WB | Adaptive MultiRate-WideBand |
| AP | Access Point |
| ARIB | Association of Radio Industries and Businesses |
| ARQ | Automatic Repeat-reQuest |
| ATDMA | Advanced Time Division Mobile Access |
| ATIS | Alliance for Telecommunications Industry Solutions |
| AWGN | Additive White Gaussian Noise |
| BC | Band Category |
| BCCH | Broadcast Control Channel |
| BCH | Broadcast Channel |
| BER | Bit-Error Rate |
| BES | Best Effort Service |
| BFN | Node B Frame Number |
| BLER | Block-Error Rate |
| BM-SC | Broadcast/Multicast Service Center |
| BMC | Broadcast and Multicast Control |
| BPSK | Binary Phase-Shift Keying |
| BS | Base Station |
| BSC | Base Station Controller |
| BTC | Block Turbo Code |
| BTS | Base Transceiver Station |
| CA | Core Network and Terminals |
| CACLR | Cumulative Adjacent Channel Leakage Ratio |
| CAPEX | Capital Expenditure |
| CC | Convolutional Code or Chase combining |
| CCCH | Common Control Channel |
| CCE | Control Channel Element |
| CCS | Channelization Code Set |

| | |
|---|---|
| CCSA | China Communications Standards Association |
| CDF | Cumulative Density Function |
| CDM | Code-Division Multiplex |
| CDMA | Code Division Multiple Access |
| CEPT | European Conference of Postal and Telecommunications Administrations |
| CIO | Cell Individual Offset |
| CL | Closed-Loop |
| CLTD | Closed-Loop Transmit Diversity |
| CN | Core Network |
| CODIT | Code-Division Testbed |
| COST | Cooperation in the field of Scientific and Technical Research |
| CPC | Continuous Packet Connectivity |
| CPICH | Common Pilot Channel |
| D-CPICH | Demodulation Common Pilot Channel |
| P-CPICH | Primary Common Pilot Channel |
| S-CPICH | Secondary Common Pilot Channel |
| CQI | Channel Quality Indicator |
| CRC | Cyclic Redundancy Check |
| C-RNTI | Cell-RNTI |
| CS | Circuit Switched or Capability Set |
| CSG | Closed Subscriber Group |
| CV | Constellation Version |
| CW | Continuous Wave |
| D2D | Device to Device |
| DB | Dual Band |
| DC | Dual-Carrier or Duall-Cell |
| DCCH | Dedicated Control Channel |
| DCH | Dedicated Channel |
| DDI | Data Description Indicator |
| DF | Dual-Frequency |
| DF-3C | Dual-Frequency Three-Cell |
| DF-DC | Dual-Frequency Four-Cell |
| DFE | Decision Feedback Equalization |
| DL | Downlink |
| DL-SCH | Downlink Shared Channel |
| DPCCH | Dedicated Physical Control Channel |
| DPCH | Dedicated Physical Channel |
| DPDCH | Dedicated Physical Data Channel |
| DRX | Discontinuous Reception |
| DSSS | Direct Sequence Spread Spectrum |
| DTCH | Dedicated Traffic Channel |
| DTX | Discontinuous Transmission |
| D-TxAA | Dual Transmit-Diversity Adaptive Array |
| E-AGCH | E-DCH Absolute Grant Channel |
| E-AI | Enhanced Acquisition Indicator |
| E-AICH | Enhanced Acquisition Indicator Channel |
| E-DCH | Enhanced Dedicated Channel |

| | |
|---|---|
| EDGE | Enhanced Data Rates for GSM Evolution and Enhanced Data Rates for Global Evolution |
| E-DPCCH | E-DCH Dedicated Physical Control Channel |
| E-DPDCH | E-DCH Dedicated Physical Data Channel |
| E-HICH | E-DCH Hybrid Indicator Channel |
| EM | Electromagnetic |
| E-RGCH | E-DCH Relative Grant Channel |
| E-RNTI | E-DCH Radio Network Temporary Identity |
| E-ROCH | E-DCH Rank and Offset Channel |
| ESCC | Enhanced Serving Cell Change |
| E-TFC | E-DCH Transport Format Combination |
| E-TFCI | E-DCH Transport Format Combination Index |
| ETSI | European Telecommunications Standards Institute |
| EUL | Enhanced Uplink |
| E-UTRA | Evolved UTRA |
| E-UTRAN | Evolved UTRAN |
| EV-DO | Evolution-Data Optimized (of CDMA2000 1x) |
| EV-DV | Evolution-Data and Voice (of CDMA2000 1x) |
| EVM | Error Vector Magnitude |
| FACH | Forward Access Channel |
| FBI | Feedback Indicator |
| FCC | Federal Communications Commission |
| FDD | Frequency-Division Duplex |
| FDM | Frequency-Division Multiplex |
| FDMA | Frequency-Division Multiple Access |
| F-DPCH | Fractional DPCH |
| FEC | Forward Error Correction |
| FET | Frame Early Termination |
| FHSS | Frequency Hopping Spread Spectrum |
| FIR | Finite Impulse Response |
| FP | Frame Protocol |
| FPLMTS | Future Public Land Mobile Telecommunications Systems |
| FRAMES | Future Radio Wideband Multiple Access Systems |
| FTP | File Transfer Protocol |
| F-TPICH | Fractional Transmitted Precoding Indicator Channel |
| GERAN | GSM EDGE RAN |
| GHz | Giga Hertz |
| GMSK | Gaussian Minimum Shift Keying |
| GPRS | General Packet Radio Services |
| GPS | Global Positioning System |
| G-RAKE | Generalized RAKE |
| GSM | Global System for Mobile Communications |
| HAP | Hybrid-ARQ Process Indicator |
| HARQ | Hybrid-ARQ |
| H-FDD | Half-Duplex FDD |
| HHO | Hard Handover |
| H-RNTI | HSDPA Network Temporary Identity |
| HRPD | High Rate Packet Data |

| | |
|---|---|
| HSDPA | High-Speed Downlink Packet Access |
| HS-DPCCH | High-Speed Dedicated Physical Control Channel |
| HS-DSCH | High-Speed Downlink Shared Channel |
| HSPA | High-Speed Packet Access |
| HS-PDSCH | High-Speed Physical Downlink Shared Channel |
| HS-SCCH | High-Speed Shared Control Channel |
| HSUPA | High-Speed Uplink Packet Access |
| ICIC | Inter-Cell Interference Cancellation |
| ID | Identity |
| IE | Information Element |
| IEEE | Institute of Electrical and Electronics Engineers |
| ILPC | Inner Loop Power Control |
| IMB | Integrated Mobile Broadcast |
| IMS | IP Multimedia Subsystem |
| IMT-2000 | International Mobile Telecommunications 2000 |
| IP | Internet Protocol |
| IPsec | Internet Protocol security |
| IPv4 | IP version 4 |
| IPv6 | IP version 6 |
| IR | Incremental Redundancy |
| IRC | Interference Rejection Combining |
| ISDN | Integrated Services Digital Network |
| ITU | International Telecommunications Union |
| ITU-R | International Telecommunications Union-Radiocommunication Sector |
| Iu | The interface used for communication between the RNC and the core network. |
| Iu_ant | The interface used for communication between the RNC and an RET unit. |
| Iu_cs | The interface used for communication between the RNC and the GSM/WCDMA circuit-switched core network. |
| Iu_ps | The interface used for communication between the RNC and the GSM/WCDMA packet-switched core network. |
| Iub | The interface used for communication between the Node B and the RNC. |
| Iur | The interface used for communication between different RNCs. |
| J-TACS | Japanese Total Access Communication System |
| KPI | Key Performance Indicator |
| L1 | Layer 1 |
| L2 | Layer 2 |
| LAN | Local Area Network |
| LMMSE | Linear Minimum Mean-Square Error |
| LPN | Low-Power Node |
| LSB | Least Significant Bit |
| LTE | Long-Term Evolution |
| M2M | Machine to Machine |
| MAC | Medium Access Control |
| MBB | Mobile Broadband |
| MBMS | Multimedia Broadcast/Multicast Service |
| MBS | Multicast and Broadcast Service |
| MBSFN | Multicast Broadcast Single Frequency Network |

| | |
|---|---|
| MBWA | Mobile Broadband Wireless Access |
| MC | Multi-Carrier |
| MCCH | MBMS Control Channel |
| MCH | Multicast Channel |
| MCS | Modulation and Coding Scheme |
| MDHO | Macro-Diversity Handover |
| MHz | Mega Hertz |
| MIB | Master Information Block |
| MICH | MBMS Indicator Channel |
| MIMO | Multiple-Input Multiple-Output |
| ML | Maximum Likelihood |
| MLD | Maximum Likelihood Detection |
| MLSE | Maximum Likelihood Sequence Estimation |
| MME | Mobility Management Entity |
| MMS | Multimedia Messaging Service |
| MMSE | Minimum Mean Square Error |
| MPR | Maximum Power Reduction |
| MRC | Maximum Ratio Combining |
| MS | Modulation Scheme |
| MSB | Most Significant Bit |
| MSC | Mobile Switching Center |
| MSCH | MBMS Scheduling Channel |
| MSR | Multi-Standard |
| MTC | Machine Type Communications |
| MTCH | MBMS Traffic Channel |
| NACK | Negative Acknowledgment (in ARQ protocols) |
| NAS | Non-Access Stratum (a functional layer between the core network and the terminal that supports signaling and user data transfer). |
| NCL | Neighbor Cell List |
| NDI | New Data Indicator |
| NHR | Number of hybrid-ARQ Retransmissions |
| NMT | Nordisk MobilTelefon (Nordic Mobile Telephony) |
| Node B | Node B, a logical node handling transmission/reception in multiple cells. Commonly, but not necessarily, corresponding to a base station. |
| NRPM | Normalized Remaining Power Margin |
| OLPC | Outer-Loop Power Control |
| OLTD | Open-Loop Transmit Diversity |
| OOB | Out-of-Band (emissions) |
| OOK | On–Off Keying |
| OPEX | Operating Expenditure |
| OVSF | Orthogonal Variable Spreading Factor |
| PA | Power Amplifier |
| PAM | Phase Amplitude Modulation |
| PAPR | Peak-to-Average Power Ratio |
| PAR | Peak-to-Average Ratio (same as PAPR) |
| PARC | Per-Antenna Rate Control |
| PBCH | Physical Broadcast Channel |
| PCCH | Paging Control Channel |

| | |
|---|---|
| P-CCPCH | Primary Common Control Physical Channel |
| PCG | Project Coordination Group (in 3GPP) |
| PCH | Paging Channel |
| PCI | Precoding Control Indication |
| PCS | Personal Communications Systems |
| PDC | Personal Digital Cellular |
| PDCP | Packet-Data Convergence Protocol |
| PDSN | Packet-Data Serving Node |
| PDN | Packet-Data Network |
| PDU | Protocol Data Unit |
| PF | Proportional Fair (a type of scheduler) |
| PHY | Physical Layer |
| PHS | Personal Handy-phone System |
| PI | Page Indicator |
| PICH | Paging Indicator Channel |
| PL | Puncturing Limit |
| PMI | Precoding Matrix Indicator |
| PoC | Push to Talk Over Cellular |
| PRACH | Physical Random Access Channel |
| PRB | Physical Resource Block |
| PS | Packet Switched |
| PSC | Primary Scrambling Code |
| PSD | Power Spectral Density |
| PSK | Phase Shift Keying |
| PSD | Power Spectral Density |
| PSTN | Public-Switched Telephone Network |
| PWI | Precoding Weight Indicator |
| QAM | Quadrature Amplitude Modulation |
| QoS | Quality-of-Service |
| QPSK | Quadrature Phase-Shift Keying |
| RAB | Radio Access Bearer |
| RACE | Research and development in Advanced Communications technologies in Europe |
| RACH | Random Access Channel |
| RAN | Radio Access Network |
| RA-RNTI | Random Access RNTI |
| RAT | Radio Access Technology |
| RB | Resource Block |
| RBS | Radio Base Station |
| RET | Remote Electrical Tilt |
| RF | Radio Frequency |
| RI | Rank Indicator |
| RIT | Radio Interface Technology |
| RL | Radio Link |
| RLC | Radio Link Protocol |
| RM | Rate Matching |
| RNC | Radio Network Controller |
| RNTI | Radio Network Temporary Identifier |

| | |
|---|---|
| RoT | Rise over Thermal |
| RR | Round Robin (a type of scheduler) |
| RRC | Radio Resource Control |
| RRC | Root Raised Cosine |
| RSCP | Received Signal Code Power |
| RRM | Radio Resource Management |
| RSN | Retransmission Sequence Number |
| RSPC | IMT-2000 Radio Interface Specifications |
| RSRP | Reference Signal Received Power |
| RSRQ | Reference Signal Received Quality |
| RTP | Real-Time Protocol |
| RTT | Round Trip Time |
| RTWP | Received Total Wideband Power |
| RV | Redundancy Version |
| RX | Receive |
| SA | Services & System Aspects |
| SAE | System Architecture Evolution |
| S-CCPCH | Secondary Common Control Physical Channel |
| SCH | Synchronization Channel |
| SCM | Spatial Channel Models |
| SCRI | Signaling Control Release Indicator |
| SDM | Spatial Division Multiplexing |
| SDMA | Spatial Division Multiple Access |
| SDO | Standards Developing Organization |
| S-DPCCH | Secondary-Dedicated Physical Control Channel |
| S-DPCH | Secondary-Dedicated Physical Channel |
| SDU | Service Data Unit |
| S-E-DPCCH | Secondary-Enhanced Dedicated Physical Control Channel |
| S-E-DPDCH | Secondary-Enhanced Dedicated Physical Data Channel |
| SEM | Spectrum Emissions Mask |
| S-ETFC | Secondary-Enhanced Transport Format |
| S-ETFCI | Secondary-Enhanced Transport Format Indicator |
| SFN | Single Frequency Network |
| SF | Spreading Factor |
| SF-DC | Single-Frequency Dual-Cell |
| SFN | Single-Frequency Network or Cell System Frame Number (in 3GPP) |
| SG | Serving Grant |
| SGSN | Serving GPRS Support Node |
| SHO | Soft Handover |
| SI | Scheduling Information |
| SIB | System Information Block |
| SIC | Successive Interference Cancellation |
| SIC | Successive Interference Combining |
| SID | Study Item Description |
| SIM | Subscriber Identity Module |
| SINR | Signal-to-Interference-and-Noise Ratio |
| SIR | Signal-to-Interference Ratio |
| SISO | Single Input Single Output |

| | |
|---|---|
| SMS | Short Message Service |
| SNR | Signal-to-Noise Ratio |
| SOHO | Soft Handover |
| SRB | Signaling Radio Bearer |
| S-RNC | Serving RNC |
| SRNS | Serving Radio Network Subsystem |
| STBC | Space-Time Block Coding |
| S-TBS | Secondary Transport Block Size |
| STC | Space-Time Coding |
| STTD | Space-Time Transmit Diversity |
| T2TP | Traffic to Total Pilot Power Ratio |
| TACS | Total Access Communication System |
| TBS | Transport-Block Size |
| TCP | Transmission Control Protocol |
| T-CPICH | TDM CPICH |
| TC-RNTI | Temporary C-RNTI |
| TD-CDMA | Time Division-Code Division Multiple Access |
| TDD | Time Division Duplex |
| TDM | Time Division Multiplexing |
| TDMA | Time Division Multiple Access |
| TD-SCDMA | Time Division-Synchronous Code Division Multiple Access |
| TF | Transport Format |
| TFC | Transport Format Combination |
| TFCI | Transport Format Combination Index |
| TIA | Telecommunications Industry Association |
| TM | Transparent Mode (RLC configuration) |
| TPC | Transmit Power Control |
| TPI | Transmitted Precoding Indicator |
| TR | Technical Report |
| TrCH | Transport Channel |
| TS | Technical Specification |
| TSG | Technical Specification Group |
| TSN | Transmission Sequence Number |
| TSTS | Time-Switched Transmit Diversity |
| TTA | Telecommunications Technology Association |
| TTA | Total Transmission Attempts |
| TTC | Telecommunications Technology Committee |
| TTI | Transmission Time Interval |
| TTT | Time to Trigger |
| TX | Transmit |
| UE | User Equipment, the 3GPP name for the mobile terminal |
| UEM | Unwanted Emissions Mask |
| UGS | Unsolicited Grant Service |
| UL | Uplink |
| UM | Unacknowledged Mode (RLC configuration) |
| UMB | Ultra Mobile Broadband |
| UMTS | Universal Mobile Telecommunications System |
| UPH | Uplink Power Headroom |

| | |
|---|---|
| USIM | UMTS SIM |
| US-TDMA | U.S.TDMA standard |
| UTRA | Universal Terrestrial Radio Access |
| UTRAN | Universal Terrestrial Radio Access Network |
| Uu | The interface used for communication between the Node B and the UE. |
| VAM | Virtual Antenna Mapper |
| VF | Version Flag |
| VoIP | Voice over Internet Protocol |
| VPN | Virtual Private Network |
| WAN | Wide Area Network |
| WARC | World Administrative Radio Congress WARC'92 |
| WCDMA | Wideband Code Division Multiple Access |
| WG | Working Group |
| WI | Work Item |
| WiMAX | Worldwide Interoperability for Microwave Access |
| WLAN | Wireless Local Area Network |
| WMAN | Wireless Metropolitan Area Network |
| WP8F | Working Party 8F |
| WRC | World Radiocommunication Conference |
| ZC | Zadoff–Chu |
| ZF | Zero Forcing |

# Index

## E

## I



## K



## L



## M

## N



## O



## P

## V



## W

Printed in the United States
By Bookmasters